Complex Analysis

Mathematische Lehrbücher und Monographien
Herausgegeben von der Akademie der Wissenschaften der DDR
Institut für Mathematik

II. Abteilung
Mathematische Monographien
Band 61

Complex Analysis
Methods, Trends, and Applications

Herausgegeben von E. Lanckau und W. Tutschke

Complex Analysis

Methods, Trends, and Applications

Edited by Eberhard Lanckau and Wolfgang Tutschke

With 28 figures

Akademie-Verlag Berlin
1983

Autoren:

Prof. Dr. Mark Benevič Balk, Smolenskiĭ pedagogičeskiĭ institut
Prof. Dr. Heinrich Begehr, Fachbereich Mathematik der FU Berlin (West)
Prof. Dr. Bogdan Bojarski, Uniwersytet Warszawa, Instytut Matematyczny
Prof. Dr. David L. Colton, University of Delaware, Department of Mathematics
Akademik Abduhamid Džuraev, Matematičeskiĭ institut s vyčislitel'nym centrom Akademii Nauk TadžSSR, Dušanbe
Prof. Dr. Helmut Florian, Institut für angewandte Mathematik der Technischen Universität Graz
Prof. Dr. Otto Forster, Mathematisches Institut der Universität München
Prof. Dr. Robert P. Gilbert, University of Delaware, Department of Mathematics
Prof. Dr. Klaus Habetha, RWTH Aachen, Lehrstuhl II für Mathematik
Prof. Dr. Rudolf Heersink, Institut für angewandte Mathematik der Technischen Universität Graz
Dr. rer. nat. Bernd Jesiak, Sektion Mathematik der Karl-Marx-Universität Leipzig
Prof. Dr. Eberhard Lanckau, Sektion Mathematik der TH Karl-Marx-Stadt
Prof. Dr. Julian Ławrynowicz, Instytut Matematyczny Polskiej Akademii Nauk, Oddział w Łodzi
Prof. Dr. Georgiĭ Ferapontovič Mandžavidze, Institut prikladnoĭ matematiki im. I. N. Vekua Tbilisskogo Gosudarstvennogo Universiteta
Prof. Dr. Erwin Mues, Institut für Mathematik der Universität Hannover
Prof. Dr. Elena Irodionovna Obolašvili, Matematičeskiĭ institut im. A. M. Razmadze Akademii Nauk GruzSSR, Tbilisi
Prof. Dr. Siegfried Prößdorf, Institut für Mathematik der Akademie der Wissenschaften der DDR, Berlin
Prof. Dr. Hans-Joachim Roßberg, Sektion Mathematik der Karl-Marx-Universität Leipzig
Prof. Dr. Grigorios Tsagas, University of Thessaloniki
Prof. Dr. Wolfgang Tutschke, Sektion Mathematik der Martin-Luther-Universität Halle-Wittenberg
Akademik Zafar Džuraevič Usmanov, Matematičeskiĭ institut s vyčislitel'nym centrom Akademii Nauk TadžSSR, Dušanbe
Prof. Dr. Vladimir Sergeevič Vinogradov, Matematičeskiĭ institut im. V. A. Steklova Akademii Nauk SSSR, Moskva
Prof. Dr. Wolfgang L. Wendland, Fachbereich Mathematik der TH Darmstadt
Prof. Dr. Lothar v. Wolfersdorf, Sektion Mathematik der Bergakademie Freiberg

Erschienen im Akademie-Verlag, DDR-1086 Berlin, Leipziger Straße 3–4

Printed in the German Democratic Republic
Lizenznummer: 202 · 100/401/83
Gesamtherstellung: VEB Druckerei „Thomas Müntzer", 5820 Bad Langensalza
Lektor: Dipl.-Math. Gesine Reiher
Gestaltung: Dietmar Kunz
LSV 1035
Bestellnummer: 763 109 6 (6699)
DDR 68,– M

Prefatory note

There are some culminating points in the development of complex analysis. A first one characterizing the meaning of complex analysis for the whole science of mathematics is the theory of holomorphic functions. This theory is based on leading papers by Cauchy, Riemann and Weierstrass. Furthermore there are essential investigations continuing those papers to the middle of our century. They led to a theory characterized by surprising harmony, suitability and perfection.

The connections between real and complex analysis were mainly given by the Cauchy Riemann differential equations. On the other hand the "holomorphic" function theory developed into a typical complex theory giving up extensively real interpretations. Thus the so called function theory became a theory, independent of the real analysis.

In the middle of our century a new development of complex analysis led to a modification of the connections between real and complex analysis. This development was clearly demonstrated by I. N. Vekua's papers on generalized analytic functions, especially. Generalized analytic functions are solutions of the system

$$\frac{\partial u}{\partial x} - \frac{\partial v}{\partial y} + au + bv = f\,, \qquad \frac{\partial u}{\partial y} + \frac{\partial v}{\partial x} + cu + dv = g$$

that is equivalent to the partial complex differential equation

$$\frac{\partial w}{\partial \bar{z}} + Aw + B\bar{w} = F\,, \tag{*}$$

where $z = x + \mathrm{i}y$, $\bar{z} = x - \mathrm{i}y$, $w = u + \mathrm{i}v$, $4A = a + d - \mathrm{i}c - \mathrm{i}b$, $4B = a - d + \mathrm{i}c - \mathrm{i}b$, $2F = f + \mathrm{i}g$. The partial complex differentiations $\partial/\partial z$, $\partial/\partial \bar{z}$ are defined by

$$\frac{\partial}{\partial z} = \frac{1}{2}\left(\frac{\partial}{\partial x} - \mathrm{i}\frac{\partial}{\partial y}\right), \qquad \frac{\partial}{\partial \bar{z}} = \frac{1}{2}\left(\frac{\partial}{\partial x} + \mathrm{i}\frac{\partial}{\partial y}\right),$$

such that the Cauchy Riemann system may be written as $\partial w/\partial \bar{z} = 0$. The theory of I. N. Vekua's partial differential equation (*) brought about new applications of complex analysis, such that complex analysis became more important for the whole science of mathematics. This new applications of complex analysis concern, for instance, boundary value problems, geometry, mechanics, quasiconformal mappings, infinitesimal bending of surfaces, the shell theory in architecture, gas dynamics, hydrodynamics and the theory of elasticity. The generalization of the complex differentiation $\mathrm{d}/\mathrm{d}z$ by partial complex differentiations led, finally, also to the concept of the so called τ-derivatives $g^\tau = \dfrac{\partial g}{\partial z} + \gamma \dfrac{\partial g}{\partial \bar{z}}$, where in each case γ must be chosen suitably (cf. Naas [2], Naas and Tutschke [5]). Another concept of complex differentiation is made use of in L. BERS' theory of pseudoanalytic functions [1]. L. BERS defines a complex difference quotient in dependence of two auxiliary functions $F = F(z)$ and $G = G(z)$ (thus the corresponding complex derivative is called the (F, G)-derivative). Another generalization of holomorphic functions is given by G. N. POLOŽIĬ's (p, q)-analytic ones that are solutions of the system

$$\frac{\partial v}{\partial y} = p\,\frac{\partial u}{\partial x} + q\,\frac{\partial u}{\partial y}, \qquad -\frac{\partial v}{\partial x} = -q\,\frac{\partial u}{\partial x} + p\,\frac{\partial u}{\partial y},$$

where $p > 0$.

The generalizations of the complex differentiation regarded above allow not only linear differential equations to be investigated. Recent results concern nonlinear differential equations (and differential inequalities), too. Also in the case of nonlinear (elliptic) differential equations there are 1—1-mappings between solutions and holomorphic functions. These mappings make it possible to reduce the construction of special solutions to those of holomorphic functions. Complex analysis leads, consequently, to a new linearization method based on the concept of holomorphy. Also the global construction of nonholomorphic functions is possible by using properties of holomorphic functions. For instance the global construction of nonholomorphic functions is possible by a modification of the well-known Cousin method.

Essentially, the disconnection between real and complex analysis mentioned above is overcome in these days. Complex analysis of today may be characterized by a systematic utilization of the concept of holomorphy. Holomorphy became a basic structural element of analysis. The concept of holomorphy produced new results, methods for proofs, and far-reaching applications in mathematics as well as in natural science. The enlarged effectiveness of the concept of holomorphy inaugurates new possibilities for numerical mathematics, too. Complex analysis leads, consequently, to new aspects of computerisation, which infiltrates the production and the administration and their various ramifications. Thus the field of complex analysis showing still many open questions became a necessity in these days.

The authors of the separate chapters characterize the problems, the methods and the results of complex analysis in these days. They describe the development of this branch in mathematics and, moreover, they demonstrate various applications of complex analysis, especially to partial differential equations.

J. NAAS

From the editors

As any scientific development the development of complex analysis is connected with an increasing specialization, too. Simultaneously, the significance of each individual scientific result has to be measured against the contribution it makes to the whole science of mathematics or at least to a sufficiently general branch of mathematics. For this reason, the organizers of the conference on "Complex Analysis and its Applications to Partial Differential Equations", regularly held at the Halle-Wittenberg Martin-Luther-University, devised an additional series of surveys aiming at a comprehensive explanation of complex analysis. From these surveys the present book originated.

The contents of such a comprehensive explanation as it is expressed in the disposition of our book has been discussed with many colleagues over a longer period of time. We are very much obliged to them. In addition to this, we obtained valuable hints by persons not included among the authors of the chapters of the book. It is not possible to name them all here, but the following persons should be mentioned representing them: First, we name Prof. H. Grunsky, who gave us the first idea to expand the comprehensive explanation of complex analysis in a series of surveys in order to make a book of it. Second, we thank Prof. V. S. Vladimirov of the Steklov Institute of the Academy of Sciences of the Soviet Union for his valuable advices. We are especially grateful to Prof. J. Naas, who permantly followed the concept of the book at issue and has enriched it by many contributions. For instance he contributed the idea to improve the homogenity of the present explanation by an integral chapter on "Classical and modern methods of complex analysis". We are also very grateful to the authors of the chapters for contributing to the elaboration of the whole concept by many ideas.

But the present book is not only the result of team-work with respect to its contents. The high number of authors necessarily required an adaption and co-ordination which was achieved by colleagues working at the Departments of Mathematics of the Halle-Wittenberg Martin-Luther-University and the Karl-Marx-Stadt Technische Hochschule who are interested in complex analysis. These are Dipl.-Math. S. Carl, Dr. sc. B. Goldschmidt, Dr. F. Heynig, Dr. sc. Lê hùng Són, Dr. W. Rüprich, and Dipl.-Math. H.-P. Schneider. The co-ordination of the so many activities necessary at the final stage of elaboration was in the hands of Dr. H. Malonek and Dipl.-Math. V. Pluschke, to whom we are very grateful. The common list of references was compiled by Mrs. Dipl.-Math. M. Wolf. Dr. H.-H. Buchsteiner of the Department of Mathematics of the Halle-Wittenberg Martin-Luther-University supported the elaboration of this book as he did in other cases before. We were given essential advices concerning the English version by Mrs. I. ECKARDT. We are also grateful to the staff of Akademie-Verlag, especially to Miss Dipl.-Math. G. REIHER and Dr. R. HÖPPNER, for the constructive co-operation.

E. LANCKAU and W. TUTSCHKE

Contents

Introduction

The present book deals with a discussion, with generalizations, and with applications of holomorphy, where holomorphy will be regarded as an autonomous concept in mathematics, especially in mathematical analysis. The concept of holomorphy will be interpreted not only from the view-point of complex analysis or even from the view-point of the more special complex function theory, but will also be discussed from the view-point of the whole area of mathematics.

Those who learned complex function theory from a usual lecture for mathematicians or physicists could get the impression, that this theory is closed in essence. Such an impression could arise, because many problems of complex function theory are very natural and, moreover, the most of the problems are solved by standard methods and formulated in natural theorems, too. Thus one could get an opinion that there are no problems in complex function theory so that this theory could lead to a new furious development of mathematics.

A deeper look at the continual growth of problems and knowledge in mathematics shows, that complex methods still play a significant role and contribute to new ideas and results in these days. One important reason for a new revival of complex analysis lies in the fact, that the relations between complex analysis and the theory of partial differential equations have been coming up to a new quality. Whereas the applications of complex analysis to partial differential equations earlier were side results, today one of the main aims of complex analysis is its systematic application to general classes of partial differential equations. Such new tendencies were started by S. BERGMAN's theory of *integral operators* [1], I. N. VEKUA's theory of *generalized analytic* functions [5] and L. BERS' theory of *pseudo-analytic* functions [1]. The theories of generalized analytic and pseudo-analytic functions are applicable to uniformly elliptic linear systems of first order for two unknown real-valued functions in the plane. Enlarging these aims today the complex analysis is applicable to systems for more than two unknown real-valued functions, to systems in $\mathbb{R}^n$ and, finally, to nonlinear differential equations.

There are three aspects of this enlargement of the aims of complex analysis: First it is possible to interpret the pecularities of holomorphic functions as properties of solutions of special systems of partial differential equations. Second complex analysis becomes applicable to general classes of differential equations, not only to special ones. And third the new general complex analysis is able to construct solutions and to describe the properties of given solutions with the help of solutions of corresponding problems for holomorphic functions. This third aspect is significant, since in the case of nonlinear equations it means their reduction to linear problems. In addition this third aspect shows that the general complex analysis is able to make use of results of classical function theory, even of such results which originally have not been connected with partial differential equations.

In order to realize the enhancement of the aims of complex analysis described above it is necessary to supply its methods. The most important integral representation in classical function theory is Cauchy's integral formula for holomorphic functions. In the general complex analysis additionally integral representations for nonholomorphic functions are used. As it is known Cauchy's integral representation is based on curve integrals with the Cauchy kernel $1/(\zeta - z)$ (where ζ is the integration variable). On the contrary in the general complex analysis double integrals with the kernels $1/(\zeta - z)$ and $1/(\zeta - z)^2$ are needed. Using these integrals it is possible to solve inhomogeneous Cauchy-Riemann partial differential equations. The solution of these equations makes it possible to include non-holomorphic functions into the considerations of complex analysis (as it is known the real and the imaginary part of a holomorphic function form a solution of the homogeneous Cauchy-Riemann system). On the other hand the concept of a solution in the general complex analysis can be interpreted in Sobolev's sense, so that it turns out to be possible to regard differential equations with weak assumptions on the coefficients. Using solutions in Sobolev's sense it will be possible to apply functional analytic and topological methods for solving partial differential equations (this means on one hand proofs for the existence of solutions; on the other hand such methods lead to approximate and exact solutions).

The consideration of inhomogeneous Cauchy-Riemann equations is important not only for applications of complex methods in the non-holomorphic case. Inhomogeneous Cauchy-Riemann equations, moreover, lead to new methods in the theory of holomorphic functions themselves. The theory of holomorphic functions of several complex variables, based on the systematic use of inhomogeneous Cauchy-Riemann equations, had been developed by L. Hörmander [2].

The present book introduces new methods of complex analysis and, moreover, it compares and connects these methods with classical ones. The book cannot give a complete description of all results in complex analysis, even not of all of the essential ones. Also it is not possible to consider all different directions in complex analysis proportionally to their "extend". Especially the new results in complex analysis are not equally covered by the book. In the main such new tendencies in complex analysis have been taken into the book, which express new relations between complex analysis and the theory of partial differential equations.

This book is not only written for specialists in complex analysis. Moreover it is directed to all mathematicians, physicists and to all scientists, who are interested in analysis, in general. The reader will find out, that the specific method in complex analysis is much more than only the formal connection of the real variables with one complex one. It is the systematic application of properties of holomorphic functions. So the main aim of this book is to illustrate the contributions of complex analysis to the progress in science and technic.

1. Classical and Modern Methods of Complex Analysis

W. Tutschke

1.1. The starting points of complex analysis

1.1.1. Complex numbers guarantee that every algebraic equation (of positive degree) in one variable possesses one solution at least (cf. 1.8.1.). On the other hand complex numbers may be interpreted as ordered pairs $z = (x, y) = x + iy$ of real numbers and, consequentely, as points in the plane. Thus functions f defined in a set G of the x, y-plane are functions of the only (complex) variable z. In order to express the real coordinates x, y by the complex variable z we associate to the complex number z the so-called *conjugate complex* number $\bar{z}$ that is defined by $\bar{z} = x - iy$. Then we have $x = (z + \bar{z})/2$ and $y = \mathrm{i}\,(\bar{z} - z)/2$.

1.1.2. Assume now that G is a domain in the z-plane and f is continuously differentiable in G. Let $z_0 = x_0 + iy_0$ be a fixed point of G. Then the linearization of f at z_0 is given by

$$\tilde{f}(z) = f(z_0) + c_1(x - x_0) + c_2(y - y_0)\,,$$

where c_1 and c_2 are the derivatives of f with respect to x and y respectively at the point (x_0, y_0). Expressing $x - x_0$ and $y - y_0$ by $z - z_0$ we get

$$\tilde{f}(z) = f(z_0) + d_1(z - z_0) + d_2(\bar{z} - \bar{z}_0)\,,$$

where

$$d_1 = \tfrac{1}{2}\,(c_1 - \mathrm{i}c_2)\,, \qquad d_2 = \tfrac{1}{2}\,(c_1 + \mathrm{i}c_2)\,. \tag{1}$$

The coefficients d_1, d_2 are called the values of the *partial complex derivatives* of f at the point z_0 with respect to z and $\bar{z}$, respectively. Denote these derivatives by $\partial f/\partial z$ and $\partial f/\partial \bar{z}$, resp. Thus in view of (1) we define

$$\frac{\partial f}{\partial z} := \frac{1}{2}\left(\frac{\partial f}{\partial x} - \mathrm{i}\frac{\partial f}{\partial y}\right), \qquad \frac{\partial f}{\partial \bar{z}} := \frac{1}{2}\left(\frac{\partial f}{\partial x} + \mathrm{i}\frac{\partial f}{\partial y}\right).$$

1.1.3. In the following we assume, that f is complex-valued (at least we must assume, that it is defined the product of the values of f with complex numbers). Let G_0 be a subdomain of G possessing a smooth boundary ∂G_0, $\overline{G_0} = G_0 \cup \partial G_0 \subsetneq G$ (see fig. 1). Using the definition of partial complex derivatives (see 1.1.2.) the *Ostrogradskii-Gauss formula* may be written in the complex form

$$\iint\limits_{G_0} \frac{\partial f}{\partial \bar{z}}\,\mathrm{d}x\,\mathrm{d}y = \frac{1}{2\mathrm{i}} \int\limits_{\partial G_0} f\,\mathrm{d}z\,. \tag{2}$$

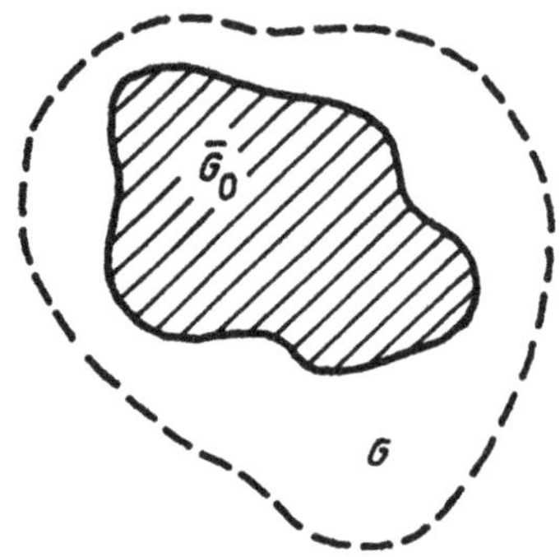

Fig. 1

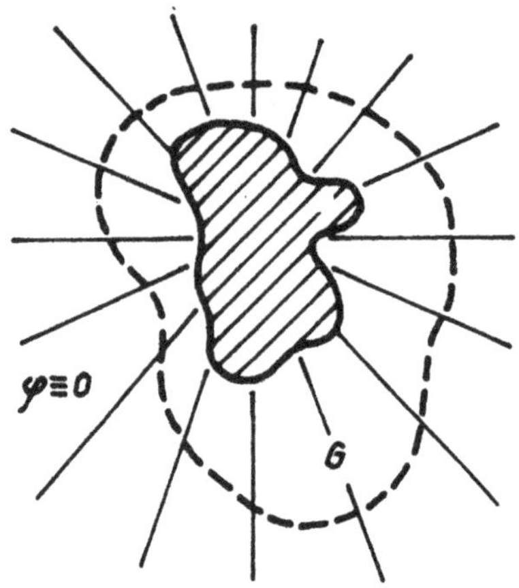

Fig. 2

1.1.4. Applying the integral formula (2) to the product $f\varphi$ we get

$$\iint_G \left(f \frac{\partial\varphi}{\partial\bar{z}} + \frac{\partial f}{\partial\bar{z}}\varphi\right) dx\, dy = 0\,, \tag{3}$$

where φ is any (complex-valued) test function belonging to $\mathscr{C}_0^1(G)$ (that means, that φ is continuously differentiable and vanishes identically outside of a compact subset of G, cf. fig. 2). The characterization (3) of classical derivatives leads to the following generalization:

The function g is called *Sobolev derivative* of f with respect to $\bar{z}$, if

$$\iint_G \left(f \frac{\partial\varphi}{\partial\bar{z}} + g\varphi\right) dx\, dy = 0 \quad \text{for every} \quad \varphi \in \mathscr{C}_0^1(G)\,.$$

Also the Sobolev derivative g is denoted by $\partial f/\partial\bar{z}$. We remark that analogously the relation

$$\iint_G \left(f \frac{\partial\varphi}{\partial z} + g\varphi\right) dx\, dy = 0$$

defines a derivative in Sobolev's sense with respect to z.

1.2. The inverses of partial complex differentiations[1])

1.2.1. In order to reduce differential equations to integral or operator equations one needs an inverse to the given differential operator. In the case of the differential operator $\partial/\partial\bar{\zeta}$ such an inverse is given by the integral operator

$$T_G h[\zeta] := -\frac{1}{\pi}\iint_G \frac{h(z)}{z-\zeta}\, dx\, dy\,, \tag{4}$$

[1]) cf. VEKUA [5].

where h is a given (continuous) function in the (bounded) domain G. The T_G-operator is right-inverse to $\partial/\partial\bar{\zeta}$, that means

$$\frac{\partial}{\partial\bar{\zeta}} T_G h = h\,, \tag{5}$$

where $\partial/\partial\bar{\zeta}$ is understood in Sobolev's sense. The function $w_0 = T_G h$ is, consequently, a special solution of the so-called *inhomogeneous Cauchy-Riemann equation*

$$\frac{\partial w}{\partial\bar{\zeta}} = h\,. \tag{6}$$

The proof of the relation (5) is based on Fubini's theorem. The integral in (4) exists as ordinary integral; this is true for every integral with singular integrand, if the integrand can be estimated by

$$\frac{\text{const}}{|z-\zeta|^\beta} \tag{7}$$

with $0 \leqq \beta < 2$. For $\beta = 2$ the corresponding integral does not exist as an ordinary integral, in general. For instance the integral

$$\Pi_G h[\zeta] := -\frac{1}{\pi}\iint\limits_G \frac{h(z)}{(z-\zeta)^2}\,\mathrm{d}x\,\mathrm{d}y \tag{8}$$

does not exist in the usual sense. But the following assertion can be proved:

If h is Hölder-continuous in G, then the integral (8) exists as *Cauchy's principal value*. That means that the integral is defined as limit by $\delta \to 0$ of the integrals about $G \setminus U_\delta(z)$, where $U_\delta(z)$ denotes the δ-neighbourhood of z. The function h, moreover, is called *Hölder-continuous* with exponent α, $0 < \alpha \leqq 1$, if there exists a constant H, such that $|h(z_1) - h(z_2)| \leqq H\,|z_1 - z_2|^\alpha$ for arbitrary points z_1, z_2 of $\bar{G}$. The set of all Hölder-continuous functions in $\bar{G}$ is denoted by $\mathscr{C}^\alpha(\bar{G})$. The existence of the integral (8) for $h \in \mathscr{C}^\alpha(\bar{G})$ follows immediately, if we write it in the form

$$-\frac{1}{\pi}\iint\limits_G \frac{h(z)-h(\zeta)}{(z-\zeta)^2}\,\mathrm{d}x\,\mathrm{d}y - \frac{h(\zeta)}{\pi}\iint\limits_G \frac{1}{(z-\zeta)^2}\,\mathrm{d}x\,\mathrm{d}y\ :$$

in view of the Hölder-continuity of h the integrand of the first integral satisfies the inequality (7) with $\beta = 2 - \alpha < 2$; by virtue of (2) the second integral is equal to

$$\frac{1}{2\mathrm{i}}\int\limits_{\partial G} \frac{\bar{z}}{(z-\zeta)^2}\,\mathrm{d}z = \frac{1}{2\mathrm{i}}\frac{\mathrm{d}}{\mathrm{d}\zeta}\left(\int\limits_{\partial G} \frac{\bar{z}}{z-\zeta}\,\mathrm{d}z\right)$$

(cf. 1.5.3.).

1.2.2. The functions $T_G h$ and $\Pi_G h$ are defined for $h \in \mathscr{C}^\alpha(\bar{G})$. In addition it can be proved, that (for $0 < \alpha < 1$) T_G and Π_G are bounded operators mapping $\mathscr{C}^\alpha(\bar{G})$ into itself. The norms in $\mathscr{C}^\alpha(\bar{G})$ we define by

$$||h||_{\mathscr{C}^\alpha(\bar{G})} := \max\left(\sup_G |h|,\ \sup_{z_1 \neq z_2} \frac{|h(z_2)-h(z_1)|}{|z_2-z_1|^\alpha}\right).$$

Then $\mathscr{C}^\alpha(\bar{G})$ is proved to be a Banach space. There is a very important connection between the T_G- and the Π_G-operator, namely for $h \in \mathscr{C}^\alpha(\bar{G})$ we have

$$\frac{\partial}{\partial\zeta} T_G h = \Pi_G h \tag{9}$$

in Sobolev's sense. The proof of this relation is similar to that of (5); it is also based on Fubini's theorem. Both relations (5) and (8) show, that $T_G h$ possesses Hölder-continuous first order derivatives. Let $\mathscr{C}^{1,\alpha}(\bar{G})$ be the Banach space of Hölder-continuously differentiable functions equipped with the norm

$$||h||_{\mathscr{C}^{1,\alpha}(\bar{G})} := \max\left(||h||_{\mathscr{C}^{\alpha}(\bar{G})}, \left\|\frac{\partial h}{\partial z}\right\|_{\mathscr{C}^{\alpha}(\bar{G})}, \left\|\frac{\partial h}{\partial \bar{z}}\right\|_{\mathscr{C}^{\alpha}(\bar{G})}\right).$$

If $0 < \alpha < 1$, then T_G is also proved to be a bounded operator mapping $\mathscr{C}^{1,\alpha}(\bar{G})$ into $\mathscr{C}^{\alpha}(\bar{G})$.

1.3. Weyl's lemma and its meaning

1.3.1. Let Φ be in G any (continuous) function, for which the derivative $\partial\Phi/\partial\bar{z}$ exists in Sobolev's sense and vanishes identically. In view of (3) this assumption means, that

$$\iint_G \Phi \frac{\partial \varphi}{\partial \bar{z}} \, dx \, dy = 0 \quad \text{for every} \quad \varphi \in \mathscr{C}_0^1(G) . \tag{10}$$

Let $\tilde{\varphi}$ be an arbitrary twice continuously differentiable test function. Then $\varphi = \dfrac{\partial \tilde{\varphi}}{\partial z}$ belongs to $\mathscr{C}_0^1(G)$. In view of

$$4\frac{\partial^2}{\partial z \, \partial \bar{z}} = 4\frac{\partial^2}{\partial \bar{z} \, \partial z} = \frac{\partial^2}{\partial x^2} + \frac{\partial^2}{\partial y^2} := \Delta \tag{11}$$

one gets from (9)

$$\iint_G U \, \Delta\tilde{\varphi} \, dx \, dy = 0 \,, \qquad \iint_G V \, \Delta\tilde{\varphi} \, dx \, dy = 0$$

for every $\tilde{\varphi}$, where $\Phi = U + iV$. From these relations Weyl [1] had deduced, that U and V resp. are *harmonic* in the following sense: in every disk the mean values of U and V resp. are equal to the values of U and V resp. at the center of the disk. On the other hand every harmonic function is proved to be an (arbitrarily often differentiable) solution of the Laplace equation

$$\Delta U = 0 \,, \qquad \Delta V = 0 \,. \tag{12}$$

Thus Φ must be arbitrarily often differentiable and the following theorem (Weyl's lemma) is proved:[1] *Let Φ be any (continuous) function fulfilling the relation*

$$\iint_G \Phi \frac{\partial \varphi}{\partial \bar{z}} \, dx \, dy = 0 \quad \textit{for every} \quad \varphi \in \mathscr{C}_0^1(G) .$$

Then Φ is an arbitrarily often differentiable solution of the so-called (homogeneous) Cauchy-Riemann equation

$$\frac{\partial \Phi}{\partial \bar{z}} = 0 \,. \tag{13}$$

1.3.2. Continuously differentiable functions fulfilling equation (13) are called *holomorphic*. Thus every function Φ fulfilling the relation (10) is proved to be holomorphic.

[1]) The book Tutschke [11] contains an elementary and complete proof.

1.3.3. As consequence of Weyl's lemma we get, for instance, that the *general solution* w of the inhomogeneous Cauchy-Riemann equation (6) is given by

$$w = w_0 + \Phi\,,$$

where w_0 is the special solution (4) and Φ is an arbitrary holomorphic function.

The complex equation (13) is equivalent to the real system

$$\begin{aligned} \frac{\partial U}{\partial x} - \frac{\partial V}{\partial y} &= 0\,, \\ \frac{\partial V}{\partial x} + \frac{\partial U}{\partial y} &= 0\,, \end{aligned} \tag{14}$$

from which it follows again, that U and V are solutions of the Laplace equation (12).

Let Φ be holomorphic, $\Phi \neq 0$. Then $\log \Phi$ is holomorphic, too. Since $\mathrm{Re}\,[\log \Phi] = \log |\Phi|$ the function $\log |\Phi|$ is a solution of the Laplace equation. Thus from the *maximum principle* for the Laplace equation one gets immediately, that the maximum principle holds for $|\Phi|$, too.

1.3.4. To a given solution U of the Laplace equation in simply connected domains by the real (homogeneous) Cauchy-Riemann system (14) a function V is defined that is uniquely determined up to a constant. The function V is called *conjugate harmonic* to U. Then $U + \mathrm{i}V$ is holomorphic.

1.3.5. Finally we remark, that a continuously differentiable function fulfilling equation (13) possesses an ordinary complex derivative at every point z_0. That means, that

$$\lim_{z \to z_0} \frac{\Phi(z) - \Phi(z_0)}{z - z_0}$$

exists. Furthermore this limit is equal to the value of $\partial\Phi/\partial z$ at z_0. Thus for a holomorphic function Φ the derivative $\partial\Phi/\partial z$ is denoted by $\mathrm{d}\Phi/\mathrm{d}z$, too.

1.3.6. The concept of a harmonic function (see 1.3.1.) may be generalized in the following way, too: A real-valued function is called *subharmonic* (*superharmonic*), if the value at the center of a disk is not larger (not smaller) than the mean value in the disk.

1.4. Basic properties of holomorphic functions[1])

1.4.1. Let Φ be a holomorphic function in G. Let, moreover, G_0 be a subdomain of G with $\overline{G}_0 \subset G$. In view of (2) and (13) one gets

$$\int_{\partial G_0} \Phi(z)\,\mathrm{d}z = 0 \tag{15}$$

(*Cauchy's integral theorem*).

Let ζ be a fixed point inside of G_0 (see fig. 3). Then $\Phi(z)/(z - \zeta)$ defines a holomorphic function in $G \setminus \{\zeta\}$. Applying (15) to this function in $G_0 \setminus \overline{U_\delta(\zeta)}$, where $U_\delta(\zeta)$ denotes the δ-neighbourhood of ζ,

$$\int_{\partial G_0} \frac{\Phi(z)}{z - \zeta}\,\mathrm{d}z = \int_{|z-\zeta|=\delta} \frac{\Phi(z)}{z - \zeta}\,\mathrm{d}z \tag{16}$$

1) Cf., for instance, Ahlfors [4], Dinghas [1], Knopp [1], A. I. and L. A. Markuševič [1], Privalov [3], Rühs [1], Šabat [1], Tutschke [1].

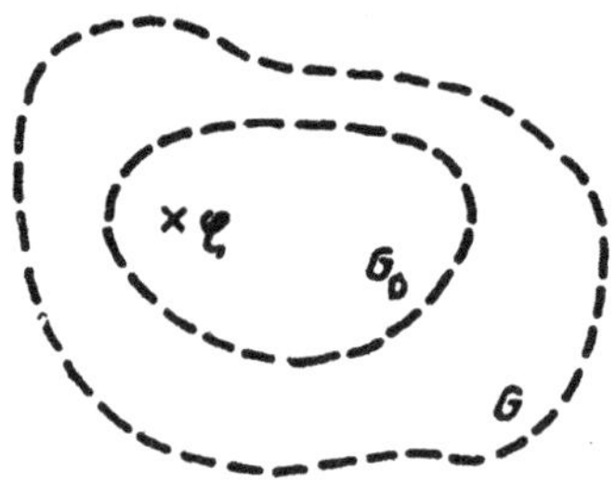

Fig. 3

follows. On the other hand the integral on the right-hand side can be written as

$$\int_{|z-\zeta|=\delta} \frac{\Phi(z) - \Phi(\zeta)}{z - \zeta} \, dz + \Phi(\zeta) \int_{|z-\zeta|=\delta} \frac{dz}{z - \zeta} .$$

The second integral in the last equation is equal to $2\pi i$. In view of the continuity of Φ the first integral tends to zero if $\delta \to 0$. Taking the limit $\delta \to 0$ from (16) one gets, consequentely,

$$\Phi(\zeta) = \frac{1}{2\pi i} \int_{\partial G_0} \frac{\Phi(z)}{z - \zeta} \, dz , \tag{17}$$

where integration is carried out in the positive direction (*Cauchy's integral formula*).

1.4.2. From Cauchy's integral formula one gets many important consequences, for instance the following one: Let $\mathcal{H}(G)$ be the set of all holomorphic functions in G, $\Phi_k \in \mathcal{H}(G)$. Assume, that 'the functions Φ_k converge locally uniformly in G (that means, that they converge uniformly on every compact subset of G). Then inside of subdomains G_0 the limit function is also representable by (17). Thus the limit function belongs to $\mathcal{H}(G)$. In consequence the space $\mathcal{H}(G)$ equipped with the topology of locally uniform convergence is complete (*Weierstrass' convergence theorem*). For a bounded domain G we regard, additionally, the set $\mathcal{H}(G) \cap \mathcal{C}(\bar{G})$ of all in G holomorphic functions that are continuous in the closure $\bar{G}$ of G. This set equipped with the maximum norm is proved to be a Banach space.

1.4.3. Another consequence of Cauchy's integral formula is an estimate of the derivative of a bounded holomorphic function. Let Φ be a bounded holomorphic function in G. Let, moreover, K be a compact subset of G possessing at least the distance $\delta > 0$ from the boundary of G. Then the disk with center at $\zeta \in K$ and radius r, $r < \delta$, belongs to G (see fig. 4). Differentiating (17) with respect to ζ one gets

$$\Phi'(\zeta) = \frac{1}{2\pi i} \int_{|z-\zeta|=r} \frac{\Phi(z)}{(z - \zeta)^2} \, dz$$

and, consequently,

$$|\Phi'(\zeta)| \leq \frac{1}{2\pi} \sup_G |\Phi| \frac{1}{r^2} 2\pi r = \frac{1}{r} \sup_G |\Phi| .$$

Taking the limit $r \to \delta$ it follows

$$\sup_K |\Phi'| \leq \frac{1}{\delta} \sup_G |\Phi| . \tag{18}$$

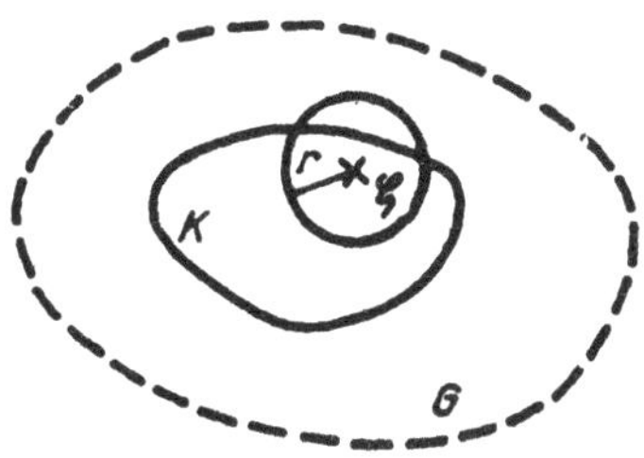

Fig. 4

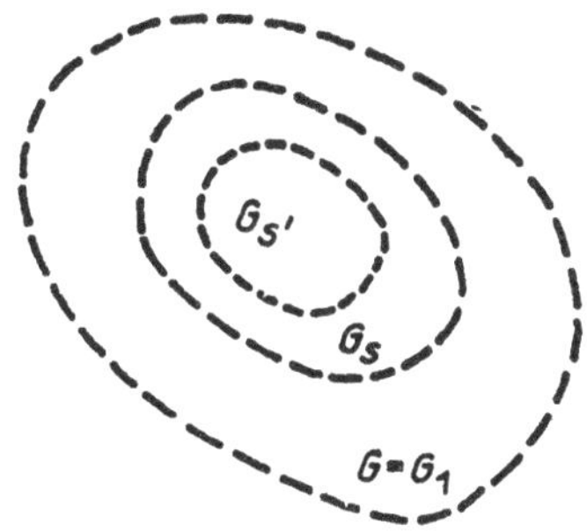

Fig. 5

Let G_s, $0 < s \leqq 1$, be a family of subdomains of G possessing the following properties (cf. fig. 5)

a) $G_{s'} \subset G_s$, if $s' < s$,

b) for $s' < s$ the distance of $\bar{G}_{s'}$ from ∂G_s is at least equal to const $\cdot\ (s - s')$,

c) $G_1 = G$.

Let $\mathscr{B}_s$ be the Banach space $\mathscr{H}(G_s) \cap \mathscr{C}(\bar{G}_s)$ equipped with the maximum norm $\|\cdot\|_s$. If $s' < s$, then $\mathscr{B}_s \subset \mathscr{B}_{s'}$. The norm in $\mathscr{B}_s$ is not smaller than the norm induced on $\mathscr{B}$ by $\mathscr{B}_{s'}$. Thus the $\mathscr{B}_s$ form a scale of Banach space. In view of (18) *the complex differentiation* $\mathrm{d}/\mathrm{d}\zeta$ *is a bounded operator mapping* $\mathscr{B}_s$ *into* $\mathscr{B}_{s'}$, where the operator norm is not larger than const $\cdot\ (s - s')$ (see TREVES [1]).

1.4.4. Let ζ_0 be a fixed point belonging to G_0 (see 1.4.1.) and ζ a point in a neighbourhood of ζ_0. Using the geometric series the kernel $1/(z - \zeta)$ of Cauchy's integral formula can be represented by

$$\frac{1}{z - \zeta} = \frac{1}{z - \zeta_0} \frac{1}{1 - \dfrac{\zeta - \zeta_0}{z - \zeta_0}} = \sum_{\nu=0}^{\infty} \frac{(\zeta - \zeta_0)^\nu}{(z - \zeta_0)^{\nu+1}} .$$

Substituting this representation into (17) term-by-term integration yields

$$\Phi(\zeta) = \sum_{\nu=0}^{\infty} a_\nu (\zeta - \zeta_0)^\nu , \tag{19}$$

where

$$a_\nu = \frac{1}{2\pi i} \int_{\partial G_0} \frac{\Phi(z)}{(z - \zeta_0)^{\nu+1}} \mathrm{d}z .$$

Thus every in G holomorphic function is *representable as power-series* with repect to $(\zeta - \zeta_0)$. The series (19) converges at least in the largest open disk of center ζ_0 contained in G (cf. fig. 6).

1.4.5. Functions permitting power-series representations possess many special properties. For example we regard the *distribution of the* w_0*-points*, in which a holomorphic function Φ takes a fixed value w_0. Let ζ_0 be any inverse image of w_0. Using the power-series representation of Φ in a neighbourhood of ζ_0 we get

$$\Phi(\zeta) - w_0 = (\zeta - \zeta_0)^k \left(a_k + a_{k+1}(\zeta - \zeta_0) + \ldots\right) ,$$

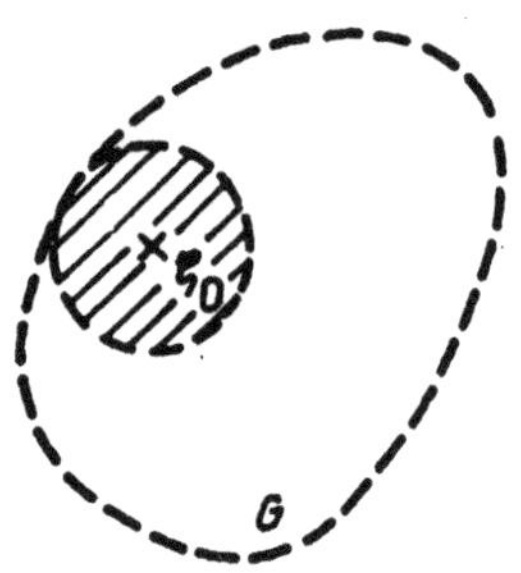

Fig. 6

if $a_0 = \ldots = a_{k-1} = 0$, $a_k \neq 0$. Since $a_k + a_{k+1}(\zeta - \zeta_0) + \ldots \neq 0$ for sufficientely small $|\zeta - \zeta_0|$, we get $\Phi(\zeta) \neq w_0$ in a neighbourhood of ζ_0 (unless Φ is constant). That means, that *the w_0-points of a non-constant holomorphic function are isolated.*

Not only such theorems on local value distribution are valid for analytic functions. Since the global behaviour of a holomorphic function is determined by its local behaviour (see 1.4.7. and 1.6.), theorems on global value distribution hold, too (see 1.4.15. and chapter 2., cf. also R. NEVANLINNA [4]).

1.4.6. In 1.4.4. it has been proved, that every holomorphic function is locally representable as power-series. Conversely let

$$\sum_{\nu=0}^{\infty} a_\nu(\zeta - \zeta_0)^\nu \tag{20}$$

be any given power-series. It is easy to prove, that there exists a real number $R \geqq 0$, such that the given series converges for all ζ with $|\zeta - \zeta_0| < R$, whereas it is divergent for ζ with $|\zeta - \zeta_0| > R$ (the number R is called the *radius of convergence*; it is uniquely determined by the given coefficients a_ν).

The convergence is uniform in compact subsets of the domain of convergence. Hence by virtue of Weierstrass' convergence theorem (see 1.4.2.) the power-series defines a holomorphic function (the partial sums are polynomials in z, therefore they are holomorphic).

1.4.7. Assume that the power-series (20) converges in the disk defined by $|\zeta - \zeta_0| < R$. The power-series defines, consequently, a holomorphic function Φ in the regarded disk. Let ζ_1 be any point of the disk. In the disk $\{\zeta : |\zeta - \zeta_1| < R - |\zeta_1 - \zeta_0|\}$ the function Φ may be represented as power-series with respect to $\zeta - \zeta_1$:

$$\Phi(\zeta) = \sum_{\nu=0}^{\infty} b_\nu(\zeta - \zeta_1)^\nu .$$

One gets this new power-series from the originally given one by the substitution $\zeta - \zeta_0 = (\zeta - \zeta_1) + (\zeta_1 - \zeta_0)$. It is possible that the new power-series converges in a larger disk with center at ζ_1 (cf. fig. 7). Then the new power-series leads to a (uniquely determined) so-called *analytic* (or *holomorphic*) *continuation* of Φ. This method can be repeated. It is possible, that continuations different from each other lead to different holomorphic functions in a neighbourhood of the same point (cf fig. 8; for instance continuation of any branch of the complex logarithm can lead to different branches). Thus analytic continuation leads to *many-valued* functions, in general. In consequence the global behaviour of holomorphic functions is connected with many-valued functions (see 1.6.).

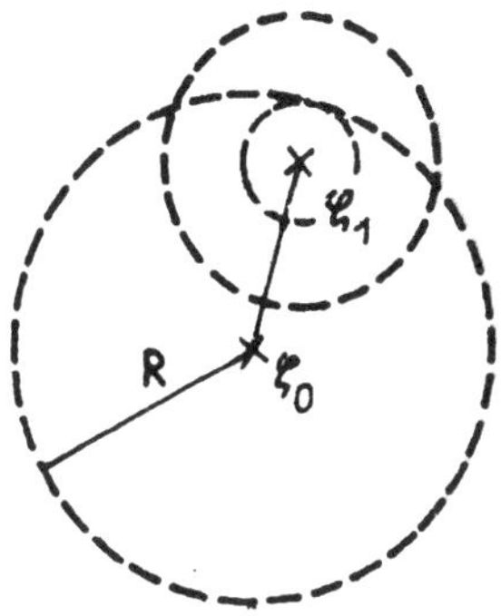

Fig. 7

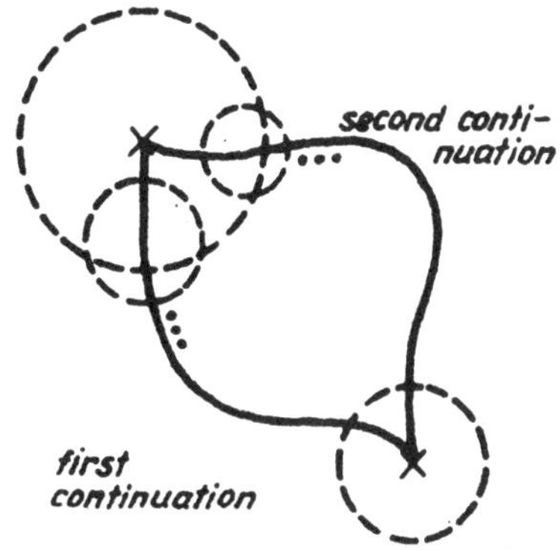

Fig. 8

1.4.8. There are also other methods in order to continue analytically a given holomorphic function. One of them is based on the following theorem:

Let Φ_1 be holomorphic in the domain G_1, Φ_2 in G_2, where G_1, G_2 do not possess common points. Assume, however, that both domains possess a common (smooth) boundary curve γ (see fig. 9). *We assume, moreover, that Φ_1, Φ_2 possess on γ the same continuous boundary values h. Then the function Φ defined in $G := G_1 \cup \gamma \cup G_2$ by*

$$\Phi = \begin{cases} \Phi_1 & in \quad G_1 , \\ h & on \quad \gamma , \\ \Phi_2 & in \quad G_2 \end{cases}$$

is proved to be holomorphic in the whole domain G.

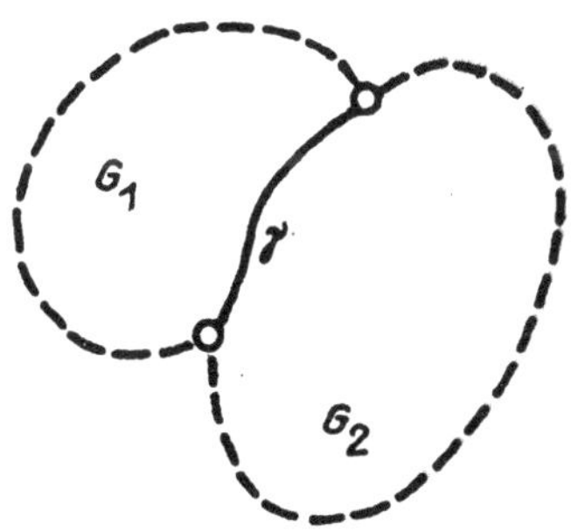

Fig. 9

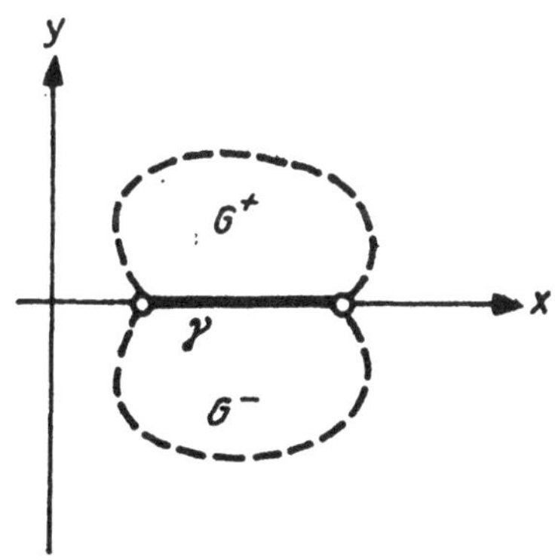

Fig. 10

The theorem follows immediately from Cauchy's integral theorem and Cauchy's integral formula. Hence such a continuation theorem holds always, if corresponding generalizations of Cauchy's integral theorem and Cauchy's integral formula are valid.

1.4.9. A constructive variant of the theorem 1.4.8 is the following reflection principle: Let G^+ be a domain in the upper half-plane, where the boundary ∂G^+ contains a segment γ of the real axis. Let G^- be the reflection image of G^+ with respect to the real axis (see fig. 10). Let $\Phi = \Phi(z)$ be defined in $G^+ \cup \gamma$. Assume, that Φ is holomorphic in G^+ and continuous and real-valued on γ. Then define by

$$w(z) = \begin{cases} \Phi(z) & \text{for} \quad z \in G^+ \cup \gamma , \\ \overline{\Phi(\bar{z})} & \text{for} \quad z \in G^- \end{cases} \tag{21}$$

a function in $G := G^+ \cup \gamma \cup G^-$. Since Φ is real-valued on γ the function w is continuous in the whole domain G. For $z \in G^-$ we have

$$\frac{\partial w}{\partial \bar{z}}(z) = \frac{\partial \overline{w}}{\partial \zeta}(\zeta) = \overline{\left(\frac{\partial w}{\partial \bar{\zeta}}\right)}[\zeta],$$

where $\zeta := \bar{z} \in G^+$. Applying the theorem 1.4.8 the function w is proved to be holomorphic in G. The method to continue Φ by (21) is called *Schwarz' reflection principle.* It is also applicable to holomorphic solutions of partial differential equation (see chapter 17.) and also to generalizations of holomorphic functions.

1.4.10. Let Φ be holomorphic in G. Assume, that the intersection of G with the real axis is not empty. Let x_0 be a point belonging to this intersection (see fig. 11). Then in view of 1.4.4. in a neighbourhood of x_0 the given function Φ is representable as power-series with respect to $z - x_0$. Especially we get the representation

$$\Phi(x) = \sum_{\nu=0}^{\infty} a_\nu (x - x_0)^\nu \tag{22}$$

or points x on the real axis.

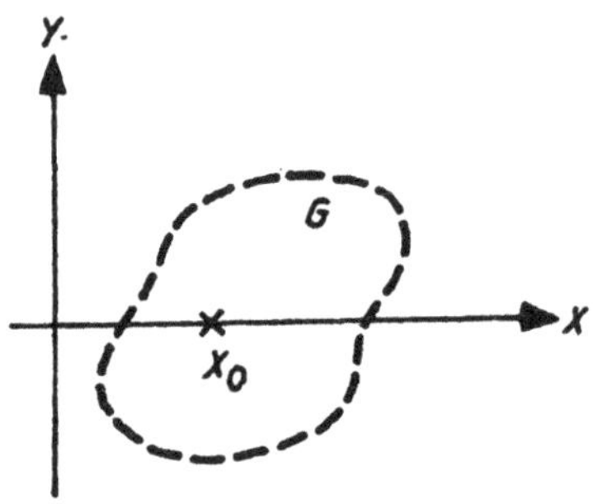

Fig. 11

Let, conversely, be given a real power-series (22). Applying the theorem formulated in 1.4.6. this power-series converges for all complex z with $|z - x_0| < R$, if it converges for all real x with $|x - x_0| < R$. That means: every function given locally as power-series can be continued into a neighbourhood of the real axis as holomorphic function.

Assume now, that the coefficients in (22) are real. Then the continuation into the complex domain leads to a uniquely determined holomorphic function with prescribed real values on the real axis. If $\Phi_0(x)$ is any power-series in $x - x_0$ possessing real values on the real axis, then $\Phi(z)$ and $\Phi(z) + i\Phi_0(z)$ define two holomorphic functions with the same real part on the real axis. There are also power-series in $z, \bar{z}$ (see 1.4.11.) with prescribed real part on the real axis, for instance $\Phi(z) + (z - \bar{z})\, \Phi_0(z)$.

From the continuation of a real power-series into the complex domain one gets consequences for the given function itself: assume, for instance, that the power-series (22) converges, if $|x - x_0| < R$. Let $M = \sum_{\nu=0}^{\infty} |a_\nu|\, r^\nu$, where $0 < r < R$. Then from (18) it follows $\sup_{|x-x_0| \leqq r/2} |\Phi'(x)| \leqq 2M/r$. Let, finally, $\Phi(x)$ be a real power-series fulfilling the inequality $|\Phi(x)| \leqq K$, if $|x - x_0| \leqq r$. Then, in general, it is not possible to deduce the estimate $|\Phi(z)| \leqq K$ for $|z - x_0| \leqq r$ (the modul of $\Phi(x) = 1 - x^2 + x^4 - + \dots = 1/(1+x^2)$ is bounded by 1, if $|x| < 1$; however $\Phi(z)$ is not bounded in $|z| < 1$). The last example also demonstrates the following fact: there are real-valued functions defined on the whole x-axis that are only *locally* (but not globally) *representable as power-series.* The uniquely determined holomorphic function that is equal to $\Phi(x)$ for $z = x$ is given by $\Phi(z) = 1/(1 + z^2)$. Since this function possesses singularities at the points $z = \mathrm{i}$ and $z = -\mathrm{i}$ it is not possible, that power-series representations for $1/(1 + x^2)$ converge globally (the representation $1/(1 + x^2) = \sum_{\nu=0}^{\infty} a_\nu (x - x_0)^\nu$ converges, if $|x - x_0| < \sqrt{1 + x_0^2}$, see fig. 12). Thus the complex point of view explains why power-series representations of $\Phi(x)$ in some cases are valid locally only.

1.4.11. Every power-series in z can be interpreted, of course, as power-series in x, y. Every power-series in x, y may be interpreted, conversely, as power-series in z and $\bar{z}$. For instance we have

$$\sum_{\nu=0}^{\infty} \sum_{\mu=0}^{\nu} \binom{\nu}{\mu} x^{2\mu} y^{2\nu-2\mu} = \sum_{\nu=0}^{\infty} (x^2 + y^2)^\nu = \sum_{\nu=0}^{\infty} z^\nu \bar{z}^\nu .$$

Thus complex methods can be used in the theory of functions possessing local power-series representations in x, y (see, for instance, Komin [1]). An important special case is the case of *polyanalytic* functions of n-th order. Such functions are polynomials in $\bar{z}$ of degree n with coefficients that are holomorphic (see chapter 3). Hence polyanalytic functions of n-th order are solutions of the differential equation

$$\frac{\partial^{n+1}\Phi}{\partial \bar{z}^{n+1}} = 0 .$$

Another possibility for a complex interpretation of real power-series is the following one: all of the given real variables $x_1, \dots, x_n$ are replaced by complex variables z_k independent from each other. In this way one gets a function that depends holomorphically on every of its complex variables (see 1.7.). The new function depending on $2n$ real variables is called the continuation of the given function into the complex domain.

1.4.12. Let Φ be a holomorphic function defined in the domain G of the z-plane. Then general theorems on the boundary behaviour of Φ are valid. Most simple is the behaviour of Φ at isolated boundary points z_0. In the case of isolated boundary points z_0 of G a neighbourhood $U_\delta(z_0)$ of z_0 exists, such that Φ is holomorphic in $U_\delta(z_0) \setminus \{z_0\}$. Then the function defined by $\Phi(z)/(z - z_0)$ is holomorphic in $U_\delta(z_0) \setminus \{z_0\}$, too. Choose r_1, r_2 with $0 < r_1 < r_2 < \delta$. Applying Cauchy's integral formula (17) with

$$G_0 = \{z : r_1 < |z - z_0| < r_2\} \text{ and } \zeta \in G_0$$

one gets (cf. fig. 13)

$$\Phi(\zeta) = \frac{1}{2\pi \mathrm{i}} \int\limits_{|z-z_0|=r_2} \frac{\Phi(z)}{z - \zeta}\, \mathrm{d}z - \frac{1}{2\pi \mathrm{i}} \int\limits_{|z-z_0|=r_1} \frac{\Phi(z)}{z - \zeta}\, \mathrm{d}z , \tag{23}$$

where integration is carried out in the positive direction. Using the method described in 1.4.4. it follows a power-series representation for the first integral. Analogously for z with $|z - z_0| = r_1$ we have $|\zeta - z_0| > |z - z_0|$, such that one gets

$$\frac{1}{z - \zeta} = -\frac{1}{\zeta - z_0} \cdot \frac{1}{1 - \dfrac{z - z_0}{\zeta - z_0}} = -\sum_{\mu=0}^{\infty} \frac{(z - z_0)^\mu}{(\zeta - z_0)^{\mu+1}} .$$

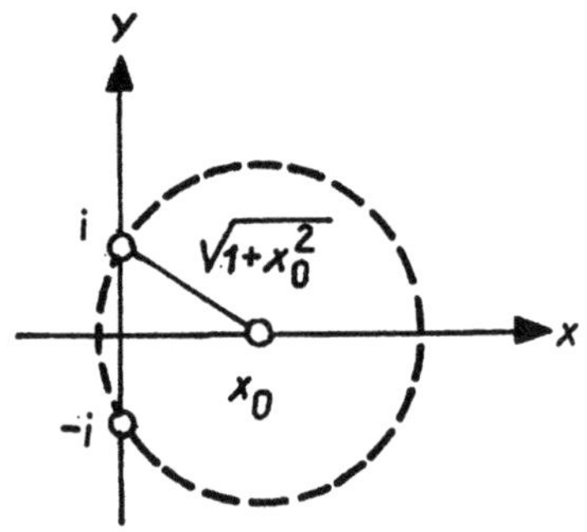

Fig. 12

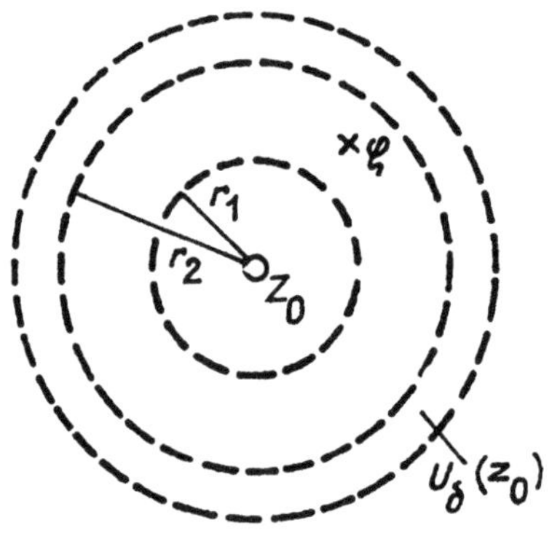

Fig. 13

Therefore the second integral in (23) possesses a representation of the form ($\mu + 1 = -\nu$)

$$\sum_{\nu=-\infty}^{-1} a_\nu(\zeta - z_0)^\nu ,$$

where

$$a_\nu = \frac{1}{2\pi i} \int\limits_{|z-z_0|=r_1} \frac{\Phi(z)}{(z - z_0)^{\nu+1}} dz$$

(we remark, that the last integral is not depending on r_1). For Φ we get, consequently, in $U_\delta(z_0) \setminus \{z_0\}$ the representation by a so-called Laurent's series

$$\Phi(\zeta) = \sum_{\nu=-\infty}^{+\infty} a_\nu(\zeta - z_0)^\nu . \tag{24}$$

The coefficient

$$a_{-1} = \frac{1}{2\pi i} \int\limits_{|z-z_0|=r} \Phi(z)\, dz$$

is called the *residue* of Φ at the point z_0 and is denoted by $\operatorname{Res}_{z_0} \Phi$. From the representation (24) follows immediately, that there are three possibilities for the behaviour of Φ at z_0:

a) All a_ν with negative index are zero,
b) only a finite number of coefficients a_ν with negative index are different from zero,
c) we have $a_\nu \neq 0$ for infinitely many negative indices.

In the case a) the function Φ is proved to be holomorphic in the whole neighbourhood $U_\delta(z_0)$ including the point z_0 (*removable singularity*). In the case b) (*pole*) the values $\Phi(\zeta)$ tend to infinity, if ζ tends to z_0 (cf. 1.6.3.). In the case c) (*essential singularity*), finally, the *Casorati-Weierstrass' theorem* holds: *To every complex number a and to every $\varepsilon > 0$ in every neighbourhood of z_0 at least one point $z \neq z_0$ exists with*

$$|\Phi(z) - a| < \varepsilon .$$

Regard now the case of a more general boundary, for instance, the case of a boundary curve. Then the representation (24), of course, is not valid. On the other hand it is possible to classify the behaviour of Φ at boundary points like in the case of isolated boundary points (see chapter 2, cf. also PRIVALOV [2] and NOSHIRO [1]). A holomorphic function possessing no other isolated singularities than poles is called *meromorphic.*

1.4.13. Another consequence of the holomorphy concerns the boundary behaviour of 1-1-mappings:

Let γ_1 and γ_2 be Jordan curves in the z-plane. The interior of γ_j is denoted by G_j. Assume that Φ is a holomorphic function in G_1 realizing a 1-1-mapping of G_1 onto G_2 (see fig. 14). Then Φ possesses limits at every point of γ_1, such that a natural extension of Φ into γ_1 exists. *This extension realizes a homeomorphism of* $G_1 \cup \gamma_1$ *onto* $G_2 \cup \gamma_2$.

Also the limits of the derivatives up to a certain order of Φ exist on γ_1, if γ_1 and γ_2 possess sufficiently differentiable representations.

We remark that the existence of holomorphic 1-1-mappings of G_1 onto G_2 follows from the *Riemann mapping theorem* (see 1.6.5. and 1.9.3.).

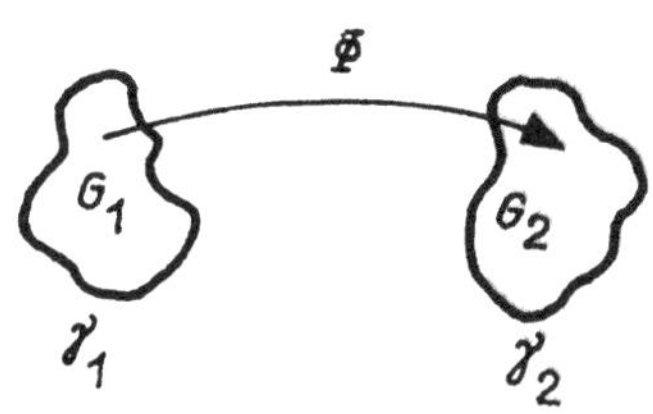

Fig. 14

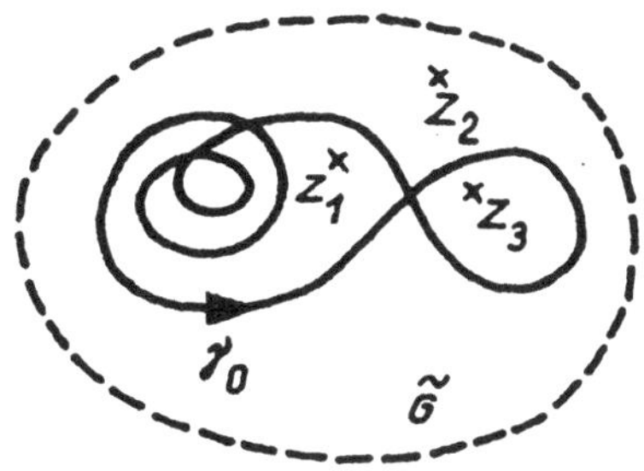

Fig. 15

1.4.14. Let G be a simply connected domain and $z_1, z_2, \ldots$ isolated points in G. Then the boundary of $\tilde{G} := G \setminus \{z_1, z_2, \ldots\}$ consists of the boundary of G and, moreover, of the isolated boundary points $z_1, z_2, \ldots$

Let Φ be a holomorphic function in $\tilde{G}$. Then a representation of type (24) holds in a neighbourhood of every z_j. Especially all the residues $\operatorname{Res}_{z_j} \Phi$ are defined. Finally we regard a closed (smooth or rectifiable) curve γ_0 in $\tilde{G}$ (see fig. 15). In view of the local uniform convergence the series (24) can be integrated term by term. On the other hand for $\nu \neq -1$ the function defined by $(\zeta - z_0)^\nu$ possesses a primitive function, namely $(\zeta - z_0)^{\nu+1}/(\nu + 1)$, such that

$$\int_{\gamma_0} \Phi(z)\, \mathrm{d}z = 2\pi \mathrm{i} \sum_k n(\gamma_0, z_k) \operatorname*{Res}_{z_k} \Phi$$

(*residue theorem*). The entire number $n(\gamma_0, z_k)$ is called *winding number* of γ_0 with respect to z_k. It can be represented by

$$n(\gamma_0, z_k) = \frac{1}{2\pi \mathrm{i}} \int_{\gamma_0} \frac{\mathrm{d}z}{z - z_k} = \frac{1}{2\pi \mathrm{i}} \int_{\gamma_0} \mathrm{d} \log (z - z_k) = \frac{1}{2\pi \mathrm{i}} \int_{\gamma_0} \mathrm{d} \arg (z - z_k)$$

and describes the increment of $\arg (z - z_k)$ along γ_0.

1.4.15. A simple, indeed, but important result of global value distribution theory for holomorphic functions is the *argument principle*:

Let Φ be in G holomorphic and different from a constant. Let, moreover, γ_0 be a closed (smooth) Jordan curve in G. The interior of γ_0 is denoted by G_0 (cf. fig. 16). The function Φ does not take all complex values on γ_0, since Φ is bounded on γ_0. Let w_0 be any complex number, such that $\Phi(z) \neq w_0$ on γ_0. Since the w_0-points of Φ are isolated (see 1.4.5.) the number of w_0-points of Φ in G_0 is finite. Let z_0 be any inverse image of w_0 in G_0. Its order is denoted by k. Then from the power-series-representation of Φ in a neighbourhood of z_0 it follows immediately, that

$$\operatorname*{Res}_{z_0} \frac{\Phi'}{\Phi - w_0} = k\,.$$

In view of the residue theorem *the number of w_0-points* counting according to order *is given by*

$$\frac{1}{2\pi i} \int_{\gamma_0} \frac{\Phi'(z)}{\Phi(z) - w_0}\, dz = \frac{1}{2\pi i} \int_{\gamma_0} d \log\big(\Phi(z) - w_0\big)$$
$$= \frac{1}{2\pi i} \int_{\gamma_0} d \arg\big(\Phi(z) - w_0\big)\,. \tag{25}$$

From this representation follows, for instance, that the regarded number will be unchanged, if one replaces w_0 by another value w in a neighbourhood of w_0. The global value distribution theory is more complicated for the whole domain G. Using an exhaustion of G this case can be reduced to that one, which concerns compact subsets. A systematic introduction into the value distribution theory is given in chapter 2.

1.5. The Cauchy type integral

1.5.1. The Cauchy integral formula (17) allows to represent a holomorphic function in the interior of a domain G_0 by its values on the boundary. Now regard the integral (17) with a complex-valued function $\psi = \psi(z)$ instead of $\Phi = \Phi(z)$, where ψ is given

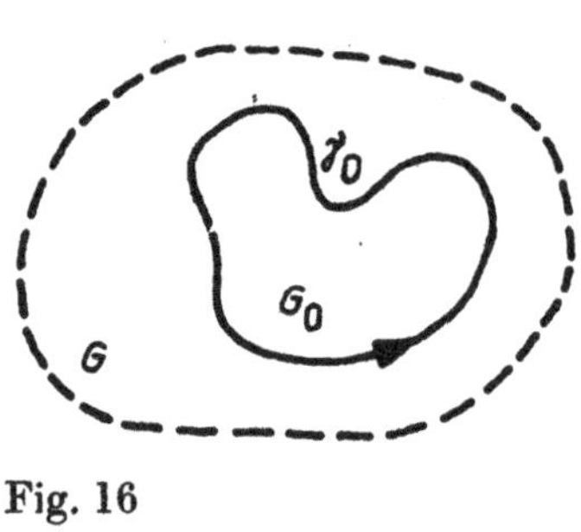

Fig. 16

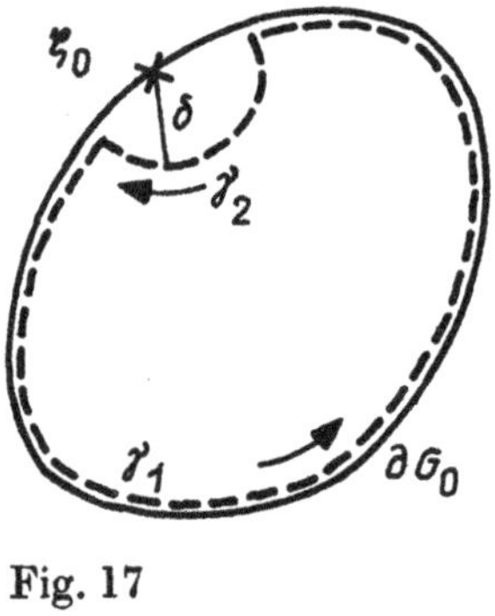

Fig. 17

only on the boundary ∂G_0 of a given (simply connected) domain G_0. That means we regard the so-called *Cauchy type integral*

$$\frac{1}{2\pi i} \int_{\partial G_0} \frac{\psi(z)}{z - \zeta}\, dz \tag{26}$$

with the *density* ψ. This integral depends on the complex parameter ζ. Immediately it exists and defines a complex-valued function Ψ for every ζ that is not belonging to the path of integration. Since the integrand depends holomorphically on ζ, the function Ψ defined by (26) is a holomorphic function in G_0 and in the outside region.

In order to study the integral (26) for points ζ_0 on the boundary, we write it in the form

$$\frac{1}{2\pi i}\int\limits_{\partial G_0}\frac{\psi(z)-\psi(\zeta_0)}{z-\zeta_0}\,dz+\frac{\psi(\zeta_0)}{2\pi i}\int\limits_{\partial G_0}\frac{1}{z-\zeta_0}\,dz\,. \tag{27}$$

Regard the closed curve $\gamma_1+\gamma_2$, where γ_1 is the part of ∂G_0 outside of the δ-neighbourhood of ζ_0 (see fig. 17). In view of Cauchy's integral theorem (15) we get

$$\int\limits_{\gamma_1+\gamma_2}\frac{1}{z-\zeta_0}\,dz=0\,. \tag{28}$$

Since $\lim\limits_{\delta\to 0}\int\limits_{\gamma_2} dz/(z-\zeta_0)=-\pi i$ we get from (28)

$$\lim_{\delta\to 0}\int\limits_{\gamma_1}\frac{1}{z-\zeta_0}\,dz=+\pi i \tag{29}$$

Regard an integral of type (26) with $\zeta_0\in\partial G_0$. If the limit

$$\lim_{\delta\to 0}\int\limits_{\gamma_1}\frac{\psi(z)}{z-\zeta_0}\,dz$$

exists, it is called the *Cauchy principal value* of the integral in question (cf. 1.2.1., where had been defined the Cauchy principal value for double-integrals). So one gets from (29), that

$$\int\limits_{\partial G_0}\frac{1}{z-\zeta}\,dz=\pi i \tag{30}$$

holds for $\zeta_0\in\partial G_0$ in the sense of Cauchy's principal value. Assume now that ψ is Hölder-continuous with exponent α, $0<\alpha\leqq 1$. Then we have

$$\left|\frac{\psi(z)-\psi(\zeta_0)}{z-\zeta_0}\right|\leqq\frac{\text{const}}{|z-\zeta_0|^{1-\alpha}}$$

and, consequently, the first integral in (27) exists in the ordinary sense. Thus we have proved:

If ψ is Hölder-continuous, then the Cauchy type integral (26) with $\zeta_0\in\partial G_0$ exists as Cauchy's principal value.

1.5.2. For the second integral in (27) we have

$$\int\limits_{\partial G_0}\frac{1}{z-\zeta}\,dz=\begin{cases}2\pi i\,, & \text{if}\quad \zeta\in G_0\,,\\ \pi i\,, & \text{if}\quad \zeta\in\partial G_0\,,\\ 0\,, & \text{if}\quad \zeta\notin\overline{G}_0\end{cases} \tag{31}$$

(the second equation (31) is identic with (30); the first and the third equation follow, for instance, from the residue theorem 1.4.14.). Using the Hölder-continuity of ψ it is possible to prove, that

$$\int_{\partial G_0} \frac{\psi(z) - \psi(\zeta_0)}{z - \zeta} \, dz \to \int_{\partial G_0} \frac{\psi(z) - \psi(\zeta_0)}{z - \zeta_0} \, dz , \tag{32}$$

if $\zeta \to \zeta_0 \in \partial G_0$ (to this and for the following see, for instance, MUSHELIŠVILI [2], POGORZELSKI [1]). Denote by $\Psi^+(\zeta_0)$ $(\Psi^-(\zeta_0))$ the limit of Ψ at $\zeta_0 \in \partial G_0$, if $\zeta \to \zeta_0$ and $\zeta \in G_0$ $(\zeta \notin \overline{G}_0)$. Then in view of (27), (31) and (32) one gets immediately the following *Sohockii-Plemelj formulas*:

$$\Psi^+(\zeta_0) = \Psi(\zeta_0) + \tfrac{1}{2}\,\psi(\zeta_0) , \qquad \Psi^-(\zeta_0) = \Psi(\zeta_0) - \tfrac{1}{2}\,\psi(\zeta_0) , \tag{33}$$

such that

$$\Psi^+(\zeta_0) - \Psi^-(\zeta_0) = \psi(\zeta_0) .$$

1.5.3. If ψ is Hölder-continuous on ∂G_0 with the Hölder exponent α, $0 < \alpha < 1$, then the Cauchy principal values of Ψ on ∂G_0 are also Hölder-continuous with the same Hölder exponent α. By virtue of (33) in this case the boundary values Ψ^+, Ψ^- are Hölder-continuous, too.

An analogous assertion holds for the derivatives.

1.5.4. If the given values ψ on ∂G_0 are the boundary values of a holomorphic function Ψ in G, then we have $\psi = \Psi^+$ on ∂G_0. In view of (33) that means

$$\Psi(\zeta_0) = \tfrac{1}{2}\,\psi(\zeta_0)$$

and, consequentely, $\Psi^-(\zeta_0) = 0$ at every point ζ_0 of G_0. Conversely we get from $\Psi^- \equiv 0$, that $\psi \equiv \Psi^+$. Thus we have proved:

Let ψ be a given (complex-valued and Hölder-continuous) function on ∂G_0. Its values are the boundary values of a holomorphic function in G_0, iff the boundary values Ψ^- of the Cauchy type integral with the density ψ vanish identically.

We remark, that in this case on ∂G_0 the values of the Cauchy type integral with the density ψ are equal to $\frac{1}{2}\,\psi$ (this follows immediately from the second equation (33)).

1.5.5. In the following we shall demonstrate how properties of holomorphic functions can be used for solving singular integral equations.

Regard on ∂G_0 the singular integral equation

$$\frac{1}{2\pi i} \int_{\partial G_0} \frac{\psi(z)}{z - \zeta} \, dz = f(\zeta) , \tag{34}$$

where f is a given (Hölder-continuous) function on ∂G_0 and ψ is the (Hölder-continuous) solution looked for.

Let ψ be any solution of the homogeneous equation ($f \equiv 0$). Then we regard the Cauchy type integral Ψ with the density ψ. Since $\Psi \equiv f$ on ∂G_0 we have $\Psi \equiv 0$ on ∂G_0, such that in view of (33) we get

$$\Psi^+ = \tfrac{1}{2}\psi , \qquad \Psi^- = -\tfrac{1}{2}\psi . \tag{35}$$

From the first equation (35) we get, that ψ are the boundary values of a holomorphic function. Using the theorem in 1.5.4. we get, consequently, $\Psi^- \equiv 0$. Then the second equation (35) shows, that $\psi \equiv 0$. Thus the solution of the inhomogeneous equation (34), if any, is uniquely determined.

Let F be the Cauchy type integral with density f. By virtue of the first equation (33) we have

$$F = F^+ - \tfrac{1}{2} f . \tag{36}$$

According to 1.5.4. the Cauchy type integral with the density F^+ is equal to $\frac{1}{2}F^+$ on ∂G_0. Hence the Cauchy type integral with the density (36) is equal to

$$\tfrac{1}{2}F^+ - \tfrac{1}{2}F = \tfrac{1}{4}f .$$

Thus *the Cauchy type integral with the density* $4f$ *is the uniquely determined solution of the singular integral equation* (34).

1.5.6. In the last section 1.5.5. it had been shown, that the concept of holomorphy leads to the solution of certain singular integral equations. Conversely many special problems for holomorphic functions (boundary problems and so on) can be deduced to singular integrals and to singular integral equations (see chapter 10 and chapter 20). Analogously one needs singular double-integrals in the theory of non-holomorphic complex-valued functions in the plane (see 1.2., 1.10.).

The Cauchy type integral can be studied also with respect to the boundary of multiply connected domains and with respect to open arcs. There are also integrals in $\mathbb{R}^n$, $n \geqq 3$, that may be regarded as higher-dimensional generalizations of the Cauchy type integral (concerning the corresponding Cauchy integral formula see chapter 16). Such integrals possess similar properties like (33).

In the sense of the distribution theory (see Schwartz [1]) one has

$$\frac{\partial}{\partial \bar{z}} \frac{1}{z} = \delta$$

where δ is the Dirac δ-distribution. Thus the kernel $1/(z - \zeta)$ of the Cauchy type integral may be regarded as fundamental solution of the differential equation $\partial w/\partial \bar{z} = 0$ (with the singularity at ζ). So the Cauchy type integral is a complex version of the Newtonian potential.

1.6. Global points of view in complex analysis

1.6.1. Complex analysis is essentially founded on the concept of complex differentiation (see 1.1.2.). Thus the complex analysis is based on local properties of the regarded functions (this concerns derivatives in classical sense as well as derivatives in Sobolev's sense). On the other hand the global existence of functions can depend on the structure of the domain, in which we look for it. This is a general fact in the theory of (partial) differential equations. We shall illustrate it in the case of complex analysis by two simple examples.

Example 1. Let G be a simply connected domain in the z-plane and Φ holomorphic in G. Then a holomorphic function w exists in G that is a solution of the differential equation

$$\frac{\partial w}{\partial z} = \Phi . \tag{37}$$

This solution is uniquely determined up to an arbitrary complex constant. The solution is given by

$$w(z) = \text{const} + \int_{\gamma(z_0,z)} \Phi(\zeta)\, d\zeta\,,$$

where z_0 is a fixed point in G and $\gamma(z_0, z)$ is any (smooth) curve connecting z_0 and z in G. Regard now instead of a simply connected domain the annulus

$$G = \{z : r_1 < |z| < r_2\}\,. \tag{38}$$

Then a globally defined solution of (37) exists in G, if and only if the given function f fulfills the side condition

$$\int_{|\zeta|=r} \Phi(\zeta)\, d\zeta = 0\,,$$

where r is any number fulfilling the inequality $r_1 < r < r_2$.

Example 2. The real part of a holomorphic function is a solution of the Laplace equation (12). Thus in simply connected domains holomorphic functions with prescribed real part on the boundary exist (the imaginary part is the conjugate harmonic function (see 1.3.4.); it is uniquely determined up to an arbitrary constant). In the case of the annulus (38) this (Dirichlet) boundary value problem for holomorphic functions is not solvable for all boundary values. For instance the function looked for is given by

$$w(z) = c_1 + (c_2 - c_1)\left(\log \frac{r_2}{r_1}\right)^{-1} \log \frac{z}{r_1} + i\,\text{const}\,, \tag{39}$$

if we prescribe constant boundary values c_1, c_2 on both boundary components $|z| = r_1$ and $|z| = r_2$, respectively (this representation of $w(z)$ is obvious, since $\log z = \log r + i\,\vartheta$, where r, ϑ are polar coordinates of z with the center at the origin). The function $w = w(z)$ given by (39) is uniquely defined globally, if and only if $c_1 = c_2$.

1.6.2. The influence of the structure of a domain on the existence of (global) solutions (see 1.6.1.) is one of the connections between global and complex analysis. Another one concerns the fact, that the local behaviour of a holomorphic function determines its global behaviour. Using the method of analytic continuation (see 1.4.7.) one can calculate the values of a holomorphic function, if one knows it in a disk only.

Example 1. Regard the power-series $\Phi(z) = \sum_{\nu=0}^{\infty} z^\nu$ that converges in the disk defined by $|z| < 1$. Since

$$\Phi(z) = \frac{1}{1-z}$$

the function Φ can be continued as holomorphic function into the whole plane excluding the point $z = 1$. That means: the regarded power-series possesses the following global property: An analytic continuation exists realizing a 1-1-mapping of

$$\{z : z \neq 1\} \quad \text{onto} \quad \{w : w \neq 0\}\,.$$

Example 2. Let $G = \{z : z \neq 0\}$ and $\Phi(z) = 1/z$. Then Φ is holomorphic in G. Each branch of the many-valued function

$$\log z = \log |z| + i \arg z$$

is (local) a primitive function of Φ. The many-valued function $\log z$ takes every value

$w \neq 0$. For a power-series

$$\Phi(z) = \sum_{\nu=0}^{\infty} a_\nu (z - z_0)^\nu$$

we have

$$\Phi^{(k)}(z_0) = k!\, a_k\,,$$

such that the following result also holds: the global behaviour of a holomorphic function is uniquely determined by its derivatives at one point. Such a result does not hold for functions, that are arbitrarily often differentiable only: the function $\tilde{\Phi}$ defined by

$$\tilde{\Phi}(x) = \begin{cases} 0, & \text{if } x \leqq 0\,, \\ \exp\left(-\dfrac{1}{x^2}\right), & \text{if } x > 0 \end{cases}$$

is infinitely differentiable and all its derivatives vanish at $x = 0$. On the other hand this function does not vanish identically. Therefore the principle of unique analytic continuation cannot be true for infinitely differentiable functions without power-series representation.

1.6.3. The method of analytic continuation leads to *many-valued* holomorphic functions (see 1.4.7. and 1.6.2., example 2). In order to interpret such functions as uniquely defined ones, one regards several exemplars of the z-plane. Then every *branch* of the many-valued function will be regarded as *single-valued* function on the corresponding sheet. The totality of needed sheets forms a *manifold* in the sense of differential geometry (see, for instance, Hodge [1]). As local coordinates in the seperate branches one can introduce the usual complex variable z, if the regarded manifold arises from the analytic continuation of a given power-series. On the other hand there are other cases in complex analysis, in which it is necessary to introduce other local parameters. To it we regard the following two examples.

Example 1. Regard the function Φ defined by

$$\Phi(z) = \frac{1}{1 - z}\,.$$

It is a meromorphic function in the whole z-plane taking all values $w_0 \neq 0$. On the other hand we have $\Phi(z_k) \to 0$ for $k \to +\infty$, if $|z_k| \to +\infty$. This suggests to complete the z-plane by one infinite point ∞, such that Φ takes the value zero at this new point $z = \infty$. A neighbourhood of ∞ is given by the exterior of a (closed) disk. In order to introduce local coordinates in a neighbourhood of $z = \infty$ we use the mapping $z = 1/\zeta$. It maps the set $\{\zeta : 0 < |\zeta| < r\}$ onto $\{z : 1/r < |z|\}$, such that $\zeta = 0$ may be interpreted as inverse image of $z = \infty$. Thus the variables ζ with $|\zeta| < r$ are local coordinates in a neighbourhood of $z = \infty$ in the so-called closed (or extended) z-plane. If a point $(\neq \infty)$ belongs to a neighbourhood of ∞ then both z and ζ are admissible coordinates. Since we have $z = 1/\zeta$ and $\zeta = 1/z$ resp., we see, that admissible coordinates are connected by a univalent holomorphic function (a univalent function takes different from each other values at different points). We remark, moreover, that the completed z-plane may be interpreted as *Riemann z-sphere*. Introducing the new coordinate ζ the given function Φ may be represented in a neighbourhood of $z = \infty$ as

$$\Phi(z(\zeta)) = \frac{1}{1 - \dfrac{1}{\zeta}} = \frac{\zeta}{\zeta - 1}\,.$$

So the function Φ is proved to be holomorphic with respect to the variable ζ in a neighbourhood of $z = \infty$ including the point $z = \infty$. On the other hand we have $|\Phi(z_k)| \to +\infty$ for $k \to +\infty$, if $z_k \to 1$. Thus we can put $\Phi(1) = \infty$. Then Φ is a (meromorphic) 1-1-mapping of the z-sphere onto the ζ-sphere.

Example 2. Regard the function Φ defined by

$$w = \Phi(z) = z^2 .$$

It maps the z-plane onto the w-plane. Using the definition of the point ∞ (see example 1) the function may be interpreted as mapping of the z-sphere onto the w-sphere, too. However the mapping is not univalent. Every point w with $w \neq 0$ and $w = \infty$ possesses two inverse images $z_1 = \sqrt{w}$ and $z_2 = -\sqrt{w}$ different from each other whereas the points $w = 0$ and $w = \infty$ resp. possess the uniquely determined inverse images $z = 0$ and $z = \infty$ resp. In order to interpret the inverse function as a uniquely defined function we introduce two exemplars of the w-sphere, $w \neq 0$ and $w \neq \infty$. In a neighbourhood of $w = 0$ and $w = \infty$ resp. we introduce $\omega = \sqrt{w}$ and $\omega = 1/\sqrt{w}$ resp. as local coordinates. Then we get

$$z = \Phi^{-1}(w(\omega)) = \omega$$

and

$$z = \Phi^{-1}(w(\omega)) = \frac{1}{\omega}$$

resp. So the inverse function Φ^{-1} is proved to be a uniquely determined, meromorphic function on a covering surface of the w-sphere. For points $w \neq 0$ ($w \neq \infty$) in a neighbourhood of $w = 0$ ($w = \infty$) we can introduce both w and $\omega = \sqrt{w}$ ($w = 1/\sqrt{w}$) as local coordinates. Then the transformation of local coordinates is given by a univalent holomorphic function.

Generalizing the two regarded examples we define now *abstract Riemann surfaces*. A Riemann surface is an orientable two-dimensional manifold, for which the transformations of local coordinates are given by univalent holomorphic functions (see, for instance, HODGE [1], R. NEVANLINNA [5]).

A complex-valued function Φ defined on a Riemann surface is called *holomorphic* (*meromorphic*), if the function depends holomorphically (meromorphically) on all local coordinates.

We remark, that the composition of two holomorphic functions is again holomorphic. Indeed, for any two complex-valued functions $\Phi = \Phi(z)$ and $z = z(\zeta)$ we have

$$\frac{\partial \Phi}{\partial \bar{\zeta}} = \frac{\partial \Phi}{\partial z} \frac{\partial z}{\partial \bar{\zeta}} + \frac{\partial \Phi}{\partial \bar{z}} \frac{\partial \bar{z}}{\partial \bar{\zeta}} \tag{40}$$

such that it follows $\partial\Phi/\partial\bar{\zeta} = 0$, if $\partial z/\partial\bar{\zeta} = 0$ and $\partial\Phi/\partial\bar{z} = 0$. Therefore Φ is proved to be holomorphic with respect to ζ, if it is holomorphic with respect to z and if, moreover, z and ζ are admissible coordinates in a neighbourhood of the same point of the regarded Riemann surface.

A Riemann surface (without boundary) is compact, if and only if it is the corresponding Riemann surface of an *algebraic function* (an algebraic function is many-valued; it is defined by an equation of the type $P(z, w) = 0$, where $P(z, w)$ is a polynomial in z and w). Every holomorphic function on a compact Riemann surface must be constant (this follows immediately from the maximum principle for holomorphic functions (see 1.3.3.). The existence of meromorphic functions on a given compact Riemann surface follows from the important *Riemann-Roch theorem* (see 15.4., p. 111, and for instance, WEYL [2], BEHNKE and SOMMER [1], FORSTER [2]). From this theorem follows, for instance, the existence of a meromorphic function with one pole only.

Every compact Riemann surface is topologically equivalent to a sphere with p handles (see, for instance, SEIFERT and THRELFALL [1]). The non-negative entire number p is called the *genus* of the Riemann surface. The value of p influences the existence of meromorphic functions on a given Riemann surface. If $w = w(z)$ is an algebraic function, then an integral of type

$$\int_\gamma R(z, w)\, \mathrm{d}z$$

is called *Abelian integral*, if the integrand is a rational function in z and w and γ is a curve on the Riemann surface. The integrand is proved to be a meromorphic function on the Riemann surface.

1.6.4. A general notion describing the solvability of an operator equation is that of the *index* of the operator. Let $\mathscr{L}$ be an operator mapping the linear space $\mathscr{R}_1$ into the linear space $\mathscr{R}_2$. Then the index of $\mathscr{L}$ is defined by

$$\operatorname{ind} \mathscr{L} = \dim \operatorname{Ker} \mathscr{L} - \dim \mathscr{R}_2/\mathscr{L}(\mathscr{R}_1)\,,$$

where

$$\operatorname{Ker} \mathscr{L} = \{x \in \mathscr{R}_1 : \mathscr{L}x = 0\}\,.$$

The global analysis aims *to express the index* of an operator *by topological invariants* of the manifold, on which the regarded operator is defined. In the following it will demonstrated, that some theorems of complex analysis may be interpreted as contributions to global analysis.

Let Φ be a meromorphic function defined on a Riemann surface. If $z, \tilde{z}$ are any local coordinates that are admissible in the same neighbourhood of a given point, then we have

$$\frac{\mathrm{d}\Phi}{\mathrm{d}\tilde{z}} = \frac{\mathrm{d}\Phi}{\mathrm{d}z} \cdot \frac{\mathrm{d}z}{\mathrm{d}\tilde{z}}\,.$$

Thus $\mathrm{d}\Phi/\mathrm{d}z$ is an expression, the values of which depend on the choice of local coordinates. Any object φ depending on the choice of local coordinates is called a *differential* on a given Riemann surface, if in new admissible coordinates $\tilde{z}$ its values $\varphi(z)$ are replaced by

$$\tilde{\varphi}(\tilde{z}) = \varphi(z)\, \frac{\mathrm{d}z}{\mathrm{d}\tilde{z}}\,.$$

Thus the derivative of a function is proved to be a differential. A differential is called *holomorphic*, if its values $\varphi(z)$ define a holomorphic function in z.

For a closed curve γ_0 the integral $\int_{\gamma_0} \varphi(z)\, \mathrm{d}z$ is called the *period* of φ along γ_0 (the value of the integral does not depend on the choice of local coordinates). A holomorphic differential is the derivative of a (holomorphic) function, if and only if all its periods are equal to zero. On the other hand all closed curves on a compact Riemann surface of genus p (see 1.6.3.) are linear combinations of $2p$ linearly independent closed curves $\gamma_1, \dots, \gamma_{2p}$ (see fig. 18), such that

$$\int_{\gamma_0} \varphi(z)\, \mathrm{d}z = \sum_{k=1}^{2p} \lambda_k \pi_k[\varphi]\,,$$

where $\pi_k[\varphi]$ denotes the period of φ along γ_k. Thus a holomorphic differential φ is the derivative of a function, if and only if all its periods $\pi_k[\varphi]$ vanish.

One can prove (see, for instance, WEYL [2], BEHNKE and SOMMER [1], R. NEVANLINNA [5]), that a uniquely determined holomorphic differential exists, such that its periods $\pi_k[\varphi]$ possess arbitrarily prescribed real parts. Also one can prove, that there exist meromorphic differentials with two first order poles at two fixed points, for instance. With respect to local coordinates z_j in neighbourhoods of the singularities the differential can be represented by

$$\varphi(z_j) = \frac{c_{-1}^{(j)}}{z_j - z_{0j}} + c_0^{(j)} + c_1^{(j)}(z_j - z_{0j}) + \dots,$$

where z_{0j} are the coordinates of the singularities. The coefficient $c_{-1}^{(j)}$ is called the residue of φ at the point z_{0j}. It is independent of the choice of the local coordinates. In view of the residue theorem (for differential on Riemann surfaces) one has $c_{-1}^{(1)} + c_{-1}^{(2)} = 0$, such that the regarded differentials can be characterized by the residue $c_{-1}^{(1)}$ at the first singular point and, moreover, by the real parts of its periods $\pi_k[\varphi]$, $k = 1, \dots, 2p$. Let $\mathcal{L}$ be the operator that associates the residue at the first singular point to the regarded differential. Thus we take $\mathcal{R}_2 = \mathbb{C}$, whereas $\mathcal{R}_1$ is the space of all differentials of regarded type. We have $\dim \operatorname{Ker} \mathcal{L} = 2p$, because a differential with residue c is uniquely determined up to a holomorphic differential. Since every residue c can be realized one gets $\dim \mathcal{R}_2/\mathcal{L}(\mathcal{R}_1) = 0$ and, consequently,

$$\operatorname{ind} \mathcal{L} = 2p .$$

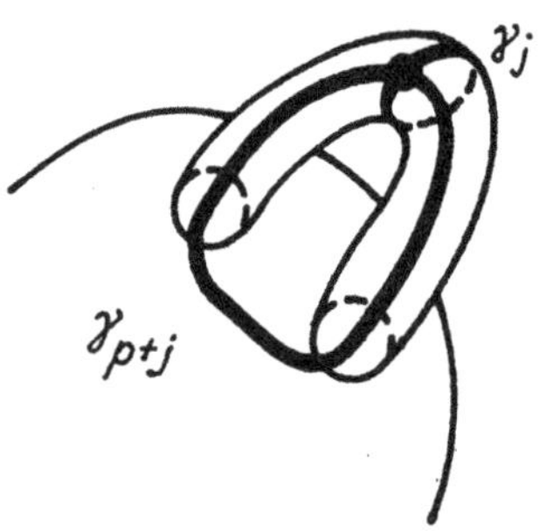

Fig. 18

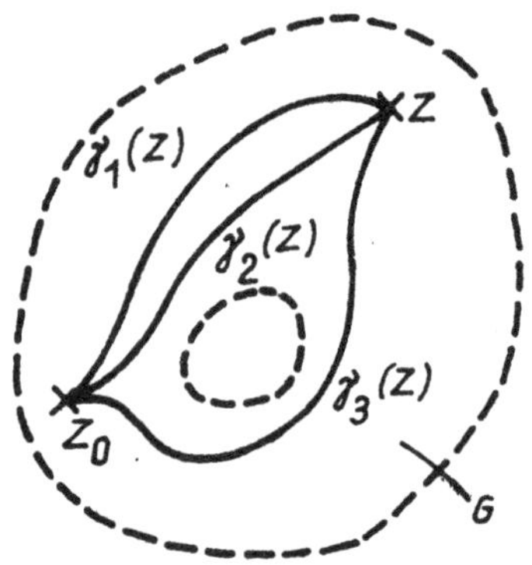

Fig. 19

1.6.5. Let $\mathcal{F}$ be any compact or noncompact Riemann surface. The *uniformization theory* aims to introduce for $\mathcal{F}$ globally admissible coordinates. To this we define the *universal covering surface* $\hat{\mathcal{F}}$ of $\mathcal{F}$:

Let z_0 be a fixed point of $\mathcal{F}$ and, moreover, $\gamma(z)$ be any curve with initial point z_0 and end point z. Two curves $\gamma_1(z)$ and $\gamma_2(z)$ are called *homotopic*, if $\gamma_2(z)$ is a continuous deformation of $\gamma_1(z)$ in $\mathcal{F}$ (in fig. 19 the curves $\gamma_1(z)$ and $\gamma_2(z)$ are homotopic, whereas $\gamma_1(z)$ and $\gamma_3(z)$ are not homotopic). The points of $\hat{\mathcal{F}}$ are homotopy classes of curves $\gamma(z)$. It is easy to prove, that $\hat{\mathcal{F}}$ is simply connected. Additionally there is

Riemann's mapping theorem. *Every simply connected Riemann surface can be mapped by a univalent holomorphic function onto one of the following sets:*

a) *the unit disk* $|\zeta| < 1$,
b) *the* ζ*-plane or*
c) *the* ζ*-sphere.*

Then ζ is a global coordinate for $\hat{\mathcal{F}}$ and, consequently, for $\mathcal{F}$, too.

One gets a natural projection of $\hat{\mathcal{F}}$ onto $\mathcal{F}$, if the point z of $\mathcal{F}$ is associated to all points of $\hat{\mathcal{F}}$, which are represented by curves $\gamma(z)$ with end-point z. Now regard 1-1-mapping of $\hat{\mathcal{F}}$ onto itself (so-called *covering transformations*), for which every point of $\hat{\mathcal{F}}$ and its transformed point possess the same projection. The set of all such transformations forms a so-called *transformation group*. Therefore the uniformization theory is connected with transformation groups of a Riemann surface.

A function defined on $\hat{\mathcal{F}}$ that is invariant under the transformation group is called *automorphic*. Such a function is also a uniquely defined function on $\mathcal{F}$ itself. Thus the construction of certain automorphic functions on $\hat{\mathcal{F}}$ proves the existence of corresponding functions on $\mathcal{F}$ itself (if $\hat{\mathcal{F}}$ is the ζ-plane and the transformations are given by $\tilde{\zeta} = \zeta + m_1 \cdot 2\omega_1 + m_2 \cdot 2\omega_2$ with entire numbers m_1, m_2 (where $\operatorname{Im} \omega_2/\omega_1 \neq 0$), the automorphic functions are *double periodic*).

1.7. Methods of the theory of functions of several complex variables[1])

1.7.1. Let G be a domain in the space $\mathbb{C}^n$ of n complex variables $z_1, \dots, z_n$. A continuous (complex-valued) function defined in G is called *holomorphic*, if it depends holomorphically on every separate variable z_j (cf. 1.3.2.). We remark, that the assumption of continuity is unnecessary, because by the important *Hartogs theorem every separately holomorphic function must be continuous.*

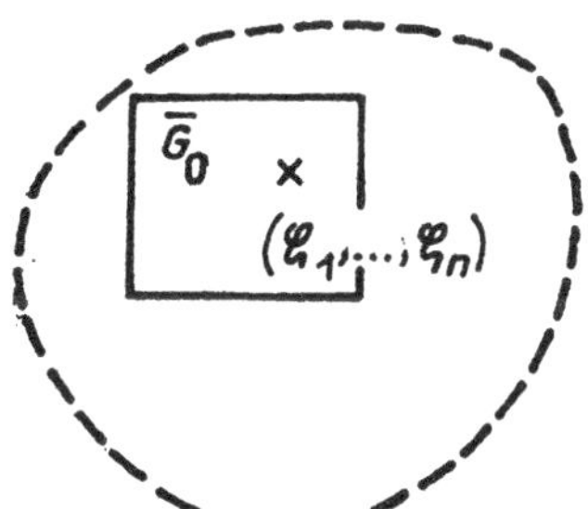

Fig. 20

Applying methods of the case of one complex variable it is easy to see, that holomorphic functions of both one complex variable and several complex variables possess many common properties. For instance from Cauchy's integral formula (17) one gets immediately the following many-dimensional variant:

Let Φ be holomorphic in G. Assume, that the closure of the so-called polycylinder (see fig. 20)

$$G_0 = \{(z_1, \dots, z_n) : |z_j - z_{0j}| < r_j\}$$

is contained in G. Let $(\zeta_1, \dots, \zeta_n)$ be a fixed point of G_0. By virtue of (17) we get

$$\Phi(\zeta_1, \dots, \zeta_n) = \frac{1}{2\pi i} \int\limits_{|z_1 - z_{01}| = r_1} \frac{\Phi(z_1, \zeta_2, \dots, \zeta_n)}{z_1 - \zeta_1} \, dz_1 .$$

1) Cf., for instance, Forster [1], Fuks [1, 2], Hörmander [2], Osgood [1], Vladimirov [1].

On the other hand the integrand may be expressed by

$$\Phi(z_1, \zeta_2, \ldots, \zeta_n) = \frac{1}{2\pi i} \int\limits_{|z_2 - z_{02}| = r_2} \frac{\Phi(z_1, z_2, \zeta_3, \ldots, \zeta_n)}{z_2 - \zeta_2} \, dz_2$$

and so on. Substituting into one another it follows, finally, the *n-dimensional Cauchy integral formula*

$$\Phi(\zeta_1, \ldots, \zeta_n) = \frac{1}{(2\pi i)^n} \int\limits_S \frac{\Phi(z_1, \ldots, z_n)}{(z_1 - \zeta_1) \ldots (z_n - \zeta_n)} \, dz_1 \ldots dz_n\,, \tag{41}$$

where $S = \{(z_1, \ldots, z_n) : |z_j - z_{0j}| = r_j,\ j = 1, \ldots, n\}$ is an n-dimensional part of the boundary ∂G_0 and is called the *distinguished surface* of G_0 (because the values of Φ in G_0 are uniquely determined by its values on S).

1.7.2. From (41) one gets similar consequences as in the one-dimensional case (17), for instance the local *power-series representation* for holomorphic functions (cf. 1.4.4.). Conversely a convergent power-series of type

$$\sum_{\nu_1, \ldots, \nu_n} a_{\nu_1 \ldots \nu_n} (\zeta_1 - \zeta_{01})^{\nu_1} \ldots (\zeta_n - \zeta_{0n})^{\nu_n}$$

defines a holomorphic function (cf. 1.4.6.). Using power-series representations, moreover, one can construct analytic continuations (cf. 1.4.7.). Replacing real variables x_j by complex ones z_j power-series in real variables

$$\sum_{\nu_1, \ldots, \nu} a_{\nu_1 \ldots \nu_n} (x_1 - x_{01})^{\nu_1} \ldots (x_n - x_{0n})^{\nu_n}$$

may be continued into a domain of $\mathbb{C}^n$ as holomorphic functions (cf. 1.4.10.). Thus holomorphic functions in n complex variables are not only special functions in $\mathbb{R}^{2n}$, but also continuations of functions in $\mathbb{R}^n$.

1.7.3. For power-series in both one and several complex variables the following *identity theorem* holds: Two power-series define the same function, if and only if all their corresponding coefficients are equal.

From this identity theorem one gets immediately the following important *permanence principle*: Let $\Phi = \Phi(z, w)$ be a holomorphic function in z and w and, moreover, $w = w(z)$ a holomorphic function defined in the domain G of the z-plane. If $\Phi(z, w(z))$ vanishes identically in any open subset of G, then we have $\Phi(z, w(z)) \equiv 0$ in the whole domain G.

1.7.4. Formula (41) is valid for polycylinders only. But there is another n-dimensional generalization of (17), namely the *Bochner-Martinelli integral formula*. It is valid for arbitrary (smoothly bounded) domains G_0 in $\mathbb{C}^n$. This formula expresses the values of a holomorphic function in the interior of G_0 by the boundary values on the whole boundary ∂G_0. We will formulate it for two complex variables:

Let Φ be holomorphic in G, $\bar{G}_0 \subset G$ and $(\zeta_1, \zeta_2) \in G_0$. Then

$$\Phi(\zeta_1, \zeta_2) = \frac{1}{4\pi^2} \int\limits_{\partial G_0} \Phi(z_1, z_2) \frac{(\bar{\zeta}_1 - \bar{z}_1)\, dz_1 \wedge d\bar{z}_2 \wedge dz_2 + (\bar{\zeta}_2 - \bar{z}_2)\, dz_2 \wedge d\bar{z}_1 \wedge dz_1}{(|z_1 - \zeta_1|^2 + |z_2 - \zeta_2|^2)^2}\,. \tag{42}$$

The proof is similar to that of (17), if one replaces the differential form

$$\Phi(z)\,\frac{dz}{z-\zeta}$$

by the integrand of (42). This integrand may be written as

$$\Omega := \Phi\left(\frac{\partial g_1}{\partial z_1}\,dz_1 \wedge d\bar z_2 \wedge dz_2 + \frac{\partial g}{\partial z_2}\,dz_2 \wedge d\bar z_2 \wedge dz_1\right),$$

where

$$g(z_1, z_2, \zeta_1, \zeta_2) = \frac{1}{|z_1-\zeta_1|^2 + |z_2-\zeta_2|^2}$$

is (up to a constant factor) a fundamental solution of $\Delta u = 0$ in $\mathbb{R}^4$ (see VLADIMIROV [3]). The differential form Ω possesses an isolated singularity at (ζ_1, ζ_2). Since Φ is holomorphic one gets

$$d\Omega = 0, \text{ if } (z_1, z_2) \neq (\zeta_1, \zeta_2),$$

where $d\Omega$ denotes the exterior differential of Ω (see, for instance CARTAN [2], KÄHLER [1]). In view of the Stokes-Cartan theorem we get

$$\int_{\partial U_\delta(\zeta_1,\zeta_2)} \Omega = \int_{\partial G_0} \Omega\,.$$

On the other hand we have

$$\int_{\partial U_\delta(\zeta_1,\zeta_2)} \frac{(\bar\zeta_1 - \bar z_1)\,dz_1 \wedge d\bar z_2 \wedge dz_2 + (\bar\zeta_2 - \bar z_2)\,dz_2 \wedge d\bar z_1 \wedge dz_1}{(|z_1-\zeta_1|^2 + |z_2+\zeta_2|^2)^2} = 4\pi^2$$

(this follows, for instance, by a direct computation with the help of polar coordinates in $\mathbb{R}^4$ with center (ζ_1, ζ_2)). Since Φ is continuous we get, finally,

$$\lim_{\delta\to 0} \int_{\partial U_\delta(\zeta_1,\zeta_2)} \Omega = 4\pi^2\Phi(\zeta_1, \zeta_2)\,.$$

1.7.5. On the one hand from the n-dimensional Cauchy integral formula (41) one concludes, that holomorphic functions of several complex variables possess many common properties with holomorphic functions of one complex variable. On the other hand one can derive such properties that holomorphic functions of one complex variable do not possess.

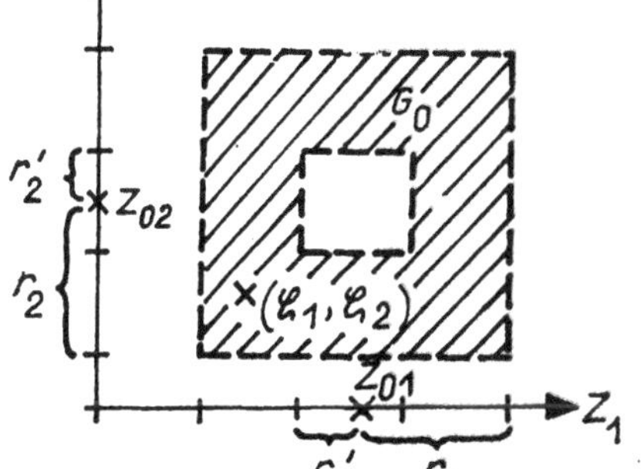

Fig. 21

Fig. 22

Let G_0 be defined by ($0 < r'_j < r_j$, see fig. 21)

$$G_0 = \{(z_1, z_2) : |z_j - z_{0j}| < r_j, j = 1, 2; |z_1 - z_{01}| > r'_1 \text{ or } |z_2 - z_{02}| > r'_2\}$$

and Φ be holomorphic in a neighbourhood of $\overline{G}_0$. Using the same argument as in 1.7.1. one gets for $(\zeta_1, \zeta_2) \in G_0$ the representation

$$\Phi(\zeta_1, \zeta_2) = \frac{1}{(2\pi i)^2} \int\limits_{|z-z_{01}|=r_1} \int\limits_{|z-z_{02}|=r_2} \frac{\Phi(z_1, z_2)}{(z_1 - \zeta_1)(z_2 - \zeta_2)} dz_1 \, dz_2 .$$

And the integral on the right-hand side of the last equation defines a holomorphic function in the whole domain

$$\tilde{G}_0 = \{(z_1, z_2) : |z_j - z_{0j}| < r_j, j = 1, 2\} ,$$

that is equal to the holomorphic function Φ given in G_0. *Thus every holomorphic function Φ in G_0 can be continued holomorphically into the larger domain $\tilde{G}_0$.*

A corresponding theorem does not hold in the case of one complex variable:

Let G be any domain in the z-plane. Every domain $\tilde{G} \supsetneqq G$ contains at least one boundary point z_0 of G (see fig. 22). Then $1/(z - z_0)$ defines a holomorphic function in G, which can not be continued holomorphically into $\tilde{G}$.

Let G be a given domain in $\mathbb{C}^n$. The largest domain into which one can continue holomorphically every in G holomorphic function, is called the *envelope of holomorphy* of G. For example the domain $\tilde{G}_0$ defined above is proved to be the envelope of holomorphy of G_0.

1.7.6. In view of the continuation theorem 1.7.5 holomorphic functions of several complex variables can not possess isolated singularities. Analogously *the zeroes of a holomorphic function are not isolated.* This can be deduced with the help of the argument principle 1.4.15 in one of the complex z_j-plane, where the other variables z_k, $k \neq j$, are regarded as parameters.

1.7.7. An autonomous method due to Hörmander [2] is the *reduction* of problems for holomorphic functions in several complex variables *to the inhomogeneous Cauchy-Riemann equations.* We will illustrate this method by the continuation problem 1.7.5. Let Φ be a holomorphic function in $G \setminus K$, where G is a domain in $\mathbb{C}^n$, K is a compact subset of G and $G \setminus K$ is assumed to be connected. We will prove that Φ can be continued holomorphically into the subset K of G.

To it let φ be any $\mathcal{E}^\infty$-function in $\mathbb{C}^n$ that vanishes identically outside of a compact set $\tilde{K}(\subset K)$ in G and that, moreover, is equal to 1 in a neighbourhood of K (cf. fig. 23). The functions

$$h_j = \frac{\partial}{\partial \bar{z}_j} [(1 - \varphi)\Phi]$$

are defined in the whole $\mathbb{C}^n$, vanish identically outside ot $\tilde{K}$ and fulfil the *integrability conditions*

$$\frac{\partial h_j}{\partial \bar{z}_k} = \frac{\partial h_k}{\partial \bar{z}_j}, \qquad j, k = 1, \dots, n .$$

Then a special solution of the inhomogeneous Cauchy-Riemann equations

$$\frac{\partial w}{\partial \bar{z}_j} = h_j, \quad j = 1, \dots, n ,$$

is given by

$$H(z_1, \dots, z_n) = -\frac{1}{\pi} \iint\limits_{C} \frac{h_1(\zeta_1, z_2, \dots, z_n)}{\zeta_1 - z_1} \, d\xi_1 \, d\eta_1 ,$$

where $\zeta_1 = \xi_1 + i\eta_1$. Hence

$$\Phi_0 = (1 - \varphi)\Phi - H$$

is holomorphic in G. Since $h_1 \equiv 0$ outside of $\tilde{K}$ the function H vanishes identically for sufficientley large $z_2, \dots, z_n$. Using the uniquess of the holomorphic continuation it follows, that $H \equiv 0$ outside of $\tilde{K}$. Therefore $\Phi \equiv \Phi_0$ outside of $\tilde{K}$. On the other hand the function Φ_0 is holomorphic in the whole domain G, such that Φ_0 is the sought holomorphic continuation of Φ into K.

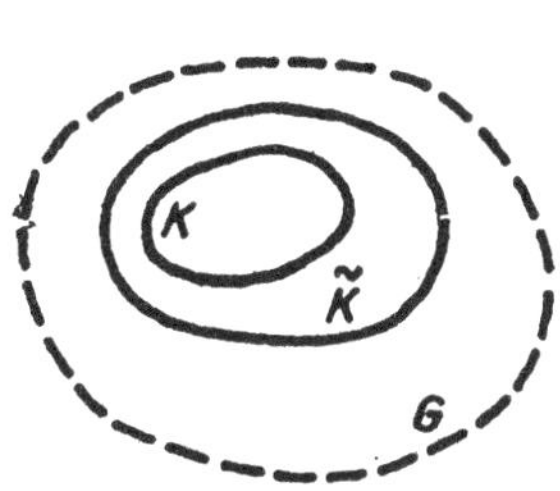

Fig. 23

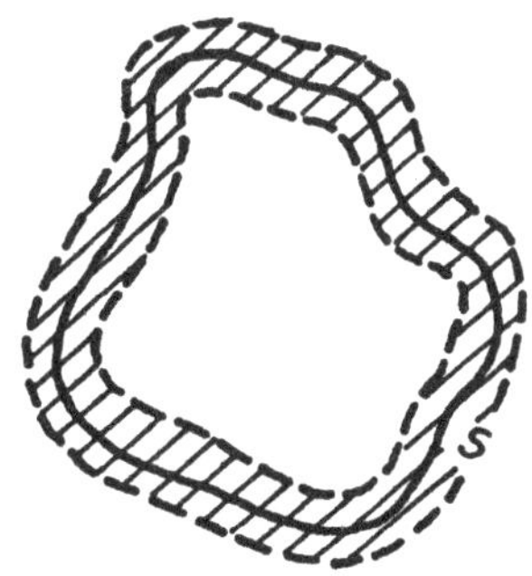

Fig. 24

1.7.8. The continuation theorem 1.7.7 may be interpreted in the following way:

Let S be a closed surface in $\mathbb{C}^n$ (which is the boundary of a domain). Let, moreover, Φ be a holomorphic function defined in a neighbourhood of S (cf. fig. 24). Then Φ can be continued holomorphically into the interior of S.

Now assume that a complex-valued function is given on S only. Then we look for conditions, such that the given values on S can be continued to a holomorphic function in the interior of S. That means, we regard the prescribed values as boundary values of a holomorphic function.

In the case of one complex variable the condition is given by the boundary behaviour of the Cauchy type integral (see 1.5.4.). In the case of at least two complex variables one must regard the *tangential Cauchy-Riemann equations* (see, for instance, Aĭsenberg and Južakov [1]). A reduction of the n-dimensional case to the case of one complex variable is given by Stout [1].

1.7.9. Complex methods allow to solve many global problems, for instance the following one:

Let G be a domain in the z-plane and let, moreover, $z_{01}, z_{02}, \dots$ be a sequence of isolated points in G. Then we look for a *holomorphic function* in G possessing *zeroes of prescribed order* at the points z_{0k}. Since zeroes of holomorphic functions of several complex variables are not isolated, the above-mentioned problem cannot be generalized directly to higher dimensions. Such a generalization needs to modify the posed problem. This had been done by P. Cousin: Let G be a domain in $\mathbb{C}$ or in $\mathbb{C}^n$. Let a covering of G by open sets U_α be given (cf. fig. 25). Then assume that in every U_α a holomorphic function Φ_α is given. In $U_\alpha \cap U_\beta$ two functions Φ_α and Φ_β are defined (if $U_\alpha \cap U_\beta \neq \emptyset$). Both functions possess the same zeroes, if a nowhere vanishing holomorphic function

$g_{\alpha\beta}$ exists in $U_\alpha \cap U_\beta$, such that

$$\Phi_\alpha = g_{\alpha\beta}\Phi_\beta .$$

Then the (multiplicative) Cousin problem is to find a uniquely determined function Φ in G possessing in every U_α the same zeroes like Φ_α (cf. 4.1.4.).

It is easy to see, that the one-dimensional problem formulated in the beginning is a special case of the general (multiplicative) Cousin problem: Take $\Phi_\alpha(z) = (z - z_{0k})^{n_k}$, if U_α is a neighbourhood of z_{0k}, otherwise $\Phi_\alpha \equiv 1$ (the neighbourhoods U_α are taken so small, that they contain no more than one of the points z_{0k}).

This one-dimensional problem can be solved directly by infinite products (see, for instance, KNESER [1]). The solution of the general case requires adequate general methods (it is possible, for instance, to reduce Cousin problems to the inhomogeneous

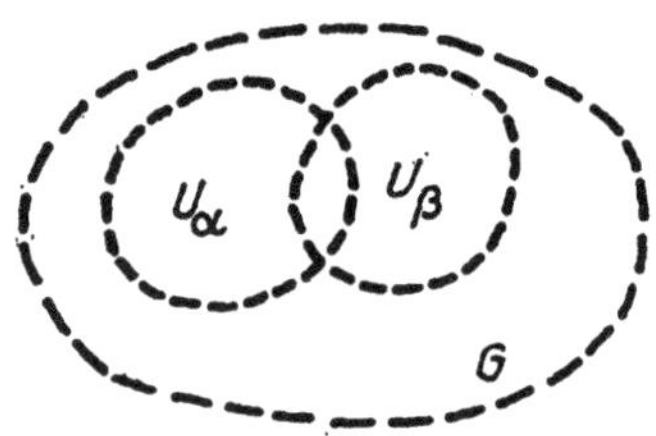

Fig. 25

Cauchy-Riemann equations, cf. 1.7.7.). The general Cousin problems are solvable for a given domain (or manifold, cf. 1.7.10.) G, if certain conditions concerning the topological structure of the regarded domain G are fulfilled. These conditions may be expressed with the help of the cohomology groups of G (see chapter 4).

1.7.10. Riemann surfaces (see 1.6.3.) had been introduced in order to study the global behaviour of holomorphic functions of one complex variable. Analogously in the case of several complex variables one introduces so-called n-dimensional complex manifolds:

A manifold of real dimension $2n$ is called an *n-dimensional complex manifold*, if any two local coordinate systems $z_1, \dots, z_n$ and $\zeta_1, \dots, \zeta_n$ are biholomorphically equivalent (that means: they are connected by holomorphic functions $z_j = z_j(\zeta_1, \dots, \zeta_n)$ and $\zeta_j = \zeta_j(z_1, \dots, z_n)$ resp., $j = 1, \dots, n$).

Many global problems concerning holomorphic functions of several comples variables (Cousin problems, for instance) can be solved for manifolds, too. Complex manifolds possessing many common properties with Riemann surfaces are the Stein *manifolds* (see 4.3.2. and, for instance, HÖRMANDER [2]).

1.7.11. An *analytic set* in $\mathbb{C}^n$ is defined as the set of all zeroes of a (uniquely defined) holomorphic function. In the case of two complex variables z, w, such a set can be interpreted as Riemann surface (that means: either z or w can be introduced as local coordinates). An analogous result is not valid in the case of more than two complex variables. Regard, for instance, the analytic set defined by

$$w^2 - z_1 z_2 = 0$$

in $\mathbb{C}^3$. Then one can prove that this analytic set is not uniformizable at (0, 0, 0) (that means: it is not possible to introduce two complex variables as local coordinates). In order to study analytic (and, in particular, algebraic) sets in higher dimensions with complex methods it is necessary, consequently, to generalize the conception of complex manifolds. This generalization leads to the conception of *complex spaces* (see, for instance, FUKS [1]).

1.7.12. Holomorphic functions Φ in $\mathbb{C}^n$ fulfil the Cauchy-Riemann system

$$\frac{\partial \Phi}{\partial \bar{z}_j} = 0\,, \qquad j = 1, \dots, n\,.$$

In the case $n \geqq 2$ this system is *overdetermined*. Many properties of holomorphic functions are valid for solutions of more general overdetermined systems, too. For instance, the continuation theorem 1.7.5, 1.7.7 holds also for systems of type

$$\frac{\partial w}{\partial \bar{z}_j} = A_j(z_1, \dots, z_n)\, w + B_j(z_1, \dots, z_n)\, \bar{w}, j = 1, \dots, n$$

(see LÊ HÙNG SÓN [1, 2], MAGOMEDOV and PALAMODOV [1]).

1.7.13. Another generalization of the concept of holomorphy to higher dimensions is based on *differential forms*. Let Ω be a given differential form and $*\Omega$ its dual, then Ω is called *harmonic*, if Ω and its dual are closed $d\Omega = 0$, $d{*\Omega}$ (see, for instance, HODGE [1], DE RHAM [1]). The operator $*$ generalizes the conception of the conjugate harmonic function (cf. 1.3.4.). Differential forms can be used also in the theory of holomorphic functions itself (cf. 1.7.4.).

1.7.14. Holomorphic functions are closely connected with mathematical physics. The first applications had beendone in the case of one complex variable. In the last years many new applications of holomorphic functions of severalcomplex variables had been originated. In the following we shall refer to three of it:

a) Regard the continuation theorem 1.4.8 in the special case, that γ is an interval on the real axis. One can generalize this special case to the case of several complex variables. This generalization is known as *edge-of-the-wedge-theorem* (the boundary function must be a distribution only). It is used in the dispersion theory in order to prove a crossing property (see TARSKI [1]).

b) Convolution equations of typ $Z * u = f$ arise, for instance, from the theory of *passive systems* (u is the sought solution in $\mathbb{R}$ or $\mathbb{R}^n$; Z and f are given; Z fulfils a so-called passivity condition). The Laplace transform of Z leads to a holomorphic function $\tilde{Z}$.

If one regards n-dimensional linear passive systems ($n \geqq 2$), then $\tilde{Z}$ is proved to be a holomorphic function of several complex variables in a tubular domain. Using properties of such holomorphic functions VLADIMIROV [2, 4, 5] had founded and developed the theory of n-dimensional linear passive systems. In this way he gets, for instance, a characterization of passive operators.

c) One of the basic problems of mathematical physics is the global construction of solutions of field equations, for instance of the (time-depending) Maxwell equations. On the other hand complex analysis allows the global construction of holomorphic functions (cf. 1.6., 1.7.9., chapter 4.). In many cases it is possible to construct solutions of field equations with the help of complex methods. For instance one gets integral representations of solutions of the Maxwell equations by the so-called *Penrose trans-*

form. It *transforms cohomology classes* (see 1.7.9., chapter 4) of a certain complex manifold *into solutions* (the corresponding complex manifold is based on a complexification of the compactified Minkowski space; see, for instance, WELLS [2]).

1.7.15. Let $\Phi_1, \dots, \Phi_m$ be holomorphic functions depending on n complex variables. Then $\Phi = (\Phi_1, \dots, \Phi_m)$ realizes a so-called *holomorphic mapping* from $\mathfrak{R}_1 = \mathbb{C}^n$ into $\mathfrak{R}_2 = \mathbb{C}^m$. One can define holomorphic mappings also in the case that $\mathfrak{R}_1$ and $\mathfrak{R}_2$ are general linear (complex) vector spaces (possibly of infinite dimensions; cf., for instance, MATOS [1]). In this case holomorphic mappings are defined as limits of polynomial mappings (so to say the definition of holomorphy is a conversion of Weierstrass' convergence theorem 1.4.2.).

An important special case is the following one: $\mathfrak{R}_1 = \mathbb{C}$ and $\mathfrak{R}_2$ is a Banach space. This case is used in functional analysis: if the complex number λ does not belong to the spectrum of the operator A, then the solution x of the equation $\lambda x - Ax = f$ depends holomorphically on λ (see, for instance, RIESZ and SZ.-NAGY [1]). In this case the holomorphy may be defined with the help of the usual complex derivative (see 1.3.5.).

1.7.16. Real part and imaginary part of a holomorphic function in several complex variables $z_1, \dots, z_n$ are harmonic in every z_j-plane. Thus they are called plurisubharmonic. *Plurisubharmonic (plurisuperharmonic)* functions are the generalization of subharmonic (superharmonic) functions (cf. 1.3.6.) to the class of several complex variables.

1.8. Complex methods in algebraic equations

1.8.1. We look for solutions of the algebraic equation

$$\sum_{\nu=0}^{n} a_\nu x^\nu = 0\,, \tag{43}$$

where a_ν are given (real) constants, $a_n \neq 0$. There are equations of this type without (real) solutions (example: $1 + x^2 = 0$ does not possess any real solution). On the other hand a uniquely determined holomorphic function exists taking the values $\sum_{\nu=0}^{n} a_\nu x^\nu$ on the real axis, namely $\Phi(z) = \sum_{\nu=0}^{n} a_\nu z^\nu$ (cf. 1.4.10.). Then it is easy to prove that Φ *possesses at least one zero in the z-plane.* All zeroes of Φ are interpreted as complex solutions of (43).

The existence of zeroes of Φ can be proved in the following way:

Let Φ_0 and Φ_1 holomorphic in a neighbourhood of $\overline{G_0}$, $\Phi_0 \neq 0$ on $\gamma_0 = \partial G_0$ and $|\Phi_1| < |\Phi_0|$ at every point of γ_0. In view of (25) the number of zeroes counting according to order of $\Phi_0 + \lambda\Phi_1$, $0 \leqq \lambda \leqq 1$, is given by

$$k_\lambda = \frac{1}{2\pi i} \int\limits_{\gamma_0} \frac{\Phi_0'(z) + \lambda\Phi_1'(z)}{\Phi_0(z) + \lambda\Phi_1(z)}\, dz\,.$$

Since k_λ is an entire number for every λ, one gets $k_0 = k_1$ (Rouché's theorem). On the other hand we have

$$\left|\sum_{\nu=0}^{n-1} a_\nu z^\nu\right| < |a_n| \cdot |z|^n\,,$$

if $|z| = r$ is sufficiently large. Let, in particular, G_0 be the domain $\{z : |z| < r\}$, and $\Phi_0(z) = a_n z^n$, $\Phi_1(z) = \sum_{\nu=0}^{n-1} a_\nu z^\nu$. Then Φ_0 and $\Phi_0 + \Phi_1$ possess in G_0 the same number of zeroes counting according to order. Thus we get $k_1 = k_0 = n$ (*fundamental theorem of algebra*).

1.8.2. In the last section it was proved that every real algebraic equation (43) possesses at least one complex solution. This complex solution is, in practice, a solution of a real system, by which the given equation (43) is replaced. So the well-known solutions $z = \pm i$ of $1 + x^2 = 0$ are solutions $z = (x, y)$ of the system

$$\mathrm{Re}\,[1 + z^2] = 1 + x^2 - y^2 = 0\,,$$

$$\mathrm{Im}\,[1 + z^2] = 2xy = 0\,.$$

Taking both real and imaginary part of a complex polynomial in z one gets only special real systems of type

$$\begin{aligned} p_1(x, y) &= 0\,, \\ p_2(x, y) &= 0\,, \end{aligned} \tag{44}$$

where p_1, p_2 are polynomials in x and y with real coefficients. The *general system* (44) is equivalent to a complex equation

$$P(z, \bar{z}) = 0\,, \tag{45}$$

where P is a polynomial in z and $\bar{z}$ (such polynomials are special polyanalytic functions, cf. 1.4.11. and chapter 3).

Whereas every equation (43) possesses at least one complex solution, the more general equation (45) is not always solvable (example: $1 + |z|^2 = 1 + z\bar{z} = 0$). On the other hand the following *generalized fundamental theorem of algebra* due to Balk [9] holds:

Let n_1 and n_2 be the degrees of $P(z, \bar{z})$ with respect to z and $\bar{z}$ resp. Let, moreover, N be the total degree of $P(z, \bar{z})$. Then the condition

$$N \geqq 2 \min(n_1, n_2) + 1$$

is sufficient for the existence of at least one solution of (45).

1.8.3. Polynomials in z and $\bar{z}$ may possess non-isolated zeroes, if the polynomial can be divided by a (so-called quasi-real) factor like $1 - |z|^2 = 1 - z\bar{z}$ or $z + \bar{z}$. Otherwise the zeroes are isolated. Applying the Euclidian algorithm to $P(z, \bar{z})$ and $\overline{P(z, \bar{z})}$ one gets, finally, a polynomial $H(z)$ in the only variable z, such that every zero of $P(z, \bar{z})$ is a zero of $H(z)$, too (the converse is not true). Thus the *calculation of isolated zeroes* of a polynomial in z and $\bar{z}$ can be reduced to the analogous problem for polynomials in z (see Heinz [1]).

1.8.4. The partial sums of a power-series in $z_1, \dots, z_n, \bar{z}_1, \dots, \bar{z}_n$ are polynomials. The calculation of their zeroes leads to an equation of type

$$P(z_1, \dots, z_n, \bar{z}_1, \dots, \bar{z}_n) = 0\,, \tag{46}$$

where $P(z_1, \dots, z_n, \bar{z}_1, \dots, \bar{z}_n)$ is a polynomial. If P depends on $z_1, \dots, z_n$ only, the equation (46) defines an analytic set in $\mathbb{C}^n$ (cf. 1.7.11.). In the special case of two complex variables $z_1 = z$, $z_2 = w$ by the corresponding equation $P(z, w) = 0$ an analytic set in $\mathbb{C}^2$ is defined, of which the projection onto the z-plane can be described by a so-called (many-valued) *algebraic function* $w = w(z)$ (cf. 1.6.3.). Excluding a finite number of critical points to every point z of the plane m values $w = w(z)$ exist different from each other, where m is the degree of P with respect to w (see, for instance, Behnke and Sommer [1], Jung [1], Weyl [2]). A complex equation $P(z, w) = 0$ is equivalent to special algebraic systems of two real equations in four reals x, y, u, v, where $z = x + iy$, $w = u + iv$.

1.8.5. The calculation of the zeroes of a general polynomial in $\mathbb{C}^2$ leads to an equation of the type

$$P(z, \bar{z}, w, \bar{w}) = 0 \,. \tag{47}$$

Such an equation is equivalent to a general algebraic system of two equations in four reals. The projection into the z-plane of the set defined by (47) leads to a many-valued function $w = w(z)$ that possesses some common properties with algebraic functions. This follows from the following assertion:

Let w be a solution of (47). Then w is also a solution of a simplified equation of the type

$$\tilde{P}(z, z, \bar{w}) = 0 \,.$$

This equation can be deduced by a Euclidian algorithm, which is analogous to that one in 1.8.3.: one eliminates $\bar{w}$ from $P(z, \bar{z}, w, \bar{w}) = 0$ and $\overline{P(z, \bar{z}, w, \bar{w})} = 0$.

If P fulfils a certain algebraic condition, then the following theorem holds (see Czerner [1, 2]):

To a given equation (47) *a finite number of critical curves exists in the z-plane* (instead of a finite number of critical points in the case of algebraic functions, cf. 1.8.4.). *In the interior (exterior resp.) of these curves a fixed number exists counting according to order of solutions $w = w(z)$ that are representable locally as power-series in z and $\bar{z}$.*

1.9. Construction of global homeomorphisms by complex methods

1.9.1. In mathematics it is often needed to *transform coordinates.* A system $f = (f_1, \ldots, f_n)$ of real-valued (continuously differentiable) functions defined in a domain G of $\mathbb{R}^n$ realizes *local homeomorphisms,* if its Jacobian is different from zero everywhere in G. It is more difficult to verify that a given function system realizes also a *global homeomorphism.* Complex methods allow (especially in the case of $\mathbb{R}^2 = \mathbb{C}$) to deduce sufficient conditions for global homeomorphisms. Furthermore they make it possible to construct global homeomorphisms (cf. 1.9.3., 1.9.4. and 1.9.5.).

Global homeomorphisms are needed in order to simplify the form of a domain or to simplify a given differential operator. Often one looks for homeomorphisms that are differentiable up to a certain order (such homeomorphisms are called *diffeomorphisms*). The *Jacobian* is given by

$$\left|\frac{\partial \zeta}{\partial z}\right|^2 - \left|\frac{\partial \zeta}{\partial \bar{z}}\right|^2,$$

if $\zeta = \zeta(z)$ is a map from the z-plane into the ζ-plane. If the mapping is given by a holomorphic function Φ, then its Jacobian is equal to $|\Phi'|^2$. Such mappings possess the following property:

Let $\mathrm{d}s$ be the length of any line element with the initial point z_0. Let, moreover, $\mathrm{d}\sigma$ be the length of its image (cf. fig. 26). Then we get

$$\frac{\mathrm{d}\sigma}{\mathrm{d}s} = |\Phi'(z_0)| .$$

That means: *the quotient* $\mathrm{d}\sigma/\mathrm{d}s$ *is independent of the direction on the given line element*; the quotient $\mathrm{d}\sigma/\mathrm{d}s$ depends, consequently, on the initial point z_0 only. A holomorphic function realizes a local homeomorphism, if $\Phi' \neq 0$. Then the mapping preserves the angle between curves. Such mappings are called *conformal*. For many applications (cf. 1.9.8.) it is important, in addition, that conformal mappings transform holomorphic functions onto holomorphic functions (see 1.6.3., particularly (40)).

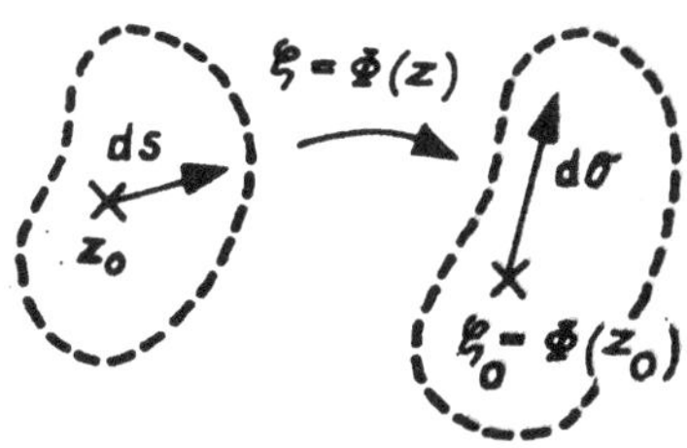

Fig. 26

1.9.2. Let ζ be a new variable depending holomorphically on the variable z. Then $\mathrm{d}\sigma/\mathrm{d}s$ is independent of the direction of $\mathrm{d}s$ (cf. 1.9.1.). Conversely assume now that $\zeta = \zeta(z)$ is any (continuously differentiable) transformation of coordinates, for which $\mathrm{d}\sigma/\mathrm{d}s$ depends only on the initial point of the line element $\mathrm{d}s$. Since

$$\mathrm{d}s^2 = \mathrm{d}x^2 + \mathrm{d}y^2$$

and

$$\mathrm{d}\sigma^2 = \mathrm{d}\xi^2 + \mathrm{d}\eta^2$$

$$= \left[\left(\frac{\partial \xi}{\partial x}\right)^2 + \left(\frac{\partial \eta}{\partial x}\right)^2\right] \mathrm{d}x^2 + 2\left[\frac{\partial \xi}{\partial x}\frac{\partial \xi}{\partial y} + \frac{\partial \eta}{\partial x}\frac{\partial \eta}{\partial y}\right] \mathrm{d}x\,\mathrm{d}y + \left[\left(\frac{\partial \xi}{\partial y}\right)^2 + \left(\frac{\partial \eta}{\partial y}\right)^2\right] \mathrm{d}y^2$$

from $\mathrm{d}\sigma = \mathrm{const} \cdot \mathrm{d}s$ it follows

$$\frac{\partial \xi}{\partial x}\frac{\partial \xi}{\partial y} + \frac{\partial \eta}{\partial x}\frac{\partial \eta}{\partial y} = 0 , \tag{48}$$

$$\left(\frac{\partial \xi}{\partial x}\right)^2 + \left(\frac{\partial \eta}{\partial x}\right)^2 = \left(\frac{\partial \xi}{\partial y}\right)^2 + \left(\frac{\partial \eta}{\partial y}\right)^2 . \tag{49}$$

If $\partial\eta/\partial y \neq 0$, then from (48) we get

$$\frac{\partial \eta}{\partial x} = C\,\frac{\partial \xi}{\partial x},$$

where C is defined by $\frac{\partial \xi}{\partial y} = C \frac{\partial \eta}{\partial y}$. Substituting the two last equations into (49) one gets $\left(\frac{\partial \xi}{\partial x}\right)^2 = \left(\frac{\partial \eta}{\partial y}\right)^2$, such that

$$\frac{\partial \xi}{\partial x} = \varepsilon_1 \frac{\partial \eta}{\partial y}, \tag{50}$$

where $\varepsilon_1^2 = 1$.

If $\partial\eta/\partial y = 0$, then $\partial\xi/\partial x$ must be equal to zero, too. Otherwise in view of (48) we would get $\partial x/\partial y = 0$ and, consequentely, a contradiction to (49) would be deduced. Hence (50) holds in all cases.

Analogously we get

$$\frac{\partial \xi}{\partial y} = \varepsilon_2 \frac{\partial \eta}{\partial x}, \tag{51}$$

where $\varepsilon_2^2 = 1$. Substituting (50) and (51) into (48) it follows $\varepsilon_2 = -\varepsilon_1$ and, finitely,

$$\frac{\partial(\xi, \eta)}{\partial(x, y)} = \varepsilon_1 \left[\left(\frac{\partial \eta}{\partial x}\right)^2 + \left(\frac{\partial \eta}{\partial y}\right)^2\right].$$

That means, that ε_1 must be equal to one, if the mapping $\zeta = \zeta(z)$ is assumed to be orientable.

In this way the following theorem is proved:

Let $\zeta = \zeta(z)$ be any (continuously differentiable) orientable mapping, for which $d\sigma/ds$ is independent of the direction of ds. Then $\zeta = \zeta(z)$ fulfils the Cauchy-Riemann equation (13) *and is, consequentely, a holomorphic function.*

1.9.3. Let G be in the z-plane a simply connected domain possessing at least two boundary points. Then in view of Riemann's mapping theorem (cf. 1.6.5.) a univalent holomorphic function $\zeta = \zeta(z)$ exists mapping G onto the unit disk $|\zeta| < 1$. If the boundary of G is a Jordan curve, then the mapping in question can be extended to a homeomorphism of $\overline{G}$ onto the closed unit disk (see 1.4.13.).

The unit disk can be regarded, consequently, as *canonical domain* of simply connected domains with at least two boundary points. Analogously multiply connected domains can be mapped conformly onto the closed ζ-plane (i.e. the ζ-sphere) minus a finite number of rectilinear slits parallel to the real axis (cf., for instance, SCHIFFER and SPENCER [1]).

1.9.4. Using local coordinates with certain additional properties the uniformization theory (cf. 1.6.5.) allows to introduce globally coordinates possessing globally the same additional properties:

Let G be a domain in the z-plane or in a Riemann surface. Assume that $\zeta, \tilde{\zeta}, \ldots$ are local coordinates, where two local coordinates are connected by a univalent holomorphic function (in view of 1.9.2. this assumption is fulfilled, if the change of variables preserves the orientation and if, moreover the line element $d\sigma$ with respect to arbitrary coordinates $\zeta = \xi + i\eta$ is given by const $\sqrt{d\xi^2 + d\eta^2}$). Then the domain G equipped with the local coordinates ζ is a Riemann surface, that is different from the given one, in general. The uniformization of this Riemann surface allows the global introduction of the regarded coordinates (cf. also BERS [3]).

1.9.5. For many applications it is important to replace a global coordinate z by another global coordinate ζ, where z and ζ are connected by the *Beltrami equation* (cf.

also NAAS [1])

$$\frac{\partial \zeta}{\partial \bar{z}} = \mu(z) \frac{\partial \zeta}{\partial z}. \tag{52}$$

The coefficient $\mu(z)$ is a given measurable function, which almost everywhere fulfils the inequality

$$|\mu(z)| \leqq \mu_0 < 1 \tag{53}$$

with a certain constant μ_0. The following theorem holds (see AHLFORS [5]):

A (uniquely determined) solution of (52) *exists that realizes a* 1-1-*map of the z-sphere onto the ζ-sphere, where* $\zeta(0) = 0$, $\zeta(1) = 1$ *and* $\zeta(\infty) = \infty$.

If $\mu(z)$ is not given in the whole z-plane, we define it by 0 outside of the given domain.

The proof of the theorem formulated above is based on a reduction of (52) to a singular integral equation containing the Π-operator (cf. 1.2.1.; the coefficient μ is approached by functions with compact support). The condition (53) effects, that (52) is *uniformly elliptic.*

Solutions of the Beltrami equation (52) possess the following geometric interpretation: their *dilatation* is bounded by

$$K = \frac{1 + \mu_0}{1 - \mu_0}$$

(the dilatation of a mapping is defined in the following way: the linearization of a mapping at the point z_0 transforms a disk onto an ellipse. The quotient of the length of the major axis and the length of the minor axis is called the dilatation at z_0). Mappings, for which the dilatation is bounded by K, are called *K-quasiconformal* (cf. AHLFORS [5], LEHTO and VIRTANEN [1]), since they possess many common properties with conformal ones. Quasiformal mappings had been introduced by GRÖTZSCH [1] and LAVRENT'EV [1].

1.9.6. The simplification of the form of a domain is important in many cases, since many problems can be solved more easily for simplified domains. So the solution of the Dirichlet boundary value problem for $\Delta u = 0$ in a disk can be represented explicitly by the *Poisson integral* (see, for instance, CARATHÉODORY [3]). On the other hand a conformal mapping transforms solutions of $\Delta u = 0$ onto solutions of the same equation. Therefore the Dirichlet boundary value problem for $\Delta u = 0$ in simply connected domains can be reduced always to the Poisson integral formula (the boundary of the regarded domain is assumed to be a Jordan curve, cf. 1.4.13., 1.9.3.).

1.9.7. The application of complex methods to partial differential equations starts with the introduction of the complex variable z instead of two reals x, y. In this way one gets, for instance, one complex equation for $w = u + iv$ instead of a system of two real equations for two unknown u, v. This complex equation contains, in general, both partial complex derivatives $\partial w/\partial z$, $\partial w/\partial \bar{z}$. Using a transformation of the coordinate z by a solution of the Beltrami equation 1.9.5. uniformly elliptic linear systems can be reduced to the *canonical form*

$$\frac{\partial w}{\partial \bar{z}} = A(z)\, w + B(z)\, \overline{w} + C(z) \tag{54}$$

(see VEKUA [5] and 11.2). This canonical form corresponds to the canonical form

$$\Delta u + a(x, y) \frac{\partial u}{\partial x} + b(x, y) \frac{\partial u}{\partial y} + c(x, y)\, u = f(x, y)$$

of a uniformly elliptic linear differential equation of second order.

1.9.8. Regard a vector field in a domain in the z-plane (see fig. 27). The field will be interpreted as velocity field of a stationary flow of an incompressible fluid. The flow does not possess any sources or whirls, if $(u, -v)$ fulfils the Cauchy-Riemann equations, such that $\Phi = u - iv$ is proved to be holomorphic. Conversely every holomorphic function Φ can be interpreted as incompressible fluid flow.

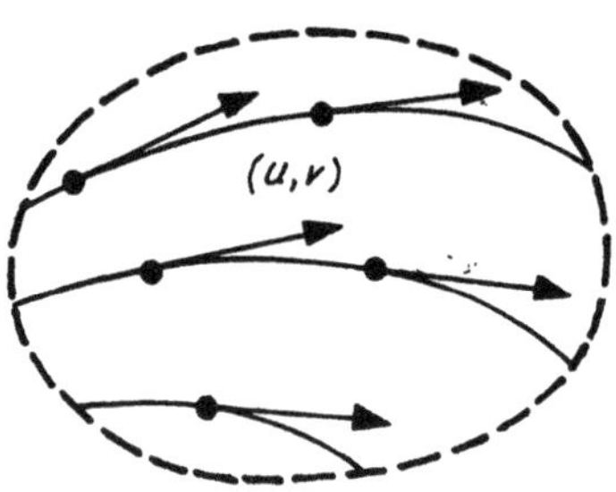

Fig. 27

Now regard any primitive function $\Psi = U + iV$ of Φ, such that

$$u - iv = \Phi = \Psi' = \frac{\partial U}{\partial x} + i\frac{\partial V}{\partial x} = \frac{\partial V}{\partial y} + i\frac{\partial V}{\partial x}.$$

Define a family of curves by $V = \operatorname{Im} \Psi = \text{const}$. The direction of normals is given by $(\partial V/\partial x, \partial V/\partial y)$, such that the direction of tangents is given by $(\partial V/\partial y, -\partial V/\partial x)$. Since we have

$$u = \frac{\partial V}{\partial y} \quad \text{and} \quad v = -\frac{\partial V}{\partial x},$$

the lines of flow (i.e. the paths of the particles of the fluid) are given by $\operatorname{Im} \Psi = \text{const}$.

On the other hand a conformal mapping $\zeta = \zeta(z)$ transforms holomorphic functions onto holomorphic functions (cf. 1.6.3.). Thus *conformal mappings transform incompressible fluid flows onto one another.* In this way it is possible to calculate special flows.

Example. With the help of the Žukovskiĭ (Joukowski) transformation

$$z = \frac{1}{2}\left(\zeta + \frac{h^2}{\zeta}\right), \qquad h = \text{const},$$

one gets the flow around a disk and, moreover, the so-called Žukovskiĭ airfoil in aerodynamics (see, for instance, DÖRRIE [1], SEDOV [1]). Other applications of complex methods in classical physics concerns the elasticity theory (see chapter 20) and the shell theory (see chapter 21).

1.9.9. Finally we hint at a connection between the existence of global homeomorphisms and uniformization theory (cf. 1.6.5.).

Let $\mathcal{F}_0$ be any Riemann surface, whose universal covering surface can be represented by the disk $\{\zeta : |\zeta| < 1\}$. Then $\mathcal{F}_0$ may be characterized by a subgroup of the group of all (linear) transformations of the disk onto itself.

Now regard any other Riemann surface $\mathcal{F}$, which can be mapped onto $\mathcal{F}_0$ by an orientable quasiconformal 1-1-mapping f (cf. 1.9.5.). Two pairs $(\mathcal{F}_1, f_1)$ and $(\mathcal{F}_2, f_2)$ are called *equivalent*, if $f_2 \circ f_1^{-1}$ can be continuously deformed into a conformal mapping of $\mathcal{F}_1$ onto $\mathcal{F}_2$. The set of all equivalence classes is said to be a *Teichmüller space*. These spaces are closely connected with transformation groups (see AHLFORS [5]).

1.10. Complex methods in differential equations[1])

1.10.1. For ordinary differential equations and for partial differential equations of elliptic type the following is true: If the coefficients of the differential equation are representable by power-series, then every solution possesses necessarily a power-series representation, too. Hence it is natural to apply properties of holomorphic functions in differential equations. However this is only a first connection between differential equations and complex analysis.

The general complex analysis makes use of the T_G- and the Π_G-operator, which are connected with partial complex differentiations in Sobolev's sense. So complex analysis is *applicable also to differential equations with coefficients not possessing power-series representations* (cf. also NAAS and TUTSCHKE [1]).

1.10.2. There are two possibilities to apply complex methods to partial differential equations:

a) First one can define certain *operators transforming* immediately *holomorphic functions onto solutions* of a given differential equation. Such transformations can be given by an integral operator (see chapter 15) or by a differential operator (see chapter 14). The transformations lead to solutions possessing explicit representations. In the main this method is advantageous in the case of special differential operators (the method of integral operators, for instance, needs the construction of the corresponding kernels).

b) A second complex method for solving partial differential equations is based on the properties of the T_G- and the Π_G-operators. Using these operators the given *differential equations* will be *reduced to an operator equation* that can be solved with the help of fixed-point theorems (see 1.10.5.). The T_G- and the Π_G-operator possess elementary kernels (namely $1/(z - \zeta)$ and $1/(z - \zeta)^2$ resp.), such that this second method does not need the construction of kernels.

1.10.3. Regard the following (real) first order system for $2m$ unknown real-valued functions $u_1, \ldots, u_{2m}$

$$H_j\left(x, y, u_1, \ldots, u_{2m}, \frac{\partial u_1}{\partial x}, \ldots, \frac{\partial u_{2m}}{\partial y}\right) = 0, \qquad j = 1, \ldots, 2m \tag{55}$$

Introduce complex variables $x + iy =: z, \qquad u_k + iu_{m+k} =: w_k, \qquad k = 1, \ldots, m,$

[1]) See BERGMAN [1], BERS [1], BERS, JOHN, and SCHECHTER [1], COLTON [13], DŽURAEV [1], GILBERT [1, 2], NAAS and TUTSCHKE [4], POLOŽIĬ [1], VEKUA [3, 4, 5], WENDLAND [4], TUTSCHKE [8, 10, 11]; with reference to ordinary differential equations cf. INCE [1].

then all derivatives $\partial u_1/\partial x, \dots, \partial u_{2n}/\partial y$ can be expressed by the derivatives $\partial w_k/\partial z =: \zeta_k$ and $\partial w_k/\partial \bar{z} =: \zeta_{m+k}$.

Now assume that the real system (55) can be solved for the real and imaginary parts of ζ_{m+k}, $k = 1, \dots, m$. Then we get

$$\zeta_{m+k} = F_k(z, w_1, \dots, w_m, \zeta_1, \dots, \zeta_m) .$$

Denote $(w_1, \dots, w_m)$ by w and $(F_1, \dots, F_m)$ by F. Taking into consideration the definition of ζ_k and ζ_{m+k} we get, finally, the following *complex normal form* of (55):

$$\frac{\partial w}{\partial \bar{z}} = F\left(z, w, \frac{\partial w}{\partial z}\right). \tag{56}$$

The most important boundary value problems for partial complex differential equations of type (56) are the following ones:

a) We look for a solution with prescribed real part on the boundary (and with prescribed imaginary part at one point). This boundary condition is called *Dirichlet's* one (cf. also the example in 1.10.5.).

b) One prescribes linear combinations of the real part and the imaginary part on the boundary. Such a boundary condition is called a *Hilbert* or a *Riemann-Hilbert* one (cf. 10.3., 10.4., 10.5., 11.3., 12.6.).

c) Finally the *Riemann boundary value problem* consists of finding a piecewise continuous solution satisfying a *jump discontinuity condition* on a given curve. That means that the boundary values w^+ and w^- corresponding to the sides of the curve satisfy a prescribed relation. This boundary value problem is also called the *Riemann-Hilbert transmission problem* as well as the problem of linear conjugation (if the jump discontinuity condition is linear). Cf. 5., 10.2., 10.4., 12.7., 20.4.3.

All boundary value problems can be solved for complex-valued solutions w as well as for vectors $w = (w_1, \dots, w_m)$. If $F = 0$ then one gets the corresponding boundary value problem for holomorphic functions and holomorphic vectors, respectively (cf. 10.2., 10.3., 5. and also 1.5.2.).

In special cases the normal form (56) can be replaced by a simpler one (see 11.2.):

In the case of linear uniformly elliptic systems for two real-valued unknown the equation (56) can be replaced by (54). Further in some cases solutions of linear uniformly elliptic systems for $2m$ real-valued unknown may be interpreted also as hyperanalytic functions (see chapter 12).

1.10.4. Complex analysis is connected closely with partial differential equations. Above all this connection concerns elliptic equations. In order to clarify the reason of it, we regard in a domain of $\mathbb{R}^n$ a linear differential operator of order k with principal part

$$\sum_{|\alpha|=k} a_\alpha(x) \frac{\partial^\alpha u}{\partial x_1^{\alpha_1} \dots \partial x_n^{\alpha_n}}, \tag{57}$$

where u is a unknown real-valued function, $x = (x_1, \dots, x_n)$, $\alpha = (\alpha_1, \dots, \alpha_n)$ and $|\alpha| = \alpha_1 + \dots + \alpha_n$. The *symbol* of this differential operator at the point x is given by the polynomial

$$\sum_{|\alpha|=k} a_\alpha(x)\, \xi_1^{\alpha_1} \dots \xi_n^{\alpha_n} . \tag{58}$$

Then the regarded operator is called *elliptic* at x, if the polynomial (58) is different from zero for any real variables ξ_j that do not vanish simultaneously (see, for instance,

Agmon [2]). The meaning of the symbol (58) for the behaviour of a differential operator with principal part (57) arises from the fact that the symbol is connected with the Fourier transform of the principal part (57):

Let u be defined in the whole R^n. Then its *Fourier transform* is defined by

$$\hat{u}(\xi) = \int_{R^n} \exp(-ix\xi)\, u(x)\, dx\,,$$

where $x\xi$ means the scalar product $\sum_{j=1} x_j\xi_j$, $\xi = (\xi_1, \dots, \xi_n)$. The derivative $\partial u/\partial x_j$ possesses the Fourier transform

$$i\xi_j\hat{u}\,.$$

Assume, in addition, that the coefficients $a_\alpha(x)$ are constant. Then the Fourier transform of the principal part (57) is given by

$$i^k \sum_{|\alpha|=k} a_\alpha \xi_1^{\alpha_1} \dots \xi_n^{\alpha_n} \cdot \hat{u}\,.$$

Hence the meaning of the symbol (58) appears.

Regard now the case of two independent variables x_1, x_2. Then the symbol (58) possesses a factorization

$$\text{const} \prod_{\varkappa=1}^{k} (\xi - \xi_{0\varkappa})\,, \tag{59}$$

where the $\xi_{0\varkappa}$ may be complex. In the case of elliptic differential operators all $\xi_{0\varkappa}$ are complex, such that complex analysis enters into the theory of elliptic operators. Another connection between the Fourier (or Laplace) transform and complex analysis is that the Fourier (or Laplace) transform is proved to be holomorphic, if the real variables ξ_j are replaced by complex variables (cf. 1.7.14.b); a characteristic function in probability theory is the Fourier transform of a distribution function, cf. chapter 22.).

A differential operator is said to be of *composed type*, if some of the $\xi_{0\varkappa}$ in (59) are complex, whereas some of them are real. Complex methods are applicable also to differential operators of composed type (see chapter 13.; using a transformation of coordinates the principal part $\mathcal{L}$ of an operator of composed type is representable as product $\mathcal{L} = \tilde{\mathcal{L}}\Delta$, where Δ is the Laplacian, cf. Malonek [3]).

In order to illustrate the definition of ellipticity for the system (56) we will regard the special case (52)

$$\frac{\partial w}{\partial \bar{z}} = \mu(z)\, \frac{\partial w}{\partial z}\,,$$

where $w = u + iv$ is complex-valued. This complex differential equation is equivalent to the real system

$$\frac{\partial v}{\partial y} = \frac{|1 - \mu|^2}{1 - |\mu|^2} \frac{\partial u}{\partial x} + 2i\, \frac{\mu - \bar{\mu}}{1 - |\mu|^2} \frac{\partial u}{\partial y}\,,$$

$$-\frac{\partial v}{\partial x} = 2i\, \frac{\mu - \bar{\mu}}{1 - |\mu|^2} \frac{\partial u}{\partial x} + \frac{|1 + \mu|^2}{1 - |\mu|^2} \frac{\partial u}{\partial y}\,,$$

if $|\mu(z)| < 1$. Eliminating the function v from the last system one gets an elliptic equation for u. Altogether from the assumption (53) the (uniform) ellipticity of the equation (52) follows. Thus in the case of the general equation (56) ellipticity means that a Lipschitz constant of the right-hand side $F(z, w, h)$ with respect to h $(= \partial w/\partial z)$ is small enough. Then a corresponding operator that will be defined with the help of $F(z, w, h)$ (cf. 1.10.5.), will be proved to be contractive.

1.10.5. Now describe a *general method* for solving the (nonlinear) equation (56). This method allows to construct the general solution as well as special solutions like those of boundary value problems.

At first assume that $w = w(z)$ is a known solution in the domain G. Then define the function Φ by

$$\Phi := w - T_G F(\cdot, w, h)\,, \tag{60}$$

where $h := \partial w/\partial z$. In view of (5) from the differential equation (56) follows that Φ is holomorphic. Differentiating (60) with respect to z we get

$$\Phi' = h - \Pi_G F(\cdot, w, h)\,. \tag{61}$$

That means that (w, h) is a solution of the system (60), (61) of singular integral equations, if w is a solution of (56). Conversely it is easy to prove that w is a solution of (56), if (w, h) is a solution of the system (60), (61), where Φ is any holomorphic function.

Now we look for a solution w of (56) fulfilling a given side condition

$$\mathscr{B}w = g\,. \tag{62}$$

Assume that a uniquely determined holomorphic function exists fulfilling this boundary condition. Define by

$$W = \Phi_{(w,h)} + T_G F(\cdot, w, h)\,, \qquad H = \Phi'_{(w,h)} + \Pi_G F(\cdot, w, h)$$

an operator $\tilde{T}$, where $\Phi_{(w,h)}$ is the uniquely determined holomorphic function fulfilling the side-condition

$$\mathscr{B}(\Phi_{(w,h)} + T_G F(\cdot, w, h)) = g$$

(we remark, that $\Phi_{(w,h)}$ is also a vector, if the solution w looked for is a vector, $w = (w_1, \ldots, w_m)$). It is sufficient to define (W, H) for every (w, h) belonging to a (closed) subset $\mathscr{D}$ of a certain function space. From the definition of $\tilde{T}$ follows immediately that every $\tilde{W}$ fulfils the side condition $\mathscr{B}W = g$. Furthermore it is clear that every fixed point (w, h) of $\tilde{T}$ fulfils the equations (60), (61), such that w is a solution of the differential equation (56).

In this way we have proved the following theorem:

Every fixed point of $\tilde{T}$ is a solution of the differential equation

$$\frac{\partial w}{\partial \bar{z}} = F\left(z, w, \frac{\partial w}{\partial z}\right)$$

and fulfils the side-condition $\mathscr{B}w = g$.

The existence of fixed points can be proved by the help of one of the usual fixed-point theorems. To it the operator $\tilde{T}$ must be estimated in $\mathscr{D}$. Let $\mathscr{D}$, for instance, be the bicylinder

$$\mathscr{D} = \{(w, h) : ||w|| \leqq R_1, ||h|| \leqq R_2\}$$

in the space of Hölder-continuous pairs (w, h). In order to apply Banach's fixed-point theorem the operator $\tilde{T}$ must be a contractive operator, mapping $\mathscr{D}$ into itself. Using well-known a priori estimates for holomorphic functions and, moreover, the boundness of the operators T_G and Π_G (cf. 1.2.1.) one gets the following result in the case of boundary value problems:

If the norm of the given boundary values and, moreover, if the diameter of the regarded domain are small enough, then the operator $\tilde{T}$ is contractive and, consequently, the boundary value problem (56), (62) *is uniquely solvable in the bicylinder $\mathfrak{D}$.*

Example. Regard the boundary value problem

$$\operatorname{Re} w = g \quad \text{on} \quad \partial G\,, \qquad \operatorname{Im} w[z_0] = c\,,$$

where z_0 is a point of the boundary ∂G (*Dirichlet's boundary value problem*). Assume, that G is simply connected. Then the regarded boundary value problem is uniquely solvable for holomorphic functions Φ: the real part $\operatorname{Re} \Phi$ is that uniquely determined solution of the Laplace equation, which possesses the boundary values g. The imaginary part $\operatorname{Im} \Phi$ is a conjugate harmonic function to $\operatorname{Re} \Phi$. Since a conjugate harmonic function is uniquely determined up to an arbitrary constant only (cf. 1.3.4.), we can prescribe the value of $\operatorname{Im} \Phi$ at z_0 (if the boundary values g are Hölder-continuous on ∂G and, moreover, if the boundary curve ∂G is sufficiently smooth, then Φ is Hölder-continuous in $\bar{G}$ (*theorem of Privalov*). Detailed estimates of the corresponding operator $\tilde{T}$ in the space of Hölder-continuous functions are given in Tutschke [11].

The same boundary value problem had been investigated by Seif-Mshimba [1] in the Sobolev space $\mathscr{W}_p^1(G)$. His result: the solution belongs to $\mathscr{W}_p^1(G)$, if the boundary values g belong to the Slobodeckiĭ space $\mathscr{W}_p^s(\partial G)$, $s = 1 - 1/p$, $p > 2$. Kaden [1] studies the operator $\tilde{T}$ in locally convex vector spaces. In this way he can solve the Dirichlet boundary value problem also in such cases, in which the boundary values g possess singularities. Rüprich [1] solves boundary value problems for differential equations of type

$$F\left(z, w, \frac{\partial w}{\partial z}, \frac{\partial w}{\partial \bar{z}}\right) = 0\,.$$

1.10.6. Complex methods allow to construct solutions of differential equations. Further they can be used to deduce *qualitative properties* of solutions of differential equations and differential inequalities. For example we will prove a factorization theorem for (continuous) solutions $w = w(z)$ of the inequality

$$\left|\frac{\partial w}{\partial \bar{z}}\right| \leqq K(z)\,|w|$$

in a bounded domain G of the z-plane, where $K(z) \geqq 0$ is given. Define

$$g = \begin{cases} \dfrac{1}{w}\dfrac{\partial w}{\partial \bar{z}}, & \text{if} \quad w \neq 0\,, \\ 0 & \text{otherwise,} \end{cases}$$

then $|g|$ is bounded by $K(z)$. Further we regard $\omega := T_G g$ and $\Phi := w \exp(-\omega)$. Then it is easy to check that Φ is holomorphic. Therefore we get for w the factorization (cf. also 11.1., theorem 11.2 in 11.2. and 12.4.)

$$w = \Phi w_0\,,$$

where $w_0 = \exp \omega \neq 0$. From this representation theorem (cf. Bers and Nirenberg [1], I. N. Vekua [8]) follows that the zeroes of $w = w(z)$ are isolated, unless w vanishes identically. Generally spoken such representation theorems allow to generalize theorems on the qualitative behaviour of holomorphic functions. Using similar representation theorems MALONEK [1, 2] has generalized the Phragmén-Lindelöf theorem to the case of systems of partial differential equations. MÄCHLER [1] has proved a continuation theorem for systems of the type

$$\frac{\partial w}{\partial z} = f_1(z, w) ,$$

$$\frac{\partial w}{\partial \bar{z}} = f_2(z, w) ,$$

where $|f_2| \leqq \mu_0 |f_1|$, $0 < \mu_0 < 1$. This theorem is an analogous theorem to Kaplan's one (see KAPLAN [1]).

1.10.7. A starting point of the complex analysis (in the plane) is the following one: problems concerning several independent variables are regarded formally like problems for one (complex) variable. In order to realize an analogous point of view in *higher* dimensions, it is helpful to introduce so-called function algebras and a certain differential operator D acting in them (see chapter 16.). They allow to generalize Cauchy's integral formula and its consequences to higher dimensions. Many important first order systems in $\mathbb{R}^n$ can be written as $Dw = 0$, such that their solutions may be interpreted as holomorphic functions (in $\mathbb{R}^n$). If the unknown functions itself enter into a given (first order) system in $\mathbb{R}^n$, then its solution are called *generalized analytic functions* in $\mathbb{R}^n$ (see OBOLAŠVILI [4, 5, 6]). Using the function-algebraic notation such systems can be written in the form $Dw = f(\cdot, w)$ (see SPRÖSSIG [1] and GOLDSCHMIDT [4, 5, 6]).

On the other hand the space $\mathbb{R}^{2n}$ can be interpreted as $\mathbb{C}^n$, such that differential equations in $2n$ real variables $x_1, \ldots, x_{2n}$ can be written as equations in n complex variables $z_1, \ldots, z_n$, where $z_j = x_j + ix_{n+j}$. The complex equations contain, in general, the differentiations with respect to the z_j as well as those with respect to the $\bar{z}_j$ (see, for instance, KOOHARA [1], MIHAÏLOV [2], MAGOMEDOV and PALAMODOV [1], TUTSCHKE [10], cf. also 1.7.12. and 1.7.13.).

1.11. Complex methods in the theory of instationary processes

1.11.1. Let u be a function depending on the variable $x = (x_1, \ldots, x_n) \in G$ and on the time t, where G is a domain in $\mathbb{R}^n$. The function u is said to describe an evolution process, if it is a solution of a differential equation of the type

$$\frac{\partial u}{\partial t} = Lu , \tag{63}$$

where L is a differential operator with respect to $x_1, \ldots, x_n$. The initial-value problem for (63) is the following one: it is to construct a solution taking arbitrarily prescribed values u_0 for $t = 0$:

$$u(x, 0) = u_0(x) . \tag{64}$$

A solution $u = u(x, t)$ of the initial-value problem (63), (64) satisfies the integro-differential equation

$$u(x, t) = u_0(x) + \int_{\tau=0}^{t} Lu[x, \tau] \, d\tau .$$

Since a differential operator is unbounded, it is not possible, in general, to use this relation for the proof of the existence and the uniquess of the solution of the initial value problem (63), (64). That means that a sequence $\{v_k\}_{k=1,2,\dots}$ defined by

$$v_{k+1}(x, t) = u_0(x) + \int_{\tau=0}^{t} Lv_k[x, \tau] \, d\tau \tag{65}$$

does not converge to the solution, in general. In some cases, however, the sequence (65) converges to the solution. For instance, the convergence can be proved, if u depends holomorphically on $x = (x_1, \dots, x_n)$ (then the x_j can be replaced by complex variables z_j).

1.11.2. Let G be a domain in $\mathbb{C}$ (or in $\mathbb{C}^n$) and, moreover, let w_0 be a holomorphic function in G. The initial-value problem

$$\frac{\partial w}{\partial t} = F\left(t, z, w, \frac{\partial w}{\partial z}\right), \tag{66}$$

$$w(z, 0) = w_0(z) \tag{67}$$

possesses a uniquely determined *holomorphic* solution $w = w(z, t)$, if F depends holomorphically on its variables. If K is any compact subset of G, then the solution exists for all t with $|t| < T_K$, where T_K (> 0) depends on K (Cauchy-Kowalewski theorem). The solution can be represented in the form of a power-series. If the right-hand side F depends linearly on w and $\partial w/\partial z$, then the solution is also the limit of the sequence (65). For the convergence of (65) it must not be assumed that the coefficients of the right-hand side of (66) depend holomorphically on t. It is sufficient that they are continuous functions with respect to (real) t. The convergence of (65) is based on the estimate (18):

Let G be a bounded domain in $\mathbb{C}$. Let, moreover, G_s be a family of subdomains like in 1.4.3. Then regard the scale of Banach spaces $\mathcal{B}_s = \mathcal{H}(G_s) \cap \mathcal{C}(\overline{G}_s)$ again (cf. 1.4.3.). A linear right-hand side F of the differential equation (66) defines a linear operator mapping $\mathcal{B}_s$ into $\mathcal{B}_{s'}$ (if $s' < s$). The norm of this operator can be estimated by

$$\frac{C}{s - s'} . \tag{68}$$

Hence the *initial value problem* (66), (67) is reduced (for linear right-hand sides) to the following *abstract* one:[1])

Let $\mathcal{B}_s$ be any scale of Banach spaces. Let, moreover, $A(t)$ be for every t ($0 \leqq t \leqq T$) a (homogeneous and) linear operator mapping $\mathcal{B}_s$ into $\mathcal{B}_{s'}$, where the norm can be estimated by (68). Assume, that u_0 and $f(t)$ (for every t) belong to every $\mathcal{B}_s$. Then we look for a solution $u = u(t)$ of the initial value problem

$$\frac{du}{dt} = A(t) \, u + f(t) , \qquad u(0) = u_0 . \tag{69}$$

---

[1]) Concerning applications and the history of the method of scales of Banach spaces cf. OVSJANNIKOV [1].

1.11.3. In order to solve the abstract initial value problem (69) we define (cf. TREVES [1] and also (65))

$$v_0(t) = u_0 + \int_{\tau=0}^{t} f(\tau)\, d\tau\,, \qquad v_{k+1}(t) = u_0 + \int_{\tau=0}^{t} A(\tau)\, v_k(\tau)\, d\tau + \int_{\tau=0}^{t} f(\tau)\, d\tau\,.$$

Denote by $||\cdot||_s$ the norm in $\mathscr{B}_s$. Then the estimate

$$||v_k(t) - v_{k-1}(t)||_s \leqq c\left(\frac{Cet}{1-s}\right)^k \tag{70}$$

holds, where c is a constant. This estimate can be proved by induction: First from the definition of the v_k we get

$$v_{k+1}(t) - v_k(t) = \int_{\tau=0}^{t} A(\tau)\left(v_k(\tau) - v_{k-1}(\tau)\right) d\tau\,. \tag{71}$$

Let s be any fixed number, $0 < s < 1$. Using (70) with $\tilde{s} = s + (1-s)/(k+1)$ and (68) with s and $\tilde{s}$ (instead of s' and s) from (71) one gets immediately

$$||v_{k+1}(t) - v_k(t)||_s \leqq \frac{C}{\tilde{s} - s}\, c\left(\frac{Ce}{1-s}\right)^k \left(1 + \frac{1}{k}\right)^k \frac{t^{k+1}}{k+1}\,.$$

Since $(1 + 1/k)^k \leqq$ e the last inequality shows, that (70) holds for $k + 1$, too.

In view of (70) the v_k converges in B_s, uniformly for all t with $0 \leqq t \leqq Ce/2(1-s)$, such that the limit of the v_k is the solution looked for.

The uniquess of the solution can be proved by a similar estimate (cf. TREVES [1])

1.11.4. By virtue of the Cauchy-Kowalewski theorem the initial value problem (66), (67) possesses a uniquely determined solution that is holomorphic in z for every fixed t. Then the *Holmgren theorem* excludes the possibility of other nonholomorphic solutions (cf. TREVES [1] and also COLTON [13]). The abstract version of the Cauchy-Kowalewski theorem allows to interpret the Holmgren theorem as a dual form of the classical Cauchy-Kowalewski theorem.

1.11.5. From the Cauchy-Kowalewski theorem the existence of a holomorphic solution of the initial value problem follows, if the initial values and the coefficients are holomorphic. The same is true, if the holomorphy is replaced by a more general (linear) differential equation (cf. TUTSCHKE [17]).

1.11.6. Finally we remark that complex methods allow also to study evolution equations with singular coefficients. In such cases the initial value problem must be replaced by a boundary value problem (cf. TUTSCHKE [18]).

Further references

NAAS and SCHMIDT [1].

2. Value Distribution Theory[1)]

H. BEGEHR

2.1. Introduction

The value distribution theory of meromorphic functions was created by the ingeneous ideas of R. NEVANLINNA more than fifty years ago and is still one of the main topics under research in complex analysis. Modern function theory starts at the end of the last century with the considerations of PICARD [1], POINCARÉ [1], HADAMARD [1, 2], BOREL [2] and — a couple of years later — of VALIRON (see VALIRON [1, 2]) on the a-points of entire functions w, that are the zeroes of $w(z) - a$, for different values of a. More precisely they recognize connections between the growth of an entire transcendental function w and the distribution of its zeroes. For rational functions it is known that for every complex value a (∞ included) $w(z) - a$ has the same number of roots in the extended complex plane $\hat{\mathbb{C}}$. Roughly speaking it turns out from Nevanlinna's theory (first main theorem) that the behaviour of every non-constant meromorphic function in $\hat{\mathbb{C}}$, relatively to different complex numbers a including infinity, is nearly balanced. More detailed information about the terms arising in the first main theorem and especially on exceptional values a are contained in the second main theorem. The results are even valid in the vicinity of an essential isolated singularity (local Nevanlinna theory) and they hold for analytic functions in a circle too, let us say, the unit disc, although there are important differences in this case.

Besides from the monographs [3, 4, 6] of R. NEVANLINNA, one can get an introduction into the theory also from those of DINGHAS [1, 2], GOL'DBERG and OSTROVSKIĬ [1] HAYMAN [3] and WITTICH [8]. Because of the variety of important contributions within the last six decades, it is impossible to give a complete overlook. Those who are interested in more detailed informations we refer to the proceedings of the 10th Nevanlinna Colloquium published in Annales Academiae Scientiarum Fennicae AI 7 (1982), especially the articles of FUCHS [2], HAYMAN [5], LEHTO [3] and WITTICH [14]. The forthcoming obituary notes for R. NEVANLINNA by AHLFORS in Acta Mathematica and by WITTICH and LOUHIVAARA in Jahresberichte der Deutschen Mathematiker Vereinigung will be of equal interest. For an application of value distribution theory to ordinary differential equations systematically developed by WITTICH (see WITTICH [8]), see the article of MUES in this collection (chapter 9).

2.2. Nevanlinna's main theorems

Let w be a non-constant meromorphic function in $|z| < R$ ($R \leqq +\infty$) and $a \in \hat{\mathbb{C}}$. Denote the number of zeroes of $w(z) - a$ in $|z| \leqq r$ ($0 \leqq r < R$) for $a \in \mathbb{C}$ by $n(r, a)$

[1)] Here I want to thank Professor I. S. LOUHIVAARA for his suggestions and hints at literature.

and the corresponding number of poles of $w(z)$ for $a = \infty$. Finally let $(a \in \mathbb{C})$

$$g(z) := z^{-\lambda}\,[w(z) - a],\ \lambda := n(0, a) - n(0, \infty)\,.$$

Then from Jensen's formula one has

$$\log |g(0)| = \frac{1}{2\pi} \int_0^{2\pi} \log |w(r\,e^{it}) - a|\,dt - N(r, a) + N(r, \infty)\,, \tag{1}$$

where

$$N(r, a) = n(0, a) \log r + \int_0^r \frac{n(t, a) - n(0, a)}{t}\,dt \qquad (a \in \hat{\mathbb{C}},\, 0 \leqq r < R)$$

is the counting function, an increasing function of r and a subharmonic function of a. A decomposition of the integrand in (1) due to those parts where $w(r\,e^{it})$ is close to a and those where it becomes close to infinity leads to Nevanlinna's proximity function

$$m_N(r, a) := \frac{1}{2\pi} \int_0^{2\pi} \overset{+}{\log} |w(r\,e^{it}) - a|^{-1}\,dt\,,$$

$$m_N(r, \infty) := \frac{1}{2\pi} \int_0^{2\pi} \overset{+}{\log} |w(r\,e^{it})|\,dt\,,$$

where for positive real a

$$\overset{+}{\log} a = \operatorname{Max}\,\{0, \log a\}\,,$$

and to the

Theorem 2.1 (First main theorem). *For any complex value a*

$$m_N(r, a) + N(r, a) = T_N(r, w) + \varphi(r, a) \tag{2}$$

where

$$T_N(r, w) := m_N(r, \infty) + N(r, \infty)$$

is the Nevanlinna characteristic function of w and

$$|\varphi(r, a)| \leqq \overset{+}{\log} |a| + |\log |g(0)|| + \log 2\,.$$

If the characteristic function turns out to be unbounded when r tends to R, this result describes the remarkable fact that the sum of counting and proximity functions, considered as a function of r, essentially is the same for every $a \in \hat{\mathbb{C}}$ up to $O(1)$. While the counting function is a measure of the distribution in the mean of the a-points of w in the disc $|z| < R$, which grows the faster with r the larger the number of a-points are, the proximity function is a measure for the mean deviation of w from a on the circle $|z| = r$, namely the mean value of $\overset{+}{\log} |w - a|^{-1}$ and of $\overset{+}{\log} |w|$, respectively, on the circle $|z| = r$, which receives its significant contribution from those arcs where w comes close to a and ∞, respectively.

Functions with bounded characteristic do occur only in the case of a finite disc $(R < +\infty)$, see section 2.3. If $R = +\infty$, the class of functions with bounded characteristic coincides with the constant functions. More general, R. Nevanlinna proved the following characterization of rational functions.

Theorem 2.2. *In order that a function meromorphic in the plane is a rational function the condition*

$$\varliminf_{r\to\infty} \frac{T_N(r, w)}{\log r} < \infty$$

is necessary and sufficient.

Because $O(1)$-terms in general are neglectable, Nevanlinna's proximity as well as the characteristic function may be replaced by functions which differ only by $O(1)$-terms.

If instead of the Euclidean distance the spherical metric is used in the definition of the proximity function, one gets the Shimizu-Ahlfors proximity function

$$m_A(r, a) := \frac{1}{2\pi} \int_0^{2\pi} \log\, [w(r\, e^{it}), a]^{-1}\, dt \qquad (a \in \hat{\mathbb{C}})$$

and the Shimizu-Ahlfors characteristic function

$$T_A(r, w) := m_A(r, \infty) + N(r, \infty)\,.$$

Another choice of a proximity and a characteristic function was given by DINGHAS [4].

The first main theorem as well as all the other results can be expressed by all these functions, too. For the characteristic function the following expression is very often useful. Let

$$d\omega(a) := (1 + |a|^2)^{-2}\, |a|\, d\, |a|\, d \arg a$$

be the area element of the Riemann sphere $\hat{\mathbb{C}}$, and

$$A(r, w) := \frac{1}{\pi} \int_{\hat{\mathbb{C}}} n(r, a)\, d\omega(a) = \frac{1}{\pi} \int_{|z|<r} (1 + |w(z)|^2)^{-2}\, |w'(z)|^2\, dx\, dy$$

and

$$T(r, w) := \int_0^r A(t, w) \frac{dt}{t}\,.$$

The function $T(r, w)$ differs from $T_A(r, w)$ by an additive constant and is obviously a convex function of $\log r$. That the same even holds for $T_N(r, w)$ is a conclusion from Cartan's identity.

In the following $m(r, a)$ and $T(r, w)$ mean any of the respective functions.

To describe the growth of w one introduces the order ϱ and the lower order λ,

$$\varrho := \varlimsup_{r\to R} \frac{\log T(r, w)}{\log r}\,, \qquad \lambda := \varliminf_{r\to R} \frac{\log T(r, w)}{\log r}\,.$$

While the first main theorem gives information about the growth of $m(r, a) + N(r, a)$, it does not say anything about the relative behaviour of the two single terms. Does in general either term play a dominant role? This question is answered by the second main theorem in the affirmative in the sense that only for very few a-values $N(r, a)$ fails to be comparable with $T(r, w)$.

Theorem 2.3 (Second main theorem). *Let $a_\mu \in \hat{C}$ $(1 \leqq \mu \leqq q, 3 \leqq q)$ be q different numbers. Then*

$$(q-2)\, T(r, w) \leqq \sum_{\mu=1}^{q} N(r, a_\mu) - N_1(r) + S(r) \tag{3}$$

where $N_1(r)$ is a counting function for multiple values of w defined as $N(r, a)$ by $n(r, a)$, by the number $n_1(r)$ of all values, counting each a-value $(a \in \hat{C})$ of w of multiplicity k only $k-1$ times, and where $S(r)$ is a remainder term for which

$$S(r) = O(1 + \overset{+}{\log} r) + O(1 + \overset{+}{\log} T(r, w)) + \log\,[1 - rR^{-1}]^{-1} \tag{4}$$

outside at most a subset E of the interval $(0, R)$, such that

$$\int_E \mathrm{d} \log\,[r(1 - rR^{-1})^{-1}] \tag{5}$$

is finite.

In case of functions of finite order E is the empty set while for functions of infinite order E really can occur (Hayman [4]).

(5) means

$$\int_E \mathrm{d} \log r < +\infty\,, \qquad R = +\infty\,; \qquad \int_E \mathrm{d} \log\,(R - r)^{-1} < +\infty\,, \qquad R < +\infty.$$

There are several different methods to prove this theorem. The original proof given by R. Nevanlinna [1] is based on the estimation of the proximity function of the ogarithmic derivative of meromorphic functions (see Dinghas [4], Hayman [3], R. Nevanlinna [3, 4, 6]). A differential geometric method given by F. Nevanlinna [1] essentially uses the hyperbolic metric of Poincaré and the theory of linear polymorphic functions which were used by Picard to prove his famous theorem (Dinghas [4], R. Nevanlinna [6]). Finally Ahlfors [1] gave a proof by using more resources from potential theory (Dinghas [1], [4], R. Nevanlinna [6]). Dinghas [3] modified Ahlfors' method (Dinghas [4]).

For the conclusion from the second main theorem it is convenient to rewrite (3) Let be

$$\delta(a) := 1 - \varlimsup_{r \to R} \frac{N(r, a)}{T(r, w)} \quad \text{(deficiency or defect of } a\text{)},$$

$$\vartheta(a) := \varliminf_{r \to R} \frac{N_1(r, a)}{T(r, w)} \quad \text{(ramification index of } a\text{)},$$

$$\theta(a) := 1 - \varlimsup_{r \to R} \frac{\overline{N}(r, a)}{T(r, w)} \quad \text{(total ramification of } a\text{)}.$$

Here $N_1(r, a)$ is defined as $N_1(r)$ counting only the multiple a-values of w rather than all multiple values of w, $\overline{N}(r, a)$ counts all a-values independently of their multiplicity only once, so that $N_1(r, a) = N(r, a) - \overline{N}(r, a)$. From the first main theorem it follows

$$0 \leqq \delta(a)\,, \qquad \vartheta(a)\,, \qquad \theta(a) \leqq 1$$

and from the second

$$\sum_{a \in \hat{C}} [\delta(a) + \vartheta(a)] \leqq \sum_{a \in \hat{C}} \theta(a) \leqq 2 + \varlimsup_{r \to R} \frac{\log\,[r(1 - rR^{-1})^{-1}]}{T(r, w)}\,. \tag{6}$$

Theorem 2.4. *If*

$$\overline{\lim_{r\to R}} \frac{\log [r(1 - rR^{-1})^{-1}]}{T(r, w)} < +\infty$$

then the quantity $\theta(a)$ vanishes for all $a \in \hat{\mathbb{C}}$ except at most for a countable set.

By a more precise investigation using Borel's lemma, Hayman ([3], § 2.3) proved

Theorem 2.5. *If*

$$\underline{\lim_{r\to R}} \frac{\log [r(1 - rR^{-1})^{-1}]}{T(r, w)} = 0 \tag{7}$$

then

$$\sum_{a\in\hat{\mathbb{C}}} [\delta(a) + \vartheta(a)] \leqq \sum_{a\in\hat{\mathbb{C}}} \theta(a) \leqq 2\,. \tag{8}$$

Condition (7) in the case $R = +\infty$ is fulfilled if w has an essential singularity at infinity (see theorem 2.2). In the case $R < +\infty$ ($R = 1$!), R. Nevanlinna ([3], p. 149) mentioned that automorphic functions which map $|z| < 1$ onto the universal covering surface of the plane punctured at $q \geqq 3$ points $a_1, a_2, \dots, q_{q-1}, a_q = \infty$ and therefore avoid the values $a_1, \dots, a_q$ fulfil

$$T(r, w) \sim \frac{1}{q-2} \log \frac{1}{1-r}$$

so that

$$\sum_{k=1}^{q} \delta(a_k) = q\,.$$

In this sense (6) is sharp. An example for an entire function with maximal defect sum was given by Hille [1] using ordinary differential equations in the complex plane. The deficiency and remification relation

$$\sum_{a\in\hat{\mathbb{C}}} [\delta(a) + \vartheta(a)] \leqq 2 \tag{9}$$

is a deep generalization of Picard's theorem, which states that in the vicinity of an essential singularity w can avoid at most two values (Picard-exceptional values). Values with positive defect $\delta(a)$ are called *Nevanlinna-deficient values.* For these values $N(r, a)$ is small, compared to $m(r, a)$. Inequality (9) tells that up to at most a countable set of a-values $N(r, a)$ becomes as big as $T(r, w)$. For more detailed discussions of (9) we refer to R. Nevanlinna [4, 6], Hayman [3], Fuchs [1, 2]. It is possible to replace $\delta(a)$ by the so-called Valiron-deficiency

$$\Delta(a) := \overline{\lim_{r\to R}} \frac{m(r, a)}{T(r, w)} = 1 - \underline{\lim_{r\to R}} \frac{N(r, a)}{T(r, w)}\,.$$

But here much weaker results are to be attained (see R. Nevanlinna [4], Hayman [4], Shea [1]). According to (6) in a finite disc the set of deficient values may be uncountable. Frostman proved (see R. Nevanlinna [4])

Theorem 2.6. *If the characteristic function of w meromorphic in $|z| < 1$ is unbounded, then the set of its deficient values has vanishing inner harmonic measure.*

More detailed considerations show (R. Nevanlinna [4], p. 276) for functions not of bounded characteristic

$$\lim_{r\to R} \frac{N(r, a)}{T(r, w)} = 1$$

except at most an a-set of inner capacity zero.

While $\vartheta(a)$ is a measure of the ramification of the image of $|z| < R$ under w over a, $\theta(a)$ gives some informations for totally ramified values, i.e. values a for which no simple a-points exist, so that $\theta(a) \geqq \frac{1}{2}$. Therefore a meromorphic function in $|z| < R$ fulfilling (7) can have at most four totally ramified values as for example the Weierstraß $\wp$-function. An important application of this remark to the uniformization of algebraic curves is a proof of the following result of PICARD (R. NEVANLINNA [4], p. 281, SELBERG [2]).

Theorem 2.7. *If the algebraic curve $f(x, y) = 0$ is of genus $p > 1$, then there is no pair of meromorphic functions $x(t)$, $y(t)$ such that $f(x(t), y(t)) \equiv 0$.*

The inverse problem of value distribution theory is to find a meromorphic function with prescribed deficiencies and ramification indices for given a-values. After first results of R. NEVANLINNA, FUCHS and HAYMAN, and GOL'DBERG, DRASIN [1] finally proved by using quasimeromorphic functions (see FUCHS [2], LEHTO [2], R. NEVANLINNA [7])

Theorem 2.8. *For given finitely or countable many complex values a_ν and non-negative numbers δ_ν, ϑ_ν such that*

$$0 \leqq \delta_\nu + \vartheta_\nu \leqq 1\,, \qquad \sum_\nu (\delta_\nu + \vartheta_\nu) \leqq 2$$

there exists a meromorphic function in the plane of arbitrarily slow growth for which for all ν

$$\delta(a_\nu) = \delta_\nu\,, \qquad \vartheta(a_\nu) = \vartheta_\nu$$

and

$$\delta(a) = \vartheta(a) = 0$$

for all $a \neq a_\nu$.

The concept of value distribution theory is founded on an exhaustion of $|z| < R$ by concentric circles. If $R = +\infty$, this exhaustion is no more uniquely given. DUGUE [1] had shown that the notation of deficiency really is dependent on the choise of the origin (see also HAYMAN [1]). For functions of finite order more can be said about deficiencies. So FUCHS, HAYMAN (see HAYMAN [3]) and WEITZMAN [1] have proved

Theorem 2.9. *Suppose w is meromorphic and of positive and finite lower order λ. Then*

$$\sum_{a \in \hat{C}} \delta^\alpha(a)$$

is bounded for $1/3 \leqq \alpha$, where the bound depends on α and λ only and unbounded for $\alpha < 1/3$.

Likewise the concept of deficiency PETRENKO [1, 4] introduced the deviations

$$\beta(a) := \varliminf_{r \to R} \frac{\log^+ M(r, (w - a)^{-1})}{T(r, w)}$$

where $M(r, w) := \max\limits_{|z| = r} |w(z)|$ (see also FUCHS [2], GOL'DBERG [3], GRIŠIN [1]).

Besides the deficiency and ramification theorem, another consequence of the second main theorem are improvements of results of BOREL, HADAMARD and PICARD (R. NEVANLINNA [4], DINGHAS [1]).

Theorem 2.10 (Theorem of Hadamard-Borel-Nevanlinna). *If for a meromorphic function in the unit disc and for $k > 0$*

$$\int_0^1 T(r, w)\,(1 - r)^{k-1}\,dr \tag{10}$$

converge, then for all $a \in \hat{\mathbb{C}}$

$$\int_0^1 N(r, a)\,(1 - r)^{k-1}\,\mathrm{d}r\,, \qquad \int_0^1 n(r, a)\,(1 - r)^k\,\mathrm{d}r\,, \qquad \sum_{n=1}^{+\infty} (1 - |z_n(a)|)^{k+1} \quad (11)$$

are convergent.

Theorem 2.11 (Theorem of Picard-Borel-Nevanlinna)

1. *For a non-constant meromorphic function in the plane, there are at most two a-values for which the counting functions* $N(r, a)$ *as well as* $n(r, a)$ *are of lower order than the characteristic function* $T(r, w)$. *Similarly only for two a-values the exponent of convergence for*

$$\sum |z_n(a)|^{-k-1}$$

can be less than the order of $T(r, w)$.

2. *If for a meromorphic function in the unit disc the quantities in* (11) *converge for three different a-values, then* (10) *is convergent, too.*

Here $z_n(a)$ *are the a-points of* w, *i.e.*

$$w(z_n(a)) = a\,.$$

2.3. Functions of bounded characteristic

As was mentioned in 2.2. a meromorphic function in the plane with bounded characteristic function reduces to a constant. But in the case of the unit disc, there exist functions of this type. Since $T(r, w)$ is increasing,

$$T(1, w) := \lim_{r \to 1} T(r, w)$$

exists and is finite. In this case, the remainder terms in the main theorems are no more neglectable, so that the above conclusions do not hold any more. R. Nevanlinna showed these functions to be representable as ratios of bounded functions (R. Nevanlinna [4], Hayman [3]). More precisely the following theorem holds.

Theorem 2.12. *If w is meromorphic in the unit disc and has a bounded characteristic, it can be represented in the form*

$$w(z) = z^p \frac{\Pi_1(z)}{\Pi_2(z)} \exp\left\{\frac{1}{2\pi} \int_0^{2\pi} \frac{\mathrm{e}^{\mathrm{i}t} + z}{\mathrm{e}^{\mathrm{i}t} - z}\,\mathrm{d}\mu(t) + \mathrm{i}\lambda\right\}$$

where p is an integer, Π_1 *and* Π_2 *are Blaschke products defined by the zeroes and poles of w respectively,* λ *is a real constant, and* μ *is a function of bounded variation.*

By this result Fatou's theorem on the boundary behaviour of bounded functions can be generalized to functions of bounded characteristic:

Theorem 2.13. *A meromorphic function in the unit disc of bounded characteristic has well-defined finite non-tangential limits almost everywhere on the unit circle.*

The theorem of F. and M. Riesz can be extended, too:

Theorem 2.14. *A meromorphic function in the unit disc of bounded characteristic which has a constant radial limit on a point set of positive measure on the unit circle is identical to this constant.*

Theorem 2.15. *If the measure of a given set E_z on the unit circle is positive, then the set E_w of radial boundary values of a nonconstant meromorphic function of bounded characteristic in $|z| < 1$ has positive inner capacity.*

Let G be a schlicht multiple connected domain, G^∞ its universal covering surface, and w the conformal map from $|z| < 1$ onto G^∞ which is an automorphic function. G is called of bounded type if w has bounded characteristic.

Theorem 2.16. *G is of bounded type if and only if its boundary has positive capacity.*

The boundary behaviour of the automorphic function w is described by

Theorem 2.17. *The measure of the set of values φ, $0 \leqq \varphi \leqq 2\pi$, for which the angular limit*

$$w(e^{i\varphi}) = \lim_{z \to e^{i\varphi}} w(z)$$

exists, is 2π or zero, depending as to G is of bounded type or not.

See R. Nevanlinna [4], Lehto [1].

The boundary behaviour of arbitrary meromorphic functions in the unit disc is characterized by results of Plessner and Meier (see Collingwood and Lohwater [1]).

A point e^{it} of the unit circle is called a *Plessner point* if the angular cluster set $C_\Delta(w, e^{it})$ is total, i.e. covers $\hat{\mathbb{C}}$ for every angle Δ with vertex in e^{it}. Let $I(w)$ denote the set of all Plessner points of w on $|z| = 1$.

A point e^{it} is called a *Fatou point* of w if the union of $C_\Delta(w, e^{it})$ over all angles Δ with vertex in e^{it} is degenerate, i.e. is a single point and if

$$\lim_{z \to e^{it}} w(z)$$

exists uniformly in any angle Δ. Let $F(w)$ denote the set of all Fatou points of w on $|z| = 1$.

A point e^{it} is called a *Meier point* of w if the complete cluster set $C(w, e^{it})$ is subtotal, e.i. it contains more than one point but not all of $\hat{\mathbb{C}}$ and the chordal cluster set $C_{\varrho(\tau)}(w, e^{it})$ is equal to $C(w, e^{it})$ for all τ, $-\pi/2 \leqq \tau \leqq \pi/2$. Let $M(w)$ denote the set of all Meier points of w on $|z| = 1$.

There is no inclusion relation between $F(w)$ and $M(w)$.

Theorem 2.18. *For each meromorphic function w in $|z| < 1$ almost all points of $|z| = 1$ belong to $F(w) \cup I(w)$.*

Theorem 2.19. *For each meromorphic function w in $|z| < 1$ all points of $|z| = 1$ except a set of first category belong to $M(w) \cup I(w)$.*

For further results see e.g. Džrbašjan and Zaharjan [1].

2.4. Generalizations of Nevanlinna theory

Value distribution theory is generalized in various directions which here only can be mentioned.

1. Generalization of the classical main theorems by modifying proximity, counting and characteristic functions (Dinghas [2], Begehr [2, 3], Rohde [1], Toda [2], Džrbašjan [1, 2]).

2. Value distribution in angular domains (R. Nevanlinna [2], Habetha [1], Ostrovskiĭ [1]).

3. Value distribution for functions meromorphic in the vicinity of a set of vanishing capacity or in a multiply connected domain with a Green function (AF HÄLLSTRÖM [1], see also DINGHAS [1, 4]).

4. Value distribution of algebroid functions (SELBERG [1], TODA [1]).

5. Ahlfors' theory of covering surfaces (AHLFORS [2], see also R. NEVANLINNA [4, 5]).

6. Value distribution for analytic mappings between Riemann surfaces (CHERN [1], OHTSUKA [1], SARIO and NOSHIRO [1], WU [1]).

7. Value distribution for pseudomeromorphic functions (AF HÄLLSTRÖM [2], PETRENKO [2]), for pseudoanalytic and approximately analytic functions (BEGEHR [4, 5], HABETHA [2]).

8. Value distribution of quasiregular mappings (RICKMAN [1], RICKMAN and MATTILA [1]).

9. Value distribution for meromorphic curves (AHLFORS [3], PETRENKO [3], H. WEYL and J. WEYL [1]).

10. Value distribution theory on complex analytic manifolds (BOTT and CHERN [1], GRIFFITHS and KING [1], KUJALA and VITTER [1], STOLL [1—3], WU [2]).

Further references

AHLFORS [8], ANDERSON [1], ANDERSON and CLUNIE [2], BAERNSTEIN [1], BAKER [1] BEGEHR [1], CLUNIE [1], FRANK [1], GACKSTATTER [1, 2], GEHRING [1], GOL'DBERG [1, 2], HAYMAN [2], HUCKEMANN [1], JULIA [1], A. Z. MOHONKO and V. D. MOHONKO [1], MUES [1], F. NEVANLINNA and R. NEVANLINNA [1], R. NEVANLINNA [7], NGOAN and OSTROVSKIĬ [1], NOSHIRO [1], PFLUGER [1], WINKLER [1], YANG and CHANG [1, 3].

3. Polyanalytic Functions

M. B. Balk

3.1. Introduction

Just as the information about the cross-section of a tree permits to tell a great deal about the tree as a whole, the information about the behaviour of an analytic function $F(z, w)$ (of two independent complex variables z, w) in the nonanalytic plane $S = \{(z, w) : w = \bar{z}\}$ (or on another nonanalytic surface) permits to judge about the global properties of this function, about its behaviour out of this surface. In that case we can get such an information which is impossible to derive from similar knowledge about the behaviour of the same function in any analytic plane (say, in the plane $\{(z, w) : w = z\}$).

This issue is at most investigated when $F(z, w)$ is a pseudopolynomial.

The restriction of a pseudopolynomial of the form

$$F(z, w) = \sum_{k=0}^{n-1} h_k(z)\, w^k \tag{1}$$

$\big(h_k(z)$ is an analytic function in a given region D, $D \subset \mathbb{C}$; $k = 0, 1, \dots, n-1\big)$ on the non-analytic plane S, i.e. a function of the form

$$f(z) = \sum_{k=0}^{n-1} h_k(z)\, \bar{z}^k \tag{2}$$

is called a *polyanalytic* (in short: p.a.) function of the n-th order (or n-analytic function) in the region D, or an areolar polynomial of the order $n - 1$ (see Théodorescu [1]). Thus, a p.a. function in D is a polynomial in $\bar{z}$ whose coefficients are analytic (in D) functions of z. We shall restrict ourselves to the case that all functions $h_k(z)$ $(k = 0, 1, \dots, n-1)$ are holomorphic (one-valued analytic) in D. A p.a. function of the second order $(n = 2)$ is called *bianalytic* (in short: b.a.). The function $h_k(z)$ is called the *k-th* (analytic) *component* of the p.a. function (2).

Note that the above mentioned definition of a p.a. function is also applicable in the case of a complex space of arbitrary dimension (p) (i.e. $z \in \mathbb{C}^p$, $p \geqq 1$), provided that $k, n-1, 0$ are interpreted as p-dimensional multiindices: $k = (k_1, \dots, k_p)$, $n - 1 = (n_1 - 1, \dots, n_p - 1)$; $0 = (0, \dots, 0)$; summation in (2) is spread over all multiindices k, for which $0 \leqq k \leqq n - 1$ (that means: $0 \leqq k_1 \leqq n_1 - 1, \dots, 0 \leqq k_p \leqq n_p - 1$). Henceforth we shall consider only p.a. functions of one variable.

P.a. functions are closely related to polyharmonic functions (which have important applications — specifically, in elasticity): a function $u(x, y)$ $(x, y$ are real variables) is a real polyharmonic function of order n in a simply connected region D if and only if $u(x, y)$ is the real (or imaginar) part of a function polyanalytic in D. In 1908 Kolosov [1] drew attention to the fact that bianalytic functions are useful for solving problems of plane elasticity. The important and fruitful investigations of Kolosov, Muscheliš-

VILI and their numerous followers concerning applications of p.a. functions to elasticity are well-known (see MUSHELIŠVILI [3] and the bibliography given there).

Every function which is n-analytic in some region may be considered as a solution of the generalized Cauchy-Riemann equation $\partial^n w/\partial \bar{z}^n = 0$; the Cauchy-Riemann operator used here (the operator of formal differentiation) may be defined by the formula

$$\partial/\partial \bar{z} = \tfrac{1}{2}(\partial/\partial x + \mathrm{i}\, \partial/\partial y).$$

The derivative $\partial w/\partial \bar{z}$ may also be defined as the areolar derivative (in the sense of Pompeiu-Théodorescu — see I. N. VEKUA [5]) or as the derivative in the sense of Sobolev-Vekua (see I. N. VEKUA [5]). In all these cases we arrive at the same class of functions (because of the ellipticity of the operator $\partial^n/\partial \bar{z}^n$). Equations with the Cauchy-Riemann operator $\partial/\partial \bar{z}$ where studied for the first time by KOLOSOV [2] (he callĕd them *conjugate equations*). KOLOSOV proposed a general way for solving such equations. I. N. VEKUA [3] worked out an effective method for solving such equations. Every equation of the form $\partial^n w/\partial \bar{z}^n = 0$ is obviously equivalent to some system of two partial differential equations for two real functions $u(x, y)$ and $v(x, y)$. E.g. for $n = 2$ we get *the system of Bicadze*:

$$\frac{\partial^2 u}{\partial x^2} - \frac{\partial^2 u}{\partial y^2} - 2\frac{\partial^2 v}{\partial x\, \partial y} = 0\,, \qquad \frac{\partial^2 v}{\partial x^2} - \frac{\partial^2 v}{\partial y^2} + 2\frac{\partial^2 u}{\partial x\, \partial y} = 0$$

(that system is cited in BICADZE [1] as an example of an elliptic system of the second order, for which the homogeneous Dirichlet boundary value problem has, in the case of the unit disk, an infinity of linear independent solutions).

Several interesting papers concerning p.a. functions have been published between 1922 and 1940 by BURGATTI [1], THÉODORESCU [1], FEDOROV [1]. But the bulk of papers devoted to p.a. functions appeared after World War II, mainly during the last two decades. These papers may be roughly divided into two parts. Some substantial papers are directly connected with the solving of boundary value problems for p.a. functions. We ought to remind here the papers by BICADZE, GAHOV, GANIN, S. V. ROGOZIN, GABRINOVIČ, ŽEGALOV, SOKOLOV, RASULOV a.o. Another large series of articles is dedicated to investigation of *function-theoretical* properties of p.a. functions, to elaboration of the apparatus of p.a. functions and its attraction to the study of functions of several complex variables (KRAJKIEWICZ, BOSCH, GOL'DBERG, OSTROVSKIĬ, ZUEV, PETROV, MALEIN, POKAZEEV, MANUILOV, TUTSCHKE, CZERNER, SCHOPF, HEINZ, AVANISSIAN, TRAORE a.o.).

It may be expected that the profound investigation of different properties of p.a. functions will extend the possibilities of their employment to solve important problems of application.

Clearly, every fact proved for p.a. functions can be reformulated in terms of function theory of several complex variables. In general, the exploration of p.a. functions should be considered as an introduction to the theory of analytic functions on non-analytic surfaces.

The close connection between the published articles about p.a. functions and such topics as meromorphic curves, Riemann surfaces, value distribution, and some other urgent branches of complex analysis allows to suppose that the theory of polyanalytic functions and their generalizations will shortly emerge as an interesting and useful chapter of contemporary complex analysis.

In this survey some results concerning p.a. functions and obtained during the last two decades are included.

3.2. Elementary properties of poly analytic functions

3.2.1. Even at the first contact with p.a. functions it is easy to note that the transfer from analytic functions to p.a. functions of the order $n > 1$ is accompanied by essential qualitative changes in the properties of such functions. Thus, for example, among the p.a. functions of order $n > 1$ there are non-constant polynomials without any complex zeroes; two p.a. functions in some region may coincide in every point of some curve, but may not coincide in any other point of the region; a p.a. function can have a bounded but infinite set of zeroes; a p.a. function in a disk is not determined by its values on the circumference; the maximum modulus of a non-constant p.a. function in some domain may be achieved in an inner point of the domain; and so on. It is of some interest to learn how "stable" some important properties of analytic functions turn out to be when we transfer to the wider class of p.a. functions.

Let us remark that every property of a p.a. function can be paraphrased as some property of an analytic function of two independent complex variables. Such a possibility emerges on the basis of the following theorem of multi-dimensional complex analysis (VLADIMIROV [1, p. 55]): *If a function of two complex variables z, w is holomorphic in some neighbourhood of the origin* $(0, 0)$ *and if $F(z, w)$ vanishes at the manifold* $w = \bar{z}$ (in the bounds of the mentioned neighbourhood), *then $F(z, w)$ is identically zero.*

3.2.2. For p.a. functions a relaxed maximum modulus principle takes place. It is based on the following assertion.

Theorem 3.2.1. *Let the function*

$$f(z) = \sum_{k=0}^{n-1} h_k(z) (\bar{z} - \bar{a})^k \tag{3}$$

be polyanalytic (of order n) in the circle $D = \{z : |z - a| < R\}$ and satisfy the condition

$$|f(z)| < M \tag{4}$$

on n circumferences

$$\Gamma_\nu = \{z : |z - a| = R_\nu\} \ (\nu = 0, 1, \dots, n-1;\ R_0 < R_1 < \dots < R_{n-1} < R) .$$

Then there exists a constant L depending only on the ratios R_ν/R $(\nu = 0, 1, \dots, n-1)$ such that the function

$$F(z, w) = \sum_{k=0}^{n-1} h_k(z) (w - \bar{a})^k$$

satisfies, in the bicylinder $\{(z, w) : |z - a| < R_0, |w - \bar{a}| < R_0\}$, the inequality

$$|F(z, w)| < LM .$$

This theorem has some useful consequences.

Theorem 3.2.2. *If the p.a. function* (3) *satisfies the inequality* (4) *in the annulus $\{z : R_0 < |z - a| < R\}$, then there exists a constant L depending only on n and the ratio R_0/R such that in the disk $\{z : |z - a| \leqq R_0\}$ the following inequalities are true:*

$$|f(z)| < LM , \qquad |h_k(z) (z - a)^k| < LM , \qquad \left|\frac{\partial^{p+q} f(z)}{\partial z^p \, \partial \bar{z}^q}\right| < \frac{p!q!}{R_0^{p+q}} LM .$$

From this a *weak maximum modulus principle* follows:

Theorem 3.2.3. *Let D be some bounded region in the complex plane $\mathbb{C}$; Let Δ be the frontier area of D of the width d* (i.e. the set of such points z of D the distance of which from the border ∂D is less than d); *let K be a compact set in D; let $f(z) = \sum_{k=0}^{n-1} h_k(z)\,\bar{z}^k$ be a p.a. function of order n in D, satisfying the condition $|f(z)| \leqq M$ in Δ. Then there exists a constant λ depending only on D, d, K, n such that for every $z \in K$ the following inequalities are true;*

$$|f(z)| \leqq \lambda M\,, \qquad |h_k(z)| \leqq \lambda M \qquad (k = 0, 1, \dots, n-1)\,.$$

For p.a. functions of a given order n this theorem allows to prove some analoga of such facts well-known in analytic function theory as the compactness principle, the Weierstrass theorem of uniformly converging sequences of analytic functions a.o. (Balk and Zuev [1, p. 207]). For p.a. functions of several complex variables analogous results were obtained by Avanissian and Traore [1]. They proved also an analogon of Hartogs' theorem for p.a. functions of several variables.

3.2.3. Conditions of uniqueness. The *inner uniqueness theorem* well-known for analytic functions is not true for p.a. functions. But a weak form of this theorem remains valid for p.a. functions:

If two functions are polyanalytic in a region D and coincide in its subregion Δ, then these functions coincide everywhere in D.

To obtain a more subtile uniqueness condition for p.a. functions one can make use of the notion of a cluster point of order n for a given set E ($E \subset \mathbb{C}$). A set E is said *to cluster to the point a* along a straight line l, if every pair of vertical angles with the vertex a containing the line l contains such portion of E, for which a is a limit point. The point a is called a *cluster point of order n* for the set E, if E clusters to the point a not less than along n different lines.

Let $H_n(D)$ denote the set of all p.a. functions of order n in D. The following uniqueness theorem is true:

Theorem 3.2.4. *A set E having a cluster point of order n in the region D is a set of uniqueness for the class $H_n(D)$* (see Balk [5]).

Instead of considering, in D, one set E having a cluster point of order n it is possible to consider (with some reservations) several sets $E_1, \dots, E_m$ with cluster points $a_1, \dots, a_m$ of an order less than n.

We shall call the set E_k *polyanalytically detachable* from the set E_j ($j \neq k$) if a p.a. function exists which takes the value zero in every point of E_k and may vanish only in a finite number of points of E_j. Theorem 3.2.4. may be generalized in the following manner:

Suppose that a set E is the sum of several such sets $E_1, \dots, E_m$ that E_j and E_k are polyanalytically detachable for any j, k ($j = 1, \dots, m$; $k = 1, \dots, m$; $j \neq k$); let E_k ($k = 1, \dots, m$) have a cluster point of order n_k in D, and $n_1 + \dots + n_m \geqq n$. Then E is a set of uniqueness for the class $H_n(D)$.

To the last theorem the following *inner factorization theorem* may be added;

Let $f(z)$ be a p.a. function of order n in D; let $E_1, \dots, E_m$ be sets having, in D, cluster points of the orders (accordingly) $n_1, \dots, n_m$; let $G_k(z)$ ($k = 1, \dots, m$) be a p.a. function *in D of order $n_k + 1$, which vanishes in every point of E_k and only in a finite number of points of E_j for every $j \neq k$. Then*

$$f(z) = \frac{h(z, \bar{z})}{A(z)} G_1(z) \dots G_m(z) ,$$

where $h(z, \bar{z})$ is a p.a. function in D and $A(z)$ is analytic in D.

3.2.4. The question of the construction of the set of zeroes of a p.a. function in the neighbourhood of its non-isolated zero arises. Let $f(z)$ be a p.a. function in the neighbourhood of a point a. We shall call $f(z)$ *irreducible* in the point a if $f(z)$ coincides with the restriction on the plane $w = \bar{z}$ of some pseudopolynomial $F(z, w) = \sum_{k=0}^{n-1} h_k(z) (w - \bar{a})^k$ which is irreducible in the point $(a, \bar{a})$.

Theorem 3.2.5. *If a p.a. function $f(z)$ $(f(z) \not\equiv 0)$ is irreducible in some point a and if a is a non-isolated zero of $f(z)$, then the set of all zeroes of $f(z)$ in some sufficiently small neighbourhood of the point a coincides with the set of points of some open simple analytic curve* (see Balk [10]).

From this theorem it follows that the set of all zeroes of a p.a. function $f(z)$ $(f(z) \not\equiv 0)$ in a sufficiently small neighbourhood of its non-isolated zero can be represented as a sum of a finite number of open simple analytic curves. By the Weierstrass preparatory theorem the same remains true for every function $f(z)$ $(f(z) \not\equiv 0)$ which is conjugately analytic (c.a.) in the point a (i.e. $f(z)$ can be represented in the form of a convergent double series $\sum_{\nu,k=0}^{\infty} C_{\nu,k}(z - a)^\nu (\bar{z} - \bar{a})^k$). Hence it follows that the set E_i of all isolated zeroes of a function which is polyanalytic (or conjugately analytic) in some region D, is discrete in D (in other words, it has no limit points in D). Taking in account that every polyharmonic function $u(x, y)$ and every function of the form $|f(z)|^2$ (where $f(z)$ is an analytic or polyanalytic function) are c.a. functions, we can derive from theorem 3.2.5 the following corollary:

The level line $(u(x, y) = \text{const}$ *or* $|f(z)| = \text{const})$ *of a polyharmonic or polyanalytic function in some sufficiently small neighbourhood of its arbitrary point either consists only of this point or is the sum of a finite number of simple analytic curves.*

3.2.5. Non-isolated zeroes of a bianalytic function. It is not difficult to show (see Balk [3]) that if a b.a. function $B(z)$ $(B(z) = \varphi(z) + \bar{z}\psi(z) \neq 0)$ has a non-isolated zero a, then the set of all zeroes of $B(z)$ in some sufficiently small neighbourhood of the point a coincides with the set of points of a simple open regular analytic curve. Inversely, for every simple open regular analytic curve Γ there exists such an analytic (in some neighbourhood $\Delta(\Gamma)$ of Γ) function $A(z)$, that the set of all points of Γ coincides with the set of all such points in $\Delta(\Gamma)$ which satisfy the equation $\bar{z} = A(z)$.

We shall mark here the following assertion:

If the set of non-isolated zeroes of the b.a. function $B(z) = \bar{z} - A(z)$ contains a closed Jordan curve Γ, then in the interior of Γ there is at least one singular point of $A(z)$ (the proof follows from the function $|B(z)|^2$ being subharmonic).

3.3. Polyentire functions

3.3.1. A function $f(z)$ is called an *entire p.a. function* or a *polyentire function*, if it is polyanalytic in the whole finite complex plane $\mathbb{C}$. Such function can be presented in the form

$$f(z) = \sum_{k=0}^{n-1} h_k(z)\, \bar{z}^k \tag{5}$$

where all "analytic components" $h_k(z)$ ($k = 0, 1, \ldots, n-1$) are entire analytic functions. A polyentire function which is a polynomial in z and $\bar{z}$ is called a *p.a. polynomial.* A polyentire function different from a polynomial is called *transcendental.* A polyentire function of the form $\sum_{k=0}^{n-1} A_k(z)\, |z|^{2k}$ (all $A_k(z)$ are entire analytic functions; $k = 0, 1, \ldots, n-1$) is called a *reduced polyentire function.* The set of all polyentire functions clearly forms a commutative ring (with regard to the operations of addition and multiplication of functions). This ring arises from joining the transcedental element $\bar{z}$ to the ring $H(\mathbb{C})$ of all entire analytic functions. Any irreducible element (see VAN DER WAERDEN [1]) of the ring $H(\mathbb{C})\,[\bar{z}]$ will be called a *globally irreducible p.a. function.* The following assertion is especially fruitful.

Theorem 3.3.1. *A globally irreducible p.a. function has only a finite number of isolated zeroes if and only if it can be represented in the form*

$$f(z) = P(z, \bar{z}) \exp g(z) \tag{6}$$

where $g(z)$ is an entire analytic function and $P(z, \bar{z})$ is a polynomial in z and $\bar{z}$.

This theorem follows from two theorems proved in BALK [7], BALK and ZUEV [2].

Theorem 3.3.2. *Any polyentire function with a bounded set of zeroes can be represented in the form* (6) (see BALK [7]).

Theorem 3.3.3. *Any globally irreducible polyentire function with non-isolated zeroes can be represented in the form* (6) (see BALK and ZUEV [2]).

The proof of theorem 3.3.2 (which is trivial in the case of entire analytic functions) was found in BALK [7] who utilized the theory of meromorphic curves elaborated before World War II by WEIL, CARTAN, and AHLFORS (see WU [3]). The proof of theorem 3.3.2 (it can be shown that it suffices to consider only the case that $f(z)$ is a reduced polyentire function) can be reduced to the estimation of the growth of some real-valued function $\bar{N}^*(r, 0, f)$ which may be considered as some analogon of the Nevanlinna counting function. Let $f_r(z)$ denote an entire analytic function coinciding with a given reduced polyentire function $f(z)$ on the circumference $\{z : |z| = r\}$; let $\bar{n}(t, 0, f_r)$ be the number of zeroes of $f_r(z)$ (without account of multiplicity) in the disk $\{z : |z| \leqq t\}$;

$$\bar{N}^*(r, 0, f) = \int_0^r \frac{\bar{n}(t, 0, f_r) - \bar{n}(0, 0, f_r)}{t}\, dt + \bar{n}(0, 0, f_r) \ln r\,. \tag{7}$$

The following assertion is true:

Theorem 3.3.4. *A polyentire function $f(z)$ can be represented in the form* (6) *if and only if*

$$\bar{N}^*(r, 0, f) = O(\ln r)\,, \qquad r \to \infty\,. \tag{8}$$

It is possible to show that for any function $f(z)$ which satisfies the condition of theorem 3.3.2 the relation (8) is true. Hence theorem 3.3.2 is true. Another proof of theorem 3.3.2 (without use of the second main theorem for meromorphic curves, which played an essential role in BALK [7]) was found by OSTROVSKIĬ [3] in 1969.

The theorem from VLADIMIROV [1] cited above allows to reformulate the assertions of theorem 3.3.1 — theorem 3.3.3 in terms of function theory of two complex variables. The term *entire pseudopolynomial* (*with regard to* w) will denote a function of the form

$$\Phi(z, w) = \sum_{k=0}^{n-1} g_k(z)\, w^k ,$$

where all $g_k(z)$ ($k = 0, 1, \ldots, n-1$) are entire analytic functions. The set of all entire pseudopolynomials obviously forms a ring (with regard to the operations of addition and multiplication of functions); we shall denote it by $H(\mathbb{C})\,[w]$; it is isomorphic to the ring of all polyentire functions $H(\mathbb{C})\,[\bar{z}]$. The term *non-analytic plane* (*in* $\mathbb{C}^2$) will denote any manyfold in $\mathbb{C}^2$ given by a condition of the form $\{(z, w) : w = a\bar{z} + bz + c\}$, where a, b, c are constants, $a \neq 0$. We shall mark now some corrollaries from theorem 3.3.1 — theorem 3.3.3.

Theorem 3.3.5. *An entire function $\Phi(z, w)$ of two independent complex variables, which is a pseudopolynomial in w and which has in some non-analytic plane a bounded set of zeroes, can be represented in the form*

$$\Phi(z, w) = e^{g(z)}\, P(z, w) ,$$

where $P(z, w)$ is a polynomial and $g(z)$ is an entire analytic function.

Theorem 3.3.6. *If at least one analytic component of a polyentire function $f(z)$ has an unbounded set of zeroes not being identically zero, then the same is true for $f(z)$.*

Theorem 3.3.7. *If the set of all zeroes of a polyentire function is bounded, then it is algebraic.*

3.3.2. A useful means to investigate polyentire functions is the well-known *theorem about the Borel identities*:

If the entire analytic functions $G_1(z), \ldots, G_m(z)$ ($z \in \mathbb{C}^p$, $p \geqq 1$, $m \geqq 2$) deprived of zeroes are linearly dependent over the ring of polynomials (in z), then it is possible to find at least two of these functions which are linearly dependent over the ring of complex constants.

By use of this assertion and of theorem 3.3.2 it is possible to prove

Theorem 3.3.8. *If any of the polyentire functions $E_1(z), \ldots, E_m(z)$ ($z \in \mathbb{C}$, $m \geqq 2$) has a bounded set of zeroes and if these m functions are linearly dependent over the ring of p.a. polynomials, then there are at least two of these functions which are linearly dependent over the same ring.*

We shall note a useful assertion which follows from theorem 3.3.3 and from the fact well-known in algebra that any polynomial $P(z, \bar{z}) = P_1(x, y) + iP_2(x, y)$ with infinitely many zeroes has a non-constant real-valued divisor $p(x, y)$:

Theorem 3.3.9. *Any polyentire function $E(z)$ can be represented in the form*

$$E(z) = I(z, \bar{z})\, Q(z, \bar{z})$$

where $I(z, \bar{z})$ is a polyentire function with only isolated zeroes and $Q(z, \bar{z})$ is a (real-valued) polynomial (see BALK and ZUEV [2] and BALK [3]).

A theorem analogous to theorem 3.3.2 can be proved for polyentire functions of several complex variables:

Theorem 3.3.10. *If an entire pseudopolynomial* $\Phi(z, w) = \sum_{k=0}^{n-1} \Phi_k(z)\, w^k$ (here is $z = (z_1, \dots, z_p)$, $w = (w_1, \dots, w_p)$, $k = (k_1, \dots, k_p)$, $n - 1 = (n_1 - 1, \dots, n_p - 1)$, $0 = (0, \dots, 0)$) *has in the non-analytic plane* $w = \bar{z}$ (that means: $w_1 = \bar{z}_1, \dots, w_p = \bar{z}_p$) *only a bounded set of zeroes, then it can be represented in the form*

$$\Phi(z, w) = e^{g(z)} P(z, w) ,$$

where $g(z)$ *is an entire analytic function and* $P(z, w)$ *is a polynominal* (see BALK and MANUILOV [1]).

The most interesting, though simple, corollary from theorem 3.3.2. is the big Picard theorem for polyentire functions:

Theorem 3.3.11. *A polyentire function* $E(z)$ *which accepts two different values* a *and* b *only on bounded sets of points in the complex plane* $\mathbb{C}$ *is a polynomial.*

The first proof of this assertion in the case of bianalytic functions (see BALK [4]) was found by use of the method of quasi-normal families of analytic functions (see MONTEL [1]), in the case of polyentire functions of arbitrary order this theorem was proved (see BALK [8]) by use of Saxer's inequality (see SAXER [1]) in Schottky's theorem. The proof of theorem 3.3.11 is very simple by using theorem 3.3.2: it follows from the conditions of theorem 3.3.11 and from theorem 3.3.2 that

$$E(z) - a = P(z, \bar{z}) \exp g(z) , \qquad E(z) - b = Q(z, \bar{z}) \exp h(z) ,$$

where $g(z)$ and $h(z)$ are entire analytic functions and $P(z, \bar{z})$ and $Q(z, \bar{z})$ are p.a. polynomials. Hence $P \exp g - Q \exp h = b - a$, and by the theorem above mentioned about the Borel identities we conclude that $g \equiv \text{const}$.

The theorem 3.3.11 can be made more precise by use of the *main theorem of algebra for p.a. polynomials* (see BALK [9]):

Theorem 3.3.12. *If the exact degree of the p.a. polynomial* $P(z, \bar{z})$ *with regard to the pair of variables* $z, \bar{z}$ *surpasses more than twice the exact degree of the polynomial with regard to one of these variables, then* $P(z, \bar{z})$ *has at least one zero.*

A method for the calculation of isolated zeroes of a p.a. polynomial was proposed by HEINZ [1].

From theorem 3.3.11 and theorem 3.3.12 the following supplement to the big Picard theorem can be drawn:

Theorem 3.3.13. *A polyentire function of order* n, *which does not accept two different complex values* a *and* b, *is a polynomial in* z *and* $\bar{z}$ *of degree less than or equal to* $2n - 2$ (see BALK [8]).

We shall note some other corollaries from theorem 3.3.2 and theorem 3.3.3.

Theorem 3.3.14. *The set of points in which a transcendental analytic function* $A(z)$ *coincides with a given polyanalytic — but not analytic — polynomial, is unbounded and discrete in* $\mathbb{C}$.

Theorem 3.3.15. *Let* $P_k(x, y)$ $(k = 0, \dots, m)$ *be complex-valued polynomials of real variables* x, y; $m \geqq 1$, $P_m(x, y) \not\equiv 0$; *let* $f(z)$ *be a polyentire function of the complex variable* $z = x + iy$. *If the auxiliary function* $\Phi(z) = \sum_{k=0}^{m} P_k(x, y)\, [f(z)]^k$ *accepts two different complex values only on a bounded set of points, then* $f(z)$ *is a polynomial.*

Theorem 3.3.16. *Let for an entire pseudopolynomial (with regard to* w) $\Phi(z, w)$ *such two (possibly coinciding) non-analytic planes* α *and* β *and such two different polynomials*

$P(z, w)$ and $Q(z, w)$ exist in $\mathbb{C}$, that the equation $\Phi(z, w) = P(z, w)$ has only a bounded set of roots in α and the equation $\Phi(z, w) = Q(z, w)$ has only a bounded set of roots in β. Then $\Phi(z, w)$ is a polynomial.

Theorem 3.3.17. *If an entire function of two complex variables (not necessarily a pseudo-polynomial) $F(z, w)$ has a bounded set of zeroes in some non-analytic plane and in every level plane $w =$* const, *then $f(z, w)$ is a polynomial.*

3.3.3. For polyentire functions it is possible to introduce the notion of *order of growth* in the same manner as it was done for analytic functions: the order of growth of the function $f(z)$ is the order of growth of the auxiliary real-valued function

$$\theta(r) = \ln \max_{|z|=r} |f(z)| \,, \qquad r \to \infty,$$

i.e. the number

$$\overline{\lim_{r\to\infty}} \, [\ln^+ \theta(r)/\ln r] \,.$$

The following assertion is true:

Theorem 3.3.18. *The order of growth of a polyentire function coincides with the greatest order of growth of its analytic components.*

For polyentire functions other characteristics of growth (type, lower order a.o.) may be introduced. It is possible to get such a generalization of the assertion of theorem 3.3.4:

Theorem 3.3.19. *If the characteristic $\overline{N}^*(r, 0, f)$ of the polyentire function $f(z)$ has a finite order of growth α, then $f(z)$ may be represented in the form* (6), *where $P(z, \bar{z})$ is assumed to be a polyentire function of order of growth α (not necessarily a p.a. polynomial).*

3.3.4. Polymeromorphic functions. A function $f(z)$ is called a *meromorphic p.a. function* in $\mathbb{C}$ (or *polymeromorphic function*) if there exists a set E without cluster points in $\mathbb{C}$ such that $f(z)$ is polyanalytic in $\mathbb{C} \setminus E$ and every point of E is a regular point or a pole for every analytic component of $f(z)$. For such functions the Picard theorem known for meromorphic analytic functions is not true. Any b.a. function of the form

$$f(z) = \sum_{k=1}^{\infty} A_k \frac{\bar{z} - \bar{a}_k}{z - a_k}$$

($a_k \to \infty$ for $k \to \infty$, $\sum_{k=1}^{\infty} |A_k| < +\infty$) is a counter-example: this function is bounded in $\mathbb{C} \setminus \{a_1, \dots, a_k, \dots\}$ by the constant $\sum_{k=1}^{\infty} |A_k|$ and therefore omits an infinity of complex values, but is not a constant. But for some classes of meromorphic p.a. functions (and even for some classes of functions of a more general type) it is possible (by use of the theorems 3.3.2 and 3.3.3) to obtain a theorem of the Picard type:

Theorem 3.3.20. *Let $F(z, w)$ be a non-rational function of two complex variables (z, w) and let $F(z, w)$ be representable as the ratio of two relatively prime entire pseudopolynomials (with regard to w), let only a bounded set of points of indeterminacy of $F(z, w)$ belong to the non-analytic plane $S = \{(z, w) : w = \bar{z}\}$. If the function $F(z, w)$ accepts in S two values c_1 and c_2 only on bounded sets E_1 and E_2, then by any choice of a constant c_3 different from c_1 and c_2 the set E_3 of all points of the plane S, in which $F(z, w) = c_3$ is unbounded and the point (∞, ∞) is the only cluster point for E_3* (see Balk [13]).

3.4. Polyanalytic functions in the neighbourhood of their isolated singularities

Let z_0 be a point in $\mathbb{C}$ and δ its closed punctured neighbourhood (that means: $\delta = \Delta \setminus \{z_0\}$, where Δ is a closed neighbourhood of z_0). The point z_0 is called an *essential singularity* for the p.a. function $f(z)$ if z_0 is an isolated singularity or regular point for every analytic component of $f(z)$ and is an essential singularity at least for one of these analytic components. A function $f(z)$ which is holomorphic (or polyanalytic, respectively) in some punctured closed neighbourhood δ of the point z_0 and has no essential singularity in z_0 will be called a *polynomoid* (and a p.a. polynomoid, respectively) in δ. A function $f(z)$ will be called *meromorphic* in the point z_0 if $f(z)$ is a polynomoid (and a p.a. polynomoid, respectively) in some punctured neighbourhood δ of z_0. The cluster set of a p.a. function in its essential singularity is total (coincides with $\bar{\mathbb{C}}$), the inverse theorem is also true. If an isolated singularity z_0 of a p.a. function $f(z)$ is not an essential singularity for $f(z)$, then the cluster set of $f(z)$ in the point z_0 is a connected part of some rational curve (that means: a curve with the equation $w = R(t)$, where $R(t)$ is a rational function). The proof in the case of b.a. functions is given in Balk and Poluhin [1]. Employment of some ideas of Balk [7] and Ostrovskiĭ [3] permits to prove a factorization theorem analogous to theorem 3.3.2 (see Krajkiewicz [2]), Balk [11]):

Theorem 3.4.1. *If $f(z)$ is a p.a. function in a punctured neighbourhood δ of the point z_0 ($z_0 \in \bar{\mathbb{C}}$) and if z_0 is not a cluster point for the set of zeroes of $f(z)$, then $f(z)$ admits the following factorization:*

$$f(z) = P(z, \bar{z}) \exp g(z) ,$$

where $P(z, \bar{z})$ is a p.a. polynomioid in δ and $g(z)$ is a polyentire function with respect to z in the case $z_0 = \infty$ and with respect to $1/(z - z_0)$ in the case $z_0 \neq \infty$.

Another possibility for the factorization of a p.a. function in some neighbourhood of its isolated singularity z_0 arises when z_0 is a cluster point for the set of non-isolated zeroes of $f(z)$. For the sake of simplicity we shall consider the case $z_0 = \infty$.

Theorem 3.4.2. *Let the function $f(z)$ be polyanalytic in $K = K(R) = \{z : R \leqq |z| < +\infty\}$ and irreducible in $K'(R') = \{z : R' < |z| < +\infty\}$ for every $R', R \leqq R' < +\infty$. If the set E of all non-isolated zeroes of $f(z)$ belonging to K is unbounded, then the function $f(z)$ may be represented in the form*

$$f(z) = M(z, \bar{z}) H(z) ,$$

where $M(z, \bar{z})$ is a p.a. function in K meromorphic in the point infinity and $H(z)$ is an entire analytic function without zeroes in $\mathbb{C}$ (see Balk and Gol'dberg [1]).

The proof of this assertion is essentially based on the use of some properties of open Riemann surfaces having ideal boundaries with null harmonic measure (see Stoilov [1], Kuroda [1]). In the case of b.a. functions a more simple proof of theorem 3.4.2. was found by use of the Carleman-Milloux estimate for harmonic functions (see R. Nevanlinna [4]). We shall note some corollaries from the theorems 3.4.1. and 3.4.2. see Balk and Gol'dberg [1]).

Theorem 3.4.3. *Let $f(z)$ be a p.a. function in the punctured neighbourhood δ of the point infinity and let the set of non-isolated zeroes of $f(z)$ be unbounded. Then $f(z)$ is in some punctured neighbourhood δ' of the point ∞ representable in the form*

$$f(z) = I(z, \bar{z}) M(z, \bar{z}) ,$$

where $I(z, \bar{z})$ is a p.a. function having in δ' only isolated zeroes, and $M(z, \bar{z})$ is a p.a. function meromorphic in the point ∞.

By application of the method of quasi-normal families of analytic functions (used earlier for a similar purpose in BALCK [4]) BOSCH and KRAJKIEWICZ [1] proved, that for p.a. functions with essential singularities the Picard theorem (known for analytic functions) remains true. The theorems 3.4.1 and 3.4.2 permit to get a more exact version of Picard's theorem for p.a. functions:

Theorem 3.4.4. *Let $f(z)$ be a p.a. function in a punctured neighbourhood δ of its essential singularity z_0; let the point z_0 not be a cluster point for the set of roots of the equation $f(z) = a$, where $a = a(z)$ is a function polyanalytic in δ and meromorphic in z_0. Then by any choice of another function $b = b(z)$ $(b(z) \not\equiv a(z))$ polyanalytic in δ and meromorphic in z_0 it is possible to find such punctured neighbourhood δ' of z_0 that the set of all roots belonging to δ' of the equation $f(z) = b$ have the point z_0 as its only cluster point.*

For p.a. functions with essential singularities it is possible to prove some theorems analogous to the one cited in 3.3. For example:

Theorem 3.4.5. *If $A(z)$ is an analytic function with an essential singularity z_0 and $\pi(z)$ is a p.a. but not analytic) function in some neighbourhood of z_0 meromorphic in this point, then the equation $A(z) = \pi(z)$ has an infinite set of roots in every neighbourhood of z_0 and in some neighbourhood δ' of z_0 this point will be the only cluster point for the set of roots of the mentioned equation.*

3.5. On factorization of polyanalytic functions in a circle

Some results which are similar to the factorization theorems cited above can be obtained for functions, polyanalytic in a circle. Let Π_D denote the ring of all functions $\pi(z)$ which are holomorphic in the circle $D = \{z : |z| < 1\}$ and satisfy there the condition

$$T(r, \pi) = O\left(\ln \frac{1}{1-r}\right), \qquad r \to 1 \tag{9}$$

$(T(r, \pi)$ is the Nevanlinna characteristic function, cf. 2.2.). The ring Π_D may be considered as an analogon of the ring of polynomials $p(z)$ (which, as it is well-known, may be characterized in the ring of entire functions by the condition: $T(r, p) = O(\ln r)$, $r \to \infty$). Every function from the ring Π_D will be called — for brevity — *a polynomoid in D*; if every analytic component of a p.a. function $\pi(z, \bar{z})$ i sa polynomoid in D, then $\pi(z, \bar{z})$ will be called a *p.a. polynomoid* in D.

Theorem 3.5.1. *Let $h(z) = \sum_{k=0}^{n-1} h_k(z)\bar{z}^k$ be a polyanalytic function in $D = \{z : |z| < 1\}$ and let $\bar{E}$ be the closure of the set of all zeroes of $h(z)$ in D. If $\bar{E} \subset D$, then $h(z)$ is representable in the form*

$$h(z) = G(z)\, P(z, \bar{z}), \tag{10}$$

where $P(z, \bar{z})$ is a p.a. polynomoid in D and $G(z)$ is holomorphic and deprived of zeroes in D (see BALK and GOL'DBERG [2]).

The proof is based on the study of the holomorphic curve (in D) $\Omega(z) = \{h_0(z), h_1(z), \ldots, h_{n-1}(z)\}$ and its characteristic

$$T(r, \Omega) = \frac{1}{2\pi} \int_0^{2\pi} \ln ||\Omega(r\, e^{i\theta})||\, d\theta,$$

where

$$||\Omega(z)|| = \sum_{k=0}^{n-1} |h_k(z)| .$$

By use of R. Nevanlinna's asymptotic formulas for the modular function (see R. NEVANLINNA [3]) it is possible to construct an example which shows that the theorem 3.5.1 ceases to be true if we consider instead of Π_D the ring of holomorphic (in D) functions satisfying the condition

$$T(r, \pi) = o\left(\ln \frac{1}{1-r}\right), \qquad r \to 1 .$$

An assertion similar to theorem 3.5.1 was proved by KRAJKIEWICZ [3].

For p.a. functions in D with non-isolated zeroes the following statement can be proved:

Theorem 3.5.2. *If $f(z)$ is an irreducible p.a. function in the unit circle D* (i.e. $f(z)$ is an irreducible element of the ring $H(D)$ $[\bar{z}]$) *and if $f(z)$ has non-isolated zeroes in D, then $f(z)$ can be represented in the form* (4.2).

We shall note some corollaries from the theorems 3.5.1 and 3.5.2.

Theorem 3.5.3. *Let $h(z)$ be a holomorphic function in the open circle D, but not a polynomoid in D; let $\pi(z, \bar{z})$ be a polynomoid in D, but not holomorphic in D. Then the set E of roots of the equation $h(z) = \pi(z, \bar{z})$ cannot compactly belong to D* (i.e. $\bar{E} \subset D$ *does not hold*).

Theorem 3.5.4. *Every function $f(z)$ which is polyanalytic in D can be represented in the form*

$$f(z) = I(z, \bar{z})\, p(z, \bar{z}) ,$$

where $I(z, \bar{z})$ is a p.a. function which has only isolated zeroes in D, and $p(z, \bar{z})$ is a p.a polynomoid in D.

Theorem 3.5.5. *Let $f(z)$ be a p.a. function in D but not a p.a. polynomoid in D; let there exist such a p.a. polynomoid (in D) $a = a(z)$ (possibly a constant) that the set M_a of all a-points of $f(z)$* i. e. the set of all roots of the equation $f(z) = a(z)$ in D) *compactly belongs to D. Then*

1. *by any choice of another p.a. polynomoid (in D) $b = b(z) \not\equiv a(z)$ the set of all b-points (in D) of $f(z)$ will not compactly belong to D;*
2. *the set M_b' of all non-isolated b-points of $f(z)$ will compactly belong to D.*

The proofs of the theorems 3.5.2 and 3.5.5 are based on N. TODA's analogon of the *theorem about Borel's identities* (see TODA [1]).

3.6. On quasi-normal families of n-analytic functions

The retaining of Montel's definition of quasi-normality (originally given for families of analytic functions) in the case of families of p.a. functions has an essential deficiency: it is not accompanied by retaining a sufficiently convenient criterion of quasi-normality for such families. That is why in BALK [12] another definition of quasi-normality is given in the case of n-analytic functions; on the one hand, it may be considered as a generalization of Montel's definition of quasi-normality; on the other hand, in reference to the new definition a criterion of quasi-normality similar to that of Montel is true. That permits — in combination with the theorems of factorization mentioned above — to precise some theorems of the Picard type proved earlier for p.a. functions, as well as to extend some results of the classical complex analysis on p.a. functions.

Let D be a region in $\mathbb{C}$ and K a compactum in D and

$$f_\mu(z) = \sum_{k=0}^{n-1} h_{\mu,k}(z)\, \bar{z}^k \qquad (\mu \in \mathbb{N}) \tag{12}$$

a sequence of p.a. functions of order n in D. The sequence (12) is called *regular of rank* $p, p \geqq 0$, *on the compactum* K if the sequence $(h_{\mu,p})$ $(\mu \in \mathbb{N})$ is uniformly convergent on K to ∞ and for every fixed k $(k = 0, 1, \ldots, n-1)$ the sequence $(h_{\mu,k}/h_{\mu,p})$ $(\mu \in \mathbb{N})$ is uniformly convergent on K to a finite limit $\theta_k(z)$ which is zero for $k > p$; the sequence (12) is called *regular of rank* (-1) *on* K if for every fixed k $(k = 0, 1, \ldots, n-1)$ the sequence $(h_{\mu,k})$ $(\mu \in \mathbb{N})$ is uniformly convergent to a finite limit $\psi_k(z)$. The sequence (12) is called *regular of rank* p, $p \geqq -1$, *inside the region* D, if it is regular of rank p on every compactum K belonging to D. The sequence (12) is called *quasi-regular of rank* p *inside* D if in D a discrete set E exists such that (12) is regular inside the region $D \setminus E$. Finally, the family F of n-analytic (in D) functions

$$f_\alpha(z) = \sum_{k=0}^{n-1} h_{\alpha,k}(z)\, \bar{z}^k \tag{13}$$

(the set of inidces α is not necessarily denumerable) is called *quasi-normal in* D if every sequence of functions extracted from F contains a subsequence which is quasi-regular inside D. The family F is called *quasi-normal in some point* if it is quasi-normal in some neighbourhood of this point. We shall note two simple properties of quasi-normal families of n-analytic functions.

Theorem 3.6.1. *A family F of n-analytic functions in the region D is quasi-normal in D if and only if it is quasi-normal in every point of D.*

Theorem 3.6.2. *If the sequence* (12) *is regular of rank* (-1) *inside D, then inside D* (12) *uniformly converges to an n-analytic function; if* (12) *is regular of rank $p \geqq 0$ inside D, then it uniformly converges to ∞ inside an open set of the form $D \setminus U_p$, where U_p is the set of zeroes of some p.a. function of the* (*exact*) *order $p + 1$.*

By multiplicity of an isolated zero z_0 of a p.a. function $f(z)$ we shall mean the number $\operatorname{Ind}(z_0, f) = \pi/2 \int_\gamma d \arg f(z)$, γ being the circumference of a circle (with centre z_0) which does not contain any points of discontinuity and any zeroes of $f(z)$, different from z_0 (the *Poincaré index*). We shall say that *a p.a. function accepts the value a in the region D not more than p times* if the function $f(z) - a$ has in D only a finite number of zeroes $z_1, \ldots, z_m$ and

$$\sum_{k=1}^{m} |\operatorname{Ind}(z_k, f(z) - a)| \leqq p\,.$$

For families of n-analytic functions the following criterion of quasi-normality (quite analogous to the known Montel criterion) is true:

Theorem 3.6.3. *Suppose that for a family F of functions n-analytic in some region D, $D \subset \mathbb{C}$, there exist such constants a, b, p, q ($a \neq b$; p and q are non-negative integers) that in D every function of the family accepts the value a not more than p times and the value b not more than q times. Then F is a quasi-normal family* (*in* D).

The proof is essentially based on some considerations of J. DUFRESNOY on normal families of holomorphic vector-functions (see DUFRESNOY [1]). The criterion of quasi-normality (theorem 3.6.3) in combination with the factorization theorems (see **3.3** to 3.5.) permits to obtain various more precise results of the Picard type for p.a. functions (*theorems of the Julia type*). Here is one of them:

Theorem 3.6.4. *Suppose that $f(z)$ is a p.a. function in some neighbourhood of a point z_0 which is an essential singularity for $f(z)$; suppose that for some constant a the set M_a of all roots of the equation $f(z) = a$ has not z_0 as its cluster point. Then such ray L with the origin z_0 exists that any angular neighbourhood $\delta(L)$ of the ray L contains, for every b, $b \neq a$, an infinite part of the set M_b of all roots of the equation $f(z) = b$; moreover, in a sufficiently small circular neighbourhood of z_0 this point is the only cluster point for the set $M_b \cap \delta(L)$.*

We shall note some other theorems which can be proved by use of the modification of the concept of quasi-normality introduced above; these theorems generalize some classical results of analytic function theory. Below $H_n(D)$ will denote the set of all n-analytic functions in the region D.

Theorem 3.6.5 (analogon to the Vitali-Montel theorem). *Let (f_k) $(k \in \mathbb{N})$ be a sequence of functions belonging to some quasi-normal family of functions, n-analytic in the region D, and let E be a set of uniqueness for $H_n(D)$. If the sequence pointwise converges on E, then it converges uniformly inside D.*

From theorem 3.6.5. and the uniqueness theorem for p.a. functions (see 3.2.) an analogon to the Vitali-Stieltjes theorem may be deduced:

Theorem 3.6.6. *Let a sequence (f_k) of functions n-analytic in some region D pointwise converge on some set E, E having, in D, a cluster point of order n; let two such numbers a and b $(a \neq b)$ exist, that $f_k(z)$ does not accept the values a and b for any k. Then the sequence (f_k) is uniformly convergent inside D.*

For the study of the boundary behaviour of p.a. functions the following corollary from theorem 3.6.5 is useful:

Theorem 3.6.7 (analogon to the Lindelöf-Montel theorem). *Let $f(z)$ be n-analytic in a open sector S (whose angle is less than π); let in S the function $f(z)$ accept two fixed values a and b not more than p times (p is finite), let $f(z)$ tend to a limit L if the point z tends to the vertex of the sector S along any of n fixed radii, belonging to S. Then $f(z)$ tends to L, if z tends to the vertex of S arbitrarily but remains in some sector S' inner to S* (see BALK [12], a special case see in PETROV [1]).

3.7. Degenerate polyanalytic functions

Let $u(x, y)$ and $v(x, y)$ be real-valued functions defined in a region $D, D \subset \mathbb{R}^2$, let them belong to the class $\mathcal{E}^\infty(D)$. We shall call the function $f(z) = u(x, y) + iv(x, y)$ $(z = x + iy)$ a *degenerate function* in D if the set $f(D)$ has no inner points. Obviously any degenerate analytic function is a constant. This is not necessarily the case for p.a. functions. First we shall note a simple auxiliary theorem:

Theorem 3.7.1. *A p.a. function in D, which is degenerate in some subregion of D, is degenerate in D.*

One of the most simple classes of degenerate p.a. functions is the class of p.a. functions of constant modulus.

Theorem 3.7.2. *The necessary (and, clearly, sufficient) condition for an n-analytic function $f(z)$ to be of constant modulus in some region D is the representability of this function in the form $f(z) = \lambda \overline{P(z)}/P(z)$, where $\lambda = \text{const}$ and $P(z)$ is a polynomial (with regard to z) of a degree less than or equal to $n - 1$* (see Balk [6]).

It follows from theorem 3.7.2 (see Balk [6]): If $\varphi(z)$ is holomorphic in a region D and $\psi(z)$ is a p.a. polynomial, and if $|\varphi(z)| \equiv |\psi(z)|$ in D (or in some subregion of D), then there exist such polynomials (in z) $r(z)$ and $s(z)$ and such constant L, that

$$|L| = 1\,, \qquad \varphi(z) = Lr(z)\,s(z)\,, \qquad \psi(z) = r(z)\,\overline{s(z)}\,. \tag{14}$$

This assertion was essentially generalized by Zuev [1] and by Krajkiewicz and Bosch [1]:

Theorem 3.7.3. *Let $\varphi(z, \bar z)$ and $\psi(z, \bar z)$ be p.a. functions in D and let $|\varphi(z, \bar z)| \equiv |\psi(z, \bar z)|$ in D. Then there exist such p.a. function (in D) $r(z)$, such p.a. polynomial $s(z)$ and such constant L, $|L| = 1$, that* (14) *holds.*

Theorem 3.7.2. permits to give a complete characterization of all degenerate bianalytic functions (see Balk [2]):

Theorem 3.7.4. *The necessary (and sufficient) condition for a b.a. function $B(z) = \varphi(z) + \bar z\psi(z)$ to be degenerate in some region D consists in the representability of $B(z)$ in one of the forms:*

$$B(z) = A(e^{i\alpha} z + \bar z) + A_1, \qquad B(z) = A(z - c)^{\gamma}\,(\bar z - \bar c) + A_1$$

(here $A, A_1, c, \alpha, \gamma$ are constants, $\operatorname{Im}\alpha = 0$, $|\gamma| = 1$). The image $B(D)$ is one of the following sets (or its connected subset): a point, a straight line, a circumference, a logarithmic spiral (i.e. a line which can be obtained from the curve with the equation — in polar coordinates — $\varrho = a^{\theta}$ ($a = \text{const}$) by a similarity transformation).

By use of Luroth's theorem (see van der Waerden [1]) it is possible to prove the following assertion:

Theorem 3.7.5. *If $f(z)$ is a degenerate polyentire function, then $f(\mathbb{C})$ is a polynomial curve* (i.e. a curve of the form $w = P(t)$, $-\infty < t < +\infty$, where $P(t)$ is a polynomial).

A p.a. function $f(z)$ (in a region D) is said to accept the value a non-isolatedly if $(fz) - a$ has a non-isolated zero (in D).

By use of the factorization theorems (see 3.3.), the following theorem can be proved:

Theorem 3.7.6. *A polyentire function of order n, which accepts n different values non-isolatedly, is necessarily a polynomial* (see Balk and Zuev [2]).

This assertion permits to find some conditions of degeneration:

Theorem 3.7.7. *Let 1. $f(z)$ be a polyentire function of order n, 2. Γ_k ($k = 0, 1, \dots, n-1$) are pairwise different lines, any of which is either a straight line or a circumference, 3. A_k ($k = 0, 1, \dots, n-1$) are constants, among which at least two are different; 4. for every k ($k = 0, 1, \dots, n-1$) $f(z) = A_k$ when $z \in \Gamma_k$.*

Then 1. $f(z)$ is a degenerate function in $\mathbb{C}$, 2. either all Γ_k ($k = 0, 1, \dots, n-1$) are concentric circumferences or all Γ_k are parallel straight lines (if $A_0 = A_1 = \dots = A_{n-1}$, then $f(z) \equiv \text{const}$).

The 1. condition may — at least in some cases — be omitted. Thus for $n = 2$ and $n = 3$ the following assertion is true:

Theorem 3.7.8. *A polyentire function of order n accepting n different values non-isolatedly is a degenerate function.*

This theorem cannot be extended to meromorphic p.a. functions. E.g. the meromorphic b.a. function $B(z) = (z^2 + z\bar{z})/(z^2 + 1)$ accepts non-isolatedly the values 0 and 1 (on the straight line $\operatorname{Re} z = 0$ and on the circumference $|z| = 1$), but $B(z)$ is not a degenerate function. However the following fact is true:

Theorem 3.7.9. *A meromorphic b.a. function, which accepts three different values non-isolatedly, is a degenerate function.*

3.8. Boundary theorems of factorization and uniqueness

First we shall note a uniqueness theorem proved by NGUYEN THUA HOP [1] (carried-out under guidance of A. V. BICADZE). A function $\Phi(z)$, which is holomorphic in some simply connected region D with a rectifiable boundary Γ, is called a *function of the class* $E_2(D)$ if $\int_{\Gamma_r} |\Phi(z)|^2 |dz| < A = \text{const}$ for all images Γ_r of circumferences $|z| = r$ by the conformal mapping of the circle $|\zeta| < 1$ on D. A b.a. function (in D) $\varphi(z) + \bar{z}\psi(z)$ is called *regular* in D, if $\varphi, \psi \in E_2(D)$. The following theorem is true (see HOP [1], BICADZE [3]).

Theorem 3.8.1. *Let D be a bounded region having as boundary an analytic curve L, and let the function $z(\zeta)$ map the unit disk on D. If $z(\zeta)$ is not a rational function, then any regular b.a. function in D which has in every regular point of Γ angular boundary values zero, is identically zero in D. If the function $z(\zeta)$ is rational, then infinitely many linearly independent regular b.a. functions in D exist which have angular boundary values zero in almost every point of Γ.*

Let $f(z)$ be a p.a. function of given order n in some region D and let the boundary ∂D contain open arcs $\Gamma_1, \dots, \Gamma_m$; let $f(z)$ have the angular limit zero in every point of "sufficiently massive" sets belonging to these arcs. Then it turns out that $f(z)$ admits some specific factorization, which we shall discuss below. It is known that every open analytic arc Γ can be presented by an equation of the form $\bar{z} = A(z)$, where $A(z)$ is an analytic function in some neighbourhood Δ_ν of Γ_ν.

Let $f(z)$ be a p.a. function in some region D, $f(z) = \sum_{k=0}^{n-1} h_k(z)\, \bar{z}^k$, and let Γ be a simple open regular analytic arc in D presented in some neighbourhood $\Delta(\Gamma)$ of Γ in the form $\bar{z} = A(z)$, where $A(z)$ is holomorphic in $\Delta(\Gamma)$. The analytic function

$$f_\Gamma(z) = \sum_{k=0}^{n-1} h_k(z)\, [A(z)]^k \tag{15}$$

will be called *connected with $f(z)$ on Γ*. If $\Gamma \subset D$, then (15) is the only function analytic in some neighbourhood of Γ and coinciding with $f(z)$ on Γ. It is possible to prove that f_Γ can be presented in $D \cap \Delta(\Gamma)$ by the formula

$$f_\Gamma(z) = f(z) + \sum_{k=1}^{n-1} \frac{1}{k!} (A(z) - \bar{z})^k \frac{\partial^k f}{\partial \bar{z}^k}. \tag{16}$$

This representation permits to factorize the p.a. function $f(z)$ *near the arc Γ* when the connected function $f_\Gamma(z)$ becomes zero. Two analytic arcs Γ and Γ_1, respectively, satisfying the equations $\bar{z} = A(z)$ and $\bar{z} = A_1(z)$, respectively, will be called *non-related* if $A_1(z)$ is not the analytic continuation of $A(z)$ in some simply connected domain.

It turns out to be useful to attract the notion of *angular limit of order not less than p* (see BALK and ZUEV [1]), as well as some notions of cluster set theory (see COLLINGWOOD and LOHWATER [1]). Let D be a region and Γ an open simple smooth arc, $\varrho(z, \partial D)$ — the distance from the point z to the boundary ∂D, $f(z)$ — a function given in D, p — an integer, $p \geqq 1$. We shall say that the function $f(z)$ has, in the point ζ, $\zeta \in \Gamma$, the angular limit zero of an order not less than p, if the auxiliary function $f(z)/[\varrho(z, \partial D)]^{p-1}$ has in ζ the angular limit zero.

Theorem 3.8.2. *Let the function $f(z)$ be n-analytic in a simply connected region D and let the boundary ∂D contain m pairwise nonrelated analytic arcs Γ_ν $(\nu = 1, \dots, m)$ with equation $\bar{z} = A_\nu(z)$ $\big(A_\nu(z)$ is a holomorphic function in some neighbourhood of the arc Γ_ν, let on every arc Γ_ν $(\nu = 1, \dots, m)$ exist such set E_ν of positive linear measure that $f(z)$ has in every point of E_ν an angular limit zero of an order not less than p_ν (p_ν is an integer, $p_\nu \geqq 1\big)$. Then in D a discrete set d exists such that in every simply connected subregion $\tilde{D}$ of $D \setminus d$ the function $f(z)$ is representable in the form*

$$f(z) = \tilde{h}(z) \prod_{\nu=1}^{m} (\bar{z} - \tilde{A}_\nu(z))^{p_\nu} \tag{17}$$

where $\tilde{A}_\nu(z)$ $(\nu = 1, \dots, m)$ is the analytic continuation of $A_\nu(z)$ into $\tilde{D}$ and $\tilde{h}(z)$ is some p.a. function in D. If $p_1 + \dots + p_m \geqq n$ (specifically, if $m \geqq n$), then $f(z) \equiv 0$.

Further references

BALK and GOL'DBERG [3], BALK [1], BICADZE [2], GAHOV [4], KOLOSOV [3], KRAJKIEWICZ [1], NAAS and TUTSCHKE [1], PASCALI [1], SCHOPF [1].

4. Cohomological Methods in Complex Analysis

O. Forster

4.1. The Cousin problems

In this section we first describe the classical theorems of Mittag-Leffler and Weierstrass which answer the problem of finding meromorphic functions in the plane with prescribed principal parts resp. prescribed orders of zeroes and poles. Then we pass to the analogous problems in several complex variables, the so-called Cousin problems. These can be reformulated and solved cohomologically.

4.1.1. The theorems of Mittag-Leffler and Weierstrass. a) *Let $(a_j)_{j\in J}$ be a family of pairwise distinct points in $\mathbb{C}$ without point of accumulation. For every $j \in J$ let there be given a principal part*

$$f_j(z) = \sum_{\nu=-k_j}^{-1} c_{j\nu}(z - a_j)^\nu .$$

Then there exists a meromorphic function F on $\mathbb{C}$ which has at every point a_j a pole with principal part f_j and is holomorphic elsewhere (Mittag-Leffler).

b) *Let $(a_j)_{j\in J}$ be a family of pairwise distinct points in $\mathbb{C}$ without point of accumulation and $(n_j)_{j\in I}$ a family of integers. Then there exists a meromorphic function F on $\mathbb{C}$ which has order n_j at a_j for all $j \in J$ (a pole if $n_j < 0$, a zero if $n_j > 0$) and is holomorphic and different from zero elsewhere (Weierstrass).*

For the proof of the two theorems one remarks first that it is trivial to find a meromorphic function which has the prescribed principal parts (resp. prescribed orders of zeroes and poles) over any compact subset of $\mathbb{C}$. Starting from this, the theorems are then proved by an exhaustion process.

4.1.2. Generalization to several complex variables. In several complex variables the analogous problems must be formulated in a different way since the zeroes and poles of a meromorphic function of several variables are no more isolated points, but 1-codimensional analytic sets.

In order to state these problems, let us introduce some notations. Let X be a complex manifold (for example $X = \mathbb{C}^n$, a domain in $\mathbb{C}^n$ or a Riemann surface). For an open subset $U \subset X$ we denote by $\mathcal{O}(U)$ the ring of holomorphic functions and by $\mathcal{M}(U)$ the ring of meromorphic functions in U. Further $\mathcal{O}^*(U)$ denotes the multiplicative group of invertible holomorphic functions in U (i.e. holomorphic functions with values in $\mathbb{C}^* = \mathbb{C} \setminus 0$) and $\mathcal{M}^*(U)$ the multiplicative group of invertible meromorphic functions (i.e. functions $f \in \mathcal{M}(U)$ which do not vanish identically on any connected component of U).

4.1.3. The Cousin-I-problem. By a Cousin-I-data (or additive Cousin data) on a complex manifold X we mean a family $\mathfrak{C} = (U_\iota, f_\iota)_{\iota \in I}$ with the following properties:

i) $\mathfrak{U} := (U_\iota)_{\iota \in I}$ is an open covering of X.
ii) $f_\iota \in \mathcal{M}(U_\iota)$ for all $\iota \in I$.
iii) $f_\iota - f_j \in \mathcal{O}(U_\iota \cap U_j)$ for all $\iota, j \in I$.

(More precisely, in the third condition, the difference $f_\iota - f_j$ should be read as $(f_\iota \mid U_\iota \cap U_j) - (f_j \mid U_\iota \cap U_j)$. In order not to complicate the notations, we often will not write out obvious restriction maps.)

A *solution* of the Cousin-I-data $(U_\iota, f_\iota)_{\iota \in I}$ is a meromorphic function $F \in \mathcal{M}(X)$ such that

$$F \mid U_\iota - f_\iota \in \mathcal{O}(U_\iota) \quad \text{for all} \quad \iota \in I\,.$$

This means that F has the same principal parts as the given Cousin-I-data. It is clear that we have to impose condition iii) on the Cousin data; otherwise a solution could not exist.

4.1.4. The Cousin-II-problem. By a Cousin-II-data (or multiplicative Cousin data) on X we mean a family $\mathfrak{C} = (U_\iota, f_\iota)_{\iota \in I}$ such that

i) $\mathfrak{U} := (U_\iota)_{\iota \in I}$ is an open covering of X.
ii) $f_\iota \in \mathcal{M}^*(U_\iota)$ for all $\iota \in I$.
iii) $f_\iota/f_j \in \mathcal{O}^*(U_\iota \cap U_j)$ for all $\iota, j \in I$.

A *solution* of such a Cousin-II-data is a meromorphic function $F \in \mathcal{M}^*(X)$ such that

$$\frac{F}{f_\iota} \in \mathcal{O}^*(U_\iota) \quad \text{for all} \quad \iota \in I\,.$$

This condition means that F has zeroes and poles of the same orders as the given f_ι.

The Cousin problems consist in finding solutions to given Cousin data. Cousin [1] has shown in 1895 that this is always possible for $X = \mathbb{C}^n$. The generalization of these problems to domains in $\mathbb{C}^n$ and complex manifolds (where the problems do not always have a solution) has led to the development of the sheaf theoretic and cohomological methods in the modern theory of several complex variables.

4.1.5. Connection with the theorems of Mittag-Leffler and Weierstrass. Before we come to the cohomological reformulation of the Cousin problems we want to show that their solution for $X = \mathbb{C}$ contains the theorems of Mittag-Leffler and Weierstrass.

a) Mittag-Leffler theorem. With the notations of 4.1.1. a), set $U_j := \mathbb{C} \setminus \{a_i : i \neq j\}$ for $j \in J$. Then $(U_j)_{j \in J}$ is an open covering of $\mathbb{C}$. Together with the principal parts

$$f_j(z) = \sum_{\nu = -k_j} c_{j\nu}(z - a_j)^\nu \in \mathcal{M}(U_j)$$

the family $(U_j, f_j)_{j \in J}$ is a Cousin-I-data. Every solution of it is a meromorphic function $F \in \mathcal{M}(\mathbb{C})$ with the prescribed principal parts f_j at a_j.

b) Weierstrass theorem. Again let $U_j := \mathbb{C} \setminus \{a_i : i \neq j\}$ and set

$$f_j(z) := (z - a_j)^{n_j}\,.$$

Then $(U_j, f_j)_{j \in J}$ is a Cousin-II-data and every solution of it is a meromorphic function $F \in \mathcal{M}(\mathbb{C})$ having order n_j at a_j for all $j \in J$.

4.1.6. Cocycles. Let X be a complex manifold and $\mathfrak{U} = (U_i)_{i\in I}$ an open covering of X. We denote by $Z^1(\mathfrak{U}, \mathcal{O})$ the set of all families $(f_{ij})_{i,j\in I}$ of functions $f_{ij} \in \mathcal{O}(U_i \cap U_j)$ with

$$f_{ij} + f_{jk} = f_{ik} \quad \text{on} \quad U_i \cap U_j \cap U_k \quad \text{for all} \quad i, j, k \in I . \tag{1}$$

With componentwise addition, $Z^1(\mathfrak{U}, \mathcal{O})$ is a group, the first cocycle group with values in $\mathcal{O}$ with respect to the covering $\mathfrak{U}$.

Remark. The cocycle condition (1) implies

$$f_{ii} = 0 \quad \text{and} \quad f_{ij} = -f_{ji} \quad \text{for all} \quad i, j \in I .$$

Analogously we define $Z^1(\mathfrak{U}, \mathcal{O}^*)$ to be the set of all families $(f_{ij})_{i,j\in I}$ of functions $f_{ij} \in \mathcal{O}^*(U_i \cap U_j)$ with

$$f_{ij}f_{jk} = f_{ik} \quad \text{on} \quad U_i \cap U_j \cap U_k .$$

This is a group with respect to componentwise multiplication. A cocycle $(f_{ij})_{i,j\in I}$ of $Z^1(\mathfrak{U}, \mathcal{O})$ (resp. $Z^1(\mathfrak{U}, \mathcal{O}^*)$) is said to *split* or to be a *coboundary* if there exist functions $g_i \in \mathcal{O}(U_i)$, $i \in I$, (resp. $g_i \in \mathcal{O}^*(U_i)$) such that

$$f_{ij} = g_j - g_i \quad \text{on} \quad U_i \cap U_j \quad \text{for all } i, j \in I$$

(resp. $f_{ij} = g_j/g_i$). The set of all splitting cocycles forms the so-called coboundary subgroup $B^1(\mathfrak{U}, \mathcal{O}) \subset Z^1(\mathfrak{U}, \mathcal{O})$ (resp. $B^1(\mathfrak{U}, \mathcal{O}^*) \subset Z^1(\mathfrak{U}, \mathcal{O}^*)$). The quotient

$$H^1(\mathfrak{U}, \mathcal{O}) := Z^1(\mathfrak{U}, \mathcal{O})/B^1(\mathfrak{U}, \mathcal{O})$$

is the first *cohomology group* of X with values in $\mathcal{O}$ with respect to the covering $\mathfrak{U}$. Analogously we define

$$H^1(\mathfrak{U}, \mathcal{O}^*) := Z^1(\mathfrak{U}, \mathcal{O}^*)/B^1(\mathfrak{U}, \mathcal{O}^*) .$$

4.1.7. Proposition. *Let X be a complex manifold, $\mathfrak{U} = (U_i)_{i\in I}$ an open covering and $(U_i, f_i)_{i\in I}$ an additive (resp. multiplicative) Cousin data on X. Define*

$$f_{ij} := f_j - f_i \in \mathcal{O}(U_i \cap U_j)$$

(*resp.* $f_{ij} := f_j/f_i \in \mathcal{O}^*(U_i \cap U_j)$). *Then the family $(f_{ij})_{i,j\in I}$ is a cocycle of $Z^1(\mathfrak{U}, \mathcal{O})$* (*resp.* $Z^1(\mathfrak{U}, \mathcal{O}^*)$). *This cocycle splits if and only if the given Cousin data has a solution.*

Proof. We consider only the additive case. Suppose that the cocycle (f_{ij}) splits, i.e. there exist functions $g_i \in \mathcal{O}(U_i)$ such that

$$f_{ij} = g_j - g_i \quad \text{on} \quad U_i \cap U_j .$$

Then $f_i - g_i = f_j - g_j$ on $U_i \cap U_j$ for all $i, j \in I$. Therefore the family $(f_i - g_i)_{i\in I}$ glues together to a global meromorphic function on X which is a solution of the Cousin data (U_i, f_i). The converse implication is as easy to prove.

4.1.8. A Heftungslemma. We want to give a proof for the fact that $H^1(\mathfrak{U}, \mathcal{O}) = 0$, where $\mathfrak{U}$ is an open covering of $\mathbb{C}^n$. The main technical tool for this is a "Heftungslemma". To formulate it, we need some notations.

We identify $\mathbb{C}^n$ with $\mathbb{R}^{2n}$. By a *rectangle* in $\mathbb{R}^{2n}$ we understand a product of compact intervals. Let

$$Q_1 = \prod_{\nu=1}^{2n} I'_\nu , \qquad Q_2 = \prod_{\nu=1}^{2n} I''_\nu$$

be two rectangles. (Q_1, Q_2) is said to be a *Cousin pair* if there exists ν such that $I'_\mu = I''_\mu$ for all $\mu \neq \nu$ and $I'_\nu = [a, b]$, $I''_\nu = [b, c]$ with real constants $a < b < c$. In this case $Q_1 \cup Q_2$ is again a rectangle.

Heftungslemma. *Let (Q_1, Q_2) be a Cousin pair of rectangles in $\mathbb{C}^n$ and let f be a holomorphic function defined in an open neighbourhood U of $Q_1 \cap Q_2$. Then there exist holomorphic functions f_1, f_2 defined in open neighbourhoods U_i of Q_i, $i = 1, 2$, such that*

$$f = f_1 - f_2 \quad \text{on} \quad U \cap U_1 \cap U_2 .$$

Proof. For simplicity we consider only the case $n = 1$. (For arbitrary n one has to consider $n - 1$ variables as parameters.) We trace a curve $\gamma_1 - \gamma_2$ around $Q_1 \cap Q_2$ as in fig. 28. By the Cauchy integral formula one has

$$f(z) = \frac{1}{2\pi i} \int\limits_{\gamma_1 - \gamma_2} \frac{f(\zeta)}{\zeta - z} d\zeta$$

for z in a neighbourhood of $Q_1 \cap Q_2$. If we define

$$f_k(z) := \frac{1}{2\pi i} \int\limits_{\gamma_k} \frac{f(\zeta)}{\zeta - z} d\zeta , \qquad k = 1, 2 ,$$

then f_k is holomorphic in a neighbourhood of Q_k and one has the desired representation $f = f_1 - f_2$.

4.1.9. Cousin induction. Let X be an open set in $\mathbb{C}^n$. Consider a property P of rectangles $Q \subset X$. Suppose that the following axioms are satisfied:

(i) Every point $x \in X$ has a neighbourhood $U \subset X$ such that P holds for every rectangle $Q \subset U$.

(ii) Let (Q_1, Q_2) be a Cousin pair of rectangles in X. If P holds for Q_1 and Q_2, it holds also for $Q_1 \cup Q_2$.

The conditions (i) and (ii) imply that P is verified for every rectangle $Q \subset X$. This is the Cousin induction principle. It is proved by a simple geometric reasoning.

We apply now the Cousin induction to the splitting of cocycles. Let $\mathfrak{U} = (U_\iota)_{\iota \in I}$ be an open covering of $\mathbb{C}^n$ and $(f_{ij}) \in Z^1(\mathfrak{U}, \mathcal{O})$ a cocycle. We say that a rectangle $Q \subset \mathbb{C}^n$ has property P if the cocycle (f_{ij}) splits over Q, that is if there exist an open neighbourhood V of Q and holomorphic functions $g_i \in \mathcal{O}(U_i \cap V)$ such that

$$f_{ij} = g_j - g_i \quad \text{on} \quad U_i \cap U_j \cap V \quad \text{for all} \quad i, j \in I .$$

It is easy to see that this property satisfies axiom (i). We claim that it satisfies also axiom (ii). For this let (Q_1, Q_2) be a Cousin pair of rectangles, V_k open neighbourhoods of Q_k and $g_i^{(k)} \in \mathcal{O}(U_i \cap V_k)$ holomorphic functions with

$$f_{ij} = g_j^{(k)} - g_i^{(k)} \text{ on } U_i \cap U_j \cap V_k .$$

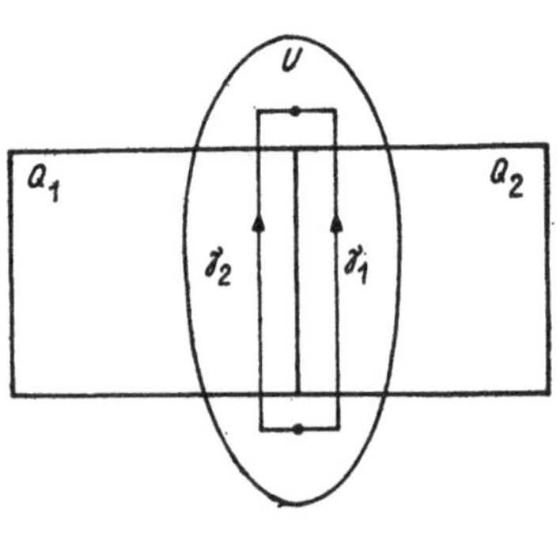

Fig. 28

Then $g_i^{(1)} - g_i^{(2)} = g_j^{(1)} - g_j^{(2)}$ on $U_i \cap U_j \cap V_1 \cap V_2$. Therefore there exists a holomorphic function $\varphi \in \mathcal{O}(V_1 \cap V_2)$ such that

$$\varphi \mid U_i \cap V_1 \cap V_2 = g_i^{(1)} - g_i^{(2)} \quad \text{for all} \quad i \in I .$$

We apply the Heftungslemma 4.1.8. to the function φ and get holomorphic functions $\varphi_k \in \mathcal{O}(V_k')$, where $Q \subset V_k' \subset V_k$, such that $\varphi = \varphi_1 - \varphi_2$ on $V_1' \cap V_2'$. We have then in particular

$$g_i^{(1)} - \varphi_1 = g_i^{(2)} - \varphi_2 \quad \text{on} \quad U_i \cap V_1' \cap V_2' .$$

$V := V_1' \cup V_2'$ is a neighbourhood of $Q_1 \cup Q_2$. There exists $g_i \in \mathcal{O}(U_i \cap V)$ with $g_i \mid U_i \cap V_k' = g_i^{(k)} - \varphi_k$ and we have

$$f_{ij} = g_j - g_i \quad \text{on} \quad U_i \cap U_j \cap V .$$

Therefore (f_{ij}) splits over $Q_1 \cup Q_2$. By the Cousin induction principle, (f_{ij}) splits over every rectangle $Q \subset \mathbb{C}^n$.

4.1.10. Theorem. *Let $\mathfrak{U} = (U_i)_{i \in I}$ be an open covering of $\mathbb{C}^n$. Then $H^1(\mathfrak{U}, \mathcal{O}) = 0$.*

Proof. We have to show that every cocycle $(f_{ij}) \in Z^1(\mathfrak{U}, \mathcal{O})$ splits. By 4.1.9. this is true over every rectangle $Q \subset \mathbb{C}^n$. We denote by Δ_k the polycylinder of multiradius k in $\mathbb{C}^n$. Then there exist functions $g_i^{(k)} \in \mathcal{O}(U_i \cap \Delta_k)$ such that

$$f_{ij} = g_j^{(k)} - g_i^{(k)} \quad \text{over} \quad U_i \cap U_j \cap \Delta_k . \tag{2}$$

We have $g_i^{(k+1)} - g_i^{(k)} = g_j^{(k+1)} - g_j^{(k)}$ on $U_i \cap U_j \cap \Delta_k$; hence these functions glue together to a global function on Δ_k, which can be approximated by functions defined on Δ_{k+1}. Therefore one can replace the functions $g_i^{(k)}$ inductively by functions $\tilde{g}_i^{(k)} \in \mathcal{O}(U_i \cap \Delta_k)$ satisfying again (2) and in addition

$$\|\tilde{g}_i^{(k+1)} - \tilde{g}_i^{(k)}\|_{U_i \cap \Delta_{k-1}} \leqq 2^{-k}$$

for all $k \in \mathbb{N}$. The functions $\tilde{g}_i^{(k)}$ converge for $k \to \infty$ to functions $g_i \in \mathcal{O}(U_i)$ such that $f_{ij} = g_j - g_i$ on $U_i \cap U_j$, q.e.d.

Remark. Similarly one proves $H^1(\mathfrak{U}, \mathcal{O}^*) = 0$. For this one needs an approximation theorem for nowhere vanishing holomorphic functions.

As proved in proposition 4.1.7., the vanishing of the cohomology groups imply that on $\mathbb{C}^n$ every additive and multiplicative Cousin problem has a solution.

4.2. Sheaves and cohomology groups

We want to place the considerations of 4.1. into a more general context. For this we give in this section a short introduction to the cohomology theory of sheaves.

4.2.1. Presheaves and sheaves. Let X be a topological space and $\mathfrak{T}$ its topology, i.e. the set of all open subsets of X. A presheaf of Abelian groups on X is a pair $(\mathcal{F}, \varrho)$ consisting of

a) a family $\mathcal{F} = (\mathcal{F}(U))_{U \in \mathfrak{T}}$ of Abelian groups,

b) a family $\varrho = (\varrho_V^U)_{U, V \in \mathfrak{T}, V \subset U}$ of group homomorphisms

$$\varrho_V^U : \mathcal{F}(U) \to \mathcal{F}(V)$$

with the following properties

i) $\varrho_U^U = \mathrm{id}_U$ for every $U \in \mathfrak{T}$,

ii) $\varrho_W^V \circ \varrho_V^U = \varrho_W^U$ for $W \subset V \subset U$.

Remark. Usually one just writes $\mathcal{F}$ instead of $(\mathcal{F}, \varrho)$. The homomorphisms ϱ_V^U are called restriction homomorphisms. For $f \in \mathcal{F}(U)$, $V \subset U$, we abbreviate $\varrho_V^U(f)$ by $f \mid V$. Analogously to presheaves of Abelian groups one can also define presheaves of vector spaces, rings etc.

A presheaf $\mathcal{F}$ on X is called a *sheaf*, if for every open set $U \subset X$ and every family of open subsets $U_i \subset U$, $i \in I$, such that $U = \bigcup U_i$ the following two conditions are satisfied:

(I) If $f, g \in \mathcal{F}(U)$ are elements such that $f \mid U_i = g \mid U_i$ for all $i \in I$, then $f = g$.
(II) Given elements $f_i \in \mathcal{F}(U_i)$ such that

$$f_i \mid U_i \cap U_j = f_j \mid U_i \cap U_j \quad \text{for all} \quad i, j \in I\,,$$

then there exists an $f \in \mathcal{F}(U)$ such that $f \mid U_i = f_i$ for every $i \in I$.

4.2.2. Examples of sheaves. a) Let X be an arbitrary topological space. For $U \subset X$ open let $\mathcal{E}(U)$ be the vector space of all continuous functions $f : U \to \mathbb{C}$. For $V \subset U$ let $\varrho_V^U : \mathcal{E}(U) \to \mathcal{E}(V)$ be the usual restriction mapping. Then $(\mathcal{E}, \varrho)$ is a sheaf of vector spaces.

b) Let X be a complex manifold. For $U \subset X$ open let $\mathcal{O}(U)$ resp. $\mathcal{M}(U)$ be the ring of holomorphic (meromorphic) functions on U, cf. 4.1.2. Together with the usual restriction mappings we get sheaves of rings $\mathcal{O}$ resp. $\mathcal{M}$ on X. Analogously we have the sheaves $\mathcal{O}^*$ and $\mathcal{M}^*$ of multiplicative groups on X.

c) Let X be an arbitrary topological space and G an arbitrary Abelian group. We define a sheaf on X, which we will also denote by G, in the following way: For $U \subset X$ open let $G(U)$ consist of all locally constant maps $U \to G$. (If U is connected, we have $G(U) = G$.) For $V \subset U$ let $G(U) \to G(V)$ be the usual restriction map. It is easy to verify all sheaf axioms.

4.2.3. Cohomology groups $H^q(\mathfrak{U}, \mathcal{F})$. Let $\mathcal{F}$ be a sheaf of Abelian groups on the topological space X and let $\mathfrak{U} = (U_i)_{i \in I}$ be an open covering of X. For $q = 0, 1, 2, \ldots$ we define the q-th cochain group of $\mathcal{F}$ with respect to $\mathfrak{U}$ as

$$C^q(\mathfrak{U}, \mathcal{F}) = \prod_{(i_0, \ldots, i_q) \in I^{q+1}} \mathcal{F}(U_{i_0} \cap \ldots \cap U_{i_q})\,.$$

$C^q(\mathfrak{U}, \mathcal{F})$ is a group with respect to componentwise addition. We set $C^q(\mathfrak{U}, \mathcal{F}) = 0$ for $q < 0$. There are *coboundary* operators

$$\delta : C^q(\mathfrak{U}, \mathcal{F}) \to C^{q+1}(\mathfrak{U}, \mathcal{F})\,.$$

defined as follows: Let $\xi = (f_{i_0 \ldots i_q}) \in C^q(\mathfrak{U}, \mathcal{F})$. Then $\delta\xi = (g_{j_0 \ldots j_{q+1}})$ with

$$g_{j_0 \ldots j_{q+1}} = \sum_{k=0}^{q+1} (-1)^k\, g_{j_0 \ldots \hat{j}_k \ldots j_{q+1}}\,.$$

Here $\hat{j}_k$ means that this index has to be omitted. In the sum it is understood that the terms have to be restricted to their common domain of definition $U_{j_0} \cap \ldots \cap U_{j_{q+1}}$.

The coboundary operators are group homomorphisms satisfying $\delta \circ \delta = 0$. Therefore, if we define the q-th cocycle resp. coboundary group by

$$Z^q(\mathfrak{U}, \mathcal{F}) := \operatorname{Ker}\left(C^q(\mathfrak{U}, \mathcal{F}) \xrightarrow{\delta} C^{q+1}(\mathfrak{U}, \mathcal{F})\right),$$

$$B^q(\mathfrak{U}, \mathcal{F}) := \operatorname{Im}\left(C^{q-1}(\mathfrak{U}, \mathcal{F}) \to C^q(\mathfrak{U}, \mathcal{F})\right),$$

we have $B^q(\mathfrak{U}, \mathcal{F}) \subset Z^q(\mathfrak{U}, \mathcal{F})$ and we can define the q-th cohomology group by

$$H^q(\mathfrak{U}, \mathcal{F}) := Z^q(\mathfrak{U}, \mathcal{F}) / B^q(\mathfrak{U}, \mathcal{F})\,.$$

The reader should convince himself that for $q = 1$ we get again the notions introduced in 4.1. for the sheaves $\mathcal{O}$ and $\mathcal{O}^*$.

For $q = 0$ we have by definition $B^0(\mathfrak{U}, \mathcal{F}) = 0$ and the sheaf axioms imply $Z^0(\mathfrak{U}, \mathcal{F}) \cong \mathcal{F}(X)$. Therefore $H^0(\mathfrak{U}, \mathcal{F}) \cong \mathcal{F}(X)$.

4.2.4. Cohomology groups $H^q(X, \mathcal{F})$. The groups $H^q(\mathfrak{U}, \mathcal{F})$ depend on the covering $\mathfrak{U}$. In order to have cohomology groups which depend only on X and $\mathcal{F}$, one has to use finer and finer coverings and then take a limit. This is done as follows. An open covering $\mathfrak{V} = (V_k)_{k \in K}$ is called *finer* than the covering $\mathfrak{U} = (U_\iota)_{\iota \in I}$, denoted $\mathfrak{V} < \mathfrak{U}$, if every V_k is contained in at least one U_ι, i.e. if there exists a map $\tau : K \to I$ such that

$$V_k \subset U_{\tau k} \quad \text{for every} \quad k \in K .$$

The map τ induces a map

$$C^q(\mathfrak{U}, \mathcal{F}) \to C^q(\mathfrak{V}, \mathcal{F})$$

by sending a cochain $(f_{i_0 \dots i_q}) \in C^q(\mathfrak{U}, \mathcal{F})$ to $(g_{k_0 \dots k_q}) \in C^q(\mathfrak{V}, \mathcal{F})$, where

$$g_{k_0 \dots k_q} = f_{\tau k_0, \dots, \tau k_q} \mid V_{k_0} \cap \dots \cap V_{k_q} .$$

This mapping maps cocycles into cocycles and coboundaries into coboundaries, hence induces a mapping

$$t^{\mathfrak{U}}_{\mathfrak{V}} : H^q(\mathfrak{U}, \mathcal{F}) \to H^q(\mathfrak{V}, \mathcal{F}) .$$

One can show that $t^{\mathfrak{U}}_{\mathfrak{V}}$ depends only on $\mathfrak{U}$ and $\mathfrak{V}$ and not on the particular map $\tau : K \to I$. For $\mathfrak{W} < \mathfrak{V} < \mathfrak{U}$ one has $t^{\mathfrak{V}}_{\mathfrak{W}} \circ t^{\mathfrak{U}}_{\mathfrak{V}} = t^{\mathfrak{U}}_{\mathfrak{W}}$. Thus we get a directed system $(H^q(\mathfrak{U}, \mathcal{F}), t^{\mathfrak{U}}_{\mathfrak{V}})$ and can form the direct limit

$$H^q(X, \mathcal{F}) := \varinjlim_{\mathfrak{U}} H^q(\mathfrak{U}, \mathcal{F}) .$$

This limit is called the *q-th Čech cohomology group* with values in $\mathcal{F}$.

4.2.5. Leray's theorem. Under certain conditions, in order to calculate $H^q(X, \mathcal{F})$ it suffices to consider one particular covering $\mathfrak{U}$. Such conditions are given by the theorem of Leray:

Let X be a metrizable topological space, $\mathcal{F}$ a sheaf of Abelian groups on X and $\mathfrak{U} = (U_\iota)_{\iota \in I}$ an open covering of X such that

$$H^m(U_{i_0} \cap \dots \cap U_{i_k}, \mathcal{F}) = 0 \quad \textit{for all } k \geqq 0 \quad \textit{and} \quad m \geqq 1 .$$

Then the natural homomorphism

$$H^q(\mathfrak{U}, \mathcal{F}) \to H^q(X, \mathcal{F})$$

is an isomorphism for every $q \geqq 0$.

Without any hypothesis one can show that the natural map $H^1(\mathfrak{U}, \mathcal{F}) \to H^1(X, \mathcal{F})$ is injective.

4.2.6. The stalks of a presheaf. Let $\mathcal{F}$ be a presheaf of Abelian groups on a topological space X and $x \in X$ a point. Let $\mathfrak{V}(x)$ be the system of all open neighbourhoods of x. Then we can form the direct limit

$$\mathcal{F}_x := \varinjlim_{U \in \mathfrak{V}(x)} \mathcal{F}(U)$$

which is called the *stalk* at x of the presheaf $\mathcal{F}$. Thus an element ("germ") $\varphi \in \mathcal{F}_x$ is represented by an element $f \in \mathcal{F}(U)$, $U \in \mathfrak{B}(x)$. Two elements $f \in \mathcal{F}(U)$, $g \in \mathcal{F}(V)$, $U, V \in \mathfrak{B}(x)$, represent the same germ in $\mathcal{F}_x$ if and only if there exists a neighbourhood $W \subset U \cap V$ of x such that $f \mid W = g \mid W$.

4.2.7. Sheaf homomorphisms. Let $\mathcal{F}$, $\mathcal{G}$ be two sheaves of Abelian groups on the topological space X. A sheaf homomorphism $\alpha : \mathcal{F} \to \mathcal{G}$ is a family of group homomorphisms

$$\alpha_U : \mathcal{F}(U) \to \mathcal{G}(U), \qquad U \text{ open in } X,$$

which are compatible with the restriction homomorphisms. Therefore we get induced homomorphisms of stalks $\alpha_x : \mathcal{F}_x \to \mathcal{G}_x$ for all $x \in X$. A sequence of sheaf homomorphisms

$$\mathcal{F}_0 \xrightarrow{\alpha_1} \mathcal{F}_1 \xrightarrow{\alpha_2} \mathcal{F}_2 \to \dots \xrightarrow{\alpha_k} \mathcal{F}_k$$

is called *exact*, if the induced sequence of stalks

$$\mathcal{F}_{0,x} \to \mathcal{F}_{1,x} \to \mathcal{F}_{2,x} \to \dots \to \mathcal{F}_{k,x}$$

is an exact sequence of Abelian groups for all $x \in X$.

4.2.8. Examples of exact sequences. a) Let X be a complex manifold. Then we have the exact sequence of sheaves

$$0 \to \mathbb{Z} \to \mathcal{O} \xrightarrow{\text{ex}} \mathcal{O}^* \to 0,$$

where ex is defined by $\text{ex}(f) = e^{2\pi i f}$. It is clear that ex is surjective on stalks and that the kernel of ex is the sheaf of locally constant integer-valued functions.

b) Let X be a real n-dimensional differentiable manifold and denote by $\mathcal{E}^q$ the sheaf of complex-valued differential forms of order q with $\mathcal{C}^\infty$ coefficients. Then we have the exact sequence

$$0 \to \mathbb{C} \to \mathcal{E}^0 \xrightarrow{\text{d}} \mathcal{E}^1 \xrightarrow{\text{d}} \mathcal{E}^2 \to \dots \xrightarrow{\text{d}} \mathcal{E}^n \to 0.$$

Here d is the exterior differentiation. The exactness is equivalent to the classical Poincaré lemma.

c) Let X be an n-dimensional complex manifold and denote by $\mathcal{E}^{0,q}$ the sheaf of differential forms of type $(0, q)$, i.e. differential forms which can be written with respect to local coordinates $z_1, \dots, z_n$ as

$$\omega = \sum_{j_1 < \dots < j_q} f_{j_1 \dots j_q} \, d\bar{z}_{j_1} \wedge \dots \wedge d\bar{z}_{j_q}.$$

For these differential forms one has an operator $\bar{\partial} : \mathcal{E}^{0,q} \to \mathcal{E}^{0,q+1}$ which is defined for ω as above by

$$\bar{\partial}\omega = \sum_{j_1 < \dots < j_q} \bar{\partial} f_{j_1 \dots j_q} \wedge d\bar{z}_{j_1} \wedge \dots \wedge d\bar{z}_{j_q},$$

where $\bar{\partial} f = \sum_{\nu=1}^{n} \partial f / \partial \bar{z}_\nu \, d\bar{z}_\nu$. One has an exact sequence

$$0 \to \mathcal{O} \to \mathcal{E}^{0,0} \xrightarrow{\bar{\partial}} \mathcal{E}^{0,1} \xrightarrow{\bar{\partial}} \dots \xrightarrow{\bar{\partial}} \mathcal{E}^{0,n} \to 0$$

(Lemma of Dolbeault-Grothendieck).

4.2.9. The long exact cohomology sequence. If $0 \to \mathcal{F}' \xrightarrow{\alpha} \mathcal{F} \xrightarrow{\beta} \mathcal{F}'' \to 0$ is a short exact sequence of sheaves on a paracompact space X, we have an induced exact sequence of cohomology groups

$$0 \to H^0(X, \mathcal{F}') \to H^0(X, \mathcal{F}) \to H^0(X, \mathcal{F}'') \xrightarrow{\delta^*} H^1(X, \mathcal{F}') \to$$

$$\ldots \to H^q(X, \mathcal{F}') \to H^q(X, \mathcal{F}) \to H^q(X, \mathcal{F}'') \xrightarrow{\delta^*} H^{q+1}(X, \mathcal{F}') \to \ldots$$

with so-called connecting homomorphisms

$$\delta^* : H^q(X, \mathcal{F}'') \to H^{q+1}(X, \mathcal{F}') .$$

We will describe these connecting homomorphisms in a special case. Let $\xi \in H^q(X, \mathcal{F}'')$ be a cohomology class which is represented by a cocycle $(f_{i_0\ldots i_q}) \in Z^q(\mathfrak{U}, \mathcal{F}'')$. Suppose that we can find $F_{i_0\ldots i_q} \in \mathcal{F}(U_{i_0} \cap \ldots \cap U_{i_q})$ such that

$$\beta(F_{i_0\ldots i_q}) = f_{i_0\ldots i_q} .$$

(This is not true in general and one has to pass to a refinement of the covering.) Then

$$(F_{i_0\ldots i_q}) \in C^q(\mathfrak{U}, \mathcal{F})$$

and

$$\delta(F_{i_0\ldots i_q}) \in Z^{q+1}(\mathfrak{U}, \operatorname{Ker} \beta) = Z^{q+1}(\mathfrak{U}, \mathcal{F}') .$$

Now $\delta^*\xi$ is represented by the cocycle $\delta(F_{i_0\ldots i_q})$.

4.2.10. The abstract de Rham theorem. *Let $\mathcal{F}$ be a sheaf of Abelian groups on a paracompact space X and*

$$0 \to \mathcal{F} \to \mathcal{A}^0 \to \mathcal{A}^1 \to \ldots \to \mathcal{A}^n \to 0$$

an acyclic resolution of $\mathcal{F}$, i.e. an exact sequence of sheaves such that $H^q(X, \mathcal{A}^k) = 0$ for all $q \geqq 1$ and $k \geqq 0$. Then for all $q \geqq 0$ we have isomorphisms

$$H^q(X, \mathcal{F}) \cong \frac{\operatorname{Ker}\left(\mathcal{A}^q(X) \to \mathcal{A}^{q+1}(X)\right)}{\operatorname{Im}\left(\mathcal{A}^{q-1}(X) \to \mathcal{A}^q(X)\right)} .$$

(One puts $\mathcal{A}^{-1} = 0$.)

Proof. Define sheaves $\mathcal{Z}^q := \operatorname{Ker}(\mathcal{A}^q \to \mathcal{A}^{q+1})$. Then we have $\mathcal{F} = \mathcal{Z}^0$ and short exact sequences

$$0 \to \mathcal{Z}^{q-1} \to \mathcal{A}^{q-1} \to \mathcal{Z}^q \to 0 , \qquad q \geqq 0 .$$

Consider the associated long exact cohomology sequence

$$H^p(X, \mathcal{A}^{q-1}) \to H^p(X, \mathcal{Z}^q) \to H^{p+1}(X, \mathcal{Z}^{q-1}) \to H^{p+1}(X, \mathcal{A}^{q-1}) .$$

The acyclicity of the sheaves $\mathcal{A}^q$ implies

$$H^p(X, \mathcal{Z}^q) \cong H^{p+1}(X, \mathcal{Z}^{q-1}) \quad \text{for all} \quad p \geqq 1$$

and

$$H^1(X, \mathcal{Z}^{q-1}) \cong \frac{H^0(X, \mathcal{Z}^q)}{\operatorname{Im}\left(H^0(X, \mathcal{A}^{q-1}) \to H^0(X, \mathcal{Z}^q)\right)}$$

$$= \frac{\operatorname{Ker}\left(\mathcal{A}^q(X) \to \mathcal{A}^{q+1}(X)\right)}{\operatorname{Im}\left(\mathcal{A}^{q-1}(X) \to \mathcal{A}^q(X)\right)} .$$

This implies the assertion.

4.2.11. The theorems of de Rham and Dolbeault. The abstract de Rham theorem can be applied to the exact sequences 4.2.8. b) and 4.2.8. c) since the sheaves $\mathscr{E}^q$ and $\mathscr{E}^{0,q}$ are acyclic, i.e.

$$H^p(X, \mathscr{E}^q) = 0 \quad \text{and} \quad H^p(X, \mathscr{E}^{0,q}) = 0$$

for all $p \geqq 1$. This can be proved by using a differentiable partition of unity (we suppose that all manifolds have countable topology). Thus we get:

a) **Theorem of de Rham.** *Let X be a differentiable manifold. Then*

$$H^q(X, \mathbb{C}) \cong \frac{\operatorname{Ker}\left(\mathscr{E}^q(X) \xrightarrow{\mathrm{d}} \mathscr{E}^{q+1}(X)\right)}{\operatorname{Im}\left(\mathscr{E}^{q-1}(X) \xrightarrow{\mathrm{d}} \mathscr{E}^q(X)\right)} \quad \textit{for all} \quad q \geqq 0\,.$$

b) **Theorem of Dolbeault.** *Let X be a complex manifold. Then*

$$H^q(X, \mathcal{O}) \cong \frac{\operatorname{Ker}\left(\mathscr{E}^{0,q}(X) \xrightarrow{\bar\partial} \mathscr{E}^{0,q+1}(X)\right)}{\operatorname{Im}\left(\mathscr{E}^{0,q-1}(X) \xrightarrow{\bar\partial} \mathscr{E}^{0,q}(X)\right)} \quad \textit{for all} \quad q \geqq 0\,.$$

4.3. Cohomology of coherent sheaves

In this last section we discuss the main theorems on the cohomology of coherent sheaves on complex manifolds and give some applications.

4.3.1. Coherent sheaves. Let X be a complex manifold and $\mathcal{O}$ the sheaf holomorphic functions on X. An *$\mathcal{O}$-module sheaf* is a sheaf $\mathcal{F}$ of Abelian groups such that every $\mathcal{F}(U)$ is equipped with the structure of an $\mathcal{O}(U)$-module compatible with the restriction maps, i.e. for open sets $V \subset U$ the diagram

$$\begin{array}{ccc} \mathcal{O}(U) \times \mathcal{F}(U) & \to & \mathcal{F}(U) \\ \downarrow & & \downarrow \\ \mathcal{O}(V) \times \mathcal{F}(V) & \to & \mathcal{F}(V) \end{array}$$

where the horizontal arrows are the module multiplications and the vertical arrows are restriction maps, is commutative.

Examples of $\mathcal{O}$-module sheaves are the direct sums $\mathcal{O}^m$ of m copies of $\mathcal{O}$. If $\mathcal{F}$ and $\mathcal{G}$ are $\mathcal{O}$-module sheaves, a sheaf homomorphism $\alpha : \mathcal{F} \to \mathcal{G}$ is called an *$\mathcal{O}$-module homomorphism* if all $\alpha_U : \mathcal{F}(U) \to \mathcal{G}(U)$, $U \subset X$ open, are $\mathcal{O}(U)$-module homomorphisms.

An $\mathcal{O}$-module sheaf $\mathcal{F}$ on the complex manifold X is called *coherent* if every point $x \in X$ possesses an open neighbourhood V such that there is an exact sequence

$$\mathcal{O}^k \to \mathcal{O}^m \to \mathcal{F} \to 0$$

of $\mathcal{O}$-module homomorphisms over V.

It follows directly from the definition that the sheaves $\mathcal{O}$ and $\mathcal{O}^m$ are coherent. Another example of a coherent sheaf is the ideal sheaf $\mathcal{J}$ of an analytic submanifold $Y \subset X$. This sheaf is defined by

$$\mathcal{J}(U) = \{f \in \mathcal{O}(U) : f \mid U \cap Y = 0\}$$

for $U \subset X$ open.

4.3.2. Stein manifolds. A complex manifold X is called a *Stein manifold* if the following conditions are satisfied:

a) X is holomorphically separable, i.e. for every pair of points $x \neq y$ on X there exists a function $f \in \mathcal{O}(X)$ with $f(x) \neq f(y)$.

b) X is holomorphically convex, i.e. for every sequence $(x_\nu)_{\nu \in \mathbb{N}}$ of points on X without accumulation point there exists a function $f \in \mathcal{O}(X)$ with $\limsup_{\nu \to \infty} |f(x_\nu)| = \infty$.

Examples of Stein manifolds. It follows at once from the definitions that $\mathbb{C}^n$ and every analytic submanifold of $\mathbb{C}^n$ is Stein. Less trivial is the Steinness of all domains of holomorphy in $\mathbb{C}^n$ (Cartan-Thullen) and all non-compact Riemann surfaces (Behnke-Stein).

4.3.3. Theorem B of Cartan-Serre. Cartan and Serre have proved in 1951/52 the following fundamental theorem on Stein manifolds (see H. Cartan [3]):

Let $\mathcal{F}$ be a coherent sheaf on a Stein manifold X. Then

$$H^q(X, \mathcal{F}) = 0 \quad \text{for all} \quad q \geqq 1 .$$

This is a strong generalization of theorem 4.1.10.

The proof uses among others Cousin induction (cf. 4.1.9.) and the following Heftungslemma of H. Cartan on invertible matrices:

Let (Q_1, Q_2) be a Cousin pair of rectangles in $\mathbb{C}^n$ and

$$A : U \to GL(m, \mathbb{C})$$

a holomorphic map on an open neighbourhood U of $Q_1 \cap Q_2$. Then there exist holomorphic maps $A_k : U_k \to GL(m, \mathbb{C})$, $k = 1, 2$, of open neighbourhoods U_k of Q_k such that

$$A = A_1 A_2^{-1} \quad \text{on} \quad U \cap U_1 \cap U_2 .$$

4.3.4. Applications of Theorem B. Let X be a Stein manifold. Then $H^1(X, \mathcal{O}) = 0$, hence $H^1(\mathfrak{U}, \mathcal{O}) = 0$ for every open covering $\mathfrak{U}$ of X. Therefore on X every Cousin-I-problem is solvable.

The exact sequence $0 \to \mathbb{Z} \to \mathcal{O} \to \mathcal{O}^* \to 0$ induces the exact cohomology sequence

$$H^1(X, \mathcal{O}) \to H^1(X, \mathcal{O}^*) \to H^2(X, \mathbb{Z}) \to H^2(X, \mathcal{O}) .$$

From Theorem B follows

$$H^1(X, \mathcal{O}^*) = H^2(X, \mathbb{Z}) .$$

Therefore we have: On a Stein manifold X with $H^2(X, \mathbb{Z}) = 0$ every Cousin-II-problem is solvable.

It is remarkable that the condition $H^2(X, \mathbb{Z}) = 0$ is purely topological (it is fulfilled in particular if X is contractible). If $H^2(X, \mathbb{Z}) \neq 0$, one can show that there exists a Cousin-II-problem on X without solution.

Another consequence of Theorem B is the following:

Let X be a Stein manifold and $Y \subset X$ an analytic submanifold. Then for every holomorphic function $f : Y \to \mathbb{C}$ there exists a holomorphic function $F : X \to \mathbb{C}$ with $F \mid Y = f$.

Proof. Let $\mathcal{O}_X$ and $\mathcal{O}_Y$, respectively, be the sheaves of holomorphic functions on X and Y, respectively, and $\mathcal{J}$ the ideal sheaf of Y. Then we have an exact sequence

$$0 \to \mathcal{J} \to \mathcal{O}_X \to \mathcal{O}_Y \to 0 .$$

(We can consider $\mathcal{O}_Y$ as a sheaf on X which is 0 on $X \setminus Y$.) The exact cohomology sequence

$$H^0(X, \mathcal{O}_X) \to H^0(X, \mathcal{O}_Y) \to H^1(X, \mathcal{J}) = 0$$

shows that the restriction map $\mathcal{O}_X(X) \to \mathcal{O}_Y(Y)$ is surjective, q.e.d.

4.3.5. Finiteness theorem of Cartan-Serre. *Let X be a compact complex manifold. Then for every coherent sheaf $\mathcal{F}$ on X and all $q \geqq 0$ the vector space $H^q(X, \mathcal{F})$ is finite-dimensional.*

The proof of this theorem by Cartan and Serre (see Cartan [4]) uses a functional analytic lemma of L. Schwartz: Let E, F be Fréchet spaces, $\sigma: E \to F$ a surjective continuous linear map and $\varkappa: E \to F$ a compact linear map. Then Coker $(\sigma - \varkappa)$ is finite-dimensional.

As one application of the finiteness theorem we want to show that it implies the existence of meromorphic functions on compact Riemann surfaces. Let X be a compact Riemann surface and $g := \dim H^1(X, \mathcal{O})$. (The number g is the genus of X.) Let a be a point of X and (U, z) a coordinate neighbourhood of a with $z(a) = 0$. Let $V := X \setminus \{a\}$. Then $\mathfrak{U} = (U, V)$ is an open covering of X and $\dim H^1(\mathfrak{U}, \mathcal{O}) \leqq g$. We have

$$Z^1(\mathfrak{U}, \mathcal{O}) \cong \mathcal{O}(U \cap V) = \mathcal{O}(U \setminus \{a\}) .$$

The functions $1/z, \dots, 1/z^{g+1} \in \mathcal{O}(U \setminus \{a\})$ can be considered as cocycles. Since $\dim H^1(\mathfrak{U}, \mathcal{O}) \leqq g$ these cocycles are linearly dependent modulo coboundaries. Therefore there exist a vector $(c_1, \dots, c_{g+1}) \in \mathbb{C}^{g+1} \setminus 0$ and functions $f \in \mathcal{O}(U)$, $g \in \mathcal{O}(V)$ such that

$$\sum_{\nu=1}^{g+1} \frac{c_\nu}{z^\nu} = f - g \quad \text{on} \quad U \cap V .$$

The equation $g = f - \sum c_\nu z^{-\nu}$ shows that g defines a meromorphic function on X which has a pole at a of order $\leqq g + 1$.

For further study we refer the reader to Grauert and Remmert [1], Forster [2], Gunning and Rossi [1] and Wells [1].

5. Connections between Complex and Global Analysis: Some Analytical and Geometrical Aspects of the Riemann-Hilbert Transmission Problem[1)]

B. Bojarski

5.1. Historical remarks

For several reasons the transmission problem belongs to the core of the index theory. Historically it was the Riemann-Hilbert problem that lead Noether [1] in 1921 to the discovery, against Hilbert's conjecture, of the phenomenon of non-vanishing indices for singular integral equations. Later, in the fundamental works of Mushelišvili, I. N. Vekua and their followers the detailed study of the Riemann transmission problem served as the main pillar of the general theory of systems of singular integral equations on piecewise smooth contours in the plane. The one-dimensional theory in turn contributed to the creation of the general abstract theory of Fredholm operators and in still further perspective, stimulated the general index theory of elliptic partial differential equations and elliptic boundary value problems. I. N. Vekua was the first who discovered and investigated (see I. N. Vekua [4]) in detail the topological nature of the two-dimensional oblique derivative problem which is closely related with the Riemann transmission problem. In the following remarks I will try to illustrate, using as an example the Riemann transmission problem, the idea repeated again and again by Prof. Vekua that the profound and detailed qualitative study of various problems naturally arising in classical complex function theory always brings us very closely to the discussion of phenomena important and genuinely interesting for the theories of most advanced mathematics, reaching far over the complex plane mathematics.

5.2. Global index and partial indices of the Riemann-Hilbert transmission problem

Let D^+ be a bounded domain in the complex plane $\mathbb{C}^1$, with the boundary Γ (for simplicity Γ is assumed to be a smooth Jordan curve), D^- — the complementary domain, so that $\Gamma = \bar{D}^+ \cap \bar{D}^-$ $\mathbb{C}^1 = D^+ \cup \Gamma \cup D^-$. For convenience of notation, assume $z = 0 \in D^+$. In what follows Ω will denote the space of all Hölder-continuous maps

$$G(t) : \Gamma \to GL(n, \mathbb{C})$$

with the natural topology. Here n is a positive integer and $GL(n, \mathbb{C})$ — is the group of invertible $n \times n$-matrices with complex entries. The condition of Hölder continuity of elements of Ω is not essential for our purposes and is introduced only to avoid some

[1)] This problem is known as *Riemann-Hilbert boundary problem*, *Riemann boundary problem* or *problem of linear conjugation*, too.

technical complications. One could work as well with the class of maps which are continuous only.

The Riemann-Hilbert transmission problem for vector functions can be formulated as follows: describe the totality of piecewise holomorphic vector functions $\Phi(z)$ in $D^+ \cup D^-$, which admit continuous boundary values on Γ

$$\Phi^{\pm}(t) = \lim_{z\to t,\, z\in D^{\pm}} \Phi(Z)\,, \qquad t \in \Gamma\,,$$

and satisfy the *transmission condition*[1]) on Γ

$$\Phi^+(t) = G(t)\,\Phi^-(t) + h(t)\,, \qquad t \in \Gamma\,, \tag{1}$$

and some condition at infinity, e.g. $\Phi(t)$ is of finite order at ∞. Usually for our purposes we shall assume this condition as the normalizing condition: $\Phi(z)$ is regular at ∞ and

$$\Phi(\infty) = 0\,. \tag{2}$$

The vector function $h(t)$ in (1), for simplicity, will be assumed to be Hölder-continuous on Γ. It is convenient to write Φ and h in (1) as column $n \times 1$-matrices.

The regularity condition at ∞ is especially suitable for the geometric interpretations of solutions of the homogeneous problem (1) ($h \equiv 0$) as description of holomorphic sections of holomorphic vector bundles over the Riemann sphere S^2, (interpreted also as the complex projective line $\mathbb{P}^1$ or a compact Riemann surface of genus $g = 0$).

Let us recall the main features of the theory of the Riemann problem (1) as presented in Mushelišvili [2]. First the homogeneous problem is discussed

$$\Phi^+(t) = G(t)\,\Phi^-(t) \tag{1'}$$

with the condition of finite order at ∞. An $n \times n$-matrix solution $X(z)$ of (1′) is called *canonical* if it has the form

$$\begin{aligned} X(z) &= X_0(z)\,, & z &\in D^+\,,\\ X(z) &= X_0(z)\,\Lambda^{-1}(z)\,, & z &\in D^-\,, \end{aligned} \tag{3}$$

where $X_0(z)$ — is a holomorphic matrix function in $D^+ \cup D^-$, admitting a continuous inverse $X_0^{-1}(z)$ in $\bar{D}^+$ and $\bar{D}^-$, respectively, including the point $z = \infty$ (thus $\det X_0(z) \neq 0$ for $z \in D^+ \cup \Gamma$ and $z \in D^- \cup \Gamma$) $\det X_0(\infty) = 1$ and $\Lambda(z)$ is a diagonal matrix with the entries $\lambda_j(z) = z^{\varkappa_j}$, $\Lambda = \{\lambda_j\}$ and $\{\varkappa_j\}$ is a sequence of integers such that

$$\varkappa_1 \geqq \varkappa_2 \geqq \dots \geqq \varkappa_n \tag{4}$$

Theorem 5.2.1 (see Mushelišvili [2]). *For every $G(t) \in \Omega$ the canonical solution always exists. The sequence* (3) *does not depend on the considered canonical solution.*

The integers in (3) are called the *partial indices* of the homogeneous transmission problem (1′) or the matrix function $G(t) \in \Omega$. In Mushelišvili [2] (and for the general case in the literature at all) there is no effective procedure for determining the partial indices — in this sense theorem 5.2.1 is a pure existence theorem. Deeper reasons for this will be discussed later. However Mushelišvili gives the following formula

$$\varkappa = \varkappa_1 + \varkappa_2 + \dots + \varkappa_n \tag{5}$$

with

$$\varkappa = \frac{1}{2\pi}\,\Delta_\Gamma \arg\det G(t) \tag{6}$$

[1]) Cf. also the chapters 10 and 20 of this book in which is regarded the special case $n = 1$ (then the solution Φ looked for is a complex-valued function).

an integer — basically calculable — which is called the *global index* of the problem (1). The case $n = 1$ is much simpler than the general $n > 1$: the *only* partial index $\varkappa_1$ is then given by the effective formula (6). In this case the canonical solution $X_0(z)$ is also given by effective Cauchy type integral formulas (see MUSHELIŠVILI [2]) and the general case (1′) is reducible by the piecewise holomorphic substitution

$$\Phi = X_0(z)\,\Psi\,, \qquad z \in D^+ \cup D^-\,, \tag{7}$$

to the special case

$$\Psi^+ = t^\varkappa \Psi^-\,, \qquad t \in \Gamma\,. \tag{8}$$

As is immediately seen, all solutions of the Riemann problem (8), normalized by condition (2) are given by the formulas

$$\begin{aligned} \Phi(z) &= z^{-\varkappa}[c_0 + \dots + c_{\varkappa-1} z^{\varkappa-1}]\,, \quad z \in D^-\,, \\ \Phi(z) &= c_0 + \dots + c_{\varkappa-1} z^{\varkappa-1}\,, \qquad z \in D^+\,, \end{aligned} \quad \text{for} \quad \varkappa \geqq 0\,. \tag{9}$$

For $\varkappa < 0$ no non-identically zero solution of (8), (2) exists. Thus in the one-dimensional case the global index $\varkappa$ completely describes the space of solutions of the homogeneous problem (1′). The general non-homogeneous problem (1) is reducible by the substitution (6) to

$$\Psi^+ = t^\varkappa \Psi^- + \tilde{h} \tag{10}$$

and one easily sees that (10)—(2) is always solvable for $\varkappa \geqq 0$ and admits $l = \varkappa$ linearly independent solutions. For $\varkappa < 0$ it is so iff the function $\tilde{h}$ satisfies $l' = |\varkappa| = -\varkappa$ linearly independent solvability conditions

$$\int_\Gamma t^k \tilde{h}\, dt = 0\,, \qquad k = 0, 1, \dots, l' - 1\,. \tag{11}$$

Notice that in both cases the formulas

$$l - l' = \varkappa, \tag{12}$$

$$l = \max(0, \varkappa) \qquad l' = \max(0, -\varkappa) \tag{12'}$$

hold. In the general case $n > 1$, once we know the canonical solution $X(z)$, the substitution (7) reduces the vector Riemann transmission problem to the *direct sum* of n scalar problems and, as easily seen, the formulas (12′) lead to the formulas (also given by Mushelišvili)

$$l = \sum_{\varkappa_i \geqq 0} \varkappa_i\,, \qquad l' = \sum_{\varkappa_i < 0} |\varkappa_i|\,. \tag{13}$$

The formula (12) holds without change. Thus the substitution (7) "uncouples" the various components of the unknown vector and theorem 5.2.1 can be viewed as some kind of structure and classification theorem what will be seen later more clearly. However, contrary to the scalar case, for $n > 1$ there is no general effectively calculable way to get the canonical solution from $G(t)$. More comments in this respect will be given later. The above discussion shows two remarkable features:

1. In the analytical discussion of the problem the integer $\varkappa$ — the global index — plays an essential role.

2. The *model* scalar case (8) has an explicite solution, which leads to simple effective formulas for the solution of the general scalar case. These formulas are very flexible and easily adjustable to discuss qualitative properties of the solutions (e.g. zeroes, behaviour at infinity, dependence on parameters etc.) and to various natural con-

structions on the *space of transmission problems* (direct and tensor products sums, dual or adjoint problems, glueing on cobordism classes). These constructions are naturally suggested by the geometric interpretations of the transmission problems in terms of fibre bundles. Thus a general *structure theorem* is established (theorem 5.2.1 the general case essentially reduces to the model case).

5.3. Relations of the Riemann-Hilbert transmission problem with other problems in analysis and geometry

The Riemann-Hilbert transmission problem (1) — as it stands — is related with a series of important problems in various branches of analysis and geometry. Here, for the completeness case, we give a list of some of these problems, naturally most of these topics are presented only in a signal form. In the next paragraph selected topics will be discussed in some detail.

First we recall that the classical and simplest way relating transmission problems with systems of one-dimensional singular integral equations goes via the Cauchy type integral representation of piecewise holomorphic functions in $D^+ \cup D^-$, normalized by (2)

$$\Phi(z) = \frac{1}{2\pi i} \int_\Gamma \frac{\varrho(t)}{t - z} \, dt \tag{14}$$

and the Sohockiĭ-Plemelj formulas for boundary values

$$\Phi^+(t) = \frac{\varrho}{2} + \frac{1}{2} S\varrho \qquad \Phi^- = -\frac{\varrho}{2} + \frac{1}{2} S\varrho \tag{15}$$

with the singular integral involution operator

$$S\varrho = \frac{1}{\pi i} \int_\Gamma \frac{\varrho(t)}{t - z} \, dt . \tag{16}$$

Inserting (15) into the transmission condition (1) results in a system of singular integral equations for the vector density ϱ

$$(E + G)\varrho + (E - G) S\varrho = 2h . \tag{17}$$

Converse holds also, since the left-hand side of (17) is the general form of the *principal* or *characteristic* part of a normal singular integral operator

$$L\varrho \equiv a\varrho + bS\varrho + \int_\Gamma k(t, \tau) \, \varrho(\tau) \, d\tau = h \tag{18}$$

with the weakly singular kernel $k(t, \tau)$ (roughly $|k(t, \tau)| \leqq c/(t - \tau)^\alpha$ $0 \leqq \alpha < 1$). The normality condition in Mushelišvili [2] states that $a + b$ and $a - b$ are invertible matrix functions, i.e. $a \mp b \in \Omega$ what is equivalent with the condition $G \in \Omega$ and is just the condition of ellipticity of the one-dimensional *pseudodifferential* operator $L\varrho$.

Now the ellipticity condition $a \mp b \in \Omega$ is essentially equivalent to the fact that the left-hand side of (18) generates a Fredholm operator in various natural functional spaces (say, the Hilbert space $L^2(\Gamma)$ of square integrable functions on Γ or many other Banach type spaces or even more general topological spaces). As a matter of fact (18) is a prototype of the general Fredholm operator in abstract space and in this sense the formula (12) is the prototype of all formulas for indices of Fredholm pseudodifferential operators expressed in terms of *topological data* of the problem considered.

Historically the general abstract theory of Fredholm operators developed from the analytical experience obtained from the study of the principal part system (17) in conjunction with the classical theory of Fredholm integral equations of the second kind or the corresponding abstract Riesz-Schauder theory.

Another chapter of functional analysis intimately connected with Riemann-Hilbert transmission problem is the theory of Fredholm pairs initiated by T. KATO in his monograph [1]. We recall that a pair (M, N) of closed subspaces of a Hilbert or, more generally, Banach space X is called a *Fredholm pair* if $M + N$ is closed and $\dim(M \cap N) = \alpha(M, N)$ and $\operatorname{codim}(M + N) = \beta(M, N)$ are both finite, with $M + N$ denoting the algebraic sum of $M + N$, i.e. the set of all elements z in X representable in the form $z = m + n$ with $m \in M$, $n \in N$.

The difference

$$\varkappa(M, N) = \alpha(M, N) - \beta(M, N)$$

is then called the *index* of the pair. The basic stability result in the book KATO [1] is that the index $\varkappa(M, N)$ is invariant, under small deformations of the pair (M, N) in the naturally topologized space of all Fredholm pairs. Now it is immediate to see that the transmission problem (1)—(2) with the normality condition $G \in \Omega$ supplies a natural example of a Fredholm pair with $M = L^+(\Gamma) = \{\Phi^+(t)\}$, $N = \{G\Phi^-(t)\}$ and $X = L^2(\Gamma)$, where $L^+(\Gamma)$ and $L^-(\Gamma)$ are subspaces in $L^2(\Gamma)$ of boundary values on Γ of vector functions holomorphic in D^+ and D^- (normalized by (2)). This observation was a starting point of my work (BOJARSKI [10]), where other examples of Fredholm pairs, naturally arising in the theory of elliptic operators on compact Riemann surfaces and compact manifolds of dimension greater than two, were given. The Riemann-Hilbert transmission problem corresponds to the simplest case of Cauchy-Riemann equations. One of the geometric results obtained along this way is the fact that any elliptic operator A on a compact manifold M of dimension n, represented as a sum of two cobordant submanifolds M^+ and M^- with the common boundary Γ, assumed to be a smooth submanifold of M of codimension 1, naturally produces in a space $X(\Gamma)$ of vector functions on Γ a pair of Fredholm subspaces generated by the Cauchy data on Γ of solutions of the homogeneous system $Au = 0$ in M^+ and M^- correspondingly.

The index of this Fredholm pair is equal to the index of A on the closed manifold M. Thus the problem of determination the index of A on M is reduced to the problem of finding the index of a Fredholm pair on Γ, which is a closed manifold of dimension $n - 1$. Now in BOJARSKI [10] it is shown that generally the calculation of the index of a Fredholm pair is reducible to the discussion of a Fredholm operator in the corresponding abstract space. The constructions suggested above are performed with the help of projectors on Cauchy data of elliptic differential equations and it is rather clear that the final results does not lead out from the class of pseudodifferential operators. This creates a basis for an inductive, with respect to the dimension, approach to the general index problem. It is also clear in the constructions sketched above the cobordism classes represented by M^+ and M^- summing up to the trivial class $M = M^+ \cup M^-$ of a manifold without boundary can be extended to general case. In particular, as is well-known from the decomposition of M into *elementary cobordism*, the homotopical and homological structure of the summands M^+ and M^- can be much simpler than that of M. It should be also remarked that this program of building an inductive proof of the index theorem has been performed for the case of the elliptic de Rham complex — theorem of de Rham — in which case the index appears as Euler characteristic (see SAMELSON [1]). One more remark should be made here: in the above sketched example the Fredholm pair was produced directly from Cauchy data. In the Riemann-Hilbert transmission problem the Fredholm pair is obtained via the transformation of the Cauchy data with the homotopically non-trivial operator G, what creates an additional *twisting* of the initial pair of Cauchy data. It is clear that this *twisting* can be done by an application of much more general operators than multiplication by $G(t)$, e.g. we can use instead of G any pseudodifferential operator, elliptic and satisfying some comutator type relations as shown in the abstract setting in BOJARSKI [10]. In general the action of G will be *only partly* responsible for the index of the resulting pair. Explicite and simple examples of this situation can be already seen in classical formulas (see I. N. VEKUA [5]) for the index of transmission problems on Riemann surfaces or boundary value problems in multiconnected plane domains.[1]) Further comments on this topic will be made after we discuss the fibre bundle interpretation of the transmission problems.

We recall that the holomorphic vector function $\Phi(z)$ — solution of (1) regular at ∞ — represents the piecewise holomorphic section of the product bundle $F = (D^+ \cup D^-) \times \times \mathbb{C}^n$ over $D^+ \cup D^-$. However if we glue together $D^+ \times \mathbb{C}^n$ with $D^- \times \mathbb{C}^n$ along Γ by identifying $(t, v) = (t, G(t)\, v)$ we obtain in general a nontrivial vector bundle F_G over the projective line. The piecewise holomorphic section of F satisfying the transmission relation (1) represents a holomorphic section of F_G and conversely, any holomorphic section of F_G is uniquely represented by piecewise holomorphic section of F. Any complex vector bundle over $\mathbb{P}^1$ can be obtained by such an identification process if D^+ is the unit disc $|z| < 1$ and D^- its exterior. The above remarks, obvious to anyone familiar with rudiments of bundle theory were noticed long ago and used in RÖHRL [1]. However in my opinion the fibre bundle point of view has not been sufficiently exploited in the theory of Riemann-Hilbert problem and vice versa. It seems worthwhile to bring again into light the interrelations between the two approaches. First steps in this direction will be done in what follows.

Now recent interest in complex fibre bundles over projective spaces — caused by the physical interpretations of complex geometry — instanton bundles etc. — opens new areas of applications of the Riemann-Hilbert transmission problem. Another domain of physical interpretations of the Riemann-Hilbert problem appeared in complex quantum field models by SATO [1], where he goes into the study of the Riemann problem for piecewise continuous, transmission coefficients $G(t)$ on the real line and the investigations of the solution as functions of parameters discribing discontinuities of G. This brings the investigation close to the original Riemann's formulation of the problem as it arose from the study of ordinary differential equations in the complex domain with algebraic singularities.

Another direction of investigations of Riemann transmission problems is connected with the study of families of transmission problems. The topological aspects of these investigations — when the family is parametrized by points of a compact space X — create a convenient and elastic tool in the study of structural properties of ring $K(X)$ of virtual complex fibre bundles over X. Families of Fredholm operators in an abstract Hilbert space play an essential role in Atiyah-Bott treatment of the complex Bott periodicity (see ATIYAH [1], ATIYAH and BOTT [1]). Actually it is worthwhile to remark here that families of Fredholm pairs — systematic studies of these as far as I

[1]) These general ideas on Fredholm pairs and index problems were included in some colloquium lectures of mine given in 1978—79 at the universities of Berlin (West), Bielefeld, Konstanz and Darmstadt.

know have not been performed yet — create apparently a more natural setting for the discussion of structure of $K(X)$ than the theory of Fredholm operators. Stability properties of families of Riemann-Hilbert problems and structurally generic classifications properties will be discussed in some detail in the next section. This brings us very close to a series of recent and current investigations in complex geometry.

These stability and genericity problems will also show the relations of Riemann transmission problems with a chapter of elliptic theory of partial differential equations — namely, the Liouville type theorems for simple systems of generalized analytic vector functions and systems with piecewise continuous coefficients (which is equivalent to elliptic systems on piecewise smooth manifolds).

5.4. Stability theorems on the Riemann-Hilbert transmission problem[1])

For the study of the totality of all Riemann-Hilbert transmission problems (1′) (for D^+ and Γ fixed this is the space Ω defined above) it is natural and convenient to introduce the equivalence relation $G_1 \sim G_2$ if there exists in $D^+ \cup D^-$ a piecewise holomorphic $n \times n$-matrix $X(z)$, invertible and continuous up to the boundary Γ, regular at infinity, $X(\infty) = E$ such that for each $t \in \Gamma$ holds

$$G_2(t) = X^+(t)\, G_1(t)\, X^-(t)\,. \tag{19}$$

Theorem 5.4.1. *Every $G \in \Omega$ is equivalent to a unique diagonal matrix $\Lambda(t)$ with the diagonal entries $t^{\varkappa_i}$, $i = 1, \dots, n$, such that $\varkappa_1 \geqq \varkappa_2 \geqq \dots \geqq \varkappa_n$. Thus every $G \in \Omega$ can be always represented in the form*

$$G(t) = X^+(t)\, \Lambda(t)\, X^-(t)\,, \qquad t \in \Gamma, \tag{20}$$

with $X(z)$ and $\Lambda(z)$ satisfying the requirements formulated above.

Representation (20) is known in the literature also as *factorization* of $G \in \Omega$ (with respect to the Riemann-Hilbert transmission problem).

Theorem 5.4.1 is equivalent to the theorem 5.2.1. There are several proofs for theorems 5.4.1, 5.2.1 available in the literature varying in the methods and special detailed assumptions on Γ, G or the class of admissible $X(z)$. Various aspects of the factorization problem could be traced in an immense literature — ranging from the papers of Birkhoff [1] via the work of Georgian school and numerous papers on boundary behaviour of Cauchy-type integrals up to the newest work of Calderón and others on the topic. Only few of these are listed in the references.

For simplicity assume now that Γ is a smooth Jordan curve — say the boundary of unit disc D^+. Then the topological space Ω decomposes into a countable number of open components $\Omega^\varkappa = \{G \in \Omega, \Delta_\Gamma \arg\det G(t) = 2\pi\varkappa\}$, $\Omega = \bigcup_\varkappa \Omega^\varkappa$, $\varkappa \in \mathbb{Z}$. G_1 and G_2 equivalent under (19) belong to the same component $\Omega^\varkappa$. However by theorem 5.4.1 the equivalence class can be uniquely paramatrized by the sequence $\varkappa_i$, $\varkappa_1 \geqq \dots \geqq \varkappa_n$ — the partial indices — and the Mushelišvili formula (5) holds. Thus we can write

$$\Omega^\varkappa = \bigcup_{\{\varkappa_i\}} \Omega_{\{\varkappa_1, \dots, \varkappa_n\}} \tag{21}$$

and we will call the classes $\Omega_{\{\varkappa_1, \dots, \varkappa_n\}}$ *leaves* of the decomposition (21).

[1]) The material in 5.4. is essentially contained in Bojarski [2, 7, 8].

The essential features of the decomposition (21) are described in the following theorems.

Theorem 5.4.2. *If $\varkappa = kn + r$, with k and r integers, $0 \leqq r < n$, then the only open and dense leaf in* (21) *is the one indexed by the sequence* $(k+1, k+1, \ldots, k+1, k, \ldots, k)$ *or equivalently the one with partial indices satisfying*

$$|\varkappa_i - \varkappa_j| \leqq 1\,, \qquad i, j = 1, 2, \ldots, n\,. \tag{22}$$

Theorem 5.4.3. *For any pair of integers a, b, $a \leqq b$, the unions*

$$\bigcup_{a \leqq \varkappa_i \leqq b} \Omega_{\{\varkappa_i\}} \tag{23}$$

are open subsets of Ω (and of $\Omega^\varkappa$ if $\sum \varkappa_i = \varkappa$).

Theorem 5.4.4. *If Γ is the Jordan curve then the leaves $\Omega_{\{\varkappa_i\}}$ are connected.*

Theorems 5.4.2 and 5.4.3 are the main *stability theorems* for Riemann-Hilbert transmission problems. Theorem 5.4.2 also describes a *typical* or *generic* situation — i.e. what happens on a dense open subset.

We recall that in view of formulas (13) partial indices $\{\varkappa_i\}$ completely describe the solvability character of the Riemann transmission problem. They also describe structural properties of the set of solutions of homogeneous problem. The global index $\varkappa$ is the only topological (for D^+ simply connected, homotopical) invariant of the problem in the sense that each $\Omega^\varkappa$ is connected. Thus in view of theorem 5.4.2 one can say that *generically the topological invariant completely describes the qualitative character of the solutions of Riemann problem.* We remark also that for $n > 1$ formulas (12′) are not true. However, we have

Corollary. *For an open and dense set of G the formulas* (12′) *hold.*

In view of theorem 5.4.1 the partial indices are well-defined functions $\varkappa_i = \varkappa_i(G)$. From what was said above it follows that these functions are not continuous on Ω. Theorem 5.4.2 fully describes the cases when they are continuous. Theorem 5.4.3 puts some restrictions on the possible *jumps* of $\varkappa_i(G)$ if the argument G is deformed slightly. Just these "obligatory jumps" make it impossible to construct any effective and general formulas for partial indices or for the canonical solution. Theorem 5.4.3 may be sharpened so as to give a full description of a sufficiently *small* neighbourhood of any class $\Omega_{\{\varkappa_i\}}$. To this aim let us call the deformation $\Omega_{\{\varkappa_i'\}}$ of the class $\Omega\{\varkappa_i\}$ *elementary*, if $\varkappa_i' = \varkappa_i$ except for two indices p and q, $p < q$, for which we have $x_p' = x_p - 1$ $x_q' = x_q + 1$.

Theorem 5.4.3′. *In any neighbourhood*[1]) *of the leaf $\Omega_{\{\varkappa_i\}}$ there exist leaves, that can be obtained from $\Omega_{\{\varkappa_i\}}$ by a finite number of elementary deformations. Sufficiently small neighbourhood of $\Omega_{\{\varkappa_i\}}$ consists of such leaves only. We notice that under elementary deformations the "dispersion" of the system of partial indices* $d(\varkappa) = \sum_{i<j} (\varkappa_i - \varkappa_j) = 2 \sum_k \times ((n+1)/2 - k)\, \varkappa_k$ *does not increase and stays constant iff all $\varkappa_i$ do not change.*

Essentially the proof of the stability results reflect the following simple abstract principle:

Proposition. *In the space of Fredholm operators F in an abstract space H on the dense and open subset of F the formulas hold:*

$$l(F) = \max(0, \varkappa) \qquad l'(F) = l(F^*) = \max(0, -\varkappa)\,.$$

[1]) Here and at other places of the text the loosely used topological notions are to be understood in the simplest way, e.g. the *neighbourhood* of the leaf is represented by a neighbourhood of a representative of a leaf in Ω etc. In the necessarily rather loose discussion of this talk — this negligence should cause no trouble.

Recall that $l(F)$ — the dimension of the ker F — and $l'(F)$ — the dimension of the coker F or ker F^* of the adjoint operator — are upper semicontinuous functions of F. $\varkappa = \varkappa(F)$ is the index of F which is a continuous function of F (i.e. locally constant).

It follows that any equivalent reduction of the transmission problem to a linear operator equation for which the Fredholm theory holds can be used as tool in the proof of stability theorems. In the sketch of the proofs below a special two-dimensional system of Fredholm integral equations of the second kind is used. This enjoys some special algebraic and complex analytic properties. In the process of this proof new connections appear between Riemann problem and Liouville type theorems for certain linear elliptic systems in the plane with piecewise continuous coefficients. Thus our list of interrelations of the Riemann transmission problem will be extended.

Before going to the proof we will describe the interpretation of the stability and genericity results in terms of structure theorems for holomorphic fibre bundles and deformations of such bundles.

Theorem 5.4.1′. *Every continuous vector bundle over $\mathbb{P}^1$ has an essentially unique holomorphic structure. As a holomorphic vector bundle it is decomposable into a direct sum of line bundles. The decomposition is essentially unique.*

Strictly speaking the method of proof suggested later requires some restrictions on the assumed smoothness of the bundle. Hölder continuity will be enough anyway. If we want to work with a priori continuous structures only — the factorization problem has to be slightly modified e.g. the factors in (22) will not be continuous up to the boundary. Qualitatively however situation does not change.

The second part of theorem 5.4.1 has a rather algebraic character and it has been recognized long ago (see Gohberg and Kreĭn [1]) that it follows from a *weak* version of factorization or equivalently, canonical solution.

Lemma 5.4.1. *The homogeneous problem* (1′) *always admits a matrix solution $\tilde{X}(z)$, meromorphic in the closed complex plane $\mathbb{P}^1$, and such that*

$$\det \tilde{X}(z) \not\equiv 0 \quad in \quad D^+ \cup D^- .$$

The vector bundle formulation of this part of the theorem is known as A. Grothendieck decomposition theorem. The first part is directly connected with the full problem of factorization and is always connected with some non-trivial analysis (Fredholm weakly singular or singular integral equations, analytical properties of Cauchy type integrals).

Smooth or continuous vector bundles ξ over the sphere (or $\mathbb{P}^1$) are classified under topological bundle isomorphisms by their first Chern class or the corresponding Chern number — which in our situation may be identified with the global index $\varkappa$ (6) of the identifying isomorphism $G: \Gamma \to GL(n, \mathbb{C})$. This identifying isomorphism produces ξ from two copies of the product bundles over D^+ and D^-. For line bundles the topological and holomorphic classifications coincide and the partial indices $\varkappa_i$ appear as Chern classes of the line components in the diagonal factorization (20). For vector bundles of rank $n > 1$ the topological and holomorphic classifications do not coincide, the latter being given by the partial indices or classes $\Omega_{\{\varkappa_i\}}$. The description of the bundle ξ over $\mathbb{P}^1$ by the identifying isomorphism $G \in \Omega$ paramatrizes the set of all complex bundles over $\mathbb{P}^1$ with elements of a natural topological space Ω (or correspondingly the holomorphic isomorphism classes of bundles with elements of the quotient space $\Omega/\sim$ of Ω by the equivalence relation (5), and allows to view Ω as a *universal deformation space* of all vector bundles over $\mathbb{P}^1$. This parametrization transfers the topology of Ω to the set of vector bundles over $\mathbb{P}^1$ and identifies the set of isomorphism classes of bundles with $\Omega/\sim$. A holomorphic bundle ξ over $\mathbb{P}^1$ is said to be *stable* if it has a neighbourhood U such that every bundle $\eta \in U$ is holomorphically isomorphic with ξ. Theorem 5.4.2 then leads immediately to the

Theorem 5.4.2'. *A holomorphic bundle ξ of rank n over $\mathbb{P}^1$ is stable if the partial indices of ξ, $\{\varkappa_i(\xi)\}$ satisfy* (22) *and are defined as indicated in theorem* 5.4.2 *by the first Chern number of ξ. The set of stable bundles is open and dense.*

Theorems 5.4.3 and 5.4.3' give some further information on the stability properties of holomorphic vector bundles over $\mathbb{P}^1$. Theorem 5.4.3' in particular gives the complete description up to isomorphism of all bundles which can be obtained by an arbitrarily small deformation of a given bundle, i.e. it describes the basis of the *quotient topology* on the set of equivalence classes.

Summarizing the above discussion we can say that though for $n > 1$ the Chern class $\varkappa$ does not describe the holomorphic isomorphism class of the bundle — *generically* i.e. on an open and dense set in the space of *all possible* bundles over $\mathbb{P}^1$ — the Chern class of the bundle — i.e. the only topological invariant of the bundle — completely describes the holomorphic structure of the bundle. In particular all *stable* situations are fully described by the topological invariants. This reflects a very general principle in complex analysis — and in analysis and geometry in general — that stable situations can be described by a considerably smaller set of invariants. Later we discuss the Riemann-Roch theorem from this point of view. In more complicated situations — e.g. for bundles over more general manifolds — in holomorphically stable cases, despite *discrete* invariants, also *continuous* invariants-moduli can appear.

Theorem 5.4.2' corresponds to general qualitative part of Grauert-Müllich stability theorem in the theory of complex bundles over projective space (see Grauert and Müllich [1]). It seems that theorems 5.4.3 and 5.4.3' give more precise information on general deformations of holomorphic bundles over $\mathbb{P}^1$ than the Grauert-Müllich theory. Surely, in general the analysis of *generic* cases can give only a very rough information about the deformation structure of the problem, and in any specific situation when only a subclass of general deformations is admitted, more detailed results can be obtained. However, in these cases, as a rule, much more detailed information on the structure of the bundle under consideration is needed.

We propose to study the factorization problem with the help of a special system of first order elliptic equations of the form

$$\frac{\partial u}{\partial \bar{z}} = \mathcal{M} u \tag{24}$$

for a vector function $u = (u_1, \dots, u_n)$ and the corresponding Vekua integral equation

$$u(z) + \frac{1}{\pi} \iint_D \frac{\mathcal{M}(t)\, u(t)}{t - z} \, dG_t = \varphi(z) \qquad (D \equiv D^+) . \tag{25}$$

Here $\mathcal{M}$ is an $n \times n$-matrix function with entries in $L_p(D)$, $p > 2$, and $\varphi(z)$ is a holomorphic vector in $L_p(D)$. System (24) is a special case of systems of elliptic equations for a generalized analytic vector in the sense of I. N. Vekua [5]. The transition from (24) to (25) and vice versa is performed by the Pompeiu-Vekua representation (see I. N. Vekua [5])

$$w(z) = -\frac{1}{\pi} \iint_D \frac{\omega(t)}{t - z} + \varphi(z) , \qquad \omega = \frac{\partial w}{\partial \bar{z}} , \tag{26}$$

of a vector function $w(z)$ belonging to the Sobolev class $W_p^1(D)$. $\varphi(z)$ — in (26) is a holomorphic vector function (in $W_p^1(D)$). If $\varphi(z)$ in (26) is the restriction of an entire function or a meromorphic function in the complex plane with poles outside of D, then $w(z)$ in (26) admits a holomorphic (or meromorphic) extension outside of D. Thus (26) gives the construction of solutions of non-homogeneous systems of Cauchy-Riemann equations, which have holomorphic (or meromorphic) extensions outside of D. Exactly the same holds for (24) and (25).

Notice that the system (25) is a weakly singular system of Fredholm equations of the second kind and the usual Fredholm-Riesz-Schauder theory for such systems holds without restrictions. In particular modulo finite-dimensional solvability conditions of the form $\int_D (\varphi, v_j)\, dG_t = 0, j = 1, \dots, N$, system (25) is solvable. Thus we get a convenient way of generating solutions of (24) satisfying various additional conditions. Notice also that solutions of (24) in a domain D form a module over the ring of holomorphic functions in D. Correspondingly, if u(z) is a solution of a homogeneous system of integral equations (25) ($\varphi \equiv 0$) $u(z)$ admits a holomorphic extension outside of D, vanishing at infinity and if, for a polynomial $f(z)$, $f(z)\, u(z) \to 0$ for $z \to \infty$ then $f(z)\, u(z)$ is again a solution of homogeneous integral equation (25). These facts make (25) a convenient and flexible tool in the study of some problems in complex analysis.

The following well-known lemma is useful. For simplicity let $\Gamma = \{z, |z| = 1\}$.

Lemma 5.4.2. *Let $P(t)$ be a non-singular matrix function on Γ and assume, for simplicity, that $P(t)$ is Hölder-continuous with an exponent $\alpha > \frac{1}{2}$. Then a necessary and sufficient condition for the existence of a matrix function $P(z)$, $|z| \leqq 1$, extending $P(t)$ onto the unit disc D and such that $P^{-1}(z)$ exists and $P_{\bar z} \in L_p(D)$ for some $p > 2$ is the equality*

$$\Delta_\Gamma \arg \det P(t) = 0\,. \tag{27}$$

Let $C(t)$ be any matrix function on Γ, $C(t) \in \Omega$, which has a holomorphic extension for $|t| < 1$, not necessarily non-singular everywhere, such that

$$\Delta_\Gamma \arg \det (G^{-1}C) = 0\,. \tag{28}$$

Let $P(z)$ be any extension of $G^{-1}C$ given by the lemma, and $\Phi(z)$ — any holomorphic solution of homogeneous problem (1'). Consider the substitution

$$\begin{aligned} u(z) &= P(z)\, \Phi(z)\,, \qquad & z \in D^+\,, \\ u(z) &= \Phi(z)\,, & z \in D^-\,. \end{aligned} \tag{29}$$

$\Phi(z)$ in (29) is holomorphic in $D^+ \cup D^-$ iff $u(z)$ is a solution of the system

$$\frac{\partial u}{\partial \bar z} = \mathcal{M} u \tag{30}$$

with

$$\mathcal{M} = \begin{cases} P_{\bar z} P^{-1} & \text{for } z \in D^+\,, \\ 0 & \text{for } z \in D^-\,, \end{cases} \qquad \mathcal{M} \in L_p(D)\,.$$

We remark that what was said above applies not only to vector solutions of (24) and (25) but to matrix solutions (e.g. systems of vector solutions) as well. For the case of $n \times n$-matrix solutions of (24) we notice also that $\Delta = \det u$ satisfies the Vekua system

$$\frac{\partial \Delta}{\partial \bar z} = \operatorname{Sp} \mathcal{M} \Delta \tag{31}$$

for generalized analytic functions. Here Sp $\mathcal{M}$ is the trace of $\mathcal{M}$. Thus Δ has only isolated zeroes, unless $\Delta \not\equiv 0$ identically. This leads us directly to the following proof of the weak factorization in lemma 5.4.1.

Proof of lemma 5.4.1. Let $\varphi(t)$ be any polynomial $n \times n$- matrix with $\det \varphi(t) \not\equiv 0$ such that (25) has a matrix solution $\tilde{u}(z)$. By (31) the matrix $\tilde{u}(z)$ will be not invertible only at isolated points of the complex plane. $\tilde{u}(z)$ is continuous on Γ. Produce $\tilde{\Phi}(z)$ by inversion of formulas (29). $\tilde{\Phi}(z)$ will be holomorphic in $D^+ \cup D^-$ and will satisfy on Γ the transmission condition

$$G^{-1}C\tilde{\Phi}^+ = \tilde{\Phi}^- .$$

The piecewise holomorphic vector

$$\Phi(z) = C(z)\, \tilde{\Phi}(z) , \qquad z \in D^+ ,$$

$$\Phi(z) = \tilde{\Phi}(z) , \qquad z \in D^- ,$$

will satisfy the conditions of the lemma.

To obtain the canonical matrix solution or the strong factorization lemma we should *transfer all (isolated zeroes and poles) of* $\det \Phi(z)$ *to* ∞ *and maximally reduce the dependence of the columns of* Φ *at infinity.* As was said above this can be done by a direct elementary but a little tedious process and because of lack of place, cannot be given here. In the discussion of stability we will restrict our attention to the simplest case, when global index $\varkappa = k\, n$.

In this case we take for $C(t)$ the diagonal matrix $\Lambda(z) = z^k E$ — multiple of identity. The corresponding distinguished leaf in the decomposition (21) we denote by Ω_k.

Lemma 5.4.3. $G(t) \in \Omega_k$ *if the Liouville theorem holds for system* (30).

Proof. We recall that the Liouville theorem for (30) says that the only solution u of (30) in the whole plane satisfying $u(z) \to 0$ for $z \to \infty$ is $u \equiv 0$. An equivalent statement is that the homogeneous Vekua equation (25) has only the trivial solution $u \equiv 0$. First we prove that Liouville implies $G \in \Omega_k$. Consider the matrix equation (25)

$$B(z) + \frac{1}{\pi} \iint\limits_{D} \frac{\mathcal{M}B}{t - z}\, \mathrm{d}G_t = E . \tag{32}$$

By assumption the equation (32) admits a solution. By (31) and known properties of generalized analytic functions it follows that any solution of (32) is invertible for each z. Define the piecewise continuous matrix function by the formulas

$$X(z) = B^{-1}(z) \qquad \text{for} \quad z \in D^- ,$$

$$X(z) = P^{-1}(z)\, B(z) \quad \text{for} \quad z \in D^+ ,$$

$X(z)$ is piecewise holomorphic. For $z \in D^-$ this is obvious. For $z \in D^+$ we have

$$X_{\bar{z}} = P_{\bar{z}}^{-1} B + P^{-1} B_{\bar{z}} = [P_{\bar{z}}^{-1} + P^{-1} P_{\bar{z}} P^{-1}]\, B \equiv 0 .$$

For $t \in \Gamma$ we have

$$G = P^{-1}\Lambda = P^{-1}B\Lambda B^{-1} = X^+\Lambda X^- \quad \text{or} \quad G \in \Omega_k .$$

Conversely, assume that the homogeneous integral equation (25) admits a non-zero solution $u_0(z)$. Then $u_0(\infty) = 0$. Then the vector function

$$\Phi(z) = z^{-k} u_0(z) , \qquad z \in D^- ,$$

$$\Phi(z) = P^{-1}(z)\, u_0(z) , \qquad z \in D^+ ,$$

is piecewise holomorphic and has zero of order $\geqq k+1$ at ∞. This implies that $\varkappa_1(G) \geqq k+1$ and the matrix $G(t)$ cannot be equivalent to $\Lambda(z) = z^k E$.

Lemma 5.4.2 implies the openness of Ω_k: really it is obvious that the set of coefficients $\mathcal{M}$ in (25) for which the Liouville theorem holds is open in $L_p(D)$.

Now the transition from $G \in \Omega_\pm$ to $\mathcal{M} \in L_p(D)$ can be chosen as a continuous operation. The fact that each distinguished leaf for which (22) holds is dense in Ω essentially follows from the explicite construction for the case $n = 2$: consider the diagonal matrix $\Lambda(t) = \begin{pmatrix} t^{\varkappa_1} & 0 \\ 0 & t^{\varkappa_2} \end{pmatrix}$ such that

$$\varkappa_1 > \varkappa_2 + 1 \,. \tag{33}$$

Let $\Lambda_\varepsilon(t) = \begin{pmatrix} t^{\varkappa_1} & \varepsilon t^{\varkappa_2+1} \\ \varepsilon t^{\varkappa_1-1} & t^{\varkappa_2} \end{pmatrix}$ for $\varepsilon > 0$. We have

$$\Lambda_\varepsilon(t) = \begin{pmatrix} t & \varepsilon - \dfrac{1}{\varepsilon} \\ \varepsilon & 0 \end{pmatrix} \begin{pmatrix} t^{\varkappa_1-1} & 0 \\ 0 & t^{\varkappa_2+1} \end{pmatrix} \begin{pmatrix} 1 & \dfrac{1}{\varepsilon} t^{\varkappa_2+1-\varkappa_1} \\ 0 & 1 \end{pmatrix} \tag{34}$$

that shows that an arbitrarily small deformation diminishes the gap between $\varkappa_1$ and $\varkappa_2$. After $(\varkappa_1 - \varkappa_2)/2$ steps of the form (34) we will obtain the distinguished equivalence class for which $|\varkappa_1 - \varkappa_2| \leqq 1$.

The above construction also appears as a decisive step in the full proof of theorem 5.4.3'.

The proof of theorem 5.4.4 runs via approximations by polynomials and then uses some elementary transformations of polynomial matrices. As another application of the point of view explained above we discuss now briefly the Riemann-Roch problem. Let be given a divisor $\mathcal{M}$ on a compact Riemann surface M of genus g. Riemann's original question was to determine the dimension $r(\mathcal{M})$ of all meromorphic functions φ on M such that the principal divisor (φ) is a multiple of $(\mathcal{M})$. For the general case of the pair $(M, \mathcal{M})$ Riemann gave only an inequality

$$r(\mathcal{M}^{-1}) \geqq d(\mathcal{M}) - g + 1 \tag{35}$$

for $r(\mathcal{M}^{-1})$ in terms of genus g of M and the degree $d(\mathcal{M})$ of $\mathcal{M}$.

As is known $d(\mathcal{M})$ is defined by the first Chern class of the complex holomorphic line bundle $\xi_{\mathcal{M}}$ defined by the divisor. It was recognized later that instead of an exact formula for $r(\mathcal{M}^{-1})$ in terms of topological data of the Chern class of ξ and cohomology of M it is possible to prove the Riemann-Roch formula

$$r(\mathcal{M}^{-1}) - i(\mathcal{M}) = d(\mathcal{M}) - g + 1 \tag{36}$$

where $i(\mathcal{M})$ is the dimension of meromorphic differentials on M, which are multiples of $\mathcal{M}$. (36) is an index formula for a differential elliptic operator on M. Now in accordance with the ideas discussed above it is possible to consider the space $\mathcal{D}(M)$ of all divisors (or divisor classes) on a fixed Riemann surface M, or, equivalently, the space $\xi(M)$ of all (equivalence classes) of holomorphic complex line bundles over M, which can be *naturally* given a structure of a topological space (say, via cocycle definition of line bundles) $r(\mathcal{M}^{-1})$ and $i(\mathcal{M})$ — can be studied then as functions on $\xi(M)$. We have then the theorem

Theorem 5.4.5. *Generically, i.e. on an open and dense set on $\xi(M)$, the formula*

$$r(\mathcal{M}^{-1}) = \begin{cases} d(\mathcal{M}) - g + 1 & \text{for} \quad d(\mathcal{M}) \geqq g - 1\,, \\ 0 & \text{for} \quad d(\mathcal{M}) < g - 1 \end{cases} \tag{37}$$

holds.

Thus, in particular, in all *stable* cases the program of Riemann can be realized.

The proof can be obtained from the general principle formulated above. There are several known ways to reduce the Riemann problem to a one-dimensional singular integral equation, to a Fredholm operator on M, or to a Riemann-Hilbert transmission problem. The details can not be given here. Analogous remarks are valid for Riemann-Roch type problems on higher-dimensional manifolds etc. but a detailed proof in each case can be given only in a framework of a proper theory of the problem discussed.

This is especially clearly seen in the study of the part of the theorem reffering to *density*.

We remark at the end that the *twisting* refered to in 5.3. when we discussed the various generalizations of the Riemann-Hilbert transmission problem, can also be performed with the independent variable t. Thus we can discuss from the point of view propagated above the transmission problem *with a shift*, say

$$\Phi^+(t) = G(t)\, \Phi^-(\alpha(t)) . \tag{38}$$

If $\alpha = \alpha(t)$ is an orientation preserving diffeomorphism of Γ the result of the discussion are analogous to the case $\alpha(t) \equiv t$ and, generally speaking, the study of the problem (38) can be reduced to the study of Riemann transmission problem without shift for the elliptic system of the form

$$U_{\bar{z}} - q(z)\, U_z = A\mathcal{M}$$

with $q(z)$ the multiple of the identity matrix $q(z) \equiv \tilde{q}(z)\, E$, $|\tilde{q}(z)| \leqq q_0 < 1$.

Simplest generalizations to several dimensions or to cases mentioned in 5.3. modulo various technical difficulties, are rather immediate.

Further references

Mushelišvili and N. P. Vekua [1], I. N. Vekua [2, 9], Plemelj [1], Bojarski [5, 7, 8], Gahov [4], Hvedelidze and Mandžavidze [1], Okonek, Schneider, and Spindler [1].

6. On Some Complex-Analytic Methods in Differential Geometry

J. Ławrynowicz and G. Tsagas

6.1. Introduction

The possibility and importance of employing complex-analytic methods in differential geometry is so wide that it presents a real difficulty to give a survey of them. They play an important role in the monograph by Kobayashi and Nomizu [1], and some aspects of them are treated in a greater detail or more generally in the monographs by Wells [1] and Yano [1]. The global aspects are treated specially by Chern [2] and a very successful method of biholomorphic invariants is given by Kobayashi [2]. For the problem of deformations one may consult the books by Morrow and Kodaira [1], and by Sundararaman [1]. Of course, this list of books is far from being complete.

In this exposition we make a special choice of material following the personal investigations of the authors. In fact, we confine ourselves to few examples. We start with some elementary examples and the Bernstein problem. Then we move to an important part of the theory of biholomorphic invariants, connected with a series of papers by Kobayashi; cf. [4]. After treating the Kobayashi pseudodistance as a capacity we concentrate on deformations of potentials and Kähler metrics related to biholomorphically invariant capacities, connected with the papers by Chern, Levine and Nirenberg [1], Ławrynowicz [1], and Gaveau and Ławrynowicz [1]. Finally, we discuss Pierzchalski's results on homotopic deformation of biholomorphic mappings (Pierzchalski [1]), exploring deep results on pseudoconformal mappings due to Yano and Ishihara [1].

6.2. Elementary examples and the Bernstein problem

A curve in the complex plane $\mathbb{C}$ is said to be *unicursal algebraic* if it can be represented by a parametric equation of the form

$$z(t) = \sum_{j=0}^{m} A_j t^j \,/\, \sum_{k=0}^{n} B_k t^k \equiv A(t)/B(t); \qquad A_m \neq 0,\ B_n \neq 0\,. \tag{1}$$

Eliminating t we obtain an equation $W(z, \bar{z}) = 0$; the degree of the polynomial W is called the *order* of the curve. If the polynomials A and B are relatively prime and B has p zeroes, real and conjugate complex (the mutually complex zeroes are calculated separately and a q-tuple zero is calculated q times), then the order equals $m + n - p$ for $m > n$ and $2n - p$ for $m \leqq n$ (cf. e.g. Deaux [1, pp. 89–91]). A plane algebraic curve represented by a Cartesian equation with real coefficients and passing k times through one of the circular points at infinity, also passes k times through the other circular point at infinity, the curve is said to be *k-circular*. A curve can be k-circular without being unicursal, and conversely. The curve (1) is $(n - p)$-circular.

A real or imaginary point (in the sense of homogeneous coordinates) is called a *focus* of a plane curve if the two isotropic (minimal) lines radiating from this point are tangents to the curve. When the curve is k-circular, a focus is said to be *singular* if the isotropic lines radiating from this point have their meet-points at the circular points. For example, the centre of a circle is its singular focus.

Let $B = B_0(B_1 + \mathrm{i}B_2)$, where B_0 is a polynomial of the p-th order with real coefficients, and B_1 and B_2 have real coefficients and are relatively prime; such a decomposition is always possible. If the curve (1) has no multiple points in the finite part of the plane on the isotropic lines radiating from the origin, then the origin is: a) an ordinary focus if the equation $A(t) = 0$ has a zero which is at least double and which is not a root of $B_1(t) - \mathrm{i}B_2(t) = 0$, b) a singular focus, if the equations $A(t) = 0$ and $B_1(t) - \mathrm{i}B_2(t) = 0$ have at least one common root.

Consider, for example, an equation of the form

$$z(t) = (A_0 + 2A_1t + A_2t^2)/(B_0 + 2B_1t + B_2t^2)\,, \tag{2}$$

where the numerator and the denominator are polynomials relatively prime. Let us distinguish three cases.

a) If the equation $B_0 + 2B_1t + B_2t^2 = 0$ has two real zeroes, then the curve (2) is called a *conic*. After dividing, if necessary, the numerator and the denominator by a constant, we can assume that B_0, B_1, and B_2 are real. Exclude the trivial case

$$\begin{vmatrix} A_0 & \bar{A}_0 & B_0 \\ A_1 & \bar{A}_1 & B_1 \\ A_2 & \bar{A}_2 & B_2 \end{vmatrix} = 0\,.$$

The equation (2) represents an *ellipse*, a *hyperbola* or a *parabola* according to the discriminant $\Delta_B = B_1^2 - B_0B_2$ being positive, negative, or zero. Since the ellipse (which does not reduce to a circle), the hyperbola, and the parabola are not circular curves, each of them has two ordinary foci; we derive them from the equation $\Delta_Bz^2 + \Delta_{AB}z + \Delta_A = 0$, where $\Delta_A = A_1^2 - A_0A_2$ and $\Delta_{AB} = A_0B_2 - 2A_1B_1 + A_2B_0$. The point $-\frac{1}{2}\Delta_{AB}/\Delta_B$ is called the *centre* of the conic.

b) If the equation $B_0 + 2B_1t + B_2t^2 = 0$ has exactly one (single) real root, then the equation (2) represents a unicyclic unicursal curve of the third order. This family contains e.g. the cissoid of Diokles, the strophoid, and the conchoid of Nikomedes.

c) If the equation $B_0 + 2B_1t + B_2t^2 = 0$ has two complex roots which are not real and not conjugate, then the equation (2) represents a bicircular unicursal curve of the fourth order. This family contains e.g. the limaçon of Pascal, the cardioid, and the lemniscates: of Booth and of Bernoulli.

Let us concentrate now on a mapping u of a domain D in $\mathbb{C}$ onto $u[D]$ in $\mathbb{C}$, conformal with respect to a Riemannian metric

$$\mathrm{d}s^2 = G_{11}(x)\,(\mathrm{d}x^1)^2 + 2G_{12}(x)\,\mathrm{d}x^1\,\mathrm{d}x^2 + G_{22}(x)\,(\mathrm{d}x^2)^2\,, \tag{3}$$

where $G_{11}G_{22} - G_{12}^2 > 0$ and $x = (x^1, x^2)$. The mapping is a solution of a differential equation of the form

$$\frac{\partial}{\partial \bar{x}_c} u_c(x) = \varkappa_u(x) \frac{\partial}{\partial x_c} u_c(x) \quad \text{for almost every} \quad x \in D \tag{4}$$

with

$$x_c = x^1 + \mathrm{i}x^2\,, \qquad u_c = u^1 + \mathrm{i}u^2\,,$$

$$\frac{\partial}{\partial x_c} = \frac{1}{2}\left(\frac{\partial}{\partial x^1} - \mathrm{i}\,\frac{\partial}{\partial x^2}\right), \qquad \frac{\partial}{\partial \overline{x_c}} = \frac{1}{2}\left(\frac{\partial}{\partial x^1} + \mathrm{i}\,\frac{\partial}{\partial x^2}\right)$$

and — on the other hand — transforms some infinitesimal ellipses onto infinitesimal circles (a *Q-quasiconformal* mapping if $(1 + |\varkappa_u(x)|)/(1 - |\varkappa_u(x)|)$ is essentially bounded by Q).

More precisely, let us consider the mapping $\mathfrak{D}u(x)$ of $T_x\mathbb{D}$ into $T_{u(x)}u[\mathbb{D}]$, the formal derivative of u, where $T_x\mathbb{D}$ is the tangent space of $\mathbb{D}$ at x, $\mathbb{D}$ being the domain D considered as a Riemannian manifold with the metric inherited from the Euclidean metric of $\mathbb{R}^2$. Let $\eta = \mathfrak{D}u(x)\,\xi \in T_{u(x)}u[\mathbb{D}]$ and $(\eta^1)^2 + (\eta^2)^2 = 1$. Then

$$(u^1_{|1}\xi^1 + u^1_{|2}\xi^2)^2 + (u^2_{|1}\xi^1 + u^2_{|2}\xi^2)^2 = 1\,, \quad \text{where} \quad u^k_{|j} = (\partial/\partial x^j)\,u^k$$

whenever x is a point of differentiability of u. Hence

$$[(u^1_{|1})^2 + (u^2_{|1})^2]\,(\xi^1)^2 + 2(u^1_{|1}u^1_{|2} + u^2_{|1}u^2_{|2})\,\xi^1\xi^2 + [(u^1_{|2})^2 + (u^2_{|2})^2]\,(\eta^2)^2 = 1\,. \quad (5)$$

Let us set $\varkappa_u = -\dfrac{\lambda - 1}{\lambda + 1}\,\mathrm{e}^{2\mathrm{i}\theta}$. Then, by (4), we have

$$u^1_{|1} = \frac{1}{2}\left(\frac{\partial}{\partial x_c}u_c + \frac{\partial}{\partial \overline{x_c}}u_c + \frac{\partial}{\partial x_c}\overline{u_c} + \frac{\partial}{\partial \overline{x_c}}\overline{u_c}\right)$$

$$= \frac{1}{2}\left(1 - \frac{\lambda - 1}{\lambda + 1}\,\mathrm{e}^{2\mathrm{i}\theta}\right)\frac{\partial}{\partial x_c}u_c + \frac{1}{2}\left(1 - \frac{\lambda - 1}{\lambda + 1}\,\mathrm{e}^{-2\mathrm{i}\theta}\right)\frac{\partial}{\partial \overline{x_c}}\overline{u_c}\,,$$

$$u^2_{|1} = \frac{1}{2\mathrm{i}}\left(\frac{\partial}{\partial x_c}u_c + \frac{\partial}{\partial \overline{x_c}}u_c - \frac{\partial}{\partial x_c}\overline{u_c} - \frac{\partial}{\partial \overline{x_c}}\overline{u_c}\right)$$

$$= \frac{1}{2\mathrm{i}}\left(1 - \frac{\lambda - 1}{\lambda + 1}\,\mathrm{e}^{2\mathrm{i}\theta}\right)\frac{\partial}{\partial x_c}u_c - \frac{1}{2\mathrm{i}}\left(1 - \frac{\lambda - 1}{\lambda + 1}\,\mathrm{e}^{-2\mathrm{i}\theta}\right)\frac{\partial}{\partial \overline{x_c}}\overline{u_c}\,.$$

Consequently,

$$(u^1_{|1})^2 + (u^2_{|1})^2 = \left|1 - \frac{\lambda - 1}{\lambda + 1}\,\mathrm{e}^{2\mathrm{i}\theta}\right|^2 \left|\frac{\partial}{\partial x_c}u_c\right|^2$$

$$= \frac{|(\lambda + 1)\,\mathrm{e}^{-\mathrm{i}\theta} - (\lambda - 1)\,\mathrm{e}^{\mathrm{i}\theta}|^2}{(\lambda + 1)^2}\left|\frac{\partial}{\partial x_c}u_c\right|^2$$

$$= (\cos^2\theta + \lambda^2\sin^2\theta)\,\frac{4}{(\lambda + 1)^2}\left|\frac{\partial}{\partial x_c}u_c\right|^2$$

$$= \left(\lambda\sin^2\theta + \frac{1}{\lambda}\cos^2\theta\right)\frac{(\lambda + 1)^2 - (\lambda - 1)^2}{(\lambda + 1)^2}\left|\frac{\partial}{\partial x_c}u_c\right|^2$$

$$= \left(\lambda\sin^2\theta + \frac{1}{\lambda}\cos^2\theta\right)J_u\,,$$

where J_u is the Jacobian of u. Let us set

$$G_{11}J_u \equiv (u^1_{|1})^2 + (u^2_{|1})^2 = \left(\lambda\sin^2\theta + \frac{1}{\lambda}\cos^2\theta\right)J_u\,.$$

Similarly, we have

$$G_{12}J_u \equiv u^1_{|1}u^1_{|2} + u^2_{|1}u^2_{|2} = -\left(\lambda - \frac{1}{\lambda}\right)\cos\theta \sin\theta \, J_u\,,$$

$$G_{22}J_u \equiv (u^1_{|2})^2 + (u^2_{|2})^2 = \left(\lambda \cos^2\theta + \frac{1}{\lambda}\sin^2\theta\right) J_u\,,$$

so that (5) becomes

$$\lambda(\xi^1 \sin\theta - \xi^2 \cos\theta)^2 J_u + \frac{1}{\lambda}(\xi^1 \cos\theta + \xi^2 \sin\theta)^2 J_u = 1\,.$$

This means that $\mathfrak{D}u(x)$ *transforms our ellipse* whose ratio of the lengths of the major and minor of its semiaxes is $\lambda(x)$ and the angle between the major semiaxis in the positive direction and the ξ^1-axis is $\theta + \frac{1}{2}\pi$ *onto the unit circle* whenever x is a point of differentiability of u and $J_u(x) \neq 0$.

It is clear that the above considerations can easily be transferred for the case of quadratic forms (3) on Riemann surfaces, where $u^k_{|j}$ has to be replaced by $u^k_{|j} = (u \circ x^{-1})^k_{|j} \circ x$ and x denotes a local coordinate system on the surface. For further details and results on complex-analytical methods in differential geometry on Riemann surfaces, involving quasiconformal mappings, we refer to KRUŠKAL [1] and KRUŠKAL, APANASOV, and GUSEVSKIĬ [1].

From the above remarks it is clear that the complex-analytical techniques are also natural when studying minimal surfaces, in particular the Plateau problem. We should like to quote here an advanced result of BOMBIERI, DE GIORGI, and GIUSTI [1] on the Bernstein problem. In fact, the result and the proof as well could be formulated in terms of $\mathbb{R}^{2m}$ instead of $\mathbb{C}^n$, but it seems more natural to involve the complex-analytic approach.

Let $S_n(r)$ be the sphere in $\mathbb{R}^n$ defined by $(x^1)^2 + \dots + (x^n)^2 = r^2$ and let $C_{2n}(r)$ be the truncated cone

$$(x^1)^2 + \dots + (x^n)^2 = (y^1)^2 + \dots + (y^n)^2 < r^2$$

of codimension one in $\mathbb{C}^n$ and with boundary $S_n(r) \times S_n(r) \subset S_{2n}(\sqrt{2}\,r)$. It can be checked that $C_{2n}(r)$ has mean curvature zero at every point except at the origin, which is singular. Let $m_t\colon \mathbb{C}^n \to \mathbb{C}^n$ be a Lipschitz map which reduces to the identity on $S_{2n}(\sqrt{2}\,r)$ and such that $\|m_t - \mathrm{id}\|_{\mathrm{lip}} < t$, where id is the identity map. SIMONS [1] proved that if $n \geqq 4$, there exists $t_0(n) > 0$ such that for $0 < t < t_0(n)$ the area of $C_{2n}(r)$ is smaller than the area of $m_t[C_{2n}(r)]$ unless $m_t = \mathrm{id}$. In the same paper SIMONS proved for $m = 5, 6, 7$ the non-existence of minimal cones in $\mathbb{R}^m$, i.e. that a complete minimal graph in $\mathbb{R}^m$ must be a hyperplane (the same result for $m = 3, 4$ was known before, the problem for general m is called the Bernstein problem in $\mathbb{R}^m$). BOMBIERI, DE GIORGI, and GIUSTI proved in [1] that for $m = 9, 10, \dots$ there exist complete minimal graphs in $\mathbb{R}^m$ which are not hyperplanes, giving a complete solution to the Bernstein problem in $\mathbb{R}^m$. This result is closely related to their following

Theorem 6.2.1. *Let $n = 4, 5, \dots$ The set $E \subset \mathbb{C}^n$, defined by*

$$(x^1)^2 + \dots + (x^n)^2 \geqq (y^1)^2 + \dots + (y^n)^2$$

has an oriented boundary of least area. (We say that a set $E \subset \mathbb{R}^m$ *has an oriented boundary of least area* if the characteristic function φ_E of E is locally of bounded variation in $\mathbb{R}^m$ and for every function g with compact support $K \subset \mathbb{R}^m$, locally of bounded variation in $\mathbb{R}^m$, we have

$$\int_K \|\mathfrak{D}\varphi_E\| \leqq \int_K \|\mathfrak{D}(\varphi_E + g)\|$$

where, for a function f, locally of bounded variation in $\mathbb{R}^m$, the integral of $\|\mathfrak{D}f\|$ over K is understood as the total variation over K of the gradient $\mathfrak{D}f$ of f in the sense of distribution theory.)

6.3. The Kobayashi pseudodistance as a capacity, its applications, and related questions

Let M be a complex manifold of (complex) dimension n. On M we can define different pseudodistances. The most important pseudodistances on M have been defined by CARATHÉODORY [1, 2] and KOBAYASHI [1].

Let D be a domain in $\mathbb{C}^n$. We denote by $G(D)$ the group of all holomorphic automorphisms of D (see KOBAYASHI [3]). If D is bounded, then it has a volume element w which is called the *Bergman volume.* This volume element is invariant under the action of $G(D)$, which is given by the formula

$$\omega = \mathrm{i}^{n^2} f \, \mathrm{d}z^1 \wedge \dots \wedge \mathrm{d}z^n \wedge \mathrm{d}\bar{z}^1 \wedge \dots \wedge \mathrm{d}\bar{z}^n ,$$

where $z^1, \dots, z^n$ are complex coordinates in $\mathbb{C}^n$ and f is a positive function on D. On D there is a *Kähler metric* which can be expressed as $\mathrm{d}s_D^2 = f_{|jk} \, \mathrm{d}z^j \wedge \mathrm{d}\bar{z}^k$ with the usual summation convention (the *Bergman metric* of D). The construction could be generalized for arbitrary M with positive definite ω.

A domain D is called *homogeneous* if the group $G(D)$ acts transitively on D. For the unit disc $\Delta = \{z \in \mathbb{C} : |z| < 1\}$ the group $G(\Delta)$ is defined by $\{kg_a : a \in \mathbb{C}, |k| = 1\}$, where $g_a(z) = (z - a)/(\bar{a}z - 1)$ for $z \in \Delta$. Clearly, $G(\Delta)$ acts transitively on Δ and therefore Δ is homogeneous. On Δ we consider a Kähler metric which is defined by $\mathrm{d}s_\Delta^2 = 4(1 - |z|^2)^{-2} \, \mathrm{d}z \, \mathrm{d}\bar{z}$ (the *Poincaré metric*).

Let $F(M)$ be the family of all holomorphic functions $f: M \to \Delta$. The *Carathéodory pseudodistance* $c(z, s)$ of $z, s \in M$ is defined by the formula

$$c(z, s) = c_M(z, s) = \sup \{\varrho(f(z), f(s)) : f \in F(M)\} ,$$

where

$$\varrho(a, b) = \frac{1}{2} \log \frac{|1 - \bar{a}b| + |a - b|}{|1 - \bar{a}b| - |a - b|} \tag{6}$$

is the hyperbolic (Poincaré) distance of $a, b \in \Delta$ with respect to Δ. It is easily verified that $c(z, z) = 0$, $c(z, s) = c(s, z)$, and $c(z, s) \leqq c(z, w) + c(w, s)$ for $z, s, w \in M$, so c is a pseudodistance indeed. It is not difficult to prove that if M is a bounded domain D in $\mathbb{C}^n$, then c is a distance in the usual sense. If $M = \mathbb{C}^n$ or M is a compact complex manifold, then c is identically equal to zero.

Clearly, $c_\Delta(z, s) = \varrho(z, s)$ for $z, s \in \Delta$ and c_Δ is invariant under the action of $G(\Delta)$. For the polydisc $D = \Delta^n$ in $\mathbb{C}^n$ we have $c_D(z, s) = \max \{c_\Delta(z^1, s^1), \dots, c_\Delta(z^n, s^n)\}$, where $z = (z^1, \dots, z^n)$ and s belong to D. If D and D' are bounded domains in $\mathbb{C}^n$ and $\mathbb{C}^{n'}$, respectively, then $D \times D'$ is a bounded domain in $\mathbb{C}^n \times \mathbb{C}^{n'}$ and

$$c_{D \times D'}\big((z, z'), (s, s')\big) \geqq \max \{c_D(z, s), c_{D'}(z', s')\} ,$$

where $(z, s) = (z^1, \dots, z^n, s^1, \dots, s^{n'})$ and (z', s') belong to $D \times D'$. The last example can be extended to any complex manifolds:

$$c_{M\times M'}((z, z'), (s, s')) \geqq \max \{c_M(z, s), c_{M'}(z', s')\} ,$$

where $z, s \in M$ and $z', s' \in M'$.

If $f: M \to M'$ is a holomorphic mapping, then $c_{M'}(f(z), f(s)) \leqq c_M(z, s)$ for $z, s \in M$. The proof is immediate and we also conclude directly that, with respect to c, every biholomorphic mapping $f: M \to f[M]$ is an isometry. It is also not difficult to prove

Proposition 6.3.1. *c_M is a distance if and only if for any two distinct points $z, s \in M$ there is a bounded holomorphic function f on M such that $f(z) \neq f(s)$.*

Now, let M be a metric space. It is called *complete* if, for every $p \in M$ and every positive number r, the closed ball of radius r around p is a compact subset of M. From proposition 6.3.1. we infer at once

Corollary 6.3.1 *Let M and M' be complex manifolds with complete Carathéodory distance. Then $M \times M'$ has also complete Carathéodory distance.*

For example, let D be a bounded domain in $\mathbb{C}^n$ and $f^1, \dots, f^n$ holomorphic functions defined on D. Let S be a connected component of the open subset of D defined by $|f^1(z)| < 1, \dots, |f^n(z)| < 1$. We suppose that the closure of S is compact and is contained in D. Then S is called *analytic polyhedron*. The distance c_S in S is complete.

We may proceed in a more general way. Let M be a complex manifold of (complex) dimension n. We call an arbitrary sequence of points $z_0, \dots, z_k$ of M, where $z_0 = z$ and $z_k = s$, a *holomorphic chain* joining z to s if there are points $a_1, b_1, \dots, a_k, b_k$ of Δ and holomorphic mappings $f_1, \dots, f_k$ of Δ into M such that $f_j(a_j) = z_{j-1}$ and $f_j(b_j) = z_j$ for $j = 1, \dots, k$. With each of such chains we associate the number $\varrho(a_1, b_1) + \dots + \varrho(a_k, b_k)$, where ϱ is defined by (6) and we call the infimum $d(z, s) = d_M(z, s)$, $z, s \in M$, of those numbers (for all possible choices of points and holomorphic mappings) the *Kobayashi pseudodistance* of $z, s \in M$. It is easy to verify that d is a pseudodistance indeed.

From the definition of the Kobayashi pseudodistance we conclude that this is obtained by dualizing the construction from which we obtain the Carathéodory pseudodistance. They both are continuous on $M \times M$. The following construction of the infinitesimal form of d_M, the intrinsic differential *Kobayashi pseudometric* k_M of M, makes this duality more apparent: we set

$$k_M(w) = \inf_f \{||v|| : v \in T\Delta, f_*(v) = w\} ,$$

where $||v||$ denotes the length of the tangent vector v, measured by the Poincaré metric ds^2_Δ and the infimum is taken over all holomorphic mappings f of Δ into M. For the unit ball B in $\mathbb{C}^n$, k^2_B coincides with the Bergman metric of B and therefore is of class $\mathscr{C}^\infty$. For the unit polydisc D in $\mathbb{C}^n$, k_D is not differentiable.

It is clear that if $f: M \to M'$ is a holomorphic mapping, then $d_{M'}(f(z), f(s)) \leqq d_M(z, s)$ for $z, s \in M$ and that $k_{M'}(f_*(w)) \leqq k_M(w)$ for $w \in TM$. Besides, with respect to d, every biholomorphic mapping is an isometry. Consequently, $d_\Delta = \varrho$. Moreover, the distance-decreasing properties make c and d particularly useful in studying holomorphic mappings. They are two extreme pseudodistances with respect to this property in the following sense:

Lemma 6.3.1. *Let M be a complex manifold and d' any pseudodistance on M such that $d'(f(a), f(b)) \leqq \varrho(a, b)$ for $a, b \in \Delta$ and every holomorphic mapping $f: \Delta \to M$.*

Then $d'(z, s) \leqq d(z, s)$ for $z, s \in M$. Similarly, if $d'(z, s) \geqq \varrho(f(z), f(s))$ for $z, s \in M$ and every $f \in F(M)$, then $d'(z, s) \geqq c(z, s)$ for $z, s \in M$.

Proof. We confine ourselves to proving the first statement only. Let $z_0, \dots, z_k$, $a_1, \dots, a_k, b_1, \dots, b_k, f_1, \dots, f_k$ be as in the definition of d. Then

$$d'(z, s) \leqq \sum_{j=1}^{k} d'(z_{j-1}, z_j) = \sum_{j=1}^{k} d'(f_j(a_j), f_j(b_j)) \leqq \sum_{j=1}^{k} \varrho(a_j, b_j) .$$

Hence $d'(z, s) \leqq \inf [\varrho(a_1, b_1) + \dots + \varrho(a_k, b_k)] = d(z, s)$, as desired.

In particular, following CHERN, LEVINE, and NIRENBERG [1], we may consider the following pseudodistance $d'(z, s) = \varrho_\Gamma(z, s)$. Let Γ be the family of all singular chains joining z and s on M, which contains all curves homotopic to a given one in the family. Then

$$\varrho_\Gamma(z, s) = \sup_{u \in A} \inf_{T \in \Gamma} |T[\mathrm{d}^c u]| , \tag{7}$$

where A is the class of all plurisubharmonic $\mathscr{E}^2$-functions u of M satisfying the condition $0 < u(z) < 1$ everywhere and T is interpreted as a current in the sense of de Rham (more precisely, $T = T_\gamma$ is the current corresponding to γ, the infimum in (7) being taken over all $\gamma \in \Gamma$). We shall replace the subscript Γ in ϱ_Γ by 0 if the family consists of all singular chains joining z to s on M. We obtain (for the proof, cf. CHERN, LEVINE, and NIRENBERG [1]):

Proposition 6.3.2. *Under the preceding notation and hypotheses, we have*

$$(2/\pi)\, c(z, s) \leqq \varrho_\Gamma(z, s) \leqq \varrho_0(z, s) \leqq (2/\pi)\, d(z, s) \quad \text{for} \quad z, s \in M .$$

All these pseudodistances are biholomorphic invariants.

From lemma 6.3.1 and proposition 6.3.2 we infer:

Corollary 6.3.2. *The Kobayashi pseudodistance gives an extreme form of the generalized Schwarz lemma: $d'(f(a), f(b)) \leqq \varrho(a, b)$ among all $d'(z, s) = \varrho_\Gamma(z, s)$.*

If, for a complex manifold M, the pseudodistance d is a distance, that is $d(z, s) > 0$ for $z \neq s$, then M is called a *hyperbolic manifold*. For two complex manifolds M and M' it is not difficult to check that

$$d_{M \times M'}((z, z'), (s, s')) \geqq \max \{d_M(z, s), d_{M'}(z', s')\}$$

where $z, s \in M$ and $z', s' \in M'$, if M and M' are hyperbolic, then $M \times M'$ is hyperbolic as well. A complex submanifold of a hyperbolic complex manifold is hyperbolic.

The following lemma is an immediate consequence of lemma 6.3.1.:

Lemma 6.3.2. *If a complex manifold M admits a [complete] distance d' for which $d'(f(a), f(b)) \leqq \varrho(a, b)$ for $a, b \in \Delta$ and every holomorphic mapping $f: \Delta \to M$, then it is [complete] hyperbolic.*

Because of the scope of this article we shall refer e.g. to KOBAYASHI and NOMIZU [1 for the well-known definitions of the Kähler and Hermitian metrics, as well as the Gaussian and holomorphic sectional curvatures. We take for granted the following lemma of GRAUERT and RECKZIEGEL [1] (cf. also KOBAYASHI [2, p. 40]):

Lemma 6.3.3. *Let $\Delta_a = \{z \in \mathbb{C} : |z| < a\}$. Consider it as a one-dimensional Kähler manifold with the metric $\mathrm{d}s_a^2 = 4a^2 A^{-1}(a - |z|^2)^{-2}\, \mathrm{d}z\, \mathrm{d}\bar{z}$, and let M be an n-dimensional Hermitian manifold whose holomorphic sectional curvature is bounded from above by a negative constant $-B$. Then every holomorphic mapping $f: \Delta_a \to M$ satisfies the estimate $f_*(\mathrm{d}s_M^2) \leqq (A/B)\, \mathrm{d}s_a^2$.*

We are going to prove (cf. KOBAYASHI [2, p. 61]):

Theorem 6.3.1. *Let M be a [complete] Hermitian manifold whose holomorphic sectional curvature is bounded from above by a negative constant. Then M is [complete] hyperbolic.*

Proof. By lemma 6.3.3 if we multiply ds_M^2 by a suitable constant, we obtain $f_*(ds_M^2) \leqq ds_\Delta^2$ for every holomorphic mapping $f: \Delta \to M$. If d' and ϱ denote the distance functions defined by ds_M^2 and ds_Δ^2, respectively, then this inequality implies that $d'(f(a), f(b)) \leqq \varrho(a, b)$ for $a, b \in \Delta$. By lemma 6.3.2 the proof is completed.

Let us consider now an open convex cone V in $\mathbb{R}^n$ which contains no entire straight lines. A mapping $f: \mathbb{C}^m \times \mathbb{C}^m \to \mathbb{C}^n$ is said to be *V-Hermitian* if:

(a) for every $w' \in \mathbb{C}^m$ the mapping $f_{w'}$ of $\mathbb{C}^m$ into $\mathbb{C}^n$ defined by $f_{w'}(w) = f(w, w')$ is complex linear,

(b) $f(w', w) = \overline{f(w, w')}$ for $w, w' \in \mathbb{C}^m$,

(c) $f(w, w)$ belongs to the closure of V for any $w \in \mathbb{C}^m$,

(d) $f(w, w) = 0$, if and only if $w = 0$. The set

$$S = \{(z, w) \in \mathbb{C}^n \times \mathbb{C}^m : \operatorname{im} z - f(w, w) \in V\}$$

is called the *Siegel domain of the second kind associated with* (V, f) (PJATECKIĬ ŠAPIRO [1]). We have the following

Theorem 6.3.2. *A Siegel domain of the second kind is complete hyperbolic.*

The proof is based on two lemmas (cf. KOBAYASHI [2, pp. 30, 57, and 64]).

Lemma 6.3.4. *A Siegel domain of the second kind can be expressed as the intersection of (possibly uncountably many) domains, each of which is biholomorphically equivalent to a product of balls.*

Lemma 6.3.5. *Let M and M_j, $j \in I$, be closed complex submanifolds of a complex manifold M' such that $M = \bigcap_j M_j$. If each M_j is a [complete] hyperbolic manifold, so is M.*

Hence the assertion of theorem 6.3.2 follows since a product of balls is complete hyperbolic.

For more advanced results one may study KOBAYASHI [2, 4].

6.4. Deformations of potentials and Kähler metrics related to biholomorphically invariant capacities

The concept of pseudodistances (7) can be further extended and modified as follows. Let M be a complex manifold of (complex) dimension n endowed with an Hermitian metric h and a $\mathscr{C}^1$ tensor field H of type $(1, 1)$. In particular we may let H depend on h or take as H an almost complex structure of the tangent bundle TM, for instance the complex structure of M. Let dv and ds^2 denote the measure and the geodesic distance associated with the metric, respectively. Let further D be a *condenser* on M, i.e. a domain whose component consists of two distinguished disjoint closed sets C_0 and C_1 (the *condenser plates*), $q: M \to \mathbb{C}$ a continuous mapping (the *inhomogeneity function*) and p a real number $\geqq 1$. Consider the class Adm D of all plurisubharmonic $\mathscr{C}^2$-functions u on the closure of D, satisfying the conditions $0 < u(z) < 1$ for $z \in D$, $u \mid \partial C_0 = 0$ and $u \mid \partial C_1 = 1$. Let (ŁAWRYNOWICZ [2]):

$$\operatorname{Cap}_p(D, q) = \inf_{u \in \operatorname{Adm} D} \left| \int_D q \, \|\nabla u\|_{ds^2}^{p-2} \det H \, du \wedge d^c u \wedge (dd^c u)^{n-1} \right| . \tag{8}$$

Let further Γ be a homology class of D with real coefficients and $\dim \Gamma = r$. Consider all currents of Γ (more precisely: corresponding to the elements of Γ) in the sense of de Rham, and a locally finite open covering $\mathscr{U} = \{u_j : j \in I\}$ of M. Denote by adm $(D, \mathscr{U})$ the family of all plurisubharmonic $\mathscr{E}^2$-functions u_j on $U_j \cap D$ defined in each member of the covering which satisfy the following conditions:

(i) the oscillation of u_j in $U_j \cap D$ is less then one,
(ii) $du_j = du_k$ in $U_j \cap U_k \cap D \neq \emptyset$.

Condition (ii) describes a closed real one-form in D. Similarly, $d^c u_j$ and $dd^c u_j$ are also well-defined in D. Without ambiguity, we can denote them omitting the indices. Let (ŁAWRYNOWICZ [1]):

$$\operatorname{cap}_p (D, q, \Gamma, \mathscr{U}) = \sup_{u \in \operatorname{adm}(D, \mathscr{U})} \inf_{T \in \Gamma} |T[q \, ||\nabla u||_{ds^2}^{p-2} \det H \mathfrak{D}^r u]| \,, \tag{9}$$

where

$$\mathfrak{D}^r u = \begin{cases} d^c u \wedge (dd^c u)^{(1/2)r-(1/2)} & \text{for } r \text{ odd,} \\ du \wedge d^c u \wedge (dd^c u)^{(1/2)r-1} & \text{for } r \text{ even.} \end{cases}$$

For a detailed description of the capacities (1) and (2), and their applications we refer to ŁAWRYNOWICZ [1–4].

The seminorms (8) and (9) are biholomorphic invariants generalizing the concept of CHERN, LEVINE, and NIRENBERG [1]. One can also define a (Dirichlet) norm more simple and natural, biholomorphically invariant, with help of Kähler metrics in the Dirichlet integral. Let, for the sake of simplicity, D be a domain in $\mathbb{C}^n$. If u is a function of the class $\mathscr{E}^1(D)$, we set (GAVEAU and ŁAWRYNOWICZ [1]):

$$|||u|||_D = [\sup_{G \in P_2(D)} \int_D du \wedge d^c u \wedge (dd^c G)^{n-1}]^{1/2} \,, \tag{10}$$

where $P_2(D)$ denotes the class of all plurisubharmonic $\mathscr{E}^2$-functions u in D, satisfying there the condition $0 \leqq u(z) \leqq 1$. It is clear that $du \wedge d^c u \wedge (dd^c G)^{n-1} = ||\nabla u||_{ds^2}^2 \, dv$, where $ds^2 = (1/2i) \, dd^c G$. Of course we can naturally generalize (10) replacing the integrand $||\nabla u||_{ds^2}^2$ by $q \, |||\nabla u|||_{ds^2}$, where $q: D \to \mathbb{R}^+$ is continuous and $p \geqq 1$; cf. GAVEAU and ŁAWRYNOWICZ [2].

Let further $P_0(D)$ be the class of all plurisubharmonic $\mathscr{E}^0$-functions in D, satisfying there the condition $0 \leqq u(z) \leqq 1$, and $\mathscr{E}_c^\infty(D)$ the class of $\mathscr{E}^\infty$-functions in D of compact support. After GAVEAU and ŁAWRYNOWICZ [1] we quote two lemmas.

Lemma 6.4.1. *For $u \in \mathscr{E}_c^\infty(D)$ we have $|||u|||_D < +\infty$ and*

$$|||u|||_D = [\sup_{G \in P_0(D)} \int_D du \wedge d^c u \wedge (dd^c G)^{n-1}]^{1/2} \,.$$

If, in addition, $n = 2$, one can also replace the above supremum by the supremum over the class $P(D)$ of functions plurisubharmonic in D, with values between 0 and 1, and not necessarily continuous.

Let $\mathscr{H}^1(D)$ denote the space of $u \in \mathscr{E}^1(D)$ with $|||u|||_D < +\infty$.

Lemma 6.4.2. *The space $\mathscr{H}^1(D)$ with the norm* (10) *is complete.*

We denote by $\mathscr{H}_0^1(D)$ the restriction of $\mathscr{H}^1(D)$ to functions $u \in \mathscr{E}_c^\infty(D)$. By lemmas 6.4.1 and 6.4.2 the space $\mathscr{H}_0^1(D)$ is complete with respect to the norms defined in terms of $P_2(D)$ and $P_0(D)$ as well.

Suppose now that D_0 is a strictly pseudoconvex domain in $\mathbb{C}^2$ whose boundary is of the class $\mathscr{E}^2$. Let $Q(D_0)$ be the class of all plurisubharmonic functions G in D_0, satisfying there the conditions $0 \leqq G(z) \leqq 1$ and $\limsup_{z \to \partial D_0} G(z) = 1$.

Since $Q(D_0)$ is non-empty, we may consider the norm

$$||u||_{D_0} = [\sup_{G \in Q(D_0)} \int_{D_0} du \wedge d^c u \wedge dd^c G]^{1/2} .$$

It is clear that $||u||_{D_0}$ exceeds the usual norm of Sobolev. Denote by $\mathcal{H}_0^1(D_0)$ the completion of $\mathcal{E}_c^\infty(D_0)$ with respect to this norm. Let now C_1 be a compact set in D_0 and $D = D_0 \setminus C_1$. We define the following two capacities of D:

$$\mathrm{CAP}(D) = \inf \{||u||_{D_0}^2 : u \in \mathcal{E}_c^\infty(D_0),\, u \mid C_1 \geqq 1\} .$$

$$\mathrm{CAP}'(D) = \sup_{G \in Q(D_0)} \inf \{ \int_{D_0} du \wedge d^c u \wedge dd^c G : u \in \hat{\mathcal{H}}_0^1(D_0),\, u \mid C_1 \geqq 1\} .$$

In the second case this means that for any G of $Q(D_0)$ we define a *norm of Dirichlet* (cf. BEURLING and DENY [1], FUKUSHIMA [1]) associated with the positive current dd^cG, which serves for calculating the capacity for this norm, and then we take the supremum of those capacities within $Q(D_0)$. One obtains (GAVEAU and ŁAWRYNOWICZ [1]):

Theorem 6.4.1. *For any D as above we have* $\mathrm{CAP}(D) = \mathrm{CAP}'(D)$ *and the supremum in* $\mathrm{CAP}'(D)$ *is attained for some G of $Q(D_0)$.*

Proof. A priori, we have $\mathrm{CAP}'(D) \leqq \mathrm{CAP}(D)$. The opposite inequality follows from the fundamental theorem of VON NEUMANN [1], appearing in the theory of games, or, more precisely, from a variant of this theorem. Let X denote the family of all functions $u \in \hat{\mathcal{H}}_0^1(D_0)$ with $u \geqq 1$ on C_0, hence X is convex. Denote by Y the family of all currents dd^cG for $G \in Q(D_0)$; hence Y is convex and compact in the weak topology. Let

$$I[u, dd^cG] = \int_{D_0} du \wedge d^c u \wedge dd^c G \tag{11}$$

be defined on $X \times Y$, therefore $dd^cG \mapsto I[u, dd^cG]$ is linear continuous on Y for every fixed $u \in X$. In addition, $u \mapsto I[u, dd^cG]$ is convex and any convex functional in a Banach space is weakly lower semicontinuous. Now the theorem of von Neumann says that the game (X, Y, I) is well-defined and this means exactly the expected assertion.

A direct calculation shows that in the case of C_1 being a ball of centre 0 and radius r_0 drilled in the unit ball $D_0 = B$, the value $\mathrm{CAP}(D_0 \setminus C_1)$ is attained at a saddle point (u_0, G_0) of the functional (11), where $G_0(z) = 0$ for $z \in C_1$, $G_0(z) = 1 + (1/\log r_0^{-1}) \log |z|^2$ for $z \in D$, and $u_0(z) = 1 - G_0(z)$ for $z \in D_0$. This suggests the following (cf. GAVEAU and ŁAWRYNOWICZ [1, 2]):

Theorem 6.4.2. *The functional $(u, G) \mapsto I[u, dd^cG]$, defined by* (11), *attains a saddle point (u_0, G_0) within the set $\hat{\mathcal{H}}_0^1(D_0) \times Q(D_0)$ with $u \mid C_1 \geqq 1$. Moreover u_0 is unique and it is the plurisuperharmonic solution of the complex Monge-Ampère equation* $dd^cu \wedge dd^cu = 0$ *in D, with values* 1 *on C_1 and* 0 *on $C_0 = \mathbb{C}^2 \setminus D_0$; one may take $G_0 = 1 - u_0$ (extremal capacitary potential).*

Proof. The existence of G_0 is already proved by theorem 6.4.1. A calculation of OKADA [1] shows that the Dirichlet norm, determined by the current dd^cG_0, defines a Dirichlet space in the sense of BEURLING and DENY [1] and FUKUSHIMA [1]. One can then apply the theory of "balayage" and obtain u_0 (cf. BEDFORD and TAYLOR [1] and BRELOT [1]). In order to prove the last statement of the theorem, we denote by

u_G the capacitary potential associated with the positive current dd^cG. After a standard approximation procedure we confine ourselves to the case of C_1 with boundary of the class $\mathscr{C}^2$. By definition, for any u of $\hat{\mathscr{H}}_0^1(D_0)$ with $u \mid C_1 = 1$, we have

$$\int_{D_0} du_G \wedge d^cu_G \wedge dd^cG \leqq \int_{D_0} du \wedge d^cu \wedge dd^cG . \tag{12}$$

Let $I_G[u] = \int_D du \wedge d^cu \wedge dd^c(G - 1)$, we have

$$I_G[u] = \int_D d[du \wedge d^cu \wedge d^c(G - 1)] + \int_D du \wedge dd^cu \wedge d^c(G - 1)$$

and

$$\int_D du \wedge dd^cu \wedge d^c(G - 1) = - \int_D d^cu \wedge dd^cu \wedge d(G - 1)$$

$$= \int_D d[(d^cu \wedge dd^cu) (G - 1)] - \int_D (dd^cu \wedge dd^cu) (G - 1) ,$$

whence

$$I_G[u] = \int_{\partial D} du \wedge d^cu \wedge d^c(G - 1) + \int_{\partial D} (d^cu \wedge dd^cu) (G - 1)$$

$$- \int_D (dd^cu \wedge dd^cu) (G - 1) .$$

Let us now take $u = v$ in (12), where v is the plurisuperharmonic solution of the complex Monge-Ampère equation $dd^cu \wedge dd^cu = 0$, with values 1 on C_1 and 0 on C_0. Therefore, since v is constant on ∂C_1 and ∂C_0, we have $dv \mid \partial D = 0$. Moreover $G \mid \partial C_0 = 1$, and hence

$$I_G[v] = - \int_{\partial C_1} (G - 1) \, d^cv \wedge dd^cv .$$

Let us consider also $I_{-v}[v]$. By the same reason we get

$$I_{-v}[v] = - \int_{\partial C_1} (-v) \, d^cv \wedge dd^cv .$$

Consequently,

$$I_G[v] - I_{-v}[v] = - \int_{\partial C_1} [(G - 1) + v] \, d^c(-v) \wedge dd^c(-v) . \tag{13}$$

But $G - 1 \leqq -v$ (a property of solution of the complex Monge-Ampère equation) and $dd^c(-v) \geqq 0$. Also, since v is constant on ∂C_1, $d^c(-v)$ contributes to the integral in question only through the exterior normal derivative on ∂D; by $-v \mid \partial C_1 = -1$ and $-1 \leqq -v \mid D \leqq 0$ we conclude that this derivative is $\leqq 0$. Thus, by (13), $I_G[v] - I_{-v}[v] \leqq 0$ and hence, by (12),

$$\int_{D_0} du_G \wedge d^cu_G \wedge dd^cG \leqq \int_{D_0} dv \wedge d^cv \wedge dd^c(1 - v) .$$

Since $1 - v$ belongs to $Q(D_0)$ and v satisfies the complex Monge-Ampère equation $dd^cv \wedge dd^c(1 - v) = 0$, v is the potential u_{1-v}. Hence

$$\int_{D_0} dv \wedge d^cv \wedge dd^c(1 - v) = \sup_{G \in Q(D_0)} \int_{D_0} du_G \wedge d^cu_G \wedge dd^cG$$

$$= \sup_{G \in Q(D_0)} \inf \{ \int_{D_0} du \wedge d^cu \wedge dd^cG : u \in \hat{\mathscr{H}}_0^1(D), u \mid C_1 \geqq 1 \} = \mathrm{CAP}'(D) .$$

Therefore the pair $(u_0, G_0) = (v, 1 - v)$ is indeed a saddle point of the functional I.

6.5. Homotopic deformation of biholomorphically continuable mappings

Let X be an analytic vector field in $\mathbb{R}^m$. It is called a *q-quasiconformal deformation*, if $||SX|| < q$, $q \geqq 0$, where S is the Ahlfors' operator (cf. Ahlfors [7]) of X: SX is the matrix with coordinates

$$[SX]_{jk} = X^j_{|k} + X^k_{|j} - \frac{1}{m}\operatorname{tr}[X^j_{|k}]\,, \quad 1 \leqq j, k \leqq m\,, \text{ and } ||a_{jk}|| = (\sum a^2_{jk})^{1/2}.$$

Ahlfors [6] and Reimann [1] proved that q-quasiconformal transformations of $\mathbb{R}^m$ generate a uniparametric group of $\exp(q\,|t|)$-quasiconformal diffeomorphisms $x \to f(x, t)$ of $\mathbb{R}^m$ onto itself.

One can consider the problem of biholomorphic continuability of such mappings from domains in $\mathbb{R}^m$ to domains in $\mathbb{C}^n$, where $m \leqq 2n - 1$ and $\mathbb{R}^m$ has to be properly immersed in $\mathbb{C}^n$. We recall that under a ***hypersurface*** in $\mathbb{C}^n$ we mean a $(2n-1)$-dimensional real-analytic submanifold of $\mathbb{C}^n$. An analytic homeomorphism f of a hypersurface M onto a hypersurface M' is said to be *biholomorphically continuable* if there are open sets U, U' of $\mathbb{C}^n$ such that $M \subset U$, $M' \subset U'$, and there is a biholomorphic mapping $\tilde{f}$ of U onto U' such that $\tilde{f} \mid M = f$.

In addition to the results of the first named author on biholomorphic continuability of quasiconformal mappings (Ławrynowicz [3] — methods of functional analysis, Ławrynowicz [4] — geometrical methods including the complex capacities described above in 6.3.), Pierzchalski [1] invented the following analytical method for $m = 2n - 1$: he compared the "parametric" equation $(\partial/\partial t) f(x, t) = \ldots$ for quasiconformal mappings (cf. Ahlfors [6], Bojarski and Iwaniec [2]) with an analogous equation for pseudo-conformal mappings of Tanaka [1] and proved that, for quasiconformal transformations locally continuable to biholomorphic mappings, the Ahlfors operator has a particularly simple form:

$$SX = \left[\begin{array}{c:c:c} A & 0 & B \\ \hdashline 0 & & 0 \\ \hdashline -\overset{t}{B} & 0 & A \end{array}\right],$$

where A is symmetric, B — antisymmetric, and they satisfy some additional conditions. These results could be generalized to non-degenerate hypersurfaces in $\mathbb{C}^n$, including the geometrical properties of S.

The method of Pierzchalski is based on the following facts (for the proof of lemma 6.5.2 below cf. Tanaka [1]).

Lemma 6.5.1. *Two biholomorphic continuations of an analytic homeomorphism f of a hypersurface M are identical in some neighbourhood of M.*

Lemma 6.5.2. *Let f be an analytic function on a hypersurface M. Then f is biholomorphically continuable if and only if*

$$\mathrm{d}f \wedge (\mathrm{d}z_1 \circ \iota) \wedge \ldots \wedge (\mathrm{d}z_n \circ \iota) = 0\,,$$

where ι is the immersion of M into $\mathbb{C}^n$.

More generally, let M be an odd-dimensional manifold: $\dim M = 2n - 1 \geqq 3$. An *almost contact structure* on M (a counterpart of an almost complex structure for odd-dimensional manifolds) is defined as the triple (F, ξ, θ), where F is a tensor field of type $(1, 1)$, ξ a vector field, and θ a one-form on M satisfying the conditions

$$F^2 = -I + \theta \otimes \xi\,, \qquad F\xi = 0\,, \qquad \theta \circ F = 0\,, \qquad \theta(\xi) = 1\,,$$

where I is the identity tensor field of type (1, 1). Let J be the canonical complex structure of $\mathbb{C}^n$. If M is an orientable hypersurface in $\mathbb{C}^n$, then for every point z of M the tangent space T_zM contains exactly one J-invariant linear subspace D_z of dimension $2n - 2$, so the correspondence $z \to D_z$ defines a distribution D called a *hyperdistribution with complex structure* J in M. Then we take an arbitrary one-form θ on M such that $\theta(X) = 0$ if and only if $X \in D$ for any vector field X on M. Finally, we take an arbitrary vector field ξ on M such that $\theta(\xi) = 1$ and define F by $FX = J(X - \theta(X)\,\xi)$ for any vector field X tangent to M. The triple (F, ξ, θ) is an almost contact structure on M.

Consider now two triples (M, D, J) and (M', D', J') where M and M' are analytic manifolds of dimension $2n - 1 \geqq 3$, D and D' are hyperdistributions in M and M', respectively, whereas J and J' are the corresponding complex structures. An analytic homeomorphism f of M onto M' is called a *pseudo-conformal mapping* if f_*X belongs to D' for any X belonging to D and if $f_*J = J'f_*$.

Suppose next that M is an orientable hypersurface in $\mathbb{C}^n$ and consider an almost contact structure (F, ξ, θ) on M, induced by the complex structure J of $\mathbb{C}^n$. Assume that the *Levi tensor* G, given by $G(X, Y) = \mathrm{d}\theta(X, JY)$ for X and Y belonging to D, is symmetric and positive definite. Then we can consider M as a Riemannian manifold with the Riemannian structure generated by G. Let Z be a vector field on M and φ_t, $-\varepsilon < t < \varepsilon$, the *local group of local transformations* generated by Z in an open set U of M, i.e. φ_t are solutions of the differential equation $(\mathrm{d}/\mathrm{d}t)\,\varphi_t(x) = Z(\varphi_t(x))$ with the initial condition $\varphi_0(x) = x$ in U. A necessary and sufficient condition for Z that φ_t be a pseudo-conformal mapping is given in terms of the Lie derivatives $\mathscr{L}_Z$ (cf. Ishihara [1]):

$$\mathscr{L}_Z F = -\theta \otimes V\,, \qquad \mathscr{L}_Z \xi = -a\xi - fV\,, \qquad \mathscr{L}_Z \theta = a\theta\,,$$

where a is a (smooth) function and V a vector field belonging to the hyperdistribution D, both being defined in U. This condition implies that, for the Ahlfors operator S, we have (Pierzchalski [1]):

$$SZ(X, Y) = cG(X, Y) \quad \text{for} \quad X, Y \text{ belonging to } D\,,$$

where c is a (smooth) function on M. Now the following result is important (Cartan [1], Tanaka [1], Chern and Moser [1], Yano and Ishihara [1]):

Lemma 6.5.3. *Suppose that M and M' are orientable real-analytic hypersurfaces in $\mathbb{C}^n$, and φ is a real-analytic homeomorphism of M onto M'. Then φ is pseudo-conformal if and only if for every point z of M there are in $\mathbb{C}^n$ neighbourhoods U of z and U' of $\varphi(z)$, and a biholomorphic mapping $\tilde{\varphi}$ of U onto U', such that $\tilde{\varphi} \mid U \cap M = \varphi$.*

The main result of Pierzchalski [1] may be formulated as follows:

Theorem 6.5.1. *Suppose that M is an orientable real-analytic hypersurface in $\mathbb{C}^n$ and Z is a vector field on M. If φ_t, $-\varepsilon < t < \varepsilon$, is a local group of local transformations generated by Z in an open set U of M, such that φ_t is biholomorphically continuable to a neighbourhood $\tilde{U}$ of U in $\mathbb{C}^n$, then $SZ = cG$, where c is a (smooth) function defined in U.*

Further references

Andreotti and Ławrynowicz [1, 2].

7. Complex Methods in the Theory of Infinitesimal Bending of a Surface

Z. D. Usmanov

7.1. Introduction

As is known, under an infinitesimal deformation of the surface S all geometric characteristics, which can be expressed by means of the coefficients E, F, G of the first fundamental quadratic form are unchanged, while the coefficients L, M, N of the second form undergo in general nonzero variations δL, δM, δN. Examination of the latter permits one to determine precisely, up to motions and symmetries of the space, the surface S^*, which is the infinitesimal deformation of the original surface S.

On the basis of the Gauss and Peterson-Codazzi equations, we get equations for the unknown functions δL, δM, δN (I. N. Vekua [5]):

$$L\,\delta N + N\,\delta L - 2M\,\delta M = 0\,,$$

$$\frac{\partial}{\partial x}(\delta M) - \frac{\partial}{\partial y}(\delta L) + \Gamma^1_{12}\,\delta L - (\Gamma^1_{11} - \Gamma^2_{12})\,\delta M - \Gamma^2_{11}\,\delta N = 0\,,$$

$$\frac{\partial}{\partial y}(\delta M) - \frac{\partial}{\partial x}(\delta N) - \Gamma^1_{22}\,\delta L - (\Gamma^2_{22} - \Gamma^1_{12})\,\delta M + \Gamma^2_{12}\,\delta N = 0\,,$$

where x, y are interior coordinates and $\Gamma^\alpha_{\beta\gamma}$ are the Christoffel symbols of the second kind.

Let S belong to the class $\mathscr{W}^{m+3}_p$, $p > 2$, i.e., the radius vector of the surface in coordinates x, y admits derivatives up to the order $m + 3$, which are p-integrable. We assume that the Gaussian curvature K of the surface is positive. Under these conditions we introduce on the surface new coordinates u, v, which are connected with the coordinates x, y by means of the Beltrami equation (I. N. Vekua [5]):

$$\partial_{\bar z} w - q(z)\,\partial_z w = 0\,, \tag{1}$$

where

$$q(z) = \frac{L - N + 2\mathrm{i}M}{L + N + 2\sqrt{LN - M^2}} \tag{2}$$

and $z = x + \mathrm{i}y$, $w = u + \mathrm{i}v$, $\mathrm{i}^2 = -1$, $2\partial_{\bar z} = \dfrac{\partial}{\partial x} + \mathrm{i}\dfrac{\partial}{\partial y}$, $2\partial_z = \dfrac{\partial}{\partial x} - \mathrm{i}\dfrac{\partial}{\partial y}$. It is established that $q(z) \in \mathscr{W}^{m+1}_p$ and $|q(z)| < 1$. Under these restrictions the equation (1) admits solutions, which give a homeomorphic map of a domain of the x, y-plane onto some domain of the u, v-plane. The coordinates u, v are called a *conjugate isometric parametrization* on the surface S. In those coordinates the coefficients of the second form satisfy the relations $L(u, v) = N(u, v)$, $M(u, v) = 0$ and the equations mentioned for the unknown functions $\delta L(u, v)$, $\delta M(u, v)$, $\delta N(u, v)$ will have the form:

$$\delta N = -\delta L\,,$$

$$\frac{\partial}{\partial u}(\delta M) - \frac{\partial}{\partial v}(\delta L) = (\Gamma^1_{11} - \Gamma^2_{12})\,\delta M - (\Gamma^1_{12} + \Gamma^2_{11})\,\delta L\,,$$

$$\frac{\partial}{\partial v}(\delta M) + \frac{\partial}{\partial u}(\delta L) = (\Gamma^2_{22} - \Gamma^1_{12})\,\delta M + (\Gamma^1_{22} + \Gamma^2_{12})\,\delta L\,.$$

Introducing the complex function

$$U(w) = K^{1/4}(w)\,(\delta M + \mathrm{i}\,\delta L)\,, \tag{3}$$

where $K(w)$ is the Gaussian curvature of the surface S, we write the preceding system in the following form:

$$\partial_{\bar{w}} U + \overline{B(w)}\,\overline{U} = 0\,. \tag{4}$$

The coefficient $B(w)$ is computed by the formula

$$B(w) = -\mathrm{e}^{\mathrm{i}\varphi(w)}\,\partial_{\bar{w}}\left(\operatorname{Arch}\frac{H}{\sqrt{K}}\right) - \frac{\partial_w K}{4K}\,\mathrm{e}^{2\mathrm{i}\varphi(w)} \tag{5}$$

where

$$\mathrm{e}^{\mathrm{i}\varphi(w)} = \frac{a^-(w)\sqrt{K(w)}}{2\sqrt{a(w)\,\mathcal{E}(w)}}\,.$$

Here $H(w)$ is the mean curvature of the surface, $\mathcal{E}(w) = H^2 - K$ is the Euler difference, $a(w) = EG - F^2$ is the discriminant of the first fundamental form (E, F, G are their coefficients), $a^-(w) = E - G + 2\mathrm{i}F$ (I. N. Vekua [5]).

The equation (4) is the complex notation of the generalized Cauchy-Riemann system with the coefficient $B(w)$, which belongs to the class $\mathcal{W}^m_p$. Namely, in this case I. N. Vekua constructed a complete theory of the equation (4) and gave its employments on infinitesimal deformations of positive curvature surfaces.

7.2. The equations of infinitesimal deformation of a surface with a flat point

Let us consider the new geometric object, the surface with an isolated flat point, around which the Gaussian curvature is positive. The immediate aim of this chapter is to tell about new problems of the complex analysis in connection with infinitesimal deformations of new geometric objects of the special type.

We consider in rectangular Cartesian coordinates $0xyz^*$ the surface S, given by the equation

$$z^* = (x^2 + y^2)^{n/2} f(x, y)\,, \qquad (x, y) \in D\,, \tag{6}$$

where $n > 2$ is any real number, D is a simply connected domain, bounded by a smooth, closed contour and containing the flat point $(0, 0)$ in its interior. We assume that $f(x, y) \in \mathcal{C}^3(\overline{D})$, $f(0, 0) > 0$ and for all points of $\overline{D}$ other than $(0, 0)$, the Gaussian curvature of the surface S is positive.

It is not possible to use immediately the theory of generalized analytic functions for the investigation of infinitesimal deformations of this surface. The reason for this is that the important condition in its use requires the estimation $K \geqq k_0 > 0$ for the

Gaussian curvature, but it is not satisfied for the surface (6) in the flat point, where $K = 0$.

7.2.1. In this subsection we consider the question of the existence of conjugate isometric coordinates on the surface (6). Calculating the coefficients L, M, N and substituting them in the formula (2), we get

$$q(z) = q_0\, e^{2i\varphi} + q^*(z)\,, \tag{7}$$

$$q_0 = \frac{n-2}{n + 2\sqrt{n-1}}\,. \tag{8}$$

Here $z = x + iy = r\, e^{i\varphi}$ and $q^*(z)$ is a known function, the specific form of which will not be needed in what follows. It will be sufficient for us to know that $q^*(z)$ has partial derivatives of the first order, which are continuous for $|z| > 0$ and bounded for $z = 0$. Moreover, $q(z) = O(|z|)$, $z \to 0$, and $|q(z)| < 1$ for all $z \in \overline{D}$. The validity of the inequality for $z = 0$ follows easily from (7), (8). For the remaining values of z it is also valid, since the surface S outside the flat point has positive curvature.

Thus for the study of the question of the existence of a conjugate isometric parametrization in a neighbourhood of the flat point at this time only bounded discontinuities in the coefficients of $q(z)$ can be present (cf. formula (7)). However, homeomorphic solutions of the Beltrami equation exist even for a wider class of $q(z)$, namely for bounded and measurable functions (Bojarski [1]).

Later we will have to know in addition the asymptotic behaviour of the homeomorphism and its partial derivatives as $z \to 0$.

Lemma 7.2.1. *The Beltrami equation* (1), (7), (8) *has a homeomorphic solution* $w(z)$, *whose partial derivatives satisfy a Lipschitz condition outside of* $z = 0$. *By means of this homeomorphism the domain* D *is mapped onto the disk* $|w| < R$ *in such a way that point* $z = 0$ *corresponds to* $w = 0$ *and*

$$w = z\,|z|^{\frac{2q_0}{1-q_0}} + O\left(|z|^{\frac{2}{1-q_0}}\right),$$

$$\partial_z w = \frac{1}{1-q_0}\,|z|^{\frac{2q_0}{1-q_0}} + O\left(|z|^{\frac{2q_0}{1-q_0}+\frac{1-q_0}{1+q_0}}\right), \qquad z \to 0\,.$$

The inverse function $z = z(w)$ *is such that* $\partial_{\overline{w}} z$ *and* $\partial_w z$ *satisfy a Lipschitz condition for* $|w| > 0$,

$$\partial_{\overline{w}} z + \left(q_0\, e^{2i\varphi} + q^*(z)\right)\overline{(\partial_w z)} = 0\,, \qquad |w| < R\,,$$

and the following relations hold:

$$z = w\,|w|^{-\frac{2q_0}{1+q_0}} + O\left(|w|^{\frac{2(1-q_0)}{1+q_0}}\right),$$

$$\partial_w z = \frac{1}{1+q_0}\,|w|^{\frac{-2q_0}{1+q_0}}\left\{1 + O\left(|w|^{\left(\frac{1-q_0}{1+q_0}\right)^2}\right)\right\}, \qquad w \to 0\,.$$

A proof of this lemma, also thorough calculations of the next subsection can be found in the paper Usmanov [1].

7.2.2. Let us consider the coefficient $B(w)$ for the surface (6). It has the form

$$B(w) = -\frac{1}{2}\left(\frac{w}{\overline{w}}\right)^2 \frac{\overline{b(w)}}{w}\,, \tag{9}$$

whereb (w) is a continuous function which satisfies a Hölder condition at the point $w = 0$ with the exponent $1 - \delta$, $\delta = \max(\alpha, 1 - \alpha)$, and $\alpha = (1 - q_0)/(1 + q_0) = 1/\sqrt{n-1}$,

$$b(0) = \frac{n-2}{2\sqrt{n-1}}, \quad n > 2. \tag{10}$$

Substituting (9) in (4), we get the equation of an infinitesimal deformation of the surface (6):

$$\partial_{\bar{w}} U - \frac{1}{2}\left(\frac{\bar{w}}{w}\right)^2 \frac{b(w)}{\bar{w}} \bar{U} = 0. \tag{11}$$

We note that equation (11) is given on the disk $|w| < R$, since the surface is mapped by means of the conjugate isometric system of coordinates onto this domain, cf. lemma 7.2.1.

We shall find a more natural restriction on the unknown function U. We start from the situation that the surface S and its infinitesimal deformation belong to the same class of functions. Then it follows that the unknown variation δL, δM, δN must necessarily be considered in the same class as L, M, N, i.e. δL, δM, δN are continuously differentiable in the domain D. But it is not enough. It is necessary to analyse the behaviour $U(w)$ at the point $w = 0$. Immediate calculations show that the Gaussian curvature $K(w) \in \mathscr{C}^1$ outside of $w = 0$ and

$$K(w) = O\left(|w|^{\frac{2(n-2)}{\sqrt{n-1}}}\right), \qquad w \to 0.$$

But then from (0.3) it follows that $U \in \mathscr{C}^1$ and

$$U(w) = O\left(|w|^{\frac{n-2}{2\sqrt{n-1}}}\right), \qquad w \to 0. \tag{12}$$

Some other natural restrictions can be found in Usmanov [2].

We introduce a new unknown function $U^*(w) = w^2 U(w)$. The equation for $U^*(w)$ is obtained from (11). Before describing the form of this equation, let us agree to the following: from now on the isometric system of coordinates will be denoted by $z = x + iy$, and the unknown function by $w(z)$, instead of the previous notation $w = u + iv$ and $U^*(w)$. We perform the operation mentioned in order to pass to the ordinary notation.

Now we formulate in final form the interrelated geometric and analytic problems. Let S be a simply connected surface of positive curvature with an isolated flat point, given by equation (6). To each infinitesimal deformation of this surface with continuously differentiable variations δL, δM, δN there corresponds a complex-valued function

$$w(z) = z^2 K^{1/4}(z)\,(\delta M + \mathrm{i}\,\delta L), \tag{13}$$

defined and continuously differentiable in the domain $|z| < R$, which satisfies the condition

$$w(z) = O\left(|z|^{2 + \frac{n-2}{2\sqrt{n-1}}}\right), \qquad z \to 0, \tag{14}$$

and the equation

$$\partial_{\bar{z}} w - \frac{b(z)}{2\bar{z}} \bar{w} = 0, \qquad |z| < R. \tag{15}$$

Here $b(z)$ is a continuous function which at the point $z = 0$ satisfies a Hölder condition with exponent $1 - \delta$ and (10).

Thus infinitesimal deformations of the surface (6) are described by the generalized Cauchy-Riemann system (15). This system requires independent investigation. The fact is that the coefficient of $\overline{w}$ has a first order singularity while the theory mentioned assumes a singularity of order less than one.

7.3. Some results from the theory of solutions of generalized Cauchy-Riemann system with a singular point

7.3.1. In this subsection a formula is given for the general solution of the equation

$$\partial_{\bar{z}} w - \frac{\lambda}{2\bar{z}} \overline{w} = f(z) , \qquad z \in G , \tag{16}$$

in the form of an operator on $f(z)$. It is assumed that $w(z) \in \mathscr{C}(\overline{G}) \cap \mathscr{C}^1(G - 0)$ and $f(z) \in \mathscr{C}(G - 0) \cap \mathscr{L}_q(\overline{G})$. We denote by $\Phi(z) \in \mathscr{C}(\overline{G}) \cap \mathscr{C}^1(G - 0)$ the solution of (16) for $f(z) = 0$.

Lemma 7.3.1. *The formula*

$$w(z) = \Phi(z) + S_G f , \qquad z \in G , \tag{17}$$

where

$$S_G f = -\frac{1}{\pi} \iint\limits_G \left[\frac{\Omega_1(z, \zeta)}{\zeta} f(\zeta) + \frac{\Omega_2(z, \zeta)}{\bar{\zeta}} \overline{f(\zeta)} \right] d\xi \, d\eta ,$$

gives the general solution of (16) *in the class* $\mathscr{C}(\overline{G}) \cap \mathscr{C}^1(G - 0)$.

In fact, the right-hand side of (17) is the sum of the general solution $\Phi(z)$ of the homogeneous equation (16) and a particular solution $S_G f$ of the inhomogeneous equation (16). A proof of the last assertion and a method of the introduction of (17) are described in Usmanov [6]. We note that the operator S_G is completely continuous from $\mathscr{L}_p(\overline{G})$, $p > 2$, into $\mathscr{C}(\overline{G})$.

7.3.2. Let us introduce the integral equation for continuous solutions of (15). We assume that $b(z) \in \mathscr{C}(\overline{G})$ and $|b(z) - b(0)| < M |z|^\alpha$, $\alpha > 0$. As in the previous subsection, it will be assumed that $w(z) \in \mathscr{C}(\overline{G}) \cap \mathscr{C}^1(G - 0)$. We write (15) in the following form

$$\partial_{\bar{z}} w - \frac{b(0)}{2\bar{z}} \overline{w} = \frac{b(z) - b(0)}{2\bar{z}} \overline{w} .$$

If we set $b(0) = \lambda$ and $\left(b(z) - b(0)\right) \overline{w}|_{2\bar{z}} = f(z)$, then (15) takes the form of (16), here $f(z)$ satisfies the condition of the subsection 7.2. In this connection formula (17), applied to (15) gives the following integral equation

$$w(z) = \Phi(z) + P_G \overline{w} , \qquad z \in G , \tag{18}$$

where $P_G \overline{w} = S_G \left(\frac{b(\zeta) - b(0)}{2\bar{\zeta}} \overline{w(\zeta)} \right)$ and $\Phi(z)$ is a continuous solution of the equation

$$\partial_{\bar{z}} \Phi - \frac{b(0)}{2\bar{z}} \overline{\Phi} = 0 , \qquad z \in G . \tag{19}$$

It is established that P_G is completely continuous from $\mathscr{C}(\overline{G})$ into $\mathscr{C}(\overline{G})$.

Thanks to (18) a connection between the sets of continuous solutions of systems (15) and (19) is established by means of the operator P_G. In this connection the solution of any problem for the equation (15) consists in investigating the analogous problem for the model equation (19).

As to (19), its theory is constructed on the basis of the generalized Cauchy integral formula

$$\Phi(z) = \frac{1}{2\pi i} \int\limits_{\partial G} \left(\frac{\Omega_1(z, \zeta)}{\zeta} \Phi(\zeta) \, d\zeta - \frac{\Omega_2(z, \zeta)}{\bar{\zeta}} \overline{\Phi(\zeta)} \, \overline{d\zeta} \right) \tag{20}$$

and a generalized integral of Cauchy type (it is obtained from (20) if on the right instead of $\Phi(\zeta)$ we write $\nu(\zeta)$ (USMANOV [4])).

Thanks to (18) and the theory of the model equation (19) we can describe the behaviour of continuous solutions of (15) at the point $z = 0$, investigate boundary problems and so on. For example, we shall introduce the formulation and the final results of the solution of the Riemann-Hilbert boundary value problem (USMANOV [2]).

Problem. To find a solution $w(z) \in \mathscr{C}(\overline{G}) \cap \mathscr{C}^1(G - 0)$ of the equation (15), which satisfies the boundary condition

$$\operatorname{Re} [\overline{g(z)} \, w] = h(z) \quad \text{for} \quad z \in \partial G \,. \tag{21}$$

It is assumed that $g(z)$ and $h(z)$ are given functions of the Hölder class $\mathscr{C}_\nu(\partial G)$, $0 < \nu \leqq 1$, $|g(z)| \neq 0$ on ∂G.

It is established

Theorem 7.3.1. *Let* $\operatorname{Ind}_{\partial G} g(z) = 0$. *Then for values* $b(0) > 0$ *or* $\operatorname{Im} b(0) \neq 0$, *the solution of the boundary value problem exists and is unique. For* $b(0) < 0$ *the homogeneous problem* ($h = 0$) *has only one nontrivial solution, while the nonhomogeneous one is solvable if and only if* $h(z)$ *satisfies a certain condition.*

Let $m = \operatorname{Ind}_{\partial G} g > 0$. *Then the homogeneous problem has* $2m$ *linearly independent solutions, and the nonhomogeneous one is solvable for any* $h(z)$.

Let $m = \operatorname{Ind}_{\partial G} g < 0$. *Then the homogeneous problem has only the trivial solution, while the fulfilment of* $2\,|m|$ *real conditions is necessary and sufficient for the solvability of the nonhomogeneous one.*

7.4. The influence of a flat point on infinitesimal deformations of the surface

From the geometric point of view the boundary condition (21) according to (13) indicates the determination of infinitesimal deformation of the surface with a preassigned linear combination of δL and δM.

For the investigation of the geometric problem we must restrict the set of solutions of the boundary problem (15), (21) by means of the additional condition (14). Let $\lambda_n = (n - 2)/2 \sqrt{n - 1}$. Then one has

Theorem 7.4.1. *Let* S *be a simply connected part of a surface of positive curvature with an isolated flat point, given by equation* (6). *With respect to the class of* S *and its infinitesimal deformation, we make the same assumption as in* 7.2.

We consider a deformation of the surface which has along the boundary of the surface a given linear combination of δL, δM, δN, *equivalent to the boundary condition* (21).

The image of the deformation of the surface essentially depends on the values of m and n, i.e. the index of the boundary condition and the order of tangency of the surface with its tangent plane at the flat point.

If $m \geqq 2\sqrt{1+\lambda_n}$, then S always has an infinitesimal deformation which satisfies the boundary condition (21), and, moreover, the whole manifold of deformations is generated by $2(m-k_n)-1$ linearly independent infinitesimal deformations, where k_n is defined by the inequalities

$$2\sqrt{1+\lambda_n}-1 \leqq k_n < 2\sqrt{1+\lambda_n}.$$

If $m < 2\sqrt{1+\lambda_n}$, then S admits one and only one infinitesimal deformation under the hypothesis that $h(z)$ satisfies $p_{n,m} = 2k_{n,m}+1$ relations. The number $k_{n,m}$ is defined by the inequalities

$$2\sqrt{1+\lambda_n}-m-1 \leqq k_{n,m} < 2\sqrt{1+\lambda_n}-m$$

and can range from 0 to arbitrarily large values as $n \to \infty$ (Usmanov [2]).

7.5. Generalizations

The method developed in I. N. Vekua [5] for the investigation of infinitesimal deformations of the surface given by the equation (6) can be extended to a more general case. Indeed, let the surface S be given in rectangular Cartesian coordinates $0x_1y_1z_1$ by the equation

$$z_1 = r_1^n f(\varphi) + h(x_1, y_1), \qquad (x_1, y_1) \in D,$$

where $n > 2$, $x_1 + iy_1 = r_1 e^{i\varphi}$, $f(\varphi)$ is a regular periodic function with period 2π, $h(x_1, y_1)$ is a rather regular one, $h(x_1, y_1) = O(r_1^{n+\alpha})$, $r_1 \to 0$, $\alpha > 0$, and its partial derivatives of first and second order for $r_1 = 0$ have zeroes of order $n+\alpha-1$ and $n+\alpha-2$. Moreover we assume that the surface has positive curvature in $\bar{D}-(0, 0)$.

On this surface there exists the conjugate isometric parametrization $z = x + iy$, $i^2 = -1$, in which the flat point corresponds to $z = 0$. We make correspond to each infinitesimal deformation the complex-valued function $w(z)$ by (13). It is proved, that $w(z)$ satisfies the equation (15), but

$$b(z) = \beta(\varphi) + o(|z|^\gamma), \qquad z \to 0. \tag{22}$$

Here $\beta(\varphi)$ is a regular periodic function with period 2π which is defined from $f(\varphi)$ and γ is a positive number.

It is clear that the asymptotic behaviour of $b(z)$ in a neighbourhood of $z = 0$ is more complicated than before. Particularly, when $\beta(\varphi) = \beta_m e^{im\varphi}$, where β_m is a complex number and m is any integer, some results were obtained in Usmanov [3].

But in other even the simpliest particular cases (e.g. $\beta(\varphi) = \beta_0 + \beta_1 e^{i\varphi}$ or $\beta(\varphi) = \beta_{-k} e^{-ik\varphi} + \beta_k e^{ik\varphi}$ etc.) we have no results in the theory of the equation (15), (22) at present, not mentioning the general case.

8. Approximation Methods for Solving Singular Integral Equations

S. Prössdorf

8.1. Introduction

The Riemann-Hilbert boundary value problem (also known as the coupling problem) may be transformed, by means of the Sohockiĭ-Plemelj formulas, into a one-dimensional singular integral equation with a Cauchy kernel and conversely (see Mushelišvili [3]). Accordingly, the approximation methods for solving singular integral equations can be employed for calculating the solution of the Riemann-Hilbert boundary value problem and vice versa. A procedure for solving the latter, which is based upon successive approximations, is given in Mandžavidze [1]. This method was successfully applied in Mushelišvili [3, §§ 133—134] for obtaining an approximate solution of the singular integral equations.

Recent years have witnessed extensive achievements in the theory of approximation methods for singular integral equations, particularly in the methods of Galerkin and collocation for equations with continuous (smooth) coefficients (consult the texts Gohberg and Feldman [1], Ivanov [2], Michlin and Prössdorf [1], Prössdorf and Silbermann [3] as well as the references cited therein).

In this paper we primarily consider two collocation methods for approximating the solution of singular integral equations. These two methods exhibit the property that they also converge for certain classes of singular integral equations with discontinuous coefficients. One of the methods utilizes trigonometric polynomials, whereas the other is based upon finite element methods for approximate solutions. The proofs of convergence and error estimates are accomplished with the tools of functional analysis, complex analysis, and approximation theory; however, these tools are very dissimilar for both methods. Initially, we shall prove a general convergence theorem concerning approximation methods for linear operator equations in Banach spaces. This theorem shall be the foundation of the aforementioned methods as well as of a multitude of approximation methods for a variety of equations.

8.2. A perturbation theorem for approximation methods

Let X_n and Y_n be closed subspaces of the Banach spaces X and Y, respectively, and Q_n denotes a linear projection operator, which is generally unbounded, from Y onto Y_n (i.e., Q_n is an idempotent surjection with domain of definition $D(Q_n) \subset Y$) $(n = 1, 2, \ldots)$. By P_n we denote the inclusion operator from the subspace X_n into the space X.

For a bounded linear operator $A: X \to Y$ (i.e. $A \in \mathcal{L}(X, Y)$), in addition to the equation

$$Ax = y (x \in X, y \in Y), \tag{1}$$

we shall consider a sequence of approximate equations

$$A_n x_n = y_n (x_n \in X_n, y_n \in Y_n), \tag{2}$$

where the operators $A_n \in \mathcal{L}(X_n, Y_n)$ are assumed to have (bounded) inverses for all $n \geqq n_0$. If additionally $A_n = Q_n A P_n$ and $y_n = Q_n y$ hold, then we say that (2) is the *projection method* $\{P_n, Q_n\}$ for the operator equation (1). We further denote by $K(A, A_n)$ the set of all elements $y \in Y$ with the following properties: $y \in D(Q_n)$ for $n \geqq n_1(y)$ and there exists a solution $x \in X$ of the equation (1) such that

$$\lim_{n\to\infty} ||x - A_n^{-1} Q_n y||_X = 0 .$$

The set $K(A, A_n)$ is called the *convergence manifold* of the operator A with respect to the family of operators $\{A_n\}$.

Obviously, $K(A, A_n)$ is a linear set contained in the image space im A of the operator A. In case A is invertible, then $K(A, A_n)$ is the maximal linear set on which the operators $A_n^{-1} Q_n$ strongly converge to A^{-1} as $n \to \infty$.

The subsequent theorem asserts the invariance of the convergence manifold under suitable perturbations of the operators A and A_n. Such results have far-reaching significance whenever concrete problems are investigated.

Theorem 8.2.1. *Let A and $A + T \in \mathcal{L}(X, Y)$ be invertible operators, $T_n \in \mathcal{L}(X_n, Y_n)$, $y_n \in Y_n$ $(n = 1, 2, ...)$ and the following conditions hold:*

(a) im $T \subset D(Q_n)$, $Q_n T \in \mathcal{L}(X, Y_n)$ $(n \geqq n_0)$,

(b) $\varrho_n := ||(A^{-1} - A_n^{-1} Q_n) T||_X \to 0$ $(n \to \infty)$,

(c) $\varepsilon_n := ||A_n^{-1}(T_n - Q_n T P_n)||_{X_n \to X} \to 0$ $(n \to \infty)$,

(d) $\lim_{n\to\infty} ||A_n^{-1} y_n||_X = 0$.

Then

$$K(A, A_n) = K(A + T, A_n + T_n) \tag{3}$$

and, for each $y \in K(A, A_n)$ and solutions $u \in X$ and $u_n \in X_n$ of the equations

$$(A + T) u = y , \qquad (A_n + T_n) u_n = Q_n y + y_n ,$$

respectively, the following estimates

$$||u - u_n|| \leqq c\{||u - A_n^{-1} Q_n A u|| + \varepsilon_n ||A_n^{-1} Q_n y|| + ||A_n^{-1} y_n||\} , \tag{4}$$

with constant $c > 0$, are valid.

Theorem 8.2.1 generalizes a number of results of the author and B. Silbermann (Prössdorf and Silbermann [2, 3] in particular [3, chapter 1, theorems 3.1, 3.2, and 5.1]). We shall demonstrate that theorem 8.2.1 can be obtained by composing the immediately aforementioned results. Theorem 8.2.1 was stated in Prössdorf and Silbermann [4], however, in a weaker form.

Proof. We set

$$\tilde{A} = A + T , \qquad \tilde{A}_n = A_n + Q_n T P_n , \qquad \Gamma_n = T_n - Q_n T P_n ,$$

for which $A_n + T_n = \tilde{A}_n + \Gamma_n$ holds. From the proof of theorem 3.1, chapter 1 in Prössdorf and Silbermann [3] follows $K(A, A_n) \subset K(\tilde{A}, \tilde{A}_n)$. The proof of theorem

5.1, chapter 1 in PRÖSSDORF and SILBERMANN [3] implies $K(\tilde{A}, \tilde{A}_n) \subset K(\tilde{A}, \tilde{A}_n + \Gamma_n)$ as well as the estimates

$$||u_n - \tilde{A}_n^{-1}Q_ny|| \leqq (1 - \varepsilon_n)^{-1} [\varepsilon_n ||\tilde{A}_n^{-1}Q_ny|| + ||\tilde{A}_n^{-1}y_n||] \tag{5}$$

for all $n \geqq 1$ with $\varepsilon_n < 1$ (compare with MICHLIN and PRÖSSDORF [1, remark 1, p. 445]). This demonstrates the inclusion $K(A, A_n) \subset K(A + T, A_n + T_n)$. We derive out of the self-evident relation

$$A_n^{-1} = (I + A_n^{-1}Q_nT)\, \tilde{A}_n^{-1} \tag{6}$$

the equations

$$u - \tilde{A}_n^{-1}Q_ny = (I + A_n^{-1}Q_nT)^{-1} [u - A_n^{-1}Q_nAu]\,. \tag{7}$$

On account of condition (b), the relations (5), (6), and (7) immediately confirm the validity of relation (4). The chain of inclusions

$$K(\tilde{A}, \tilde{A}_n + \Gamma_n) \subset K(\tilde{A}, \tilde{A}_n) \subset K(A, A_n)$$

is a direct consequence of the relationships $\tilde{A}_n^{-1} = (I + \tilde{A}_n^{-1}\Gamma_n)\,(\tilde{A}_n + \Gamma_n)^{-1}$ and (6). This completes the proof of theorem 8.2.1.

Remark 8.2.1. The following estimate is easily derivable:

$$||u - A_n^{-1}Q_nAu|| \leqq \varrho_n ||u|| + ||A^{-1}y - A_n^{-1}Q_ny||\,.$$

Remark 8.2.2. Condition (c) of theorem 8.2.1 is trivially satisfied for $T_n = Q_nTP_n$, in particular for the projection method $\{P_n, Q_n\}$ for the perturbed operator $A + T$

Remark 8.2.3. If there exists a Banach space $Z \subset Y$ continuously imbedded into Y, such that $Z \subset K(A, A_n)$ and the operator $T: X \to Z$ is compact, then condition (b) of theorem 8.2.1 is satisfied.

Remark 8.2.4. In case $T = 0$, then the conditions (c) and (d) of theorem 8.2.1 yield stability conditions for the approximation method (2) (compare also with (5)).

8.3. Trigonometric collocation method in the space L_2

We next introduce several conventions in notation. Γ denotes the unit circle of the complex plane, i.e. $\Gamma = \{t \in \mathbb{C} : |t| = 1\}$. $L_p = L_p(\Gamma)$ $(1 \leqq p < \infty)$ stands for the Banach space of all Lebesgue-measurable complex-valued functions $f(t)$ $(t \in \Gamma)$ on Γ, whose p-th power is Lebesgue-integrable, with norm

$$||f||_p := \left((2\pi)^{-1} \int_0^{2\pi} |f(e^{is})|^p \,ds\right)^{1/p}.$$

Further, $R(\Gamma)$ is the Banach space of all bounded and Riemann-integrable functions on Γ with norm $||f||_\infty := \sup_{t\in\Gamma} |f(t)|$ and $C(\Gamma)$ $(\subset R(\Gamma))$ is the Banach space of the continuous functions on Γ.

The symbol PC(Γ) designates the algebra of functions $f(t)$ on Γ which are piecewise continuous as well as continuous from the left on Γ (with respect to the orientation of Γ) in the following sense: for all $t \in \Gamma$ the limits $f(t \pm 0)$ exist and are finite, $f(t - 0) = f(t)$ and f is discontinuous at most at a finite number of points $t \in \Gamma$. To each function $f \in \mathrm{PC}(\Gamma)$ we assign a naturally orientable closed curve $\tilde{f}$, which we obtain by adjoining, to the image of f, the line segments

$$\mu f(t+0) + (1-\mu) f(t) \qquad (0 \leqq \mu \leqq 1)\,.$$

We call the function $f \in \mathrm{PC}(\Gamma)$ *regular*, if $0 \notin \tilde{f}$. The number of times the curve $\tilde{f}$ windes around the origin ($t = 0$) is called the *index* of the regular function $f \in \mathrm{PC}(\Gamma)$ and shall be designated by $\operatorname{ind} f$. Each regular function $f \in \mathrm{PC}(\Gamma)$ can be written in the form

$$f(t) = r(t)\, h(t)\,, \tag{7}$$

where $r(t)$ is a (trigonometric) polynomial and $h \in \mathrm{PC}(\Gamma)$ such that $\|1 - h\|_\infty < 1$; moreover, $\operatorname{ind} f = \operatorname{ind} r$ (see Gohberg and Feldman [1], 118–119).

Furthermore, we denote by $\mathcal{T}_n$ the set of all (trigonometric) polynomials of degree n — i.e. all expressions in t of the form $\sum\limits_{k=-n}^{n} \xi_k t^k$ ($t \in \Gamma$, $\xi_k \in \mathbb{C}$). To each bounded function f on Γ we assign the Lagrange interpolation polynomial

$$(L_n f)(t) = \sum_{k=-n}^{n} a_k t^k \in \mathcal{T}_n$$

with coefficients

$$a_k = (2n+1)^{-1} \sum_{k=-n}^{n} f(t_j)\, t_j^{-k}\,, \quad t_j = \mathrm{e}^{2\pi \mathrm{i} j/(2n+1)} \qquad (k, j = 0, \pm 1, \dots, \pm n)\,.$$

The polynomial $L_n f$ satisfies the conditions

$$(L_n f)(t_j) = f(t_j) \qquad (j = 0, \pm 1, \dots, \pm n)\,,$$

and these conditions determine $L_n f$ uniquely. Obviously, the operator $L_n : f \to L_n f$ is an (unbounded) projection operator in each of the spaces $L_p(\Gamma)$ ($1 \leqq p < \infty$). It is known, that $\|f - L_n f\|_p \to 0$ ($n \to \infty$) for arbitrary $f \in R(\Gamma)$ and $1 \leqq p < \infty$ (see Zygmund [1]).

By S we denote the singular integral operator with the Cauchy kernel over the path Γ,

$$(Sx)(t) := (\pi \mathrm{i})^{-1} \int_\Gamma x(\tau)\, (\tau - t)^{-1}\, \mathrm{d}\tau \qquad (t \in \Gamma)\,.$$

Since $S \in \mathcal{L}(L_p(\Gamma))$ ($1 < p < \infty$) and $S^2 = I$ hold, the operators $P = 2^{-1}(I + S)$ and $Q = 2^{-1}(I - S)$ are bounded projection operators in the space $L_p(\Gamma)$.

We shall now consider the singular integral equation of the form

$$(Ax)(t) := a(t)\,(Px)(t) + b(t)\,(Qx)(t) = y(t) \qquad (t \in \Gamma) \tag{8}$$

with coefficients $a, b \in \mathrm{PC}(\Gamma)$. As is well known, the operator $A \in \mathcal{L}(L_2(\Gamma))$ is invertible if and only if $b(t) \neq 0$ ($t \in \Gamma$), a/b is regular and $\operatorname{ind}(a/b) = 0$ (see e.g. Gohberg and Feldman [1], Gohberg and Krupnik [1]). The object which we seek is an approximate solution of equation (8) having the appearance $x_n(t) = \sum\limits_{k=-n}^{n} \xi_k t^k \in \mathcal{T}_n$, where the coefficients $\xi_k = \xi_k^{(n)}$ ($k = 0, \pm 1, \dots, \pm n$) are determined by the system of equations

$$(Ax_n)(t_j) = y(t_j) \qquad (j = 0, \pm 1, \dots, \pm n)\,,$$

which is identical to

$$a(t_j) \sum_{k=0}^{n} t_j^k \xi_k + b(t_j) \sum_{k=-n}^{-1} t_j^k \xi_k = y(t_j) \qquad (j = 0, \pm 1, \dots, \pm n)\,. \tag{9}$$

Theorem 8.3.1. *Let $a, b \in \mathrm{PC}(\Gamma)$ and $A = aP + bQ$ be an invertible operator in the space $L_2(\Gamma)$. Then system (9) possesses a unique solution for all n sufficiently large. Moreover, the solutions $x_n \in \mathcal{T}_n$ converge in L_2-norm to the solution $x \in L_2(\Gamma)$ of equation (8) for an arbitrary function $y \in R(\Gamma)$.*

Proof. It is quite apparent that equation (9) is equivalent to

$$L_n A P_n x_n = L_n y \,. \tag{9'}$$

Henceforth, P_n shall always denote the projection operator which assigns to every function $f \in L_p(\Gamma)$ the n-th partial-sum $\sum_{k=-n}^{n} f_k t^k$ of its Fourier expansion $f(t) = \sum_{k=-\infty}^{\infty} f_k t^k$. It is well-established that $\|f - P_n f\|_p \to 0$ as $n \to \infty$ for arbitrary $f \in L_p(\Gamma)$ $(1 < p < \infty)$.

Let us now consider an $f \in L_2(\Gamma)$. Since $\sum_{k=-n}^{n} t_k^m = 0$ for any integer m satisfying $0 < |m| \leqq 2n$, it is obvious that

$$\|L_n f\|_2^2 = \sum_{k=-n}^{n} |a_k|^2 = (2n+1)^{-1} \sum_{j=-n}^{n} |f(t_j)|^2 \,.$$

This justifies for an arbitrary bounded function g the estimate

$$\|L_n g f\|_2 \leqq \|g\|_\infty \, \|L_n f\|_2 \,. \tag{10}$$

We next turn our attention to the operator $G := I + gP$, where $g \in R(\Gamma)$ and $\|g\|_\infty < 1$. The operator $G \in \mathcal{L}(L_2(\Gamma))$ is invertible. In consequence of the norm-estimate (10) and $\|P\|_2 = 1$, the operators $G_n := L_n G P_n \in \mathcal{L}(\mathcal{T}_n)$ are invertible and moreover, the norm-estimates $\|G_n^{-1}\| \leqq (1 - \|g\|_\infty)^{-1}$ hold. Inequalities (10) further imply the uniform boundedness in $L_2(\Gamma)$ of the sequence of operators $\{G_n\}_{n=1}^{\infty}$, and we obtain for each $u_m \in \mathcal{T}_m$ and $n \geqq m$ the convergence property

$$G_n u_m = u_m + L_n g P u_m \xrightarrow{L_2(\Gamma)} G u_m \qquad (n \to \infty) \,.$$

Therefore, we conclude by means of the Banach-Steinhaus theorem that $G_n u \xrightarrow{L_2(\Gamma)} Gu$ for arbitrary $u \in L_2(\Gamma)$. The relation

$$G_n^{-1} L_n y = G_n^{-1} P_n [L_n y - G_n x] + P_n x \,, \qquad Gx = y \,,$$

which is more than self-evident, yields

$$\|G_n^{-1} L_n y - G^{-1} y\|_2 \leqq \frac{\|P_n\|_2}{1 - \|g\|_\infty} [\|L_n y - y\|_2 + \|Gx - G_n x\|_2] + \|P_n x - x\|_2 \xrightarrow[n \to \infty]{} 0 \tag{11}$$

for an arbitrary $y \in R(\Gamma)$.

Let the conditions of theorem 8.3.1 be fullfilled. Because of relationship (7),

$$\frac{a(t)}{b(t)} = r(t)\, h(t) \tag{12}$$

is true, where $r(t) \in \mathcal{T}_m$, $\operatorname{ind} r = 0$ and $\|g\|_\infty < 1$ with $g(t) = h(t) - 1 \in \mathrm{PC}(\Gamma)$. It is a well-established fact that the function $r(t)$ possesses the representation (see e.g. Gohberg and Feldman [1, p. 17]).

$$r(t) = r_+(t)\, r_-(t) \,,$$

where $r_{\mp}^{\pm 1}$ and $r_{-}^{\pm 1}$ are analytic functions of the inner and outer domains of Γ, respectively, with $r_-(\infty) = 1$. Hence, the equations $Pr_{\mp}^{\pm 1}P = r_{\mp}^{\pm 1}P$ and $Pr_{-}^{\pm 1}P = Pr_{-}^{\pm 1}$ are valid and they confirm the following representation of the operator A

$$A = br_+(hr_-P + r_+^{-1}Q) = br_+[G(Pr_- + Qr_+^{-1}) + T]$$

with $T = hQr_-P + Pr_+^{-1}Q$. Since the operator $T\colon L_2(\Gamma) \to R(\Gamma)$ is compact, it suffices to demonstrate theorem 8.3.1 for the operator $B := br_+G(Pr_- + Qr_+^{-1})$ instead for A (in consequence of theorem 8.2.1).

We note $L_nf = L_nfL_n$ and $(Pr_- + Qr_+^{-1})\, u_n \in \mathcal{T}_n$ for all $u_n \in \mathcal{T}_n$ and hence

$$B_n := L_nBP_n = L_nbr_+G_n(Pr_- + Qr_+^{-1})\, P_n\,,$$
$$B_n^{-1}L_n = (Pr_-^{-1} + Qr_+)\, G_n^{-1}Lb^{-1}r_+^{-1} \tag{13}$$

follows. Relations (11) and (13) imply for arbitrary $y \in R(\Gamma)$

$$B_n^{-1}L_ny \xrightarrow{L_2(\Gamma)} (Pr_-^{-1} + Qr_+)\, G^{-1}b^{-1}r_+^{-1}y = B^{-1}y \qquad (n \to \infty)\,,$$

which terminates the proof of theorem 8.3.1.

Theorem 8.3.1 was first proved for a more restricted class of coefficients a and b (e.g. a and b are Hölder-continuous) in GABDULHAEV [1] by means of a different technique. The inequality (10) may also be found in GABDULHAEV [2]. The proof of theorem 8.3.1 was carried out in JUNGHANNS [1] with assumptions as immediately aforementioned. The arguments used herein to justify theorem 8.3.1 overlap essentially with the proof stated in JUNGHANNS [1]; however, the proof given in this presentation is a natural generalization of the proof, with Hölder-continuous coefficients a and b, which appeared in PRÖSSDORF and SILBERMANN [3].

Remark 8.3.1. Obviously, theorem 8.3.1 with the above-stated proof remains valid for arbitrary bounded coefficients a and b possessing the following properties: a/b is representable in the form (12) with a non-vanishing Hölder-continuous function $r(t)$, b^{-1}, $h \in R(\Gamma)$ with $\|1 - h\|_\infty < 1$, and $aP + bQ$ is invertible in $L_2(\Gamma)$.

Remark 8.3.2. If the operator $aP + bQ$ has only a one-sided inverse, i.e. $\varkappa :=$ $:= \operatorname{ind} r \neq 0$ is true (see (12)), then the operator $A^{(\varkappa)} := t^{-\varkappa}aP + bQ$ satisfies the conditions of theorem 8.3.1. If $x_n(t) = \sum_{k=-n}^{n} \xi_k t^k$ is the solution of the system

$$(A^{(\varkappa)}x_n)\,(t_j) = y(t_j) \qquad (j = 0, \pm 1, \dots, \pm n)$$

for this particular case, then the functions $\tilde{x}_n(t) = \sum_{k=0}^{n} \xi_k t^{k-\varkappa} + \sum_{k=-n}^{-1} \xi_k t^k$ converge in $L_2(\Gamma)$-norm toward a solution of equation (8), provided this equation is solvable (also consult MICHLIN and PRÖSSDORF [1, Kap. XVII, § 3.5]).

Remark 8.3.3. A singular integral equation of the form

$$\tilde{a}(t)\, x(t) + \frac{\tilde{b}(t)}{\pi i} \int_{\tilde{\Gamma}} \frac{x(\tau)}{\tau - t}\, d\tau = y(t) \qquad (t \in \tilde{\Gamma})\,, \tag{14}$$

where $\tilde{\Gamma}$ denotes an arbitrary closed Ljapunov-curve, can be transformed, via a suitable map from $\tilde{\Gamma}$ onto the unit circle Γ, into an equation where the operator has the form $aP + bQ + T$ with operator T compact (see MICHLIN and PRÖSSDORF [1, Kap. II, Th. 4.3]). On account of this, theorems 8.2.1 and 8.3.1 lay the foundations for the development of collocation methods for equation (14).

Remark 8.3.4. Theorem 8.3.1 still remains intact if we seek an approximate solution of the form $u_n(t) = \sum_{k=-n}^{n-1} \eta_k t^k$, where the coefficients η_k are determined by the system $(Au_n)(\tau_j) = y(\tau_j)$ $(j = -n, \dots, n-1)$ with $\tau_j = e^{\pi i j/n}$. This method is occasionally referred to as the Multhopp method (consult PRÖSSDORF and SILBERMANN [3]).

8.4. Trigonometric collocation method in the space L_p

As evident from the proof of theorem 8.3.1, all of its arguments can be carried over to an $L_p(\Gamma)$-formulation $(1 < p < \infty)$ of theorem 8.3.1 with the exception of the statements concerning the operators G and G_n. To guarantee the invertibility of these operators as well as the convergence

$$||G^{-1}y - G_n^{-1}L_n y||_p \to 0 \quad \text{as} \quad n \to \infty \qquad (y \in R(\Gamma)), \tag{15}$$

a generalized version of inequality (10) — i.e., a version with $f \in \mathcal{T}_n$ — must be introduced. This requires that we allow for the introduction of a constant. Out of ZYGMUND [1], theorem X.7.5, the estimate

$$||L_n g u_n||_p \leqq C_p\, ||g||_\infty\, ||u_n||_p \qquad (u_n \in \mathcal{T}_n) \tag{16}$$

follows, where C_p is a positive constant depending upon p, for any arbitrary bounded function g. Now we assume that the function $g \in R(\Gamma)$ satisfies the inequality

$$||g||_\infty < 1/(C_p\, ||P||_p) =: m_p\,. \tag{17}$$

The operators $G = I + gP \in \mathcal{L}(L_p(\Gamma))$ and $G_n = L_n G P_n \in \mathcal{L}(\mathcal{T}_n)$ are invertible in this case and $||G_n^{-1}|| \leqq 1/(1 - ||g||_\infty\, C_p\, ||P||_p)$. After this stage we note that relation (15) follows via the reasoning utilized in the proof of theorem 8.3.1.

Remark 8.4.1. Constant C_p can be estimated by means of $C_p \leqq 9(1 + 2A_p)$, where A_p is the constant appearing in the theorem of M. RIESZ (see ZYGMUND [1, proof of theorem X.7.5]). It is well-known that $A_p = \tan(\pi/2p)$ for $1 < p \leqq 2$ (see PICHORIDES [1]) and $A_p = \cot(\pi/2p)$ for $2 \leqq p < \infty$. In consideration of $||S||_p = A_p$ (see GOHBERG and KRUPNIK [1, Kap. XIII]) we conclude $||P||_p \leqq (1 + A_p)/2$ and consequently $m_p \geqq 2/9(1 + 3A_p + 2A_p^2)$.

We let $\mathfrak{M}_p$ denote the set of all functions $f(t)$ on Γ representable in the form

$$f(t) = r(t)\, h(t)\,, \tag{18}$$

where $r(t)$ $(\in \mathcal{T}_m)$ is a polynomial and $h(t) \in R(\Gamma)$ is a function such that $g(t) = h(t) - 1$ satisfies inequality (17). The proof of theorem 8.3.1 and the foregoing considerations lead immediately to the following theorem.

Theorem 8.4.1. *Let $a(t)$ and $b(t) \in R(\Gamma)$ be functions such that $a/b \in \mathfrak{M}_p$ and the operator $aP + bQ$ is invertible in the space $L_p(\Gamma)$ $(1 < p < \infty)$. Then the collocation method* (9) *converges in L_p-norm for arbitrary $y \in R(\Gamma)$.*

Since any continuous function $f(t)$ possesses the form (18), where $||1 - h||_\infty$ can be chosen arbitrarily small, we conclude $C(\Gamma) \subset \mathfrak{M}_p$ as well as the following corollaries.

Corollary 8.4.1. *Let $a(t)$, $b(t) \in C(\Gamma)$ be continuous functions and $aP + bQ$ be an invertible operator in $L_p(\Gamma)$ $(1 < p < \infty)$. Then the collocation method* (9) *converges in the space $L_p(\Gamma)$ for arbitrary $y \in R(\Gamma)$.*

Corollary 8.4.1 was demonstrated in BOĬKOV [1] for the case $p = 2$ and for arbitrary $p > 1$ in JUNGHANNS [1].

Corollary 8.4.2. *Let the function* $f \in \mathrm{PC}(\Gamma)$ *satisfy*

$$|\arg f(t+0) - \arg f(t-0)| \leqq \alpha ,$$

$||f(t+0)| - |f(t-0)|| \leqq \Delta$ *and* $|f(t)| \geqq \gamma > 0$ *for all* $t \in \Gamma$, *where* $4 \sin^2 (\alpha/2) + \Delta^2/\gamma^2 < m_p^2$ *holds. Then* $f \in \mathfrak{M}_p$.

Proof. It is clear that there exists a function $d \in C(\Gamma)$ such that $|f(t)| \leqq |d(t)|$ and $|d(t)| - |f(t)| \leqq \Delta$ for all $t \in \Gamma$. We set $\varrho(t) := |f(t)|/|d(t)| \leqq 1$. Simple geometric considerations demonstrate

$$\left|\frac{f(t)}{d(t)} - 1\right|^2 = 4\left(\sin^2\frac{\alpha}{2}\right)\varrho(t) + (1-\varrho(t))^2 .$$

Therefore, $||1 - f/d||_\infty \leqq 4 \sin^2 \alpha/2 + \Delta^2/\gamma^2$. Since every function $d \in C(\Gamma)$ can be approximated by trigonometric polynomials as close as desired, we conclude that $f \in \mathfrak{M}_p$.

Generalizations of theorem 8.3.1 and corollary 8.4.1 to the case of matrix equations of the form (8) and (19) were obtained recently by Junghanns and Silbermann [2] by means of a local principle. For the case, if Γ is a finite interval on the real line, consult the complete paper of Junghanns and Silbermann [1].

8.5. The method of mechanical quadratures

In the present section we turn to the singular integral equation of the form

$$(Ax)(t) := a(t)(Px)(t) + b(t)(Qx)(t) + \int_\Gamma k(t,\tau)\, x(\tau)\, d\tau = y(t) \quad (t \in \Gamma) \quad (19)$$

with kernel $k(t,\tau) \in C(\Gamma \times \Gamma)$. In the above-mentioned method an approximate solution is sought in the form $x_n(t) = \sum_{k=-n}^{n} \xi_k t^k$ with coefficients ξ_k to be calculated from

$$a(t_j) \sum_{k=0}^{n} t_j^k \xi_k + b(t_j) \sum_{k=-n}^{-1} t_j^k \xi_k + \sum_{k=-n}^{n} \gamma_{jk} \xi_k = y(t_j) \qquad (j = 0, \pm 1, \dots, \pm n), \quad (20)$$

where

$$\gamma_{jk} = \frac{2\pi i}{2n+1} \sum_{r=-n}^{n} k(t_j, t_r)\, t_r^{k+1}, \qquad t = e^{2\pi i j/(2n+1)} .$$

Let $H^\mu(\Gamma)$ $(0 < \mu \leqq 1)$ denote the Banach space of all functions $f \in C(\Gamma)$ satisfying a Hölder condition with exponent μ (see e.g. Mushelišvili [3]). Further, $C^{m,\mu}(\Gamma)$ stands for the set of all functions $f \in C(\Gamma)$ whose m-th derivative belongs to $H^\mu(\Gamma)$.

Theorems 8.2.1 and 8.3.1 lead to the following theorem (for the proof see Michlin and Prössdorf [1, pp. 454—455]).

Theorem 8.5.1 (Gabdulhaev [2], Gabdulhaev and Duškov [1], Zolotarevskiĭ [1], Prössdorf and Silbermann [2, 3]). *Let* $a, b \in H^\mu(\Gamma)$ $(0 < \mu < 1)$ *satisfy*

(i) $a(t) \neq 0$, $b(t) \neq 0$ $(t \in \Gamma)$,

(ii) $\operatorname{ind} a = \operatorname{ind} b$,

and let $\dim \ker A = 0$. *Then for sufficiently large* n *and for arbitrary* $y \in R(\Gamma)$ *the system* (20) *has a unique solution and the approximate solutions* x_n *converge to the solution* $x \in L_p(\Gamma)$ *of equation* (19) *in the* $L_p(\Gamma)$*-norm* $(1/\mu < p < \infty)$ *as* $n \to \infty$.

Under the additional assumptions of $a, b, y \in C^{m,\mu}(\Gamma)$ *and* $k(t,\tau) \in C^{m,\mu}(\Gamma \times \Gamma)$ *the following estimates hold:* $||x - x_n||_p = O(n^{-m-\mu})$ *and* $||x - x_n||_{H^\nu} = O(n^{\nu-m-\mu} \ln n)$ $(\forall \nu : 0 < \nu < \mu)$.

Theorem 8.5.1 also retains its validity for matrix equations of the form (19). In this case i) and ii) must be replaced by the following:

(i') $\det a(t) \neq 0$, $\det b(t) \neq 0$ $(t \in \Gamma)$,

(ii') the left factorization indices of the matrix $b^{-1}a$ are zero.

We emphasize that conditions (i) and (ii) of theorem 8.5.1 are necessary for the convergence of the method of mechanical quadratures because of the following reason (see Prössdorf [2], Prössdorf and Schmidt [1]). Let $k(t, \tau)$ be a sufficiently smooth function on $\Gamma \times \Gamma$ and $a, b \in H^\mu(\Gamma)$ $(0 < \mu < 1)$. If for sufficiently large n and for an arbitrary $y \in C(\Gamma)$ the system (20) has a unique solution x_n which converges in L_p-norm to a solution of (19) for fixed $p > 1/\mu$, then conditions (i) and (ii) are satisfied.

As the subsequent example shows, the method of mechanical quadratures (collocation method) may converge for sufficiently smooth functions y or even for all $y \in R(\Gamma)$, although condition (i) fails to hold.

Example. Let $B = P\varrho + Q$ with $\varrho(t) = (t^{-1} - t_0^{-1})^r$, where $t_0 \in \Gamma$ is fixed and r is real such that $0 < r \leq 1$. We set $X = L_p(\Gamma)$ $(1 < p < \infty)$ and $B_n = L_nBP_n$. It can be easily verified that $R(\Gamma) \subset K(B, B_n)$ provided $r < 1/p$ and $H^\lambda(\Gamma) \subset K(B, B_n)$ for $\lambda > r - 1/p$ and $r \geqq 1/p$. Obviously, the operator B may be written in the form (19) with $a(t) = \varrho(t) \in H^r(\Gamma)$ and $b(t) \equiv 1$. We note $a(t_0) = 0$.

Approximation methods for equation (19), whose coefficients fail to satisfy the condition (i) of theorem 8.5.1, were investigated in detail in Prössdorf and Silbermann [3].

8.6. A finite element collocation method

In this method an approximate solution of equation (8) is sought in the form of a polygon

$$x_n(t) = \sum_{k=0}^{n-1} \xi_k \varphi_k^{(n)}(t) \tag{21}$$

with piecewise linear finite elements given by the formula

$$\varphi_k^{(n)}(t) = \begin{cases} (t - t_{k-1})\,(t_k - t_{k-1})^{-1} & \text{for } \ t \in t_{k-1}t_k\,, \\ (t_{k+1} - t)\,(t_{k+1} - t_k)^{-1} & \text{for } \ t \in t_kt_{k+1}\,, \\ 0 & \text{otherwise}\,, \end{cases}$$

where $t_k = e^{2\pi ik/n}$ $(k = 0, 1, \ldots, n-1)$ and $t_{k-1}t_k$ stands for the arc on Γ with the end points t_{k-1} and t_k. The coefficients $\xi_k = \xi_k^{(n)}$ have to be determined such that the equation $Ax_n = y$ is satisfied at the points t_j — i.e. such that

$$a(t_j) \sum_{k=0}^{n-1} \xi_k (P\varphi_k^{(n)})\,(t_j) + b(t_j) \sum_{k=0}^{n-1} \xi_k (Q\varphi_k^{(n)})\,(t_j) = y(t_j)\,, \qquad j = 0, 1, \ldots, n-1\,. \tag{22}$$

The approximation method (22) is called the *polygonal method*.

Lavrent'ev [1] is likely to have used the polygonal method for the first time. Ivanov [2] found, contrary to the statement of Sofronov [1], that the invertibility of the singular operator is generally not sufficient for the convergence of the polygonal method. For sufficiently small coefficients of the Cauchy singular operator S the convergence of this method in the space $H^\mu(\Gamma)$ $(0 < \mu < 1)$ was shown in Gab-

DULHAEV [1]. As the results of DANG and NORRIE [1], WASHIZU and IKEGAWA [1] demonstrate, the polygonal method is highly efficient from the numerical point of view. Recently, the following general convergence theorem for the polygonal method was proved.

Let Ω denote the set of all bounded functions on Γ satisfying the condition

$$\vartheta_n := n^{-1} \sum_{k=0}^{n-1} \omega_k^2 \to 0 \quad \text{as} \quad n \to \infty\,,$$

where ω_k stands for the oscillation of the function on the arc $t_{k-1}t_{k+1}$. Obviously, $R(\Gamma) \subset \Omega$.

Theorem 8.6.1. (PRÖSSDORF and SCHMIDT [2]). *Let $a, b \in H^\mu(\Gamma)$ $(1/2 < \mu \leqq 1)$. Then for sufficiently large n and for arbitrary $y \in \Omega$ the system* (22) *has a unique solution and the approximate polygons x_n converge to a solution $x \in L_2(\Gamma)$ of equation* (8) *in the L_2-norm if and only if*

$$a(t)\,\lambda + b(t)\,(1-\lambda) \neq 0 \qquad (t \in \Gamma, 0 \leqq \lambda \leqq 1)\,. \tag{23}$$

Remark 8.6.1. It is evident that (23) implies the conditions (i) and (ii) of theorem 8.5.1. It can be proved that (23) is equivalent to the strong ellipticity of the singular operator A — i.e. there exist a function $\theta(t) \in C(\Gamma)$ and a constant $\alpha > 0$ such that

$$\operatorname{Re} \theta(t)\, a(t) \geqq \alpha\,, \qquad \operatorname{Re} \theta(t)\, b(t) \geqq \alpha \qquad (t \in \Gamma)$$

(see WENDLAND and STEPHAN [1]), where the convergence of the Galerkin method for strongly elliptic operators was obtained by means of finite element methods).

Theorem 8.6.2 (PRÖSSDORF and SCHMIDT [2]). *The polygonal method converges in the space $L_2(\Gamma)$, if the coefficients a, b satisfy one of the following conditions:*

1. *$a, b \in C(\Gamma)$ and* (23) *is fulfilled.*
2. *$a, b \in PC(\Gamma)$, $\operatorname{Re} a(t) \geqq \alpha$, $\operatorname{Re} b(t) \geqq \alpha > 0$ $(t \in \Gamma)$ and*
$$\max_t |a(t+0) - a(t-0)| + \max_t |b(t+0) - b(t-0)| \leqq \alpha\,.$$
3. *$b = 1$, $a = 1 + g$, where $g \in \Omega$ and $\|g\|_\infty < 1$.*

Theorem 8.6.3. (PRÖSSDORF and SCHMIDT [2]). *Let $a, b \in H^\mu(\Gamma)$ satisfy the condition* (23). *Then for arbitrary $y \in H^\mu(\Gamma)$ the following estimates hold:*

$$\|x - x_n\|_2 = O(n^{-\mu} \ln n) \quad \textit{for} \quad \mu > 0$$

and

$$\|x - x_n\|_{H^\nu} = O(n^{\nu+1/2-\mu} \ln n) \quad \textit{for} \quad \mu > 1/2$$

and

$$\forall \nu : 0 \leqq \nu \leqq \mu - 1/2\,.$$

The theorems 8.6.1, 8.6.2, and 8.6.3 can be generalized to the case of matrix equations of the form (8) (see PRÖSSDORF and SCHMIDT [3]). In such a case conditions $\dim \ker A = 0$ and (I)

(I) $\det\,[a(t)\,\lambda + b(t)\,(1-\lambda)] \neq 0$ $(t \in \Gamma,\ 0 \leqq \lambda \leqq 1)$

are necessary and sufficient for the L_2-convergence of the polygonal method (for an arbitrary vector function y with components from Ω). Condition (I) is equivalent to the condition (II)

(II) there exists a continuous matrix function $\theta(t)$ such that

$$\operatorname{Re} \theta(t)\, a(t) \geqq \alpha I \quad \text{and} \quad \operatorname{Re} \theta(t)\, b(t) \geqq \alpha I > 0$$

for any continuous matrix functions a and b.

Condition (I) does not imply the invertibility of the matrix-operator $A = aP + bQ$ in general.

Unlike the trigonometric collocation method no factorization methods may be used in the polygonal method. The reason for this is the fact that Sx_n is generally not a polygon.

The proofs of the above-mentioned theorems are based on the following auxiliary propositions (see PRÖSSDORF and SCHMIDT [2]):

1°. Let K_n denote the interpolation projection given by

$$(K_nf)(t) = \sum_{k=0}^{n-1} f(t_k)\, \varphi_k^{(n)}(t)\,.$$

Then for an arbitrary $f \in \Omega$ the estimate $\|K_nf - f\|_2 \leqq c\sqrt{\delta_n}$ hold, where $c = \text{const} > 0$.

2°. Let H_n be the orthogonal projection of $L_2(\Gamma)$ onto the subspace of all polygons (21). The union over n of the eigenvalues of the n-dimensional operators K_nSH_n is a dense subset of the segment $[-1, 1]$.

3°. $\|Af - K_nAH_nf\|_2 \to 0$ as $n \to \infty$ for arbitrary $a, b \in \Omega$ and $f \in L_2(\Gamma)$.

4°. If $a, b \in H^\mu(\Gamma)$ and $f \in H^\lambda(\Gamma)$ $(0 < \mu, \lambda \leqq 1)$, then the following estimates hold:

$$\|Af - K_nAH_nf\|_2 = \begin{cases} O(n^{-\mu}) & \text{for} \quad \mu < \lambda\,, \\ O(n^{-\lambda} \ln n) & \text{for} \quad \lambda \leqq \mu\,. \end{cases}$$

5°. If $a, b \in H^\mu(\Gamma)$ with $\mu > 1/2$, then the operators K_nAH_n may be represented as $K_nAH_n = D_n + U_n$, provided condition (23) holds. Herein $|(D_nx_n, x_n)| \geqq \alpha \|x_n\|_2^2$ $(n \geqq n_0)$ and $\|U_n - U\|_2 \to 0$ as $n \to \infty$ for a compact operator $U: L_2(\Gamma) \to C(\Gamma)$.

Recently, RATHSFELD [1] proved the following generalization of theorems 8.6.1 and 8.6.2 by means of a local technique:

Theorem 8.6.4. *Let* $a, b \in \mathrm{PC}(\Gamma)$ *and condition* (III) *be fulfilled:*

(III) $b(t) \neq 0$ *and*

$$\lambda\left[\mu \frac{a}{b}(t+0) + (1-\mu)\frac{a}{b}(t-0)\right] + 1 - \lambda \neq 0$$

$$\forall\, t \in \Gamma,\ \forall\, \mu, \varepsilon \in [0, 1]\,.$$

Then the polygonal method converges in $L_2(\Gamma)$ *for arbitrary* $y \in R(\Gamma)$. *Condition* (III) *is necessary for the stability of the polygonal method.*

Collocation methods for singular integral equations and pseudodifferential equations on closed curves, which utilized piecewise polynomial splines of arbitrary degree $d \geqq 1$, are investigated in recent papers by ARNOLD and WENDLAND [1], PRÖSSDORF [3], and SCHMIDT [1].

Further references

PRÖSSDORF and SILBERMANN [1].

9. Ordinary Differential Equations in the Complex Domain

E. Mues

9.1. Introduction

The theory of ordinary differential equations in the complex domain has a long history going back to the eighteenth century. Only to list the names of famous mathematicians who contributed to this part of function theory would be a hard piece of work. There is a series of excellent monographs in this field: Bieberbach [1], Golubev [1], Hille [5] and Ince [1]. Beside these monographs also the books of Herold [1] and Hille [1] should be mentioned.

Because of the variety of theories and of results it is impossible to try to give a survey. So let us pick out a modern part of research. We shall mainly be concerned with the existence of single-valued solutions in the plane of nonlinear differential equations (abbreviated DE) and the limitations which the existence of such solutions impose on the form of the equation.

The prototype for this range of ideas is a theorem proved by Malmquist [1] in 1913 which describes the structure of the rational function $R(z, w)$, if the DE $w' = R(z, w)$ is supposed to have transcendental meromorphic solutions. This theorem has inspired a large literature. An alternative proof and an extension to more general equations were given by Yosida in 1932 [1] using the Nevanlinna value distribution theory. A systematic examination of the implications of this theory for linear and nonlinear DE's was undertaken by Wittich in the 1950's (see his monograph [8]). The theory of differential polynomials developed in the 1970's gave also a powerful tool for new research in this field.

In the next section we start with some basic and fundamental classical results in analytic continuation and singularities of the solutions connected with the works of Fuchs and Painlevé.

For the notations of value distribution theory used here see chapter 2.

9.2. Analytic continuation and singularities

Let us start with the classical initial value problem

$$w' = f(z, w)\,, \tag{1}$$

$$w(z_0) = w_0\,. \tag{2}$$

Theorem 9.1. *Let $f(z, w)$ be holomorphic in a domain $D \subset \mathbb{C}^2$ and let $(z_0, w_0) \in D$. Then there exists exactly one solution w of the initial value problem* (1), (2), *i.e., a function $w(z)$ holomorphic in a neighbourhood $U_R(z_0)$ of z_0 satisfying* (2) *and* $(z, w(z)) \in D$, $w'(z) = f(z, w(z))$ *for $z \in U_R(z_0)$.*

To emphasize the dependence of the initial parameters z_0, w_0, let us also write $w(z; z_0, w_0)$ instead of $w(z)$.

By the law of permanence of functional equations an analytic continuation $W(z)$ of a solution $w(z)$ of (1) is also a solution of (1) as long as $(z, W(z))$ remains in the domain of $f(z, w)$. However, analytic continuation of a solution $w(z; z_0, w_0)$ can lead to points (z, w) which must not be regular points of $f(z, w)$. For instance, $w = (w_0/z_0) z$ is a solution of $w' = w/z$, $w(z_0) = w_0$ $(z_0 \neq 0)$. Although the point $(0, 0)$ is a singular point of $f(z, w) = w/z$, the solution is holomorphic in the whole plane.

The question whether apart from $w(z; z_0, w_0)$ there exists another solution of (1), (2) in a punctured neighbourhood of z_0 (e.g., with an algebraic singularity at z_0 or tending to w_0 if $z \to z_0$ along a certain arc) is answered by an important theorem of PAINLEVÉ [1].

Theorem 9.2 (PAINLEVÉ). *The following assumptions shall hold: $f(z, w)$ is holomorphic in a domain $D \subset \mathbb{C}^2$, $(z_0, w_0) \in D$. Let γ be an arc starting at z_0 and ending at $z_1 \in \mathbb{C}$; $\gamma: t \to \gamma(t)$, $t_0 \leqq t \leqq t_1$, $\gamma(t_0) = z_0$, $\gamma(t_1) = z_1$. The solution $w(z; z_0, w_0)$ of (1) can be continued analytically along γ for $t_0 \leqq t < t_1$ and $(\gamma(t), w(\gamma(t); z_0, w_0)) \in D$ for $t_0 \leqq t < t_1$. Let now (ζ_ν), $\zeta_\nu \in (\gamma)$, be a sequence with $\zeta_\nu \to z_1$ and $w(\zeta_\nu; z_0, w_0) \to w_1 \in \mathbb{C}$, and let $(z_1, w_1) \in D$. Then the analytic continuation $w(z, z_0, w_0)$ along γ is holomorphic at z_1 and coincides with $w(z; z_1, w_1)$ in a neighbourhood of z_1.*

Let us draw some conclusions.

(i) If analytic continuation of $w(z; z_0, w_0)$ into the point z_1 along γ is not possible, then the arc $(\gamma(t), w(\gamma(t); z_0, w_0))$, $t_0 \leqq t < t_1$, tends to the boundary of D if $t \to t_1$.

(ii) Suppose that $f(z, w)$ is holomorphic at (z_1, w_1). Then apart from the solution $w(z; z_1, w_1)$ there exists no other solution of (1) in a punctured neighbourhood of z_1 tending to the limit w_1 if $z \to z_1$. In particular, there is no solution of (1) with an algebraic branch point at z_1 having the value w_1 at z_1.

(iii) Suppose that $w(z)$ is a solution of (1) holomorphic in a punctured neighbourhood of $z_1 \in \mathbb{C}$. Suppose further that there exists $w_1 \in \mathbb{C}$ such that $f(z, w)$ is holomorphic at (z_1, w_1). Then z_1 cannot be an essential singularity for $w(z)$.

The point $(z_0, w_0) \in \mathbb{C}^2$ is called a *regular point* of the DE $w' = f(z, w)$ if $f(z, w)$ is holomorphic at (z_0, w_0). Using substitutions $z = 1/\zeta$ or $w = 1/v$ or both and considering the transformed DE, this definition can be extended to the points (∞, w_0), $w_0 \in \mathbb{C}$, (z_0, ∞), $z_0 \in \mathbb{C}$, and (∞, ∞) resp. theorem 9.2 remains valid with corresponding modifications for regular points of this kind. A singular point of $f(z, w)$ is called a *singular point* of the DE $w' = f(z, w)$.

As an example consider a pole of $f(z, w)$: $f(z, w)$ has a pole at $(z_0, w_0) \in \mathbb{C}^2$ if $1/f(z, w)$ is holomorphic in a neighbourhood of (z_0, w_0) and has a zero at (z_0, w_0) (poles are not isolated!).

Theorem 9.3. *Suppose that $f(z, w)$ has a pole at $(z_0, w_0) \in \mathbb{C}^2$ and $1/f(z_0, w) \not\equiv 0$. Then there exists exactly one solution of* (1) *satisfying* (2). *This solution has an algebraic branch point at z_0.*

The condition $1/f(z_0, w) \not\equiv 0$ is not superfluous, as the example $w' = \alpha(w/z)$, $0 \neq \alpha \in \mathbb{R}$, shows. Here $1/f(0, w) \equiv 0$. The structure of the singularity at $z = 0$ of the solutions $w(z) = cz^\alpha$ depends essentially on the choice of α.

Theorems 9.2 and 9.3 give a first connection between the singularities of the solutions and the singularities of the DE. E.g., if z_1 is an essential singularity of a solution of (1), then for no w_1 the point (z_1, w_1) can be a regular point of (1).

If in $w' = f(z, w)$ the right-hand side is a rational function with respect to the variable w, it is easier to get a complete survey on the possible singularities occuring with analytic continuation of a local holomorphic solution $w(z; z_0, w_0)$. In

$$w' = \frac{P(z, w)}{Q(z, w)} \ (= R(z, w)) \tag{3}$$

let P and Q be polynomials with respect to w (relatively prime) the coefficients of which are holomorphic in a fixed domain $D \subset \mathbb{C}$. Let $R(z, w)$ be holomorphic at (z_0, w_0) and suppose that $w(z; z_0, w_0)$ has been continued analytically along an arc γ from $z = z_0$ to $z = z_1$, as in theorem 9.2. Now ask what will happen to the solution as $z \to z_1$. For the various possibilities see BIEBERBACH [1] or HILLE [5]. Here, it is useful to distinguish between the so-called *fixed* and *movable* singularities.

Definition. Let (Δ) be a fixed differential equation. Let S be the set of all $z \in \hat{\mathbb{C}}$ occurring as singularities of any solution of (Δ). If S has interior points the set of solutions of (Δ) is said to have *movable singularities.* In the other case one says that the solutions have only *fixed singularities.* Considering only a subset of S instead of S (e.g., S_a the set of algebraic branch points) there is an analogous definition.

Examples. (i) $w' = (1 + 1/z^2)\, w$; solutions $w(z) = c \exp(z - 1/z)$, $c \in \mathbb{C}$. There are only fixed singularities $z = 0$, $z = \infty$. Let us remark that this is a general property of linear DE's $w' = p_0(z) + p_1(z)\, w$, p_0, p_1 being holomorphic in a domain $D \subset \mathbb{C}$. The solutions have only fixed singularities.

(ii) $w' = 1/2zw$; solutions $w(z) = \sqrt{\log(z/z_0)}$, $0 \neq z_0 \in \mathbb{C}$ arbitrary. There are movable algebraic branch points, fixed logarithmic branch points at $z = 0$ and $z = \infty$.

The question for the structure of the movable singularities of the solutions of (3) is settled by the following theorem of PAINLEVÉ [1].

Theorem 9.4 (PAINLEVÉ). *The only movable singularities of the solutions of* (3) *are poles, ordinary algebraic branch points, and algebraic poles.*

In this connection, naturally the question arises under which conditions the solutions of (3) are free of movable branch points. This leads to a famous characterization of the Riccati equation.

Theorem 9.5. *If the set of the solutions of* (3) *has no movable algebraic singularities then* (3) *necessarily is a Riccati equation,*

$$w' = a_0(z) + a_1(z)\, w + a_2(z)\, w^2 , \tag{4}$$

the a_j being holomorphic in a domain $D \subset \mathbb{C}$. There is no solution possessing an algebraic branch point in D.

Let us remark that, in general, the solutions of a DE (4) are not single-valued in D, as the following example shows:

$$w' = \frac{w}{z} + \frac{w^2}{z^2}, \qquad D = \mathbb{C} \setminus \{0\} .$$

Solutions: $w(z) \equiv 0$ and $w(z) = -z/\log cz$, $0 \neq c \in \mathbb{C}$.

The Riccati equation is one of the most important nonlinear DE's. Basic and fundamental properties can be found in BIEBERBACH [1], HILLE [5], INCE [1]. We will come back to the Riccati equation in the next section.

Theorem 9.4 of Painlevé can be extended to general algebraic DE's of first order, $P(z, w, w') = 0$. Here P is a polynomial in w and w' (which can be assumed to be irreducible) with coefficients holomorphic in a domain $D \subset \mathbb{C}$. So movable singularities are only poles and algebraic singularities.

All these investigations concerning fixed and movable singularities are based on the pioneering work of L. FUCHS [1, 2]. Also the question for first order algebraic DE's without movable algebraic singularities can be settled (Fuchsian conditions), see GOLUBEV [1], especially the case of the binomial equation $w'^m = R(z, w)$ ($m \geqq 2$, R rational).

Concerning Briot-Bouquet equations $P(w, w') = 0$, see BANK [4], BIEBERBACH [1], GOLUBEV [1], HILLE [7].

All these problems are much more complicated in the case of nonlinear DE's of higher order. PAINLEVÉ and mainly GAMBIER [1] (both using the α-method due to PAINLEVÉ) investigated the case

$$w'' = R(z, w, w') . \tag{5}$$

Here, R is a rational function of w and w' with coefficients holomorphic in a fixed domain $D \subset \mathbb{C}$. The problem to find all equations (5), the solutions of which are free of movable singularities of multi-valued character, leads to 50 different types of DE's (5), 44 of which can be integrated by well-known functions or can be reduced to simpler DE's. Only six types define new functions, the so-called Painlevé transcendents. E.g., the first and second Painlevé equations are

$$\text{(I)} \quad w'' = 6w^2 + z , \qquad \text{(II)} \quad w'' = 2w^3 + zw + c .$$

All solutions are meromorphic in $\mathbb{C}$ and have an infinite number of poles. For a detailed discussion see BOUTROUX [2], GOLUBEV [1], HILLE [5], INCE [1], PAINLEVÉ [2].

9.3. Single-valued solutions in the plane

Apart from theorem 9.5 there is another famous characterization of the Riccati equation proved by MALMQUIST in 1913 [1].

Theorem 9.6 (Malmquist theorem). *If $R(z, w)$ is a rational function of z and w and if the DE*

$$w' = R(z, w) \tag{6}$$

has a solution w meromorphic in $\mathbb{C}$, then either w is a rational function or (6) *is a Riccati equation,*

$$w' = a_0 + a_1 w + a_2 w^2 , \tag{7}$$

where the a_j are rational functions of z.

The example in section 9.2 shows that the solution of a Riccati equation with rational coefficients must not necessarily be single-valued. However, the following is true.

Theorem 9.7. *If the coefficients a_j of a Riccati equation* (7) *are entire functions, then every solution is meromorphic in $\mathbb{C}$.*

The modern way to prove the Malmquist theorem is via the Nevanlinna value distribution theory. The first one using this theory as a tool to prove results on single-valued solutions of DE's was YOSIDA [1]. In 1932 he gave a new proof and also a generalization of theorem 9.6.

Theorem 9.8 (Malmquist-Yosida theorem). Let $R(z, w)$ *be a rational function of* z *and* w *and let* $m \geqq 1$ *be an integer. If the DE*

$$w'^m = R(z, w)$$

has a solution meromorphic and transcendental in $\mathbb{C}$, *then* R *is a polynomial in* w *of degree at most* $2m$, *the coefficients of which are rational functions of* z.

Wittich [6] proved in 1954 a local version of the Malmquist theorem using the local Nevanlinna theory concerning the neighbourhood of an essential singularity.

Theorem 9.9. *If* $R(z, w)$ *is a rational function of* z *and* w *and if the DE* (6) *has a solution* w *meromorphic in a punctured neighbourhood of a point* $z_0 \in \hat{\mathbb{C}}$, z_0 *being an essential singularity for* w, *then* (6) *is a Riccati equation.*

The following result is also due to Wittich [6, 7].

Theorem 9.10. *If in the DE* (7) *the coefficients* a_j *are polynomials,* $a_2(z) \not\equiv 0$, *then the solutions are either constant or rational or transcendental meromorphic. In the transcendental case the order* $\lambda(w)$ *is always finite and of the form* $\lambda(w) = 1 + k/2$, $k + 1 \in \mathbb{N}$. *Also* $m(r, w) = O(\log r)$ *in this case and so the solution has an infinite number of poles. For* $\alpha \in \mathbb{C}$ *one has always* $m(r, 1/w - \alpha) = O(\log r)$ *with two exceptional cases at most. For an exceptional* α *the following holds: The constant* α *is a solution of the Riccati equation and also a Picard value for all transcendental solutions. For every transcendental solution the equality sign holds in the second main theorem.*

Wittich also proved: If the Riccati equation (7) with rational coefficients a_j has a solution w transcendental and meromorphic in $\mathbb{C}$, then the order satisfies $\lambda(w) \geqq \frac{1}{2}$.

In order to formulate and to state more general theorems concerning single-valued solutions, let us first give some definitions.

Definition. Let f be meromorphic and transcendental in the plane. We denote by $S(r,f)$ any real-valued function of r satisfying $S(r,f) = o(T(r,f))$ for $r \to \infty$ outside possibly a set of r-values of finite Lebesgue measure. The famous theorem on the proximity function of the logarithmic derivative (which easily leads to the second main theorem) states that $m(r, f'/f) = S(r,f)$. A meromorphic function a satisfying $T(r, a) = S(r,f)$ is called a *small* function with respect to f.

A differential polynomial in f is a finite number of monomials of the form

$$M[f] = af^{n_0}f'^{n_1} \dots f^{(k)n_k}.$$

Here the $n_j \geqq 0$ are integers, and a is a constant or a small function with respect to f. $\gamma_M = n_0 + n_1 + \dots + n_k$ and $\Gamma_M = n_0 + 2n_1 + \dots + (k + 1)\, n_k$, respectively, are called the *degree* and the *weight*, respectively, of the monomial $M[f]$. The degree γ_P and the weight Γ_P, respectively, of a differential polynomial $P = \sum_j M_j$ are defined as the maximum of the degrees and the maximum of the weights, respectively, of the occuring monomials.

The following generalization of the Malmquist-Yosida theorem was proved in 1978/79 by Gackstatter and Laine [1] and independently by Strelic [2].

Theorem 9.11. *Let* $P[w]$ *be a differential polynomial in* w *with rational coefficients. If* $R(z, w)$ *is a rational function of* z *and* w *and if the DE*

$$P[w] = R(z, w)$$

has a transcendental meromorphic solution in $\mathbb{C}$, *then* R *is a polynomial in* w *of degree at most* Γ_P.

The case $P[w] = w'^m$ is the Malmquist-Yosida theorem. The most general criterion in this direction is due to STEINMETZ [1].

Theorem 9.12. *Let $P[w]$ be a differential polynomial in w and let $H(z, w)$ be the quotient of two entire functions in z and w. A function f meromorphic in $\mathbb{C}$ is called admissible with respect to (P, H) if*

(i) *every coefficient in P is a small function with respect to f,*

(ii) *there exists a set $\Omega \subset \mathbb{C}$ with a finite cluster point such that $T(r, H(z, \tau)) = S(r, f)$ for all $\tau \in \Omega$.*

If the DE $P[w] = H(z, w)$ has a solution admissible with respect to (P, H), then H is a polynomial with respect to w of degree at most Γ_P.

For entire admissible solutions the degree does not exceed γ_P.

The bounds given in theorem 9.12 are sharp. Clearly, theorem 9.12 contains theorem 9.11.

Let us also mention the following result of RELLICH [1].

Theorem 9.13. *In the DE*

$$w' = f(z, w) \tag{8}$$

let f be an entire function of z and w. If (8) has two different entire solutions u and v, then every other entire solution w has the form

$$w = u + c(v - u)\,. \tag{9}$$

If $f(z, w)$ is nonlinear in w, then there are at most countably many $c = c_n$ such that (9) is a solution of (8). The c_n have no finite cluster point.

It follows that a DE $w' = f_0(w)$ with f_0 entire and nonlinear has no nonconstant entire solutions. This result was also proved by WITTICH [2] and is also contained in theorem 9.12. If in (8) f is a polynomial in w, one can replace w' by a differential polynomial in w and use theorem 9.12.

Using the Wiman-Valiron theory WITTICH [2] proved

Theorem 9.14. *If the differential polynomial $P[w]$ (polynomial coefficients) has exactly one monomial $M_0[w]$ with $\gamma_{M_0} = \gamma_P$, then the DE $P[w] = 0$ has no transcendental entire solutions.*

There is an intimate connection between the appearance of zeroes of a differential polynomial in a meromorphic function f and the nonexistence of meromorphic solutions in $\mathbb{C}$ of certain nonlinear DE's.

Let Q be a differential polynomial with rational coefficients.

A. Suppose Q has the following property: If f is transcendental meromorphic in $\mathbb{C}$, then $Q[f](z) \not\equiv 0$ and $Q[f](z)$ has zeroes. It follows that if $g \not\equiv 0$ is an entire function, then the DE $Q[w] = 1/g$ has no transcendental meromorphic solutions in $\mathbb{C}$.

B. Suppose Q has the following property: If $g \not\equiv 0$ is entire, then the DE $Q[w] = 1/g$ has no transcendental meromorphic solutions in $\mathbb{C}$. It follows that $Q[f](z)$ has zeroes for every transcendental meromorphic f.

So theorems which guarantee the appearance of zeroes of differential polynomials yield nonexistence theorems for meromorphic solutions of the corresponding DE. Let us mention such a result (see FRANK and MUES [2]):

Theorem 9.15. *Let f be a transcendental meromorphic function in $\mathbb{C}$. Let Q and P be differential polynomials in f such that $P[f](z) \not\equiv 0$, $Q[f](z) \not\equiv 0$. If $n \geqq 3 + \Gamma_P$*

($n \in N$) then

$$f^n Q[f] + P[f]$$

has an infinite number of zeroes.

The conclusion remains true if f is supposed to be entire and $n \geqq 2 + \gamma_p$. The bounds are best possible.

For further results concerning differential polynomials see CLUNIE [1], FRANK [2], MUES [3, 4], STEINMETZ [6].

9.4. Value distribution properties of the solutions

Let us start with the linear case.

Theorem 9.16. *Let the a_j ($j = 0, 1, \ldots, n-1$) be polynomials, $a_0(z) \not\equiv 0$. Every solution of the n-th order linear DE*

$$w^{(n)} + a_{n-1}w^{(n-1)} + \ldots + a_0 w = 0$$

is an entire function of finite order.

There is a finite set of positive rational numbers $\lambda_1 > \lambda_2 > \ldots > \lambda_k \geqq 1/n$ ($k \leqq n$) such that $\lambda(w) \in \{\lambda_j\}$ for every transcendental solution w. The λ_j are uniquely determined by the degrees of the a_j. Especially $\lambda(w) = 0$ implies that w is a polynomial.

Let us remark that $\lambda_1 \in \{\lambda(w) : w \text{ a solution}\}$. A solution w with $\lambda(w) < \lambda_1$ is called *subnormal*. The problem of finding necessary and/or sufficient conditions for the existence of subnormal solutions in the case $n \geqq 3$ is still open.

For further results concerning these problems see FRANK [1, 2], FRANK and WITTICH [1], WITTICH [8, 13].

The case of the Riccati equation was treated in 9.3. Concerning Riccati equations with transcendental entire or meromorphic coefficients, see HILLE [4], LAINE [2], MUES and STEINMETZ [1].

Wishing to investigate in an analogous manner the solutions of the Yosida-type equations $w'^m = P(z, w)$, one finds only a very poor class of functions. In 1978 STEINMETZ [2] proved the following result.

Theorem 9.17. *Let $P(z, w)$ be a polynomial in w with rational coefficients. If the DE*

$$w'^n = P(z, w)$$

of minimal degree $n \geqq 2$ has a solution transcendental and meromorphic in C, then up to linear fractional transformations the following cases can only occur:

$$\text{(i) } w'^2 = a_0(z)(w-\tau_1)(w-\tau_2)(w-\tau_3)(w-\tau_4),$$
$$\text{(ii) } w'^3 = a_0(z)(w-\tau_1)^2(w-\tau_2)^2(w-\tau_3)^2,$$
$$\text{(iii) } w'^4 = a_0(z)(w-\tau_1)^2(w-\tau_2)^3(w-\tau_3)^3,$$
$$\text{(iv) } w'^4 = a_0(z)(w-\tau_1)^3(w-\tau_2)^4(w-\tau_3)^5,$$
$$\text{(v) } w'^2 = a_0(z)(w-\tau_1)(w-\tau_2)\big(w-q(z)\big)^2.$$

$\tau_j \neq \tau_k$ for $j \neq k$ (except in case (i), $\tau_3 = \tau_4$ say). a_0 and the nonconstant q are rational functions.

In the cases (i)—(iv) it is easy to give sufficient conditions for a_0 guaranteeing the existence of single-valued solutions in the plane, since in the case of constant a_0 integration leads to elliptic functions. The order of the solutions is always finite. It is worth mentioning that transcendental solutions of order zero can occur, different from the case $n = 1$. Sufficient conditions for the existence of meromorphic solutions of (v) and also the value distribution properties are known (STEINMETZ [2]).

Considering the DE only in a punctured neighbourhood of an essential singularity of the solutions, the corresponding results were found by BANK and KAUFMAN [3]; see also [1].

Concerning the order of meromorphic solutions of first order algebraic DE's, one has the following result of GOL'DBERG [2].

Theorem 9.18. *If the first order algebraic DE $P(z, w, w') = 0$ has a solution w meromorphic in $\mathbb{C}$, then $\lambda(w) < \infty$.*

See also BANK [2] and BANK and KAUFMAN [2].

Theorem 9.19. *Let the second order algebraic DE be given by*

$$w''^q + A_1(z, w, w')\, w''^{q-1} + \dots + A_q(z, w, w') = 0\,,$$

the A_j being polynomials in z, w, w'. Let w be a solution meromorphic in $\mathbb{C}$. If there exists an $\alpha \in \mathbb{C}$ such that the counting function $N(r, 1/w - \alpha)$ has finite order, then w has finite order, too.

See GOL'DBERG [4] and also BANK [6], BANK and LAINE [1], STEINMETZ [5].

Concerning the counting functions resp. the proximity functions of the solutions of a general algebraic DE, there is a simple criterion due to A.Z. and V. D. MOHON'KO [1].

Theorem 9.20. *Suppose that the algebraic DE $P[w] = 0$ has a solution w transcendental and meromorphic in $\mathbb{C}$. If the constant function $w_0(z) \equiv c$, $c \in \mathbb{C}$, is not a solution of $P[w] = 0$, then $m(r, 1/w - c) = S(r, w)$.*

Malmquist-type theorems for the DE

$$w'' = R(z, w, w')\,, \qquad R \text{ rational}\,,$$

are not known. If the right-hand side is independent of w', then some results are known (RELLICH [2], WITTICH [5]). Also theorems like 9.12 do sometimes work. For special DE's, e.g., the six Painlevé equations, there are careful investigations concerning value distribution properties and also concerning the question on the existence of common solutions of Fuchsian equations (special first order algebraic DE's) with some Painlevé equations (see GROMAK [1], KOSTJUKOVIČ [1], LUKAŠEVIČ [1, 2], [3, 4], LUKAŠEVIČ and JABLONSKIĬ [1], JABLONSKIĬ [1]).

Further references

BANK [1, 3, 5, 7, 8], BANK and KAUFMAN [1, 4], BÖHMER [1—3], BOUTROUX [1], BRIOT and BOUQUET [1—3], CHAZY [1], CLUNIE [2], FRANK [3], FRANK and HENNEKEMPER [1], FRANK, HENNEKEMPER, and POLLOCZEK [1], FRANK and MUES [1], HILLE [2, 6, 8, 9], LAINE [1, 3], MUES [1, 2, 5], MUES and REDHEFFER [1], NIKOLAUS [1], STEINMETZ [3, 4], STRELIC [1], VALIRON [3, 4], WITTICH [1, 3, 4, 9—12], YANG [1—3], YOSIDA [2, 3].

10. Boundary Value Problems for Analytic and Generalized Analytic Functions

H. Begehr

10.1. Introduction

In the classical complex analysis there are two types of boundary value problems which arise in a natural way from the Cauchy formula.

(i) Let φ be an integrable function defined on a rectifiable curve L of $\mathbb{C}$, then the function φ given by the Cauchy integral

$$\Phi(z) := \frac{1}{2\pi i} \int_L \varphi(\zeta) \frac{d\zeta}{\zeta - z}$$

for z not on L represents an analytic function which has a jump on L. If Φ^+ denotes in a suitable sense the boundary value of Φ approaching L from the left with respect to the orientation of L and Φ^- the corresponding limit from the right then on L, the formula from Plemelj-Sohockiĭ

$$\Phi^+ = \Phi^- + \varphi$$

holds. The function Φ therefore is a solution of a special Riemann boundary value problem. More general this Riemann problem demands for a given curve L in $\mathbb{C}$ and for on L given functions g, G, where G does not vanish on L, a function Φ analytic in $\hat{\mathbb{C}} - L$ the boundary values of which on L fulfil the condition

$$\Phi^+ = G\Phi^- + g\,. \tag{1}$$

(ii) A slightly modified Cauchy integral for the unit circle is the Poisson formula. Let φ be a real, integrable function defined on the unit circle, then

$$\Phi(z) := \frac{1}{2\pi i} \int_{|\zeta|=1} \varphi(\zeta) \frac{\zeta + z}{\zeta - z} \frac{d\zeta}{\zeta} + ic \qquad (c \in \mathbb{R})$$

represents a function analytic in the unit disc, the boundary values of which on $|z| = 1$ coincide with φ in all points of continuity of φ and for which Im $\Phi(0)$ is given by c (see H. A. Schwarz [1]). Here Φ turns out to be a solution of a special Hilbert boundary value problem. More general it requires to find an analytic function w in a given domain D of $\mathbb{C}$, such that for given functions g, G on $L = \partial D$

$$\operatorname{Re}\{\overline{G}w\} = g \tag{2}$$

is valid.

Both these problems which for the first time were posed by Riemann [problem (2) in [1], 35pp., and problem (1) in [2], 379pp.], and which still inspire mathematicians with new ideas are exclusively considered here. An indication of the broad interest these problems had caused is the variety of their notations. Problem (1) often is called the *boundary value problem of linear conjugation*, or the *Riemann boundary value problem*, or the *Hilbert boundary value problem*. Sometimes the names of CARLEMAN and PRIVALOV are also involved. (2) is named *Riemann-Hilbert* or only *Hilbert boundary value problem* or sometimes *Haseman problem*. Referring to this one may have a look at the different interpretations and the historical notes of GAHOV [3] and MUSHELIŠVILI [2]. We will use *Riemann* for (1) and *Hilbert* for (2).

First solutions for the Riemann problem were given by CARLEMAN [1], HILBERT [2], PLEMELJ [1], SOHOCKIĬ [1] and for the Hilbert problem by HILBERT [2] and VOLTERRA [1]. Significant contributions for (1) were given by GAHOV [1—3] and MUSHELIŠVILI [1, 2] and for (2) by F. NOETHER [1] and SOBOLEV [1].

While the first solutions of both problems were attained through the help of integral equations, F. NOETHER was the first to use singular integral equations. Owing to this relationship to the theory of singular integral equations, including the theory of Wiener-Hopf of singular integral equations of convolution type, there are many results, especially for problem (1), and still every month there are several new papers published.

An examination as well as a historical overlook can be found in the books of GAHOV [3], MUSHELIŠVILI [2] and I. N. VEKUA [2]. The survey article of HVEDELIDZE [3] on new developments for (1) presents extensive references.

In order that a boundary value problem is well-posed the assumptions on the boundary curve L (or more general on the system of curves with no common points) and on the given function defined on L have to agree with the desired boundary behaviour of the solutions. Three cases are to be distinguished.

1. The solutions including their boundary values on L are continuous (continuous boundary value problems).

2. The solutions are continuous, even on L, (from the right and from the left) with the exception of finitely many points of L (piecewise continuous boundary value problems).

3. The solutions are neither continuous nor piecewise continuous on L (discontinuous boundary value problems).

The classical assumptions considered by GAHOV (continuous case) and MUSHELIŠVILI (piecewise continuous case) correspond to 1 and 2, respectively. They include smoothness of L as well as Hölder continuity and piecewise Hölder continuity, respectively, of the coefficients in the boundary condition. The results for discontinuous boundary value problems are considered in the case of the Riemann problem by HVEDELIDZE [3] and for the Hilbert problem by the same author in [2]. SIMONENKO [4] extended essentially the class of allowable coefficients G for problem (1) and showed this extention to be maximal in the space $\mathscr{L}_2$.[1]) Other generalizations of (1) and (2) have been undertaken in several directions. They are studied e.g.

a) on Riemann surfaces (KOPPELMAN [1], RODIN [1], RODIN and JURAKULOV [1], SLIZKIĬ [3], ČIBRIKOVA [1]),

b) for vectors of analytic functions (BOJARSKI [2, 4], HVEDELIDZE [2, 3], SIMONENKO [3], N. P. VEKUA [5]),

c) for solutions of elliptic systems of first order in the plane (generalized analytic and generalized hyperanalytic functions) (BEGEHR [6], BEGEHR and GILBERT [3, 5],

---

[1]) Here I want to thank Professor PRÖSSDORF for putting my attention to the papers of SIMONENKO.

BOJARSKI [5], GAHOV [3], HAACK and WENDLAND [1], HILBERT [2], KOPPELMAN [2], TEICHERT-GAIDA [2], I. N. VEKUA [5], WENDLAND [2, 4], WEN [1]), for elliptic equations of second order (FLORIAN and HEERSINK [1], LI [1], I. N. VEKUA [4, 5], WEN [5]), and for polyharmonic as well as polyanalytic functions (GAHOV [3]),

d) in form of nonlinear boundary value problems for nonlinear elliptic systems in the plane (BEGEHR and GILBERT [1], [2], BEGEHR and HILE [1, 2], BEGEHR and HSIAO [1, 2, 5], BOJARSKI and IWANIEC [1], BUCHANAN and GILBERT [1], GILBERT [3], NAAS and TUTSCHKE [2, 3], TUTSCHKE [8, 9, 12, 14, 15, 16], I. N. VEKUA [5, 335pp], WEN [2—4], WAROWNA-DORAU [1, 2], WENDLAND [1, 2, 4], v. WOLFERSDORF [5], WOLSKA-BOCHENEK [1, 2]).

Moreover there are some special generalizations as

α. Boundary value problems including higher derivatives, e.g.

$$\sum_{\mu=0}^{m} a_\mu(\zeta)\frac{\mathrm{d}^\mu}{\mathrm{d}\zeta^\mu}w^+(\zeta) = \sum_{\nu=0}^{n} b_\nu(\zeta)\frac{\mathrm{d}^\nu}{\mathrm{d}\zeta^\nu}w^-(\zeta) + g(\zeta) \qquad (\zeta \in L)$$

or more general

$$\sum_{\mu=0}^{m}\left[a_\mu(\zeta)\frac{\mathrm{d}^\mu}{\mathrm{d}\zeta^\mu}w^+(\zeta) + \int_L A_\mu(\zeta, z)\frac{\mathrm{d}^\mu}{\mathrm{d}z^\mu}w^+(z)\,\mathrm{d}z\right] =$$

$$= \sum_{\nu=0}^{n}\left[b_\nu(\zeta)\frac{\mathrm{d}^\nu}{\mathrm{d}\zeta^\nu}w^-(\zeta) + \int_L B_\nu(\zeta, z)\frac{\mathrm{d}^\nu}{\mathrm{d}z^\nu}w^-(z)\,\mathrm{d}z\right] + g(\zeta) \qquad (\zeta \in L)$$

(GAHOV [2, p. 316], GUZ [1]), and

$$\mathrm{Re}\left\{\sum_{k=0}^{m}\overline{G_k(\zeta(s))}\,\frac{\mathrm{d}^k w(\zeta(s))}{\mathrm{d}s^k}\right\} = g(s)$$

(s parameter of arc length on L) or

$$\mathrm{Re}\,\{\sum_{k=0}^{m}\overline{G_k(\zeta)}\,w^{(k)}(\zeta) + \int_L H_k(\zeta, z)\,w^{(k)}(z)\,\mathrm{d}z\} = g(\zeta) \qquad (\zeta \in L)\,,$$

respectively (GAHOV [3, p. 291], MUSHELIŠVILI [2, p. 203]).

β. Loaded Riemann problem

$$w^+(\zeta) - G(\zeta)\,w^-(\zeta) + \sum_{k=1}^{n}\{a_k(\zeta)\,w^+(\zeta_k) - b_k(\zeta)\,w^-(\zeta_k)\} = g(\zeta) \qquad (\zeta, \zeta_k \in L)$$

(SALIMOV [1]).

γ. Nonlinear Riemann problem of potential type

$$[w^+(\zeta)]^{\alpha^+(\zeta)} = G(\zeta)\,[w^-(\zeta)]^{\alpha^-(\zeta)} + g(\zeta) \qquad (\zeta \in L)$$

(ARZANOV [1]).

δ. Shift problems

$$w^+(\alpha(\zeta)) = G(\zeta)\,w^-(\zeta) + g(\zeta) \qquad (\zeta \in L)$$

(ALEHNO [2, 5], GAHOV [3], KAC [3], KVESELAVA [1], MANDŽAVIDZE [1, 2, 4], MURTAZAEV [1], PČEL'NIK [1], SIMONENKO [3], ŽUKOVA [2]).

ε. Boundary value problems of infinite index (ALEHNO [1, 2], ALEKNA [2], ROGOZIN and TOLOČKO [1]).

ζ. Riemann boundary value problems of the type

$$w^+(\zeta) = a(\zeta)\, w^-(\zeta) + b(\zeta)\, \overline{w^-(\zeta)} + c(\zeta) \qquad (\zeta \in L)$$

or

$$w^+(\zeta) = a(\zeta)\, w^-(\zeta) + b(\zeta)\, \overline{w^-(\zeta)} + c(\zeta)\, w^-(\bar{\zeta}) + d(\zeta)\, \overline{w^-(\bar{\zeta})} + e(\zeta) \qquad (\zeta \in L)$$

[Gorodža et al. [1], Juhanonov [1], Kazidzhanova [1], Mihaĭlov [1], Murtazaev (1]).

Some other generalizations only shall be mentioned: generalizations with respect to curves L of infinite length or systems of countably many curves, and generalizations with respect to the coefficients of the problem (Alehno [4], Gahov [3], Grudskiĭ and Dybin [1], Hvedelidze [3], A. I. and S. I. Jacko [1], Kulagina [1], Kac [4], Mkojan [1], Šeško [1], Sznitko [1], Zabreĭko et al. [1]). In the following, some characteristic results of the classical problems (1) and (2) for analytic as well as for generalized analytic functions and for nonlinear boundary value problems connected with them shall be presented.

10.2. Riemann boundary value problem for analytic functions

Basic for the Riemann boundary value problem (cf. 1.10.3., chapter 5, 20.4.3.) are the formulae of Plemelj-Sohockiĭ. Let ϱ be a Hölder-continuous function defined on a smooth (open or closed) curve L, then the Cauchy integral

$$\Phi(z) := \frac{1}{2\pi i} \int_L \varrho(\zeta) \frac{d\zeta}{\zeta - z} \tag{3}$$

has the property that in every point ζ of L with — in the case of an open curve — exceptions of the end points the limits

$$\Phi^+(\zeta) := \lim_{\substack{z\to\zeta \\ \text{from right}}} \Phi(z)\,, \qquad \Phi^-(\zeta) := \lim_{\substack{z\to\zeta \\ \text{from left}}} \Phi(z)$$

exist where $z \notin L$ tends to $\zeta \in L$ in the first case from the right with respect to the orientation of L and in the second case from the left. The Cauchy integral (3) still has a significance for $z \in L$ if the integral in this case is taken as the Cauchy principal value. With this interpretation in mind one finds

$$\Phi^+ = \tfrac{1}{2}\varphi + \Phi\,, \qquad \Phi^- = -\tfrac{1}{2}\varphi + \Phi\,, \tag{4}$$

or equivalently

$$\Phi^+ - \Phi^- = \varphi\,, \qquad \Phi^+ + \Phi^- = 2\Phi\,. \tag{5}$$

Therefore (3) represents the — uniquely given — solution of the Riemann problem

$$\Phi^+ = \Phi^- + \varphi \qquad \text{on } L. \tag{6}$$

Let now L be a simply closed, positively oriented, smooth curve bounding a domain D^+ with $0 \in D^+$. Let the complement of $D^+ \cup L$ with respect to the Riemann sphere $\hat{\mathbb{C}}$ be D^-, and let G and g be Hölder-continuous functions defined on L such that G does not vanish on L, then the general Riemann problem (1) easily can be reduced to the special form (6). To this purpose the index of problem (1)

$$n := \operatorname{ind} G := \frac{1}{2\pi i} \int_L d \log G \in \mathbb{Z}$$

and the function X analytic in $\hat{C} - L$ defined by

$$X(z) := \begin{cases} \exp \Gamma(z), & z \in D^+, \\ z^{-n} \exp \Gamma(z), & z \in D^-, \end{cases}$$

where

$$\Gamma(z) := \frac{1}{2\pi i} \int_L \log \{\zeta^{-n} G(\zeta)\} \frac{d\zeta}{\zeta - z}$$

have to be considered. X gives a factorization of G in the form

$$X^+ = GX^- \quad \text{on} \quad L$$

so that for every solution Φ of (1) ΦX^{-1} is an analytic function outside L which fulfils

$$\frac{\Phi^+}{X^+} = \frac{\Phi^-}{X^-} + \frac{g}{X^+} \quad \text{on} \quad L.$$

So one gets (see Gahov [3, 98 p.]; cf. also 20.4.3.).

Theorem 10.1. *For $n \geqq 0$ (1) is soluble for every Hölder continuous function g. The solution is representable in the form*

$$\Phi(z) = \frac{X(z)}{2\pi i} \int_L \frac{g(\zeta)}{X^+(\zeta)} \frac{d\zeta}{\zeta - z} + X(z)\, P_n(z) \tag{7}$$

where P_n is an arbitrary polynomial of degree not bigger than n. For $n < 0$ (1) is unsolvable in the homogeneous case while it is solvable in the inhomogeneous case if g fulfils the $-n - 1$ conditions

$$\int_L \frac{g(\zeta)}{X^+(\zeta)} \zeta^{k-1}\, d\zeta = 0 \qquad (1 \leqq k \leqq -n - 1). \tag{8}$$

If these conditions are fulfilled then the solution is uniquely given by (7) where P_n has to be replaced by 0.

Is the solution required to vanish at infinity, then P_n has to be replaced by a polynomial of degree less than or equal to $n - 1$, and in the case $n < 0$ (8) has to be valid for $k = -n$, too. As it can be seen (1) is uniquely solvable for arbitrary (Hölder-continuous) g only for $n = -1$ and under the additional assumption at ∞ only for $n = 0$. It is obvious that the conditions (8) arise from the fact that Φ has to behave regularly at infinity.

The assumptions on G can slightly be modified so that G is allowed to have finitely many zeroes and poles on L,

$$G(\zeta) = \prod_{\mu=1}^{m} (\zeta - \alpha_\mu)^{m_\mu} \prod_{\nu=1}^{n} (\zeta - \beta_\nu)^{-n_\nu} G_1(\zeta) \qquad (\alpha_\mu, \beta_\nu, \zeta \in L)$$

where G_1 does not vanish on L and is Hölder-continuous (see Gahov [3, 109 pp.]).

The curve L can be replaced by a system of non-intersecting, smooth, closed curves (see Gahov [3, 113 pp.]). Privalov [1] has suggested to consider the Riemann problem under the following assumptions:

(i) L is a simply closed, redifiable curve.

(ii) There exists a real number $k > 1$, such that

$$k^{-1} \leqq |G(\zeta)| \leqq k \qquad (\zeta \in L),$$

and

$$g \in \mathscr{L}_p(L) \quad \text{for} \quad 1 < p.$$

(iii) Φ^+ and Φ^- are angular limit values on L of the analytic function Φ, which is assumed to be representable by a Cauchy type integral over L.

The formulae of Plemelj-Sohockiĭ are valid for $\varphi \in \mathcal{L}_p(L)$ $(1 < p)$ almost everywhere on L, where the boundary values and the Cauchy principal value of (3) exists, too (see Gahov [3, 43 p.], Hvedelidze [3, 343 p.], Zabreĭko et al. [1, 156 p.]).

This incontinuous Riemann boundary value problem first was considered by Hvedelidze (see his article [3]) for the case where L is a Ljapunov curve and G is Hölder-continuous. For the formulation of his results some definitions are needed.

Definition 10.1. Let $\mathcal{R}_p$ denote the set of simply closed, rectifiable curves γ, for which there exists a constant c_p dependent on p $(1 < p)$ only such that for all $f \in \mathcal{L}_p(\gamma)$

$$\left\| \int_\gamma f(\zeta) \frac{\mathrm{d}\zeta}{\zeta - z} \right\|_p \leqq c_p \|f\|_p .$$

Moreover let $\mathcal{R} := \bigcup_{1<p} \mathcal{R}_p$.

Definition 10.2. Let $\mathcal{K}_p(\gamma)$ denote the set of Cauchy integrals

$$\frac{1}{2\pi \mathrm{i}} \int_\gamma \varphi(\zeta) \frac{\mathrm{d}\zeta}{\zeta - z} \quad (z \notin \gamma)$$

for $\varphi \in \mathcal{L}_p(\gamma)$ and for fixed complex polynomial $q(z)$ $\mathcal{K}_{p,q(z)}(\gamma)$ the set

$$\mathcal{K}_{p,q(z)}(\gamma) := \{\Phi + q : \Phi \in \mathcal{K}_p(\gamma)\}$$

and

$$\widetilde{\mathcal{K}}_p(\gamma) := \bigcup_{q(z) \in Q} \mathcal{K}_{p,q(z)}(\gamma)$$

the union of the sets $\mathcal{K}_{p,q(z)}(\gamma)$ over all complex polynomials.

Definition 10.3. A function G defined on γ is called *factorable* in $\widetilde{\mathcal{K}}_p(\gamma)$ if there exist an $X^+ \in \widetilde{\mathcal{K}}_p(\gamma)$ and an $X^- \in \widetilde{\mathcal{K}}_{p/p-1}(\gamma)$ such that almost everywhere on γ

$$G(\zeta) = X^+(\zeta)/X^-(\zeta) .$$

Because every factor

$$X = \begin{cases} X^+ & \text{in } D^+ , \\ X^- & \text{in } D^- \end{cases}$$

of G in $\widetilde{\mathcal{K}}_p(\gamma)$ has the same order at infinity (see Hvedelidze [3, p. 379]) this order can be used to define an index of G.

Definition 10.4. If G has a factorization in $\widetilde{\mathcal{K}}_p(\gamma)$, then the order of a factor of G at infinity multiplied by -1 is the *index* of G in the set $\widetilde{\mathcal{K}}_p(\gamma)$ denoted by

$$\operatorname{ind}\,[G, \widetilde{\mathcal{K}}_p(\gamma)] .$$

Theorem 10.2. *Let $L \in \mathcal{R}$, G be factorable in $\widetilde{\mathcal{K}}_p(L)$ and $X^+ \in \mathcal{W}_p(L)$, then under the assumptions of Privalov* (1) *in the case $n := \operatorname{ind}\,[G, \widetilde{\mathcal{K}}_p(L)] \geqq 0$ is solvable in $\widetilde{\mathcal{K}}_p(L)$ for every g. The solutions are of the form* (7) *where P_n is a polynomial,* $\deg P_n \leqq n$. *For $n < 0$* (1) *is solvable if and only if the conditions* (8) *are valid.*

Solutions in $\widetilde{\mathcal{K}}_p(L)$ — those are solutions vanishing at infinity — are given as in theorem 10.1. This theorem is given in Hvedelidze [3] more generally for analytic matrices and moreover also proved in spaces $\mathcal{L}_p(\gamma, \omega)$ where ω is a weight function. There one can find (385 pp.) examples for factorizations of incontinuous functions.

Important contributions to the Riemann problem with incontinuous coefficients are due to DANILJUK [1], GOHBERG [1], IVANOV [1], MIHLIN [2], SIMONENKO [1—4]. They consider finite systems of closed Ljapunov curves L and $g \in \mathcal{L}_p(L)$. The continuity condition on G is replaced by the following assumption: G is bounded and measurable and for each $\zeta_0 \in L$ there exist a neighbourhood $U(\zeta_0)$ and a $\delta > 0$ so that $G(\zeta)$ for $\zeta \in U(\zeta_0)$ is contained in a sector with vertex at the origin and of opening angle

$$\frac{2\pi - \delta}{\max\left\{p, \dfrac{p}{p-1}\right\}}.$$

The solutions as before are surched in the class $\mathcal{K}_p(L)$. If the index of the problem is adapted to the assumptions in a suitable way by functional analytic methods, it can be shown that the Noether theorems for the Riemann problem continue to hold even in this case.

The Riemann problem for analytic vectors can be solved in $\mathcal{K}_2^n(L)$ — the space of n-vectors with components in $\mathcal{K}_2(L)$ — where $g \in L_2^n(L)$ is a vector with components in $\mathcal{L}_2(L)$ and g is a bounded measurable $(n \times n)$-matrix which can be written as a product $G_1G_2G_3$, where G_1 and G_3 are continuous non-singular matrices and G_2 is such that

$$\operatorname{Re} G_2(\zeta) = \tfrac{1}{2}\{G_2(\zeta) + G_2^*(\zeta)\} \geqq \nu > 0\,.$$

Here G_2^* denotes the conjugate (transposed matrix with conjugate complex entries) matrix of G_2, $\geqq$ compares Hermitian matrices, and ν is independent of ζ. The Noether theorems hold in this case, too. The index is invariant with respect to the representation defined by the variation of $\arg G_1G_3$ on L (see SIMONENKO [3]).

In the formulation of the theory of singular integral equations the Riemann problem reads (see e.g. PRÖSSDORF [1])

$$R_G\Phi \equiv \Phi^+ - G\Phi^- = g\,, \qquad \Phi^\pm = \tfrac{1}{2}(\pm\varphi + S\varphi) \qquad \text{on L}$$

where S is a singular operator with Cauchy kernel. In SIMONENKO [4] among others and including even results for analytic vectors and matrices with respect to the spaces $\mathcal{K}_p^n(L)$ and $\mathcal{K}_p^{n,n}(L)$ respectively the following results are proved by the author's theory of local type operators.

Theorem 10.3. *If G is a bounded measurable function which fulfils*

(i) $0 < \operatorname{ess\,inf} |G| < \operatorname{ess\,sup} |G| < +\infty$,

(ii) *for any $\zeta_0 \in L$ there exist a neighbourhood $U(\zeta_0)$, $\delta > 0$, and bounded functions $G_{\zeta_0}^+$, $G_{\zeta_0}^-$, which are boundary values of functions from $\mathcal{K}_p(L)$, such that*

$$|\arg(G_{\zeta_0}^+ G G_{\zeta_0}^-)| < \frac{\pi - \delta}{\max\left\{p, \dfrac{p}{p-1}\right\}} \quad \textit{in } U(\zeta_0)\,,$$

then R_G is a Noetherian operator from $\mathcal{L}_p(L)$ into itself.

Theorem 10.4. *For $p = 2$ the converse of theorem* 10.3. *is true.*

This is the maximal extension of the problem (1) mentioned above.

10.3. Hilbert boundary value problem for analytic functions

Contrary to the Riemann problem for the Hilbert problem there is a difference between the cases of simply and of multiply connected domains. In the second case (see GAHOV [3], KUTATELADZE [1]) the solutions in general are multi-valued analytic functions and it is not possible, in general, to find a single-valued analytic solution. In the following we will restrict the consideration to the case of simply connected domains. For the general case we refer to the book of GAHOV [3]. Let us first consider a circle, say the unit disc, where we have the formula of Schwarz

$$w(z) = \frac{1}{2\pi \mathrm{i}} \int_0^{2\pi} u(t) \frac{\mathrm{e}^{it} + z}{\mathrm{e}^{it} - z} \, \mathrm{d}t + \mathrm{i}c \qquad (|z| < 1)$$

which represents for given real function u on $|z| = 1$ and given real constant c an analytic function inside the disc, the real part of which in the points of continuity of u on $|z| = 1$ coincide with u and the imaginary part of which attains c at $z = 0$.

Let more general D be a simply connected domain with a Green's function $G(z, \zeta)$ and let $H(z, \zeta)$ be the harmonic conjugate of $G(z, \zeta)$ and

$$M(z, \zeta) := (G + \mathrm{i}H)(z, \zeta)$$

then for real functions u on $L = \partial D$ the integral

$$(Su)(z) := \frac{1}{2\pi} \int_L u(\zeta) \frac{\partial M(z, \zeta)}{\partial n_\zeta} \, \mathrm{d}s_\zeta$$

where n_ζ denotes the inner normal and s_ζ the arc length parameter of L in the point ζ is an analytic function in D. If L is smooth and u is Hölder-continuous on L then the boundary value function of Re Su is u and Im Su vanishes at a fixed point z_0 of D given by $H(z_0, \zeta) \equiv 0$.

S is called the Schwarzian operator of D. In the case where L is the real line, S is given by an integral of Cauchy type. For multiply connected domains S still has the same property but in $D - \{\zeta\}$

$$M(z, \zeta) + \log(z - \zeta)$$

is multiple-valued.

There is another representation for S. Let D again be the unit disc and denote by

$$G^{\mathrm{I}}(z, \zeta) := -\frac{1}{2\pi} \log \left| \frac{\zeta - z}{1 - \bar{\zeta} z} \right|, \quad G^{\mathrm{II}}(z, \zeta) := -\frac{1}{2\pi} \log |(\zeta - z)(1 - \bar{\zeta} z)|$$

the Green's functions of D and let d and d_n be the differential operators

$$\mathrm{d} := \frac{\partial}{\partial \zeta} \partial \zeta + \frac{\partial}{\partial \bar{\zeta}} \mathrm{d}\bar{\zeta}, \qquad \mathrm{d}_n := -i \left\{ \frac{\partial}{\partial \zeta} \mathrm{d}\zeta - \frac{\partial}{\partial \bar{\zeta}} \mathrm{d}\bar{\zeta} \right\}$$

then we have

$$(Su)(z) = -\int_L u(\zeta) \, [\mathrm{d}_n G^{\mathrm{I}} - \mathrm{i} \, \mathrm{d} \, G^{\mathrm{II}}] (\zeta, z) \, .$$

This formula is valid for all domains for which the Green's functions exist (see HAACK and WENDLAND [1]).

Definition 10.5. Let D be a domain of $\hat{\mathbb{C}}$. An operator S is called a *Schwarz operator* if for given real function u on the boundary L of D it determines an analytic function the limiting values of the real part of which on L coincide with u and the imaginary part of which vanishes at a certain point of D.

If now instead of the real part a linear combination of the real and the imaginary parts that is

$$\operatorname{Re}\{\overline{G(\zeta)}\, w(\zeta)\} = g(\zeta) \qquad (\zeta \in L) \tag{9}$$

where G, g are given functions on L is prescribed, then we are dealt with the Hilbert boundary value problem.

The classical assumptions on the coefficients G and g are Hölder-continuity and $G(\zeta) \neq 0$ on L.

As by the Riemann problem the solution of the Hilbert problem depends on the index

$$n := \operatorname{ind} G := \frac{1}{2\pi} \int_L \mathrm{d}\arg\, G(\zeta) \in \mathbb{Z}$$

and can be reduced to the simpler (Dirichlet) problem $G \equiv 1$.

In the following it is assumed that $0 \in D$. By multiplication with a regularizing factor (see GAHOV [3, 220 p.]) G can be replaced by an analytic function while the index is unchanged. Without loss of generality one can assume $|G| = 1$. Then

$$\tau := S(\arg G - n \arg \zeta)$$

where S is the Schwarz operator of D implies

$$\operatorname{Re}\{\zeta^{-n}\, \mathrm{e}^{-\mathrm{i}\tau(\zeta)}\, w(\zeta)\} = |\zeta|^{-n}\, \mathrm{e}^{\operatorname{Im}\tau(\zeta)}\, g(\zeta) \qquad (\zeta \in L)\,. \tag{10}$$

Therefore in the case of non-negative index

$$w(z) = z^n\, \mathrm{e}^{\mathrm{i}\tau(z)}\, \{S(|\zeta|^{-n}\, \mathrm{e}^{\operatorname{Im}\tau(\zeta)}\, g(\zeta)) + Q(z)\} \tag{11}$$

is the general solution of (9) where

$$Q(z) = \mathrm{i}c_0 + \sum_{k=1}^{n} \{c_k \omega^k(z) - \bar{c}_k \omega^{-k}(z)\}$$

($c_0 \in \mathbb{R}$, $c_k \in \mathbb{C}$) and ω is the conformal mapping from D onto the unit disc (normalized by $\omega(0) = 0$, $\omega'(0) > 0$). It can easily be seen that Q has a pole at zero of order not bigger than n and has $2n + 1$ free real parameters. (11) even in the case $n < 0$ represents a solution if $Q(z) \equiv \mathrm{i}c_0$. But in order that this solution behaves like a regular function in $z = 0$, c_0 has to vanish as well as

$$S(|\zeta|^{-n}\, \mathrm{e}^{\operatorname{Im}\tau(\zeta)}\, g(\zeta))$$

has to have a zero of order at least $-n$ in $z = 0$. This leads to $-2n - 1$ conditions on the coefficients of (9) which can be written down explicitely only if the Schwarz operator is known. In the case of the unit disc the conditions are

$$\int_{|\zeta|=1} |\zeta|^{-n}\, \mathrm{e}^{\operatorname{Im}\tau(\zeta)}\, g(\zeta) \frac{\mathrm{d}\zeta}{\zeta^{k+1}} = 0 \qquad (0 \leqq k \leqq -n-1)\,.$$

Theorem 10.5. *For $n \geqq 0$ the Hilbert problem (9) is soluble for any Hölder-continuous. coefficients. The homogeneous problem has $2n + 1$ over $\mathbb{R}$ linearly independent solutions. For $n < 0$ the homogeneous problem is unsolvable and the inhomogeneous problem is solvable iff $-2n - 1$ conditions are fulfilled.*

In the case of negative index there exist meromorphic solutions which must have at least $-n$ poles in D counted with respect to their multiplicity.

With respect to the homogeneous problem the theorem is valid even for multiply connected domains if multi-valued functions are accepted as solutions. For additional conditions for the inhomogeneous problem as well as for single-valued solutions for this problem see GAHOV [3, 340 pp.].

The Hilbert problem is soluble analoguously for an unbounded domain if L is bounded (see GAHOV [3, 224 pp.]).

GAHOV as well as MUSHELIŠVILI have developed connections between the Riemann and the Hilbert problem. In the case of the unit circle they are obvious because of the similarity of the Cauchy and the Schwarz integral. This can be found in GAHOV [3, 233 pp.] and MUSHELIŠVILI [2, 100 pp.].

In order to consider piecewise analytic functions in a domain instead of the whole plane $\hat{\mathbb{C}}$, one has to replace the assumptions at infinity which gives uniqueness of the solution by assumptions on boundary values on the boundary of the domain. The so-called compound problem in its simplest form is as follows.

Let D be a simply connected domain with smooth boundary L and let γ be a simply closed curve in D which surrounds the origin. On L and on γ there are given Hölder-continuous functions G_k, g_k for $k = 1$ and $k = 2$, respectively, so that the G_k do not vanish. Find a function w analytic in $D - \gamma$ such that

$$\begin{cases} w^+ = G_1 w^- + g_1 & \text{on } \gamma\,, \\ \operatorname{Re}\{\overline{G_2} w\} = g_2 & \text{on } L\,. \end{cases}$$

For problems of this type see AKBAROV [3], HAĆ and SŁYSZ [1, 2], MAMOURIAN [1], LU [1], OBOLAŠVILI [1], ROGOŽINA [1], TUTSCHKE [15], WAROWNA-DORAU [2], WOLSKA-BOCHENEK [1, 2].

10.4. Boundary value problems for generalized analytic functions

The Cauchy integral and the Schwarz operator are basic for the boundary value problems of Riemann and Hilbert for analytic functions, respectively. For the more general class of generalized analytic functions which are solution of an equation of the form (see I. N. VEKUA [5, chapter III])

$$w_{\bar{z}} = aw + b\overline{w} \qquad a, b \in \mathcal{L}_{p,2}(\mathbb{C}) \tag{12}$$

there exist analoga of the Cauchy as well as of the Schwarz integral. Therefore these problems can be solved for this class of functions also in a similar way. The same is true for hyperanalytic and generalized hyperanalytic functions (see GILBERT [2], GILBERT and HILE [1, 2], BEGEHR and GILBERT [3, 4], GOLDSCHMIDT [3] and 12.6., 12.7. and 11.3.).

For each pair of coefficients a and b of (12) there exists a singular kernel $\Omega_1(z, \zeta)$ of Cauchy type and a weakly singular kernel $\Omega_2(z, \zeta)$ both uniquely given such that the generalized Cauchy integral

$$w(z) := \int_L \{\Omega_1(z, \zeta)\, \varphi(\zeta)\, d\zeta - \Omega_2(z, \zeta)\, \overline{\varphi(\zeta)\, d\zeta}\} \qquad (z \in \hat{\mathbb{C}} - L)$$

gives a solution of (12) for which under the same assumptions as above the formulae of Plemelj-Sohockiĭ hold (see I. N. VEKUA [5, III § 8, § 10, § 14)]). So the Riemann problem for solutions of (12) are sovable analogously to the analytic case (see GAHOV [3, 393 pp.]) and the same statements hold. Here the factorization of G is not necessarily but practically taken as the piecewise analytic function X from 10.2. The polynomials P_n in theorem 10.1 have to be replaced by linear combinations of generalized analytic analoga of the powers z^n which exist because of the similarity principle. Corresponding remarks with respect to the conditions in the case of negative index hold.

Theorem 10.6. *The general solution of the Riemann problem for generalized analytic functions in the case $n \geqq 0$ has the form*

$$w = \{\omega + \sum_{k=0}^{n} [\lambda_k F_k + \mu_k G_k]\} \exp \underset{\sim}{T}\Phi ,$$

$$\omega := \begin{cases} \underset{\sim}{T}_1 \tilde{g} & \text{in } D^+ , \\ z^{-n} \underset{\sim}{T}_1 \tilde{g} & \text{in } D^- , \end{cases}$$

$$(\underset{\sim}{T}\Phi)(z) := \frac{1}{2\pi i} \int_L \{\Omega_1(z,\zeta)\, \Phi(\zeta)\, d\zeta - \Omega_2(z,\zeta)\, \overline{\Phi(\zeta)}\, \overline{d\zeta}\} \qquad (z \in \mathbb{C}) ,$$

$$\Phi(z) := \log G(z) - n \log z , \qquad \tilde{g} := g \exp\left(-\tfrac{1}{2}\Phi - \underset{\sim}{T}\Phi\right) ,$$

where $\underset{\sim}{T}_1$ is a suitable integral operator defined similar to $\underset{\sim}{T}$, λ_k and μ_k are arbitrary real constants, F_k and G_k are special solutions of a suitable associated homogeneous Riemann boundary value problem of index n for generalized analytic functions.

For $n < 0$ the inhomogeneous problem is solvable only under additional assumptions. The solution then has the above form where the λ_k's and μ_k's have to be replaced by zero. More precise informations can be found in Begehr and Gilbert [5] and in Gahov [3, p. 393]. See also Kazidzhanova [1], Kvitko [1]. Another possibility for finding a particular solution of the inhomogeneous Riemann problem is given by the representation of the solution of an inhomogeneous equation of the form

$$w_{\bar{z}} = aw + b\bar{w} + c ,$$

$$|a| + |b| \leqq \begin{cases} K, & |z| \leqq 1 , \\ K\,|z|^{-1-\varepsilon} & 1 \leqq |z| , \end{cases} \qquad c \in \mathcal{L}_{p,2}(\mathbb{C}) \quad (0 < \varepsilon,\ 2 < p) \tag{13}$$

which is

$$w(z) := (\underset{\sim}{P}c)(z) := -\frac{1}{\pi} \int_{\mathbb{C}} \{\Omega_1(z,\zeta)\, c(\zeta) - \Omega_2(z,\zeta)\, \overline{c(\zeta)}\}\, d\xi\, d\eta$$

(see Begehr and Hile [1] and I. N. Vekua [5, III, § 13]). The solution of

$$\begin{cases} w_{\bar{z}} = aw + b\bar{w} + c & \text{in } \mathbb{C} - L , \\ w^+ = w^- + g & \text{on } L , \\ w(\infty) = 0 , \end{cases} \tag{14}$$

is given by

$$\varrho(z) := \frac{1}{2\pi i} \int_L g(\zeta) \frac{d\zeta}{\zeta - z}$$

in the form

$$w = \varrho + \underset{\sim}{P}[c + a\varrho + b\bar{\varrho}] . \tag{15}$$

To this special ($G \equiv 1$) Riemann problem the general case can be transformed by

$$\omega = \frac{w}{X} + P_n \qquad (0 \leqq n)$$

where P_n is a suitable complex polynomial of degree not exceeding n and X is the analytic factorization of G. If $n < 0$ the same transformation with $P_n \equiv 0$ leads to a reduction of the special problem but now ω has to have a zero of order at least $-n$ at infinity, so that w will behave regularly there. This condition is fulfillable only for ϱ by the assumptions on the coefficients

$$\int_L \frac{g(\zeta)}{X^+(\zeta)} \zeta^{k-1} \, d\zeta = 0 \qquad (1 \leqq k \leqq -n-1)$$

but in general not for $\underset{\sim}{P}[c + a\varrho + b\bar{\varrho}]$, where only

$$|\underset{\sim}{P}[c + a\varrho + b\bar{\varrho}]| \leqq M(p, \varepsilon, K) \{||\varrho||_0 + ||c||_{p,2}\} |z|^{-\alpha} \quad \left(z \neq 0, \alpha = \frac{p-2}{p}\right)$$

is known.

The representation formula (15) leads to an a priori estimate for solutions of (14) useful e.g. for uniqueness theorems.

Lemma 10.1 (a priori estimate). *Let w be a solution of the Riemann problem* (14) *where the assumptions* (13) *are fulfilled, then there exists a constant $\varkappa$ dependent on L, p, ε, and K, but not on w and g such that*

$$||w||_{\alpha, \hat{C}} := ||w||_{\alpha, \hat{D}^+} + ||w||_{\alpha, \hat{D}^-} \leqq \varkappa \{||g||_{\alpha, L} + ||w_{\bar{z}} - aw - b\widetilde{w}||_{p,2}\} \tag{16}$$

(see Begehr and Hile [1]).

The Hilbert boundary value problem for generalized analytic functions first was considered by Hilbert [2] who solved it for (12) written as a real system by integral equation methods. For the unit disc the Schwartz operator can be generalized (I. N. Vekua [5, 309 pp.]), while for arbitrary simply or multiply connected domains this is not known up to now. But there exists an analogue to the Schwarz formula for the unit disc, which is deduced from the generalized Cauchy formula (Pompeiu formula, a consequence from Green's formula)

$$w(z) = \frac{1}{2\pi i} \int_L w(\zeta) \frac{d\zeta}{\zeta - z} + \frac{1}{2\pi i} \int_D w_{\bar{\zeta}}(\zeta) \frac{d\zeta \, d\bar{\zeta}}{\zeta - z} \qquad (z \in D).$$

It is

$$w(z) = \frac{1}{2\pi i} \int_{|\zeta|=1} \operatorname{Re} w(\zeta) \frac{\zeta + z}{\zeta - z} \frac{d\zeta}{\zeta} + \frac{i}{2\pi} \int_{|\zeta|=1} \operatorname{Im} w(\zeta) \frac{d\zeta}{\zeta}$$

$$+ \frac{1}{2\pi i} \int_{|\zeta|<1} \left\{ \frac{w_{\bar{\zeta}}(\zeta)}{\zeta - z} + \frac{\overline{z w_{\bar{\zeta}}(\zeta)}}{1 - \bar{\zeta} z} \right\} d\zeta \, d\bar{\zeta} \qquad (|z| < 1).$$

Again using Green's functions as in 10.3. one gets

$$w(z) = -\int_L \operatorname{Re} w(\zeta) \, [d_n G^{I} - i dG^{II}] \, (\zeta, z) - i \int_L \operatorname{Im} w(\zeta) \, d_n G^{II}(\zeta, z)$$

$$+ i \int_D \{ w_{\bar{\zeta}}(\zeta) \, [G^{I}_{\zeta} + G^{II}_{\zeta}] \, (\zeta, z) + \overline{w_{\bar{\zeta}}(\zeta)} \, [G^{I}_{\bar{\zeta}} - G^{II}_{\bar{\zeta}}] \, (\zeta, z) \} \, d\zeta \, d\bar{\zeta} \quad (z \in D).$$

This representation formula which under suitable assumptions on the boundary data for solutions of (12) leads to an integral equation of Fredholm type is still valid for domains D for which the Green's functions G^{I} and G^{II} exist (Haack and Wendland

[1], Wendland [4]). From a refinement of it one gets an a priori estimate for solutions of the Hilbert problem for generalized analytic functions in the case $n \geqq 0$.

Lemma 10.2 (a priori estimate). *Let $w \in \mathscr{C}^{1+\alpha}(\hat{D})$ fulfil*

$$\left.\begin{aligned} &\operatorname{Re}\{\overline{G}w\} = g \quad on\ L && (G, g \in \mathscr{C}^{1+\alpha}(L)\,, \quad n = \operatorname{ind} G \in \mathbb{N}_0)\,, \\ &\int_L \operatorname{Im}\{\overline{G}w\}\, ds = \varkappa && (\varkappa \in \mathbb{R})\,, \\ &w(z_k) = a_k && (1 \leqq k \leqq n) \qquad (z_k \in D,\, a_k \in \mathbb{C}) \end{aligned}\right\} \tag{17}$$

and let $a, b \in \mathscr{C}^\alpha(\hat{D})$ fulfil $\|a\|_\alpha + \|b\|_\alpha \leqq K < +\infty$, then there exists a constant γ dependent on D, z_k $(1 \leqq k \leqq n)$, G, α, K but independent of w, g, and $\varkappa$ such that

$$\|w\|_{1+\alpha,\hat{D}} \leqq \gamma\{\|g\|_{1+\alpha,L} + |\varkappa| + \sum_{k=1}^{n} |a_k| + \|w_{\bar{z}} - aw - b\overline{w}\|_{\alpha,\hat{D}}\}\,. \tag{18}$$

This a priori estimate first was proved by Wendland [4] who used functional analytic methods and later by Begehr and Hsiao [1], using constructive methods so that this estimate just as (16) for the Riemann problem can be used for a numerical treatment of nonlinear Hilbert boundary value problems, as will be considered in the next section.

Another conclusion of the mentioned refinement of the representation formula is the following result (Wendland [4, p. 19]).

Theorem 10.7. *Let D be a simply connected bounded domain of $\mathbb{C}$ the boundary L of which is a Ljapunov curve, say, $L \in \mathscr{C}^{1+\alpha}$ $(0 < \alpha \leqq 1)$, let $G \in \mathscr{C}^{1+\alpha}(L)$, $|G| = 1$, $n = \operatorname{ind} G \in \mathbb{N}_0$, $z_k \in D$ $(1 \leqq k \leqq n)$, and let $a, b \in \mathscr{C}^\alpha(\hat{D})$. Then the Hilbert problem* (17) *for solutions of*

$$w_{\bar{z}} = aw + b\overline{w} + c$$

is uniquely solvable in $\mathscr{C}^{1+\alpha}(\hat{D})$ for every $c \in \mathscr{C}^\alpha(\hat{D})$, every $g \in \mathscr{C}^{1+\alpha}(L)$, every $\varkappa \in \mathbb{R}$ and arbitrary $a_k \in \mathbb{C}$.

For solutions and solvability conditions in the case of negative index we refer to Haack and Wendland [1] and Wendland [4].

It is quite obvious that the results of this section can be generalized to solutions of more general elliptic systems — generalized Beltrami equations — of the form

$$w_{\bar{z}} + \mu_1 w_z + \mu_2 \overline{w}_z = aw + b\overline{w} + c\,. \tag{19}$$

One can e.g. extend the a priori estimates (see Bojarski [1], Vinogradov [1—3], Begehr and Hsiao [3—5], Begehr and Hile [2]).

Lemma 10.3. *Let $a, b, c \in \mathscr{L}_{p',p}(\mathbb{C}) \cap \mathscr{L}_{p,2}(\mathbb{C})$ $(1 < p' < 2 < p)$ and let μ_1, μ_2 be measurable and fulfil*

$$\|\mu_1\|_0 + \|\mu_2\|_0 \leqq q_0 < 1\,, \qquad |\mu_1(z)| + |\mu_2(z)| = O(|z|^{-\varepsilon}) \quad (z \to \infty, 0 < \varepsilon)\,.$$

Then for every solution of (19) *which fulfils the boundary condition of* (14) *one has*

$$\begin{aligned} \|w\| &:= \|w\|_0 + \|w_z\|_{p',p} + \|w_{\bar{z}}\|_{p',p} \leqq \\ &\leqq \beta[\|g\|_{1+\alpha} + \|w_{\bar{z}} + \mu_1 w_z + \mu_2 \overline{w}_z - aw - b\overline{w}\|_{p',p}] \end{aligned}$$

where β depends on L, p', p, α, q_0, and k where $\|a\|_{p',p} + \|b\|_{p',p} \leqq k$.

Lemma 10.4. *Let $a, b, c \in \mathscr{L}_p(\hat{D})$ $(2 < p)$ and let μ_1, μ_2 be measurable and fulfil*

$$\|\mu_1\|_0 + \|\mu_2\|_0 \leqq q_0 < 1\,.$$

Then for every solution of (19) *which fulfils the boundary condition* ($n = \operatorname{ind} G \geqq 0$)

$$\operatorname{Re}\{\bar{G}w\} = g \quad \text{on } L\,, \qquad \int_L \operatorname{Im}\{\bar{G}w\}\, ds = \varkappa\,, \qquad w(z_k) = a_k \ (1 \leqq k \leqq n)$$

one has

$$\|w\| := \|w\|_0 + \|w_z\|_p + \|w_{\bar{z}}\|_p$$
$$\leqq \beta[\|g\|_{1+\alpha} + |\varkappa| + |a_1| + \dots + |a_n| + \|w_{\bar{z}} + \mu_1 w_z + \mu_2 \overline{w_z} - aw - b\bar{w}\|_p]$$

where β depends on D, G, p, α, q_0, and k where $\|a\|_p + \|b\|_p \leqq k$.

It has to be mentioned that in the last two lemmata there are restrictions on p' and p and on p, respectively, which ask them to be close enough to 2.

For other results in connection with boundary value problems for (19) we refer to Bojarski [1], Mamourian [1], Tjurikov [1], Vinogradov [1—4] and Wen [1—4]. Generalizations to systems of equations of first order were given by Murtazaev [2]. See also Nguyen Thua Hop [2].

10.5. Nonlinear boundary value problems

Boundary value problems of Riemann as well as of Hilbert type are considered for nonlinear elliptic systems of first order not only with linear but also with nonlinear boundary conditions. In their most general form they are as follows:

10.5.1. Riemann boundary value problem

$$\left.\begin{aligned} w_{\bar{z}} &= H(z, w, w_z) && \text{in } \mathbb{C} - L\,,\\ w^+ &= Gw^- + F(z, w^+, w^-) && \text{on } L\,,\\ w(z) &= O(|z|^{m-n}) && (z \to \infty)\,. \end{aligned}\right\} \tag{20}$$

This problem, according to the considerations in the preceding section, is reducible to the special case $G \equiv 1$, $n = 0$. If, moreover, $m \in \mathbb{N}$, then a transformation of the form $w = \omega - P_m$, where P_m is a suitable polynomial of degree at most m leads to the special problem

$$\begin{aligned} w_{\bar{z}} &= H(z, w, w_z) && \text{in } \mathbb{C} - L\,,\\ w^+ &= w^- + F(z, w^+, w^-) && \text{on } L\,,\\ w(\infty) &= 0\,. \end{aligned}$$

10.5.2. Hilbert boundary value problem

$$\left.\begin{aligned} w_{\bar{z}} &= H(z, w, w_z) && \text{in } D\,,\\ \operatorname{Re}\{\bar{G}w\} &= g(z, w) && \text{on } L\,. \end{aligned}\right\} \tag{21}$$

In the case of non-negative index n the additional conditions

$$\left.\begin{aligned} &\int_L \operatorname{Im}\{\bar{G}w\}\, ds = \varkappa(w)\\ &w(z_k) = a_k(w) \qquad (z_k \in D,\ 1 \leqq k \leqq n) \end{aligned}\right\} \tag{22}$$

again will give uniqueness of the solution.

Problem (20) was solved for linear equations (12) by Warowna-Dorau [1], Hać and Słysz [2], for nonlinear but of w_z independent H by Begehr and Hile [1]. (21, 22) was solved again with H independent of w_z by Wendland [1, 4], Begehr and Hsiao [1, 2]. For equations (12) Warowna-Dorau [2], Wolska-Bochenek [1, 2], Hać and Słysz [2] considered a nonlinear compound boundary value problem.

Nonlinear equations of the general type up to now are only studied in connection with linear boundary problems. BOJARSKI and IWANIEC [1] (Dirichlet problem) used fixed point methods of Banach and Schauder, NAAS and TUTSCHKE [2, 3], TUTSCHKE [12, 16] (Hilbert problem) reduced the Hilbert problem to the Dirichlet problem and used the Banach fixed point theorem. Compare also WENDLAND [4, e.g. § 26].

The most general formulation in this connection was given by TUTSCHKE [10]. Under suitable assumptions on H (Lipschitz condition) a solution of the elliptic equation

$$w_{\bar{z}} = H(z, w, w_z) \quad \text{in } D$$

has to be found which fulfils a side condition of the form $\underset{\sim}{B}w = g$ where $\underset{\sim}{B}$ is an operator for which for arbitrary g_1, g_2

$$\underset{\sim}{B}(\Phi + g_1) = g_2$$

always is uniquely solvable by an analytic function.

Then the problem is solvable with fixed point methods (Banach or Schauder). Applying this result in TUTSCHKE [10] the compound boundary value problem for the nonlinear equation can be solved by specifying B to a suitable linear operator

Recently using the a priori estimates from lemmata 10.4 and 10.5, respectively both boundary value problems (20) and (21), (22) with boundary and side data in dependent of w have been solved (BEGEHR and HILE [2]).

Theorem 10.8. *Let p' and p, $1 < p' < 2 < p$ be close enough to 2 and*

$$0 < \varepsilon < 1 < 2\alpha < 2 .$$

Let H be a measurable function from $\mathbb{C}^3$ into $\mathbb{C}$, $H(\cdot, 0, 0) \in \mathcal{L}_{p',p}(\mathbb{C}) \cap \mathcal{L}_{p,2}(\mathbb{C})$,

$$\begin{aligned} &|H(z, w_1, v) - H(z, w_2, v)| \leqq K(z)\, |w_1 - w_2| , && K \in \mathcal{L}_{p',p}(\mathbb{C}) \cap \mathcal{L}_{p,2}(\mathbb{C}) , \\ &|H(z, w, v_1 | - H(z, w, v_2)| \leqq q(z)\, |v_1 - v_2| , && 0 \leqq q(z) \leqq q_0 < 1 , \\ &q(z) = O(|z^{-\varepsilon}|) \quad \text{as} \quad z \to \infty . \end{aligned} \tag{23}$$

Let $F \in \mathcal{C}^\alpha(L)$. Then the special form of (20) is uniquely solvable.

Theorem 10.9. *Let H satisfy (23) and*

$$\begin{aligned} &H(\cdot, 0, 0) \in \mathcal{L}_{\infty,\mathrm{loc}}(\mathbb{C}) , \qquad H(z, 0, 0) = O(|z|^{-1-\varepsilon-n}) \quad \text{as} \quad z \to \infty , \\ &q(z) = O(|z|^{-\varepsilon-m}) \quad \text{as} \quad z \to \infty , \\ &K \in \mathcal{L}_{\infty,\mathrm{loc}}(\mathbb{C}) , \qquad K(z) = O(|z|^{-1-\varepsilon-m}) \quad \text{as} \quad z \to \infty , \end{aligned}$$

where $m \geqq 0$ and n are integers. Let $F, G \in \mathcal{C}^\alpha(L)$ $(1 < 2\alpha < 2)$ and $G(z) \neq 0$ on L and $\operatorname{ind} G = n$. *Then there exists a unique solution w of* (20).

Theorem 10.10. *Let $p > 2$ be close enough to 2 and $1 < 2\alpha < 2$. Let H be a measurable function from $D \times \mathbb{C} \times \mathbb{C}$ into $\mathbb{C}$, $H(\cdot, 0, 0) \in \mathcal{L}_p(\hat{D})$, fulfilling* (23), *where $K \in \mathcal{L}_p(\hat{D})$. Let $g, G \in \mathcal{C}^\alpha(L)$, $G(z) \neq 0$ on L, $n = \operatorname{ind} G \geqq 0$, $\varkappa \in \mathbb{R}$, $a_k \in \mathbb{C}$ $(1 \leqq k \leqq n)$. Then* (21), (22) *is uniquely solvable.*

These results can be generalized to nonlinear boundary and side data under suitable Lipschitz conditions. Because the Lipschitz constants have to be very small the results are still unsatisfactory.

Further references

ABDURAHMANOV [1], AKBAROV [1, 2], AKSENT'EV, HOHLOV, and ŠIROKOVA [1], AKSENT'EV, IL'INSKIĬ et al. [1], ALEHNO [3, 6], ALEKSANDRIJA [1], ALEKNA [1], BEGEHR [3, 7], BINDERMAN [1, 2, 3], BOJARSKI [3], CALDERÓN et al. [1], ČEREDNIČENKO [1], CERNECKIĬ and DIDENKO [1], ČIBRIKOVA and SIL'VESTROV [1], CIECHANOWICZ-HALKA [1], A. and S. CHAKRABARTI [1], ČORIEV [1], DECEWICZ [1], DIDENKO and TIHONENKO [1], DYBIN and GAPONENKO [1], DŽURAEV [1], EFENDIEV [1, 2], GABRINOVIČ [1, 2], GAVDZINSKIĬ, and MEČAEV [1], GAVRILOV [1, 2, 3], GILBERT [4], GLOWACKI [1], GOGONEA [1], HASEMAN [1], HELLWIG [1], HILBERT [1], HVEDELIDZE [1], IKEBE, LI, and STENGER [1], ISAHANOV [1—4], JACENKO [1], A. I. and S. I. JACKO [2], KAC [1, 2, 5], KAKIČEV [1, 2], KAKIČEV and ŠELKOVIČ [1], KAZIEV [1], KOVERNYĬ and JAŽENKO [1], KRUGLOV [1], KOČURA [1], KULAGINA [1], KUPRIN [1], LE-DIN'ZON [1], LITVINČUK [1], LUBOWICZ and WIEPRZKOWICZ [1], MARTIROSJAN [1], MEUNARGIJA [1], MICHLIN and PRÖSSDORF [1], MONAHOV [1], MUSHELIŠVILI [1], NECAEV and PROČENKO [1], NITSCHE [1, 2], OBOLAŠVILI [2], ORTON [1, 2], OBNOSOV [1], PASCALI [2, 3], PANČENKO [1], PRIMAČUK [1—3], PRIVALOV [2], RODIN [2], ROGOZIN [1], RUSAK and ŠEŠKO [1], RYSJUK [1], SEĬFULLAEV [1], SLIZKIĬ [1, 2], SOLDATOV [1], TEICHERT-GAIDA [1], TIHONENKO [1, 2], TIMOFEEV [1, 2], VAN TRIAN [1], TUNGATAROV [1], USMANOV [1], N. P. VEKUA [2, 3], VOLEVIČ and GINDIKIN [1], WENDLAND [7], ZVEROVIČ [1].

11. Elliptic Systems in the Plane

V. S. Vinogradov

The theory of elliptic systems in the plane is at present developed in much more detail than the corresponding theory in spaces of higher dimensions. This is because of the possibility to apply the classical methods of function theory in one complex variable and the theory of one-dimensional singular integral equations. The representation of the solutions of elliptic systems by means of analytic functions in one complex variable is the fundamental device for realizing these applications. Wide literature is devoted to the establishment and application of such representation formulas and is essentially based on the papers of E. Goursat, É. Picard and T. Carleman. Here we shall use the results of the authors I. N. Vekua [3, 4, 5], Bojarski [1], Bers and Nirenberg [1, 2], Bers, John, and Schechter [1]. Hence, for the investigation of elliptic systems of first order in the plane by means of the methods of complex analysis we shall start from a special system and arrive at a general one.

11.1. Generalized analytic functions

A complex-valued function $w(z) = u(x, y) + iv(x, y)$ is called a *generalized analytic function* if it satisfies the equation

$$\frac{\partial w}{\partial \bar{z}} + aw + b\bar{w} = 0 . \tag{1}$$

We shall assume that the coefficients of the equation (1) a, b are elements of $\mathcal{L}_p(G)$ where $p > 2$ and G is some bounded domain with a sufficient smooth boundary. Then we have to find a solution of the Sobolev space $\mathcal{W}_p^{(1)}(G)$.

The following representation theorem of the solutions of (1) by means of analytic functions holds (I. N. Vekua [5, chapter 3, § 4]):

Theorem 11.1.1. *Every solution of $\mathcal{W}_p^{(1)}(G)$ of equation* (1) *is representable by*

$$w(z) = \Phi(z)\, e^{\omega(z)} , \tag{2}$$

where $\Phi(z)$ is an analytic and Hölder-continuous function in G with the exponent $p - 2/p$ and $\omega(z) \in \mathcal{W}_p^{(1)}(G)$.

The functions $\Phi(z)$ and $\omega(z)$ may be expressed by $w(z)$ in the following way. Introducing the function

$$A(z) = \begin{cases} a(z) + b(z)\,\dfrac{\overline{w(z)}}{w(z)} & \text{for} \quad w(z) \neq 0 , \\ a(z) + b(z) & \text{for} \quad w(z) = 0 \end{cases} \tag{3}$$

we have $A(z) \in \mathscr{L}_p(G)$, $p > 2$ and

$$\omega(z) = \frac{(-1)}{\pi} \iint_G \frac{A(\zeta)}{\zeta - z} \mathrm{d}\xi \, \mathrm{d}\eta , \qquad \zeta = \xi + \mathrm{i}\eta . \tag{4}$$

$\Phi(z)$ is defined by formula (2)

It is also possible to state the inverse problem. Let $\Phi(z)$ be an analytic function in G which is Hölder-continuous with exponent $p - 2/p$ in G[1]). The question now arising is, whether it is possible to find a solution of (1) which can be represented in the form (2).

If there exists a solution of the problem then it would be not unique. In fact, for any analytic function $\Omega(z)$ in G one has also the following representation of $w(z)$ in addition to the representation formula (2)

$$w(z) = \Phi_1(z) \, \mathrm{e}^{\omega_1(z)}$$

with

$$\Phi_1(z) = \Phi(z) \, \mathrm{e}^{\Omega(z)} \quad \text{and} \quad \omega_1(z) = \omega(z) - \Omega(z) . \tag{5}$$

In virtue of this result it is possible to formulate the theorem 11.1.1 more precisely, namely.

Theorem 11.1.2. *Under hypotheses of theorem* 11.1.1 *one may choose the function* $\omega(z)$ *in the representation formula* (2) *in such a way that it is pure imaginary* (*or pure real*) *on the boundary of the domain G.*

Proof. It is sufficient to take $\Omega(z)$ as a solution of the boundary value problem:

$$\frac{\partial \Omega}{\partial \bar{z}} = 0 \quad \text{in } G , \qquad \operatorname{Re} \Omega(z)|_{\partial G} = \operatorname{Re} \omega(z)|_{\partial G}$$

$$(\operatorname{Im} \Omega(z) = \operatorname{Im} \omega(z)|_{\partial G}) .$$

As it was shown in I. N. Vekua [5, chapter 3, § 7], for any analytic function $\Phi(z)$ in G which may possess arbitrary singularities in G (unique) and for an arbitrary fixed point t in G a solution $w(z)$ of (1) exists such that $w_0(z) = w(z)/\Phi(z)$ is continuous in G and it can be continuously extended into the whole plane. Furthermore $w_0(z)$ satisfies the following three conditions:

(i) $w_0 \in C_{p-2/p}(\mathbb{C})$,
(ii) $w_0(z) \neq 0$,
(iii) $w_0(t) = 1$.

(Note, that $w(z)$ satisfies the equation (1) in the points of regularity of $\Phi(z)$). From this follows the existence of a solution of the inverse problem concerning the representation of solutions by analytic functions.

We call the representation formula (2) the *representation of the first kind* of generalized analytic functions by analytic functions.

Besides, there is a second representation of generalized analytic functions by analytic ones. This representation is obtained in the following way (I. N. Vekua [5, chapter 3, § 5]). For every function of $\mathscr{W}_p^{(1)}(G)$ the formula holds

$$w(z) = \Phi(z) - \frac{1}{\pi} \iint_G \frac{\partial w}{\partial \bar{\zeta}} \frac{\mathrm{d}\xi \, \mathrm{d}\eta}{\zeta - z} , \tag{6}$$

---

[1]) One may regard this problem under the more general situation that the trace $\Phi|_{\partial G} \in \mathscr{W}_p^{\left(1-\frac{1}{p}\right)}(\partial G)$.

where $\Phi(z)$ is some analytic function in G. Analogeously we get from (6) for a solution of (1) the following representation formula

$$w(z) - \frac{1}{\pi} \iint\limits_G \frac{aw(\zeta) + b\overline{w(\zeta)}}{\zeta - z} \, d\xi \, d\eta = \Phi(z) . \tag{7}$$

The kernel of this equation has a weak singularity (less than the dimension of the plane) and therefore Fredholm's alternative may be applied to it.

Moreover one can show (cf. I. N. VEKUA [4]) that the solution of (7) can be expressed in terms of Neumann's series and one gets the representation formula of the form

$$w(z) = \Phi(z) + \iint\limits_G \Gamma_1(z, \zeta) \, \Phi(\zeta) \, d\xi \, d\eta + \iint\limits_G \Gamma_2(z, \zeta) \, \overline{\Phi(\zeta)} \, d\xi \, d\eta . \tag{8}$$

The kernels Γ_i may be expressed by series or by limits of converging iterative sequences the form of which essentially depend on the domain G where the given equation is regarded.

The representation formula (8) is called the *representation of second kind* of generalized analytic functions by analytic functions. We note that this formula is valid only for $z \in G$ whereas for the convergence of the series representing the Γ_i, $i = 1, 2$, the boundedness of the domain is essential. For establishing the representation formula (8) in unbounded domains it is necessary to impose an additional condition on the coefficients $a(z)$ and $b(z)$ concerning its decreasing behaviour at infinity. This condition should ensure the convergence of the integral

$$Kw = -\frac{1}{\pi} \iint\limits_G \frac{aw + b\overline{w}}{\zeta - z} \, d\xi \, d\eta$$

for any $w \in \mathcal{L}_p(G)$ and for Kw acting from $\mathcal{L}_p$ into $\mathcal{L}_p$.

From formula (8) it is possible to deduce Cauchy's integral formula for the system (1). To this assertion we note that $\Phi(z)$ in (6) may be written in form

$$\Phi(z) = \frac{1}{2\pi i} \int\limits_{\partial G} \frac{w(\zeta)}{\zeta - z} \, d\zeta . \tag{9}$$

Formula (9) immediately follows from (6) by multiplying with $1/2\pi i \; 1/(\zeta - z)$ and integrating over ∂G. Inserting (9) into (8) and changing the order of integration over G and ∂G we get

$$w(z) = \frac{1}{2\pi i} \int\limits_{\partial G} \Omega_1(\zeta, z) \, w(\zeta) \, d\zeta - \frac{1}{2\pi i} \int\limits_{\partial G} \Omega_2(\zeta, z) \, \overline{w(\zeta)} \, d\zeta . \tag{10}$$

If system (1) is the Cauchy-Riemann system, then $\Omega_1(\zeta, z) = 1/(\zeta - z)$ and $\Omega_2(\zeta, z) = 0$. In the general case Ω_1 has a singularity of order $1/(\zeta - z)$ for $z \to \zeta$ but Ω_2 has a logarithmic one (I. N. VEKUA [5]).

If the coefficients in (1) are constants, then the kernels are expressed by the fundamental solution of the system (1) and they do not depend on the domain G where we regard this system. Their precise expressions by McDonald functions are given in TOKIBETOV [1].

If the coefficients of (1) belong to $\mathcal{L}_{p,\mathrm{loc}}(G)$ then a method for constructing their kernels Ω_1 and Ω_2 is given in the paper GOLDSCHMIDT [2].

But, if we consider the corresponding inhomogeneous equation to (1), i.e.

$$\frac{\partial w}{\partial \bar z} + aw + b\bar w = f\,, \tag{11}$$

then for its solution the following formula holds

$$w(z) = \frac{1}{2\pi \mathrm{i}} \int_{\partial G} \Omega_1(\zeta, z)\, w(\zeta)\, \mathrm{d}\zeta - \Omega_2(\zeta, z)\, \overline{w(\zeta)}\, \overline{\mathrm{d}\zeta}$$

$$- \frac{1}{\pi} \iint_G \left(\Omega_1(\zeta, z) f(\zeta) + \Omega_2(\zeta, z) \overline{f(\zeta)}\right) \mathrm{d}\xi\, \mathrm{d}\eta \tag{12}$$

where

$$w_1(z) = -\frac{1}{\pi} \iint_G \left(\Omega_1(\zeta, z) f(\zeta) + \Omega_2(\zeta, z) \overline{f(\zeta)}\right) \mathrm{d}\xi\, \mathrm{d}\eta$$

is a solution of the inhomogeneous equation (11).

The representation formulas play an important role for studying the properties of the solution of equation (1) and for proving the solvability of boundary value problems for (1) and (11).

Carleman's theorem about the distribution of zeroes of the solution (1) (the zeroes are isolated or the solution vanishes identically), Rouché's theorem and the argument principle immediately follow from the representation formula of the first kind.

By means of the representation formula of the second kind (8) the Riemann-Hilbert boundary value problem, i.e.,

$$\mathrm{Re}\,\{g(t)\, w(t)\}|_{\partial G} = \gamma(t) \tag{13}$$

can be transformed into an equivalent singular integral equation over the boundary of the domain (I. N. Vekua [4, 5]).

To this end we have to take the representation of the analytic function $\Phi(z)$ in (8) by a Cauchy type integral with real density given by I. N. Vekua, i.e.,

$$\Phi(z) = \frac{1}{2\pi \mathrm{i}} \int_{\partial G} \frac{\mu(t)}{t - z}\, \mathrm{d}t + \mathrm{i}c\,. \tag{14}$$

After having set (8) into (13) we get a singular integral equation and thus it is possible to write down the Noether condition and a index formula for it

11.2. Transformation of an elliptic system, consisting of two real equations into equations of generalized analytic functions and an equation of special type, respectively

Let us consider the elliptic system

$$\frac{\partial \boldsymbol{u}}{\partial x} + \boldsymbol{A} \frac{\partial \boldsymbol{u}}{\partial y} + \boldsymbol{B}u = \boldsymbol{f} \tag{15}$$

where $\boldsymbol{A}$ and $\boldsymbol{B}$ are matrix functions of second order $\boldsymbol{u} = (u_1, u_2)^{\mathrm{T}}$ is the unknown vector function and $\boldsymbol{f} = (f_1, f_2)^{\mathrm{T}}$ is the given vector function. The domain of definition G of all these functions is assumed to be some bounded and simply connected domain with a smooth boundary.

Let the elements of the matrix $\boldsymbol{A}$ be continuously differentiable in $\overline{G}$ but those of $\boldsymbol{B}$ and $\boldsymbol{f}$ belong to $\mathcal{L}_p(G)$, $p > 2$. Then we can show that the equation (15) can be transformed into an inhomogeneous equation of generalized analytic functions (11) by changing the dependent and independent variables. If the coefficients are not smooth, one can not proceed in such a way. Therefore a transformation will be given such that (15) leads to an equation of special complex type.

Let be $\boldsymbol{A} \in C^1(\overline{G}) \cap C(\overline{G})$. In the following it will be shown how one can weaken that restriction on the smoothness of the coefficients.

Because of its ellipticity the matrix $\boldsymbol{A}$ may be diagonalized globally in G by applying to it a similarity transformation, i.e.,

$$\boldsymbol{A} = \boldsymbol{T}\begin{pmatrix} \lambda(z) & 0 \\ 0 & \overline{\lambda(z)} \end{pmatrix}\boldsymbol{T}^{-1}, \qquad \boldsymbol{T} = \begin{pmatrix} t_1 & \overline{t_1} \\ t_2 & \overline{t_2} \end{pmatrix},$$

$\operatorname{Im} \lambda(z) > 0$ in $\overline{G}$. $\lambda(z)$ and elements of $\boldsymbol{T}$ are of the same regularity as elements of $\boldsymbol{A}$. By introducing a new unknown complex function $w(z)$, defined by

$$u_1 = t_1 w + \overline{t_1}\overline{w}, \qquad u_2 = t_2 w + \overline{t_2}\overline{w}$$

the system (15) obtains the following form

$$\frac{\partial w}{\partial x} + \lambda \frac{\partial w}{\partial y} + \dots = f_1, \qquad \frac{\partial \overline{w}}{\partial x} + \overline{\lambda} \frac{\partial \overline{w}}{\partial y} + \dots = \overline{f_1}. \tag{16}$$

Here the second equation is the complex conjugate to the first one and therefore it can be omitted. Using the complex differentiations $\partial/\partial\bar{z}$ and $\partial/\partial z$ we can write the equation (16) in the form

$$\frac{\partial w}{\partial \bar{z}} + \mu \frac{\partial w}{\partial z} + \tilde{a} w + \tilde{b} w = f, \tag{17}$$

where $\mu = 1 + \lambda i/1 - \lambda i$. Since $\operatorname{Im} \lambda > 0$ in $\overline{G}$ it follows that

$$|\mu| \leqq |\mu_0| < 1, \tag{18}$$

i.e., the equation (17) is a uniformely elliptic system in G.

Let us consider now the Beltrami system corresponding with the principal part of (17), namely

$$\frac{\partial \zeta}{\partial \bar{z}} + \mu \frac{\partial \zeta}{\partial z} = 0. \tag{19}$$

It is well known (cf. I. N. Vekua [5]) that there exists a homeomorphism $\zeta = \zeta(z): G \to G$ satisfying the system (19). Introducing the new variable ζ into (17) defined above we get the inhomogeneous equation of generalized analytic functions in the form

$$\frac{\partial w}{\partial \bar{\zeta}} + a w + b \overline{w} = f.$$

(Here the formulas for the calculation of the coefficients may be omitted.) All the considerations carried out above remain true also in the case when $\boldsymbol{A} \in \mathcal{W}_p^{(1)}(G)$, $p > 2$.

If elements of the matrix $\boldsymbol{A}$ are not differentiable or moreover discontinuous, then it is not possible to transform equation (15) into the system (1). But introducing the unknown complex function $w = u_1 + iu_2$ one can transform the elliptic system (15) into an equation of special type, namely for $a_{21} - a_{12} > 0$

$$\frac{\partial w}{\partial \bar{z}} + \mu_1 \frac{\partial w}{\partial z} + \mu_2 \frac{\partial \overline{w}}{\partial z} + a w + b \overline{w} = f \tag{20}$$

and for $a_{21} - a_{12} < 0$

$$\frac{\partial w}{\partial z} + \mu_1 \frac{\partial w}{\partial \bar{z}} + \mu_2 \overline{\frac{\partial w}{\partial z}} + aw + b\bar{w} = f \tag{21}$$

(The condition of ellipticity implies $a_{21} - a_{12} \neq 0$.) In both cases the condition of ellipticity of equation (15) is equivalent with the inequality

$$|\mu_1| + |\mu_2| < 1 , \tag{22}$$

and uniform ellipticity is equivalent with

$$|\mu_1| + |\mu_2| \leqq \mu_0 < 1 . \tag{23}$$

Further on we shall assume that the system already has been transformed into one of equations (20) or (21) in the whole domain. We decide to deal with equation (20). In this case the following representation theorem of solutions proved by Bers, Nirenberg, and Bojarski holds (Bojarski [1], Bers and Nirenberg [1]).

Theorem 11.2. *Let $f = 0$, μ_1 and μ_2 be measurable and bounded in G and $a, b \in \mathcal{L}_p(G)$, $p > 2$. Then for any solution of the system (20) the representation formula $w(z) = \Phi(\chi(z))e^{\omega(z)}$ is valid where $\chi(z)$ and $\omega(z)$ belong to $\mathcal{W}_p^{(1)}(G)$, $p > 2$, and $\chi(z)$ is a homeomorphism $G \to G$, satisfying some Beltrami system, but $\Phi(\chi)$ is an analytic function of the complex variable χ in G.*

11.3. Investigation of the Riemann-Hilbert boundary value problem

Now let us regard the Riemann-Hilbert problem for the system (15), i.e., we are seeking solutions of the system

$$\frac{\partial \boldsymbol{u}}{\partial x} + \boldsymbol{A} \frac{\partial \boldsymbol{u}}{\partial y} + \boldsymbol{B}\boldsymbol{u} = \boldsymbol{f} \quad \text{in } G , \tag{24}$$

satisfying the boundary condition (cf. also 1.10.3., 10.4., and 12.6.)

$$\alpha u_1 + \beta u_2 = \gamma \quad \text{on } \partial G . \tag{25}$$

We shall assume that the functions α, β, and γ satisfy a Hölder condition on ∂G with exponent larger than $\frac{1}{2}$.

Applying methods of complex analysis we shall show that the problem (24), (25) can be replaced by an equivalent singular integral equation allowing us to determine the dimensions of the kernel and cokernel.

We make the following assumptions concerning the coefficients of the system (24):

(i) Let elements of the matrix $\boldsymbol{A}$ be measurable and bounded in G and the system (24) may be transformed globally in G into one of the complex equations of special type (20) or (21), i.e., in the whole domain G one has either $a_{21} - a_{12} > 0$ or $a_{21} - a_{12} < 0$.

(ii) Elements of $\boldsymbol{B}$ and $\boldsymbol{f}$ belong to $\mathcal{L}_p(G)$, $p > 2$.

(iii) The system (24) is a uniformly elliptic one in G, i.e., the condition (23) is satisfied.

We shall deal with that case, where the system (24) has been transformed into (20).

Without restriction of generality let G be the unit circle defined by

$$G = \{z : |z| < 1\}$$

and the boundary condition (25) given in the form

$$\operatorname{Re} \{t^{\varkappa} w(t)\}|_{\partial G} = 0 , \tag{26}$$

where

$$\varkappa = \frac{1}{2\pi} \Delta|_{\partial G} \{\arg (\alpha - i\beta)\} . \tag{27}$$

Let us consider separately the two cases $\varkappa > 0$ and $\varkappa \leqq 0$.

1. Let $\varkappa \leqq 0$. By setting the representation formula (6) into the boundary condition (26) and by solving the resulting boundary value problem for the analytic function $\Phi(z)$ we get the following representation of any function of $\mathscr{W}_p^{(1)}(G)$ satisfying the boundary condition (26)

$$w(z) = T\varrho + P_{-2\varkappa}(z) , \qquad \varrho = \frac{\partial w}{\partial \bar{z}} \tag{28}$$

where

$$T\varrho = -\frac{1}{\pi} \iint \left[\frac{\varrho(\zeta)}{\zeta - z} + \frac{z^{-2\varkappa+1}\overline{\varrho(\zeta)}}{1 - z\bar{\zeta}}\right] d\xi \, d\eta ,$$

$$P_{-2\varkappa}(z) = a_{-\varkappa} + \dots + a_1 z^{-\varkappa-1} + i a_0 z^{-\varkappa} - \bar{a}_1 z^{-\varkappa+1} - \dots - \bar{a}_{-\varkappa} z^{-2\varkappa} ,$$

$a_1, \dots , a_{-\varkappa}$ are arbitrary complex numbers and a_0 is an arbitrary real number. Inserting (28) into equation (20) we obtain the singular integral equation:

$$\varrho + \mu_1 S\varrho + \mu_2 \overline{S\varrho} + aT\varrho + b\overline{T\varrho} = f - \mathfrak{L}(P_{-2\varkappa}) \tag{29}$$

(here by $\mathfrak{L}(\cdot)$ we denote the application of the differential operator on the left-hand side of equation (20)). The right-hand side of equation (29) contains $-2\varkappa + 1$ independent real parameters. The singular operator $S\varrho = \partial/\partial z \, (T\varrho)$ may be separated into the sum of two operators (Vinogradov [1]), namely

$$S\varrho = S_1\varrho + T_1\varrho$$

where $T_1\varrho$ is a completely continuous operator but $S_1\varrho$ is a singular integral operator with operator norm in $\mathscr{L}_2(G)$ which is equal to one. From the Riesz-Torin theorem about the continuous dependence of operator norms for operators acting in the scale of $\mathscr{L}_p$-spaces, $1 < p < \infty$, and in virtue of uniform ellipticity (23) it follows that the singular part of the equation (29) is invertible in $\mathscr{L}_p(G)$ for p nearly to two: $2 < p < 2 + \varepsilon$. Consequently, (29) is an equation of the Fredholm type, and moreover it is an equation of the Fredholm type in the whole scale, $2 < p < \infty$, if $\mu_1(z)$ and $\mu_2(z)$ are continuous (Vinogradov [6]).

It can be proved that the corresponding homogeneous equation to equation (29) possesses only the solution identical zero (Vinogradov [1]). The proof is carried out by means of methods of the complex analysis by applying the representation formula of Bers, Nirenberg, and Bojarski (24). Hence, the equation (29) is invertible. Solving the equation (29) and setting the solution into (28) we get a multiplicity of solutions of the problem (26) in this case. The problem (26) is solvable for any right-hand side and the homogeneous problem ($f = 0$) possesses exactly $-2\varkappa + 1$ linearly independent solutions over the real number field.

Therefore, the kernel has the dimension $-2\varkappa + 1$, the cokernel has the dimension zero and the index of the problem is $1 - 2\varkappa$.

2. Now let $\varkappa > 0$. Then for any function of $\mathscr{W}_p^{(1)}(G)$ satisfying the boundary condition (26) the same procedure carried out in case 1 yields the representation

$$w(z) = T\varrho\,, \qquad \varrho(z) = \frac{\partial w}{\partial \bar{z}}\,. \tag{30}$$

Here we have

$$T\varrho = -\frac{1}{\pi}\iint\limits_G \left[\frac{\varrho(\zeta)}{\zeta - z} + \frac{\bar{\zeta}^{2\varkappa-1}\overline{\varrho(\zeta)}}{1 - z\bar{\zeta}}\right] d\xi\, d\eta\,,$$

besides $\varrho(\zeta)$ has to fulfil $2\varkappa - 1$ real conditions (after separation of the real and imaginary part)

$$\iint\limits_G [\zeta^{\varkappa-l-1}\varrho(\zeta) + \bar{\zeta}^{\varkappa+l-1}\overline{\varrho(\zeta)}]\, d\xi\, d\eta = 0\,, \tag{31}$$

$$l = 0, 1, \ldots, \varkappa - 1\,.$$

Inserting (30) into (20) we get the following singular integral equation

$$\varrho + \mu_1 S\varrho + \mu_2 \overline{S\varrho} + aT\varrho + b\overline{T\varrho} = f\,. \tag{32}$$

However, for the investigation of the boundary value problem it is necessary to find a solution of (32), not in the whole $\mathscr{L}_p(G)$ but only in the subspace of $\mathscr{L}_p(G)$ defined by conditions (31).

In the same way as it was done before in case 1 one can show that the operator defined by the left-hand side of equation (32) is invertible, i.e., for any f this equation possesses a uniquely defined solution. Determining the solution of (32) and inserting it into (28) we obtain a function satisfying (20). For fulfilling the boundary conditions the function has also to satisfy conditions (31).

From (32) follows, that its solution may be expressed by a linear operator acting on f. Inserting the solution represented in such a way into (31) we get $2\varkappa - 1$ conditions, which have to be satisfied by the function f. These requirements on f are necessary and sufficient for the solvability of the boundary value problem.

Hence, in this case we can state that the homogeneous problem only possesses the solution identical to zero whereas the inhomogeneous one is solvable only if $2\varkappa - 1$ conditions are fulfilled. These conditions are linear functionals in $\mathscr{W}_p^{(1)}(G)$. Consequently, the kernel is null-dimensional and the cokernel $2\varkappa - 1$-dimensional whereas the index is $1 - 2\varkappa$.

In both cases the index can be expressed by $\varkappa$ in the same way.

If the system (24) has been transformed globally in G into the other complex equation of the special type (21), then the investigation would be carried out in just the same way. The results of solvability would differ from the previous ones only by the fact that $\varkappa$ is replaced by $-\varkappa$.

In the case of higher-dimensional problems we have not still appropriate representation formulas of solutions of elliptic systems by analytic functions of more than one complex variable.

Further references

Vinogradov [5].

12. Recent Results and Developments in Generalized Hyperanalytic Function Theory

R. P. Gilbert

12.1. Introduction

In this chapter we shall present a brief survey of recent results concerning generalized hyperanalytic function theory in the plane. For greater detail the reader is directed to the book by Buchanan and the present author which is scheduled to appear shortly Buchanan and Gilbert [2]. The *generalized hyperanalytic* functions bear the same relation to hyperanalytic functions, as say, the generalized analytic functions of Bers [1] and I. N. Vekua [5] to analytic functions. In order to keep our discussion within the bounds of a single chapter we must restrict our discussion to $\mathbb{R}^2$ and use the definition of hyperanalytic functions introduced by Douglis [1]. For discussions concerning hyperanalytic functions in $\mathbb{R}^n$ the reader is directed to the recent literature by Delanghe [1–3], Brackx [1, 2], Hile [1–3], Hile and Protter [1, 2], Snyder [1, 2], and the book by Buchanan and Gilbert [2]. See also the habilitation thesis of Goldschmidt [4]. For topics in $\mathbb{C}^n$ there is the monograph by Tutschke [10], and also the papers Tutschke [2, 7].

The hyperanalytic functions are solutions of a system of $2r$ first-order elliptic equations in the plane, having only complex eigenvalues which are either all equal or equal to their conjugate values; this is a special case of a *Bojarski system* (cf. Bojarski [5]). We remark, moreover that there is, in general, no normal form for the first-order, elliptic system; however, under certain assumption on the smoothness of the generalized eigenvectors it is possible to introduce a normal form. The reader wishing more information is directed to the excellent treatment of this topic in chapter III of the monograph by Wendland [4].

12.2. Douglis systems

For expository purposes we treat primarily what we refer to as *Douglis systems* in what follows. In terms of real coefficients a Douglis system may be written as

$$\begin{aligned}
&u_{0,x} - v_{0,y} + \dots = g_0\,,\\
&u_{0,y} + v_{0,x} + \dots = h_0\,,\\
&\vdots\\
&u_{k,x} - v_{k,y} + a u_{k-l,x} + b u_{k-l,y} + \dots = g_k\,,\\
&u_{k,y} + v_{k,x} + a v_{k-l,x} + b v_{lk-l,y} + \dots = h_k\\
&(k = 1, 2, \dots, r-1)\,.
\end{aligned} \tag{1}$$

DOUGLIS [1] introduced the *hypercomplex algebra* $\mathfrak{A}_n$ generated by the nilpotent e ($e^n = 0$) and the imaginary number i which permits the elliptic system (1) to be written in a rather compact form. More precisely, if one introduces the hypercomplex functions $w(x, y)$ as

$$w(x, y) := \sum_{k=0}^{r-1} e^k w_k(x, y), \quad \text{with} \quad w_k(x, y) := u_k(x, y) + i v_k(x, y),$$

then after some manipulation (1) may be put into

$$Dw + \sum_{k;l=0}^{r-1} e^k (\overline{A_{kl}} w_l + B_{kl}\overline{w_l}) = F. \tag{2}$$

Here D is the differential operator defined as

$$D := \frac{\partial}{\partial \bar{z}} + q\frac{\partial}{\partial z}, \qquad q(x, y) := \sum_{k=1}^{r-1} e^k q_k(x, y),$$

and the q_k are functions of the coefficients a and b appearing in (1) above. DOUGLIS presented a fairly complete study of the equation $D_\Phi = 0$ in DOUGLIS [1], and refers to the classical solutions as hyperanalytic functions. GILBERT and HILE [1—3] investigated (2) and developed much of the so-called general theory following ideas originating out of the work of I. N. VEKUA [5] and BERS [1] for the complex equation

$$\frac{\partial w}{\partial \bar{z}} + \bar{a}w + b\bar{w} = f. \tag{3}$$

Further contributions to the theory of generalized hyperanalytic function theory were made by BEGEHR and HSIAO [2], BUCHANAN [1], BEGEHR and GILBERT [3—5], BUCHANAN and GILBERT [1], GILBERT and SCHNEIDER [1], GILBERT and WENDLAND [1], GOLDSCHMIDT [1—3], KÜHN [1], HABETHA [4], and WENDLAND [4].

There are applications of hyperanalytic, and generalized hyperanalytic functions to problems of mathematical physics, such as plate and shell problems, nonhomogeneous elasticity, Oseen flows etc., a number of these are discussed in the monograph by WENDLAND cited above.

12.3. Properties of hyperanalytic functions

Before discussing the theory of weak solutions of (1) it is necessary to collect some facts about hyperanalytic functions. DOUGLIS [1] showed that to each hyperanalytic function there corresponds a unique n-tuple of analytic functions. Indeed, if $t(z)$ is the so-called Douglis generating function, then an arbitrary solution of $D\Phi = 0$ has a representation

$$\Phi(z) = \sum_{p=1}^{r-1} f_p(t(z))\, e^p, \tag{4}$$

where the analytic functions $f_p(z)$ are uniquely determined. The function $t(z) := t(x, y)$, moreover, is a solution of $Dt = 0$, and is normalized such that its complex part is z, that is $t := z + E(x, y)$, where $E^n = 0$. Furthermore, if $q \in \mathscr{B}^\alpha(\mathbb{C})$[1]), we find that $E \in \mathscr{B}^{1,\alpha}(\mathbb{C})$. DOUGLIS showed the existence of a Green's identity for smooth functions

[1]) $\mathscr{B}^\alpha(\mathbb{C})$ is the family of Hölder continuous functions of index α, which are bounded in the entire plane. The $\mathscr{B}^{k,\alpha}(\mathbb{C})$ are the functions which are Hölder continuous up to order k, of index α, and bounded in the entire plane.

in domains G, with Hölder-smooth boundaries ∂G,

$$\mathrm{i} \iint_G t_{\bar\zeta}(\zeta)\,[wDv + (Dw)\,v]\,\mathrm{d}\xi\,\mathrm{d}\eta = \int_{\partial G} wv\,\mathrm{d}t(\zeta)\,. \tag{5}$$

From this identity one easily obtains a *Cauchy-Pompeiu type* representation for smooth functions,

$$w(z) = \frac{1}{2\pi\mathrm{i}} \int_{\partial G} \frac{w(\zeta)\,\mathrm{d}t(\zeta)}{t(\zeta) - t(z)} + T(Dw)\,(z)\,, \tag{6}$$

where

$$(Tv)\,(z) := -\frac{1}{\pi} \iint_G \frac{t_{\bar\zeta}(\zeta)\,v(\zeta)\,\mathrm{d}\xi\,\mathrm{d}\eta}{t(\zeta) - t(z)}\,. \tag{7}$$

Gilbert and Hile [1] showed that these representations could be extended to functions with Sobolev derivatives.

12.4. Lower triangular systems

Since uniqueness is lost with regard to some of the representation formulae associated with the differential system (2) we will restrict our further discussions to the lower triangular system

$$Dw + \sum_{k=0}^{r-1} \mathrm{e}^k \sum_{l=0}^{k} (\bar A_{kl} w_l + B_{kl} \bar w_l) = F\,. \tag{8}$$

Furthermore, in analogy with the Vekua theory we ask that the coefficients A_{kl}, $B_{kl} \in \mathcal{L}_{p,2}(\mathbb{C})$. In this instance, it has been shown in Gilbert and Hile [1] that the only solution of the homogeneous equation which is bounded in the entire plane is a so-called generalized constant, i.e. the solution must have the form

$$w(z) = c \exp\,[\omega(z)]\,, \tag{9}$$

where c is a hypercomplex constant, and $\omega \in \mathcal{B}^\alpha(\mathbb{C})$, $\alpha = (p-2)/p$, is a hypercomplex function. One may view (9) as a special form of the *similarity principle*, which for the complex homogeneous equation says that each solution may be written as $w(z) = \Phi(z)\,\mathrm{e}^{\omega(z)}$, where $\omega(z)$ is analytic and $\omega \in \mathcal{B}^\alpha(\mathbb{C})$, $\alpha = (p-2)/p$ (cf. 11.1., formula (2)). This is not the case for the homogeneous form of the full system (2) as Habetha [4] has shown; however, in the case of the lower triangular system (8) we are able to establish an analogous result (cf. Begehr and Gilbert [6]). To this end, we introduce the integral operator

$$T(\tau) := -\frac{1}{\pi} \iint_G \frac{t_{\bar\zeta}(\zeta)\left[\sum_{i=0}^{r-1} \sum_{j=0}^{i} c_{ij}\tau_j\,\mathrm{e}^i\right] \mathrm{d}\xi\,\mathrm{d}\eta}{t(\zeta) - t(z)}\,, \tag{10}$$

on the class of hypercomplex functions $\tau \in \mathcal{B}^\alpha(\mathbb{C})$. Since the null space of $I - T$ is the empty set, and T is compact, we may solve the following system of hypercomplex operator equations

$$\tau^{(k)} = T(\tau^{(k)}) + \mathrm{e}^k\,, \qquad k = 0, 1, \ldots, r-1\,. \tag{11}$$

Furthermore, the functions $\tau^{(k)}$ are all solutions of the generalized hyperanalytic equation

$$D\tau^{(k)} + \sum_{i=0}^{r-1} \sum_{j=0}^{i} \mathrm{e}^i c_{ij} \tau_j^{(k)} = 0\,, \qquad k = 0, 1, \ldots, r-1\,, \tag{12}$$

hence, each $\tau^{(k)}$ has a representation of the form (9),

$$\tau^{(k)}(z) = c_k \, e^{\omega_k(z)} . \tag{13}$$

As $\omega_k(z) \to 0$ as $z \to \infty$, we may also conclude $c_k = e^k$. The vector of hypercomplex solutions $[\tau^{(0)}, \tau^{(1)}, \dots, \tau^{(r-1)}]$ associated with differential equations of the form (12) can be put into one-to-one correspondence with the matrix solutions of the system

$$\boldsymbol{I} \frac{\partial \boldsymbol{T}}{\partial \bar{z}} + \boldsymbol{Q} \frac{\partial \boldsymbol{T}}{\partial z} + \boldsymbol{C}\boldsymbol{T} = 0 , \tag{14}$$

where $\boldsymbol{Q}(z) := \begin{bmatrix} 0 & & & & \\ & \ddots & & 0 & \\ q_1 & & \ddots & & \\ \vdots & \ddots & & \ddots & \\ q_{r-1} & \cdots & q_1 & & 0 \end{bmatrix}$, $c := (c_{ij})$, and $\boldsymbol{I}$ is the identity matrix which, moreover, satisfies the asymptotic condition $\boldsymbol{T}(\infty) = \boldsymbol{I}$. Clearly the system (14) is the matrix equivalent of the hypercomplex equation (12).

As the leading term in a componentwise expansion of the solutions $\tau^{(k)}$ must be $\tau_k^{(k)} e^k$, we can conclude from (14) that the matrix $\boldsymbol{T}$ is lower triangular and, moreover, non-singular. Hence, (14) takes on the simplified form

$$\frac{\partial \tau_i^{(k)}}{\partial \bar{z}} + \sum_{j=k}^{i-1} q_{i-j} \frac{\partial \tau_j^{(k)}}{\partial z} + \sum_{j=0}^{i} c_{ij} \tau_j^{(k)} = 0 , \qquad i \geqq k . \tag{15}$$

An arbitrary vector solution of (14) may be written in the form $\boldsymbol{w} = \boldsymbol{T}\boldsymbol{\Phi}$, where $\boldsymbol{T}$ is the unique matrix solution of (14) which satisfies $\boldsymbol{T}(\infty) = \boldsymbol{I}$. The components of $\boldsymbol{w}$ are then given by BEGEHR and GILBERT [6]

$$w_k = \sum_{j=0}^{k} \tau_k^{(j)} \Phi_j , \qquad k = 0, 1, \dots, r-1 ,$$

where the $\tau_k^{(j)}$ are the matrix entries of $\boldsymbol{T}$ and Φ_j are the components of the hyperanalytic function Φ. Reverting back to hypercomplex form we have

$$w = \sum_{k=0}^{r-1} e^k w_k = \sum_{j=0}^{r-1} \Big(\sum_{k=j}^{r-1} \tau_k^{(j)} e^k\Big) \Phi_j = \sum_{j=0}^{r-1} \tau^{(j)} \Phi_j . \tag{16}$$

Lemma (BEGEHR and GILBERT [6]). *If the coefficients $c_{ij} \in \mathscr{L}_{p,2}(\mathbb{C})$, then each solution of (12) may be represented in the form*

$$w(z) = \sum_{j=0}^{r-1} \tau^{(j)}(z) \, \Phi_j(z) , \tag{17}$$

where $\Phi = \sum_{j=0}^{r-1} e^j \Phi_j$ is a solution of $D^*\Phi = 0$, where D^* in matrix form is given by $D^* := \partial/\partial\bar{z} + \tilde{\boldsymbol{Q}} \, \partial/\partial z$, with $\tilde{\boldsymbol{Q}} := \boldsymbol{T}^{-1}\boldsymbol{Q}\boldsymbol{T}$.

The $\tau^{(j)}(z)$ are solutions of (12) satisfying $\tau^{(j)}(\infty) = e^j$ $(j = 0, 1, \dots, r-1)$ which are of class $\mathscr{B}^\alpha(\mathbb{C})$, $\alpha = (p-2)/p$.

Corollary. *If $A_{ij}, B_{ij} \in \mathscr{L}_{p,2}(\mathbb{C})$, $p > 2$ then the system*

$$Dw + \sum_{i=0}^{r-1} \sum_{j=0}^{i} e^i (A_{nj} w_j + B_{ij} \overline{w_j}) = 0 , \tag{18}$$

permits each solution to have a representation of the form (17), where in the present case the $\tau^{(j)}$ are now solutions of (18).

Proof. We notice that by introducing the matrix $C := (C_{ij})$

$$C_{ij} := A_{ij} + B_{ij}\frac{\overline{wj}}{wj}, \qquad i, j = 0, 1, \dots, r-1, \tag{19}$$

we may rewrite (18) in the form (12) where the coefficients C_{ij}, moreover, satisfy the hypothesis of the above theorem. See also in this regard the doctoral dissertation by KÜHN [1].

12.5. A generalized hyperanalytic Cauchy integral formula

It has been shown by GILBERT and HILE [1] that it is possible to construct fundamental singular solutions for the system (18) in the form

$$\Omega^{(1)}(z,\zeta) := X^{(1)} + \mathrm{i}X^{(2)}, \qquad \Omega^{(2)}(z,\zeta) := X^{(1)} - \mathrm{i}X^{(2)}, \tag{20}$$

where

$$X^{(1)}(z,\zeta) := \frac{\exp\left(\omega^{(1)}(z) - \omega^{(1)}(\zeta)\right)}{2\left(t(\zeta) - t(z)\right)}, \quad X^{(2)}(z,\zeta) := \frac{\exp\left(\omega^{(2)}(z) - \omega^{(2)}(\zeta)\right)}{2\mathrm{i}\left(t(\zeta) - t(z)\right)}.$$

This leads to a generalized hyperanalytic Cauchy representation (for regions with Hölder-continuous boundaries), namely

$$w(z) = C[w] := \frac{1}{2\pi\mathrm{i}}\int_{\partial G} \Omega^{(1)}(z,\zeta)\, w(\zeta)\, \mathrm{d}t(\zeta) - \Omega^{(2)}(z,\zeta)\, \overline{w(\zeta)}\, \overline{\mathrm{d}t(\zeta)}\,. \tag{21}$$

If $\Phi(z)$ is defined by the Cauchy formula as

$$\Phi(z) = \frac{1}{2\pi\mathrm{i}}\int_{\partial G} \frac{w(\zeta)\, \mathrm{d}t(\zeta)}{t(\zeta) - t(z)} \tag{22}$$

an application of the Plemelj formula shows that one may replace $w(\zeta)$ in the above integral by $\Phi(\zeta)$. This follows from the local behaviour of the *generalized Cauchy kernels*, namely

$$\Omega^{(1)}(z,\zeta) - \frac{1}{t(\zeta) - t(z)} = O(|z-\zeta|^{-2/p}), \qquad \Omega^{(2)}(z,\zeta) = O(|z-\zeta|^{-2/p}).$$

12.6. The Hilbert boundary value problem for generalized hyperanalytic functions

Next we turn to discuss several of the boundary value problems that occur by analogy with the analytic function theory. First we consider the so-called *Hilbert boundary value problem* (cf. 1.10.3., 10.3., 10.4., and 11.3.) associated with (18). Namely we seek a solution of (18) which satisfies the boundary condition

$$\operatorname{Re}(\lambda w) = \Phi\,.$$

Our investigations are greatly simplified, at least from a computational point of view if we are able to work in the hypercomplex algebra. To this end we choose a special form of (18), for which the matrices $\boldsymbol{A}$ and $\boldsymbol{B}$ are quasidiagonal. This permits us to write our boundary value problem in the hypercomplex form

$$\left.\begin{aligned} L[w] &:= Dw + \bar{a}w + b\bar{w} = f \quad \text{in } G, \\ \operatorname{Re}(\bar{\lambda}w) &= \alpha \quad \text{on} \quad \partial G. \end{aligned}\right\} \text{(A)} \tag{23}$$

In what follows we assume $\lambda \in \mathscr{C}^1(\partial G)$. From the Gauss-Green formula (5) it is clear that the proper choice for the adjoint, homogeneous problem is

$$\left.\begin{aligned} L'[w'] &:= Dw' - aw' - b^* w' = 0 \quad \text{in } G \\ \operatorname{Re}\left(\lambda \frac{\mathrm{d}t}{\mathrm{d}z}\frac{\mathrm{d}z}{\mathrm{d}s} w'\right) &= 0 \quad \text{on} \quad \partial G, \end{aligned}\right\} (\text{Å}') \tag{24}$$

where ds is an arclength differential and $b^* := \bar{b}\,\overline{t(z)}\,\big(t(z)\big)^{-1}$. We seek a solution to problem (Å′) in the form

$$w'(z) = -\frac{1}{2\pi \mathrm{i}} \int_{\partial G} \left(\Omega^{(1)}(\zeta, z)\, \lambda^{-1}(\zeta) + \Omega^{(2)}(\zeta, z)\, \overline{\lambda(\zeta)^{-1}}\right) f(\zeta)\, \mathrm{d}s \tag{25}$$

where $f(\zeta)$ is a real hypercomplex density. The Plemelj formula then implies that $f(\zeta)$ must satisfy the condition

$$\int_{\partial G} K_1(\zeta, z)\, f(\zeta)\, \mathrm{d}s = 0 \quad \text{for} \quad z \in \partial G, \tag{26}$$

where

$$K_1(\zeta, z) := -\frac{\operatorname{Re}}{2\pi}\left\{\lambda(\zeta) \frac{\mathrm{d}t}{\mathrm{d}z}\frac{\mathrm{d}z}{\mathrm{d}s}\left(\Omega^{(1)}(\zeta, z)\, \lambda^{-1}(\zeta) + \Omega^{(2)}(\zeta, z)\, \overline{(\lambda(\zeta))^{-1}}\right)\right\}. \tag{27}$$

The solutions to problem (A) may be expressed as a sum of *generalized Cauchy representations*

$$w := w_1 + w_2 := C[\partial\bar{\lambda}^{-1}](z) + C[\mathrm{i}\mu\bar{\lambda}^{-1}](z). \tag{28}$$

Again using the Plemelj relations we see that the solutions of problem (A) may be represented in the form (28) providing μ is a solution of

$$\int_{\partial G} K_1(\zeta, z)\, \mu(z)\, \mathrm{d}s(z) = \gamma_0(\zeta), \tag{29}$$

where

$$\gamma_0(\zeta) := \gamma(\zeta) - \operatorname{Re}\,[\overline{\lambda(\zeta)}\, w_1^+(\zeta)] = \operatorname{Re}\,[\overline{\lambda(\zeta)}\, w^-(\zeta)].$$

For (29) to be solvable it is necessary and sufficient that

$$\int_{\partial G} \gamma_0(\xi)\, f_j(\xi)\, \mathrm{d}s_\xi = 0 \qquad (j = 1, \ldots, k), \tag{30}$$

where the $\{f_1, \ldots, f_k\}$ form a complete system of solutions to the adjoint homogeneous, singular integral equation (26) to (27) (see Begehr and Gilbert [6]). Further analogies with the theory of singular integral equations as developed for complex equations by Muschelišvili [2] and I. N. Vekua [5] are apparent. Rather than continue in this approach at this time we choose to modify the approach of Haack and Wendland [1]. The reader interested in pursuing this analysis further is directed to a work by the author with Begehr (Begehr and Gilbert [6]).

Returning to the problem (23), we assume now that $\lambda, \gamma \in \mathscr{C}^1(\partial G)$ furthermore, we define the index of the data by

$$\operatorname{Ind}[\bar{\lambda}] := \operatorname{Ind}[\bar{\lambda}_0] := \frac{1}{2\pi} \int_{\partial G} \mathrm{d} \arg \bar{\lambda}_0, \tag{31}$$

where $\lambda_0 :=$ complex part of λ, and $\Lambda := \lambda - \lambda_0$ is the nilpotent part. If the index is zero, it has been shown that it is always possible to reduce the problem to the form

$$D\omega + \bar{a}_1\omega + b_1\bar{\omega} = 0\,, \qquad \operatorname{Re}\omega|_{\partial G} = \gamma\,, \tag{32}$$

where $\bar{a}_1 := \bar{\lambda}^{-1}D\bar{\lambda} + \bar{a}$, $b_1 := \lambda b\bar{\lambda}^{-1}$, $\lambda_0 \neq 0$ on ∂G, and there exists a harmonic extension of λ_0 to G. In this case it is easy to show, using the hypercomplex polar representation[1]) of $\lambda = \varrho e^{i\Theta}$, that λ^{-1} exists in $\bar{G}$.

Having reduced the problem to the form (32), a solution may be found using one of several imbedding methods (WENDLAND [3], BEGEHR and GILBERT [3]).

For the case of negative index, Ind $[\bar{\lambda}] = -m$, $m \in \mathbb{N}^+$, we introduce a new unknown $v(z) := \sum_{i=1}^{m} \big(t(z) - t(z_i)\big)^{-1} w(z)$, where the points z_i are taken interior to G the problem is reduced to a similar boundary value problem of index zero (BEGEHR and GILBERT [3]). In this way, it may be shown in this book that every solution of the homogeneous boundary value problem (Å) of negative index $-m$ may be written in the form

$$w_h(z) = \mu \prod_{i=1}^{m} \big(t(z) - t(z_i)\big)\, v_h(z)\,, \qquad \mu := \sum_{k=0}^{r-1} e^k \mu_k$$

being a hypercomplex constant and $v_h(z)$ being a solution to a homogeneous problem (Å) of index zero, which, moreover, has a nonvanishing complex part. In order to better classify the solutions of the homogeneous problem (Å) we introduce the concept of linear independence over A_r. We shall refer to m hypercomplex functions $w_i(z)$ $(i = 1, 2, \dots, m)$ having nonidentically vanishing complex parts w_{i0} linearly independent over A_r when for hypercomplex constants

$$\sum_{i=1}^{m} \lambda_i w_i = 0$$

implies that all the λ_i are zero. We have shown (BEGEHR and GILBERT [3]) that the homogeneous boundary value problem (Å) of negative index $-m$ has exactly $2m + 1$ linearly independent solutions over A_r (with nonvanishing complex parts).

To treat the homogeneous problem (A) of positive index $m = \operatorname{Ind}[w] > 0$, we introduce a new unknown

$$v(z) := \prod_{i=1}^{m} \big(t(z) - t(z_i)\big)\, w(z)\,,$$

where as before the $z_i \in \operatorname{int}(G)$. The differential equation becomes

$$Dv + \bar{a}v + b \prod_{i=1}^{m} [\overline{t(z)} - \overline{t(z_i)}]\, [t(z) - t(z_i)]^{-1}\, \bar{v} = 0\,, \tag{33}$$

and the boundary condition becomes

$$\operatorname{Re}\, [\bar{\lambda} v \prod_{i=1}^{m} \big(t(z) - t(z_i)\big)^{-1}] = 0\,. \tag{34}$$

As the reduced boundary value problem (33—34) has index zero all solutions w of problem (Å) with positive index **must** have at least m poles, which may be arbitrarily placed at m points $z_1, \dots, z_m \in \operatorname{int}(G)$.

In general, we may say that an arbitrary solution of $L[w] = 0$ which does not vanish nor have poles on ∂G has an index which is the difference between its number

---

[1]) The modulus ϱ, as well as the argument θ, are hyperreal quantities; θ_0 is, however, determined only up to an integral multiple of 2π.

of poles and its zeroes, where multiplicities are counted. If in contrast to the above discussion we now permit zeroes to occur on the boundary, then for the case of (Å) with negative index $-m$ we are permitted, in the case where all zeroes occur on ∂G, $2m$ arbitrarily chosen zeroes. Hence we obtain solutions of the form

$$w_h(z) = \mu v_h \prod_{i=1}^{2m} \left(t(z) - t(z_i)\right) , \tag{35}$$

where μ is a real hypercomplex constant, and v_h is a solution to another problem of type (Å), but of index zero.

If the data λ vanishes on ∂G then certain problems naturally occur as λ^{-1} no longer exists when $\lambda_0 = 0$. It is possible to overcome these problems in certain instances, and for a detailed discussion of this the reader is directed to the book by Buchanan and Gilbert [2].

Let us suppose that $\bar{\lambda}$ has a finite number of zeroes at $z_j \in \partial G$ ($j = 1, \dots, l$). In this case we define

$$\operatorname{Ind}[\bar{\lambda}] := \operatorname{Ind}[\bar{\lambda}_0] := \frac{1}{2\pi i} \sum_{j=0}^{l} \int_{s_j}^{s_{j+1}} \frac{d}{ds} (\arg \bar{\lambda}_0)\, ds , \tag{36}$$

where $z = z(s) \in \mathscr{C}^{1,\alpha}$, $z_j := z(s_j)$, and ds is the arc length differential. The jump in $\arg \lambda_0$ may be removed by multiplying $\bar{\lambda}$ by $f(z) := \prod_{j=1}^{l} [t(z) - t(z_j)]^{-1}$, which may be seen by a short computation (Begehr and Gilbert [6]). In the instance that the index is now zero the problem (Å) is thereby regularized. Otherwise, further zeroes or poles may be placed within G to achieve zero index.

It should be mentioned that we require sufficient regularity on ∂G and for λ such that tangential derivatives may be taken. We require further that

$$\arg \left(\lim_{s \to s_j^+} \frac{\bar{\lambda}}{|\bar{\lambda}|} \right) - \arg \left(\lim_{s \to s_j^-} \frac{\bar{\lambda}}{|\bar{\lambda}|} \right) = -\pi . \tag{37}$$

If ∂G and λ are $\mathscr{C}^{1,\alpha}$ this is equivalent to requiring that this condition holds for just the complex part of λ and its modulus as is the condition in the complex case (Haack and Wendland [1, p. 290–299]). We recall that $|\lambda|$ is a hyperreal quantity known as the *hypercomplex modulus* of λ. The modulus $\varrho := \sum_{k=0}^{r-1} e^k \varrho_k$, moreover, is uniquely determined when $\lambda_0 \neq 0$ as the solution of the system

$$\sum_{l=0}^{k} \varrho_l \varrho_{k-l} = \sum_{l=0}^{k} \lambda_l \overline{\lambda_{k-l}} , \quad \text{providing} \quad \varrho_0 > 0 . \tag{38}$$

The hyperreal quantity called the hypercomplex argument used in the criteria (37) above is defined for $\lambda = \sum_{k=0}^{r-1} e^k \lambda_k$ as

$$\theta := \arg \lambda := \arg [\lambda_0] + 2\pi n + i \sum_{k=1}^{r-1} (-1)^k \frac{N^k}{k} \exp (ik \arg [\lambda_0]) , \tag{39}$$

where $N :=$ nilpotent part $(\lambda/|\lambda|)$.

By introducing as a new unknown $v(z) := \prod_{i=1}^{2m+l} [t(z) - t(z_i)]\, w(z)$ where the $\{z_1, \dots, z_l\}$ are zeroes of $\bar{\lambda}$ on $\partial \dot{G}$, and the $\{z_{l+1}, \dots, z_{l+2m}\}$ are arbitrarily chosen the boundary value problem is reduced to one of index zero, that is by defining $\bar{\Lambda} := I(s)\, \bar{\lambda}/|\lambda|$

$f/|f|$, where $I(s)$ is a suitably defined jump function, such as $I(s) := (-1)^j$ for $s \in (s_j, s_{j+1})$, $(j = l+1, \dots, l+2n-1)$ we obtain the transformed boundary data

$$\operatorname{Re}(\bar{\Lambda} v) = \tau(s)\, |\lambda|^{-1}\, |f|^{-1}\, \Phi \quad \text{on} \quad \partial G\,, \tag{40}$$

and differential equation

$$Dv + \bar{a} v + (\bar{f}\,|f|^{-1}\, b)\, \bar{v} = 0 \quad \text{in } G\,. \tag{41}$$

It is not difficult to show Λ is continuous at the zeroes of λ (Buchanan [1]). however, the right-hand side of (40) causes some difficulties (Buchanan [1]), and some conditions must be prescribed concerning Φ at the zeroes of λ and f. The problem (Å), however, is clearly solvable.

12.7. The Riemann boundary value problem for generalized hyperanalytic functions

We turn next to the *Riemann boundary value problem*, for which we assume the coefficients $a, b \in \mathcal{L}_{p,2}(\mathbb{C})$ where $p > 2$. For this problem, piecewise continuous solutions of the differential equation $L[w] = 0$ are sought which satisfy on a closed, Hölder-continuous curve Γ the jump discontinuity condition (cf. 1.10.3., chapter 5, 10.2., 10.4. and 20.4.3.)

$$w^+ = g w^- + \gamma \quad \text{on } \Gamma\,. \tag{42}$$

The notation w^+, w^- refers to the solution w in the regions G^+ (inside Γ) and G^- (outside Γ), respectively the weak solutions of $L[w] = 0$ are Hölder-continuous in G^+ and G^-. For simplicity in presentation, we assume moreover, that the coefficients a, b have compact support.

The solution to the Riemann boundary value problem is dependent on the index of g, which we define as (see Begehr and Gilbert [5])

$$\operatorname{Ind}[g] := \frac{1}{2\pi i} \int_\Gamma d \log g = \frac{1}{2\pi i} \int_\Gamma d \log g_0\,. \tag{43}$$

For the homogeneous problem $\operatorname{Ind}[g] = \operatorname{Ind}[w^+] - \operatorname{Ind}[w^-] = N^+ + N^- \geqq 0$, where N^+ and N^- are the number of zeroes (counted with respect to their multiplicity) of w^+ and w^-, respectively.

If the index of g is $m \in \mathbb{N}$ then $[t(z)]^{-m}\, g(z)$ is a Hölder-continuous, non-vanishing hypercomplex function of index zero. Consequently we may introduce the single-valued, Hölder-continuous function $\Phi(z) := \log g(z) - m \log t(z)$ on Γ. Consequently, the function $(P\Phi)(z)$ defined by

$$(P\Phi)(z) := \frac{1}{2\pi i} \int_\Gamma [\Omega^{(1)}(z,\zeta)\, \Phi(\zeta)\, dt(\zeta) - \Omega^{(2)}(z,\zeta)\, \overline{\Phi(\zeta)}\, \overline{dt(\zeta)}] \tag{44}$$

is defined in $\mathbb{C}/\Gamma$. Moreover, this permits the introduction of the new unknown, $\omega := w \exp(-P\Phi)$, which is the solution of a transformed problem

$$L[\omega] := D\omega + \tilde{a}\omega + \tilde{b}\bar{\omega} = 0 \quad \text{in} \quad \mathbb{C} \setminus \Gamma\,. \tag{45}$$

and $\omega^+ = t^m \omega^- + \tilde{\gamma}$ on Γ. Here the new coefficients are given by

$$\tilde{a} := (1 - P\Phi)\, a - b\, \overline{P\Phi}\,, \qquad \tilde{b} := b \exp(-2i \operatorname{Im} P\Phi)\,, \tag{46}$$
$$\tilde{\gamma} := \gamma \exp(-\tfrac{1}{2}\Phi - P\Phi)\,.$$

This transformation is the device used in the work of BEGEHR and GILBERT [5], to permit a direct analysis of the problem through a series of elementary reductions. For the simplest case, namely that of zero index the general solution to the problem is easily obtained in the form (see BEGEHR and GILBERT [5])

$$w(z) = [C \exp \omega_0(z) + \tilde{P}\gamma] \exp [P\Phi] , \tag{47}$$

where $c \exp \omega_0$ is a *generalized constant,* and $\tilde{P}$ is the generalized Cauchy operator associated with the coefficients $\tilde{a}, \tilde{b}$.

When the index of g is positive the transformation

$$\omega_1(z) := \begin{cases} \omega^+(z) , & z \in G^+ , \\ t^m \omega^-(z) , & z \in G^- \end{cases} \tag{48}$$

permits us to find a particular solution to the reduced problem (44), namely the solution $\omega^+(z) := \omega_1(z)$ in G^+, $\omega^-(z) := t^{-m}\omega_1$ in G^-, where

$$\begin{aligned} \omega_1(z) &:= (P_1\tilde{\gamma})(z) \\ &:= \frac{1}{2\pi i} \int_\Gamma [\Omega_1^{(1)}(z, \zeta)\, \tilde{\gamma}(\zeta)\, dt(\zeta) - \Omega_1^{(2)}(z, \zeta)\, \tilde{\gamma}(\zeta)\, \overline{dt(\zeta)}] . \end{aligned} \tag{49}$$

Here the generalized Cauchy kernels are associated with the coefficients A_1, B_1 of the transformal problem

$$\begin{aligned} &D\omega_1 + A_1\omega_1 + B_1\overline{\omega}_1 = 0 \quad \text{in} \quad \mathbb{C} \setminus \Gamma , \\ &\omega_1^+ - \omega_1^- = \tilde{\gamma} \quad \text{on } \Gamma . \end{aligned} \tag{50}$$

The coefficients A_1 and B_1 are defined by

$$A_1 := \tilde{a} , \qquad B_1 := \begin{cases} b , & z \in G^+ , \\ \tilde{b}t^m(\bar{t}^{-m}) , & z \in G^- . \end{cases} \tag{51}$$

We note, moreover, that $\overline{\omega}(z)$ satisfies the asymptotic condition $\omega^-(z) = O(|z|^{-m-1})$ as $|z| \to \infty$.

Now we must seek the general form of the solution to the homogeneous problem. This is done by making use of a hypercomplex extension of Bers' concept of *generating pairs* due to the author with HILE (GILBERT and HILE [2]). Suppose that the $\{\hat{F}_k, \hat{G}_k\}$ $(k = 1, 2, \ldots, m)$ are generating pairs associated with the differential equation,

$$L_k[\omega] := D\omega + A_1\omega + B_1 t^k (\bar{t})^{-k}\, \overline{\omega} = 0 , \qquad k = 1, 2, \ldots, m . \tag{52}$$

It has been shown by BEGEHR and GILBERT [5] that any hypercomplex function W defined and bounded in $\mathbb{C}$ may be represented in the form

$$W = \Phi_k F_k + \Psi_k G_k , \qquad k = 1, \ldots, m , \tag{53}$$

where Φ_k and Ψ_k are hyperreal-valued functions. Moreover, Φ_k and Ψ_k may be solved for in the form

$$\Phi_k = \frac{W\overline{\hat{G}}_k - \overline{W}\hat{G}_k}{\hat{F}_k\overline{\hat{G}}_k - \overline{\hat{F}}_k G_k} , \qquad \Psi_k = \frac{W\overline{\hat{F}}_k + \overline{W}\hat{F}_k}{\hat{F}_k\overline{\hat{G}}_k - \overline{\hat{F}}_k G_k} . \tag{54}$$

The $\{\hat{F}_k, \hat{G}_k\}$ $(k = 1, \ldots, m)$ satisfy the equations (50), furthermore, having a pair $\{\hat{F}_k, \hat{G}_k\}$ we may compute the coefficients of (52) as (see BEGEHR and GILBERT [5])

$$A_1 = \frac{(D\hat{F}_k)\,\overline{\hat{G}}_k - (D\hat{G}_k)\,\overline{\hat{F}}_k}{\hat{F}_k\overline{\hat{G}} - \overline{\hat{F}}_k\hat{G}_k} , \qquad \frac{B_1 t^k}{(\bar{t})^k} = \frac{-(D\hat{F}_k)\, G_k + (D\hat{G}_k)\,\hat{F}_k}{\hat{F}_k\overline{\hat{G}}_k - \overline{\hat{F}}_k\hat{G}_k} .$$

The generating pairs $\{\hat{F}_k, \hat{G}_k\}$ $(k = 1, \dots, m)$, moreover, are continuous and bounded in $\mathbb{C}$, and are normalized such that $\hat{F}_k(\infty) = 1$, $\hat{G}_k(\infty) = \mathrm{i}$.

We define, next, certain solution pairs $\{\hat{F}_k, \hat{G}_k\}$ to the equation (44) by

$$\tilde{F}_k(z) := \big(t(z)\big)^k \hat{F}_k(z) \quad \text{in } G^+ , \qquad (55)$$
$$\tilde{F}_k^-(z) := \big(t(z)^{k-m} \hat{F}_k(z)\big) \quad \text{in } G^- ,$$

and

$$\tilde{G}_k^+(z) := \big(t(z)\big)^k \hat{G}_k(z) \quad \text{in } G^+ , \qquad (56)$$
$$\tilde{G}^-(z) := \big(t(z)\big)^{k-m} \hat{G}_k(z) \quad \text{in } G^- .$$

The functions $\tilde{F}_k$, $\tilde{G}_k$ $(k = 1, \dots, m)$ have zeroes of order $k - m$ at ∞, and may be seen to be independent of one another. Finally, the general solution to problem (44) with homogeneous data may be represented as a linear combination of the pairs $\{\tilde{F}_k, \tilde{G}_k\}$ where the coefficients $\{\lambda_k, \mu_k\}$ are hyperreal constants, namely

$$\omega_h(z) := \sum_{k=0}^{m} (\lambda_k \tilde{F}_k + \mu_k \tilde{G}_k)(z) . \qquad (57)$$

Hence the general solution to the nonhomogeneous problem (41) may be represented as (see Begehr and Gilbert [5])

$$\omega = [\omega + \sum_{k=0}^{m} (\lambda_k \tilde{F}_k + \mu_k \tilde{G}_k)] \exp(P\Phi) \qquad (58)$$

where $\omega(z)$ is given by (48) and (49).

If the index of g is negative we must include poles in the homogeneous problem. It may be shown that the inhomogeneous problem is given by (see Begehr and Gilbert [5])

$$w(z) = \frac{X(z)}{2\pi \mathrm{i}} \int_{\Gamma} \left\{ \Omega_1^{(1)}(z, \zeta) \left[\frac{\tilde{\gamma}(\zeta)}{X^-(\zeta)}\right] \mathrm{d}t(\zeta) - \Omega_1^{(2)}(z, \zeta) \left[\frac{\tilde{\gamma}(\zeta)}{X^-(\zeta)}\right] \overline{\mathrm{d}t(\zeta)} \right\}, \qquad (59)$$

where the transformed data must satisfy the conditions

$$\int \tilde{\gamma}(s)\, t^{k-1}(s)\, \mathrm{d}t(s) = 0 , \qquad 1 \leqq k \leqq -m - 1, \qquad (60)$$

$X(z)$ is the so-called *characteristic solution* which is given by

$$X^+ := \exp(P\Phi)^+ , \qquad X^- := t^{-m} \exp(P\Phi)^- . \qquad (61)$$

There are numerous other directions the development of generalized hyperanalytic function theory can take. The reader is referred to the book of Buchanan and Gilbert [2] for further information.

Further references

Begehr and Hile [1], Gilbert [2], Haack [1], Hellwig [1], Pascali [3], Tutschke [4], v. Wolfersdorf [6].

13. Complex Methods in the Theory of Boundary Value Problems with respect to Systems of Partial Differential Equations of Composed Type

A. DŽURAEV

13.1. Introduction

It is known that the methods of complex function theory have a wide use in many questions of mathematical analysis and its applications. A special statement deals with the application of these methods in the theory of partial differential equations, of first order, too, and in the theory of elliptic equations and systems. In this time the methods of complex analysis especially in multidimensional analysis penetrate into many parts of modern mathematics and physics. The object of this chapter is not a wide use of the methods of complex analysis in the general theory of partial differential equations. Here we will give an illustration for the use of some widely known methods of application of complex variables in some systems of partial differential equations of nonclassified type.

We consider linear systems with vector-valued functions of $n + 1$ variables $u(x, t) = (u_1, u_2, \dots, u_N)$ with

$$\partial_t^m u + \sum_{j<m} A_{\alpha,j}(x, t)\, D^\alpha\, \partial_t^j u = f \tag{1}$$

where $D^\alpha = \partial_{x_1}^{\alpha_1} \partial_{x_2}^{\alpha_2} \dots \partial_{x_n}^{\alpha_n}$, $|\alpha| = \alpha_1 + \alpha_2 + \dots + \alpha_n$, $A_{\alpha j}(x, t)$, $f(x, t)$ are given real $(N \times N)$-matrices and a vector-valued function, respectively. The system (1) is called a *system of composed type* in the point $(x, t) \in \mathbb{R}^{n+1}$ if the algebraic equation in τ

$$\det \Big(I\tau^m + \sum_{\substack{|\alpha|+j=m \\ (j<m)}} A_{\alpha,j}(x, t)\, \xi^\alpha \tau^j\Big) = 0$$

where $0 \neq \xi = (\xi_1, \dots, \xi_n)$ is fixed has in the point (x, t) real as well as complex roots.

For $n = 1$ the theory of boundary value problems for systems of partial differential equations of composed type in bounded domains is given in DŽURAEV [1] where the methods of complex variables are systematically used and we are given a connection between the boundary value problems for systems of partial differential equations of composed type and the theory of singular integral equations with shifts in the arguments.

In DŽURAEV [1] we considered the following first order systems of equations with respect to a real vector-valued function

$$u(x, y) = (u_1, u_2, \dots, u_N)$$

with

$$u_x - A(x, y)\, u_y - B(x, y)\, u = 0 \tag{2}$$

in some simply connected domains $D \subset \mathbb{R}^2$. This system is of composed type if the following algebraic equation in λ

$$\det \big(A(x, y) - \lambda I\big) = 0$$

has in the point (x, t) real as well as complex roots $\lambda_j(x, y)$ in D. In order to formulate our problem in a bounded domain D we consider for example the case $N = 3$. In this case the last algebraic equation has one real and two complex-valued roots. They are $\lambda_1(x, y)$ and $\lambda_0(x, y)$, $\overline{\lambda}_0(x, y)$. By the real characteristic which arises in this case the boundary $\Gamma = \partial D$ of the domain D is split into two parts: Γ_1 and $\Gamma_2 = \Gamma - \Gamma_1$. In this case one part of our problem is the following: Find a solution of the system (2) when the following conditions are given:

$$\big(a(t), u(t)\big) = h(t)\,, \qquad t \in \Gamma\,,$$

$$\big(a^s(t), u(t)\big) = h^s(t)\,, \qquad t \in \Gamma_s\,,$$

where

$$a(t) = \big(a_1(t)\,,\, a_2(t)\,,\, a_3(t)\big)$$

$$a^s(t) = \big(a_1^s(t)\,,\, a_2^s(t)\,,\, a_3^s(t)\big)$$

are real vector fields given on Γ and Γ_s where s is equal to 1 or 2. First this problem is reduced to the following boundary value problem with shift:

$$\operatorname{Re}\{c(t)\, w(t) + d(t)\, w[\alpha(t)]\} = y(t) \quad \text{on } \Gamma$$

for the solutions of some first order complex elliptic equations which generalize analytic functions in Vekua's sense or their generalizations. Second the last problem is reduced to the following singular integral equation with Carleman's shift $\alpha(t)$:

$$A(t)\, \varphi(t) + B(t)\, \varphi[\alpha(t)] + \frac{C(t)}{\pi i} \int_\Gamma \frac{\varphi(\tau)\, d\tau}{\tau - t} + \frac{D(t)}{\pi i} \int_\Gamma \frac{\varphi(\tau)\, d\tau}{\tau - \alpha(t)} +$$

$$+ T(\varphi) = f(t)\,, \qquad t \in \Gamma\,.$$

This complex method gives the possibility to introduce the conjugate problem to formulate the Noetherian theorems and to find an explicit formula for the index of our problem. In DŽURAEV [1] many other problems for systems of composed type in bounded domains in the plane have been formulated.

In this chapter we restrict to the consideration of special classes of systems in semispaces and cylindrical domains and we will find properly posed problems for such systems and their solutions.

13.2. Problems in semispaces

Let $\mathbb{R}^n$ be the n-dimensional Euclidian space of points $x = (x_1, \dots, x_n)$ with real coordinates and $\mathbb{R}_+^{n+1}$ be the semispace $\mathbb{R}_+^{n+1} = \{(x, t) : x \in \mathbb{R}^n, t > 0\}$. Let $K(\mathbb{R}_+^{n+1})$ be the class of the complex-valued functions $u(x, t)$ which (together with u_t) belong for every $t \geqq 0$ to the space $\mathcal{H}(\mathbb{R}^n)$ which consists of all quadratic sumable functions in $\mathbb{R}^n$ and its differentiations to x in the sense of generalized functions which with reference to t do not grow faster than a certain power of itself. Normally it is supposed that the sequence $\varphi_k(x)$ converges in $\mathcal{H}(\mathbb{R}^n)$ to zero if such an s exists that $\varphi_k = (1 - \Delta)^s\, \psi_k$, $\psi_k \in L_2(\mathbb{R}^n)$ where the ψ_k converge in the norm of $L_2(\mathbb{R}^n)$ to zero. Let $\tilde{\mathcal{H}}(\mathbb{R}^n)$ be the dual space to $\mathcal{H}(\mathbb{R}^n)$ in the sense of Fourier's transformation with reference to x.

We consider the following complex equation

$$\frac{\partial^2 u}{\partial t^2} + L(D)\,\bar{u} = f(x, t) \tag{3}$$

where $L(D) = \sum_{|\alpha|=2m} a_\alpha D^\alpha$ is a linear homogeneous elliptic operator of the order $2m$ with complex constant coefficients a_α. This equation is equivalent to a system of two real equations, where its adequate characteristic form is

$$\left(\tau^2 - |L(\xi)|\right)\left(\tau^2 + |L(\xi)|\right)$$

where $L(\xi) = \sum_{|\alpha|=2m} a_\alpha \xi^\alpha$ and therefore the equation (3) is for all (x, t) of composed type. We will find a solution of the equation (3) in the class K in the semispace $\mathbb{R}_+^{n+1}$. Which assumption must be current to the hyperplane that the problem to find such solutions in $\mathbb{R}_+^{n+1}$ is properly posed? We apply to (1) the Fourier transformation to x:

$$\tilde{u}(\xi, t) = F[u]\,(\xi, t) = (2\pi)^{-n/2} \int_{\mathbb{R}^n} u(x, t)\, e^{i(x,\xi)}\, dx. \qquad \tilde{f}(\xi, t) = F[f]\,(\xi, t)$$

then we obtain the following equation

$$\frac{\partial^2 \tilde{u}(\xi, t)}{\partial t^2} + (-1)^m L(\xi)\, \overline{\tilde{u}(-\xi, t)} = \tilde{f}(\xi, t)\,. \tag{4}$$

We introduce the new function $\tilde{v}(\xi, t)$:

$$\tilde{v}(\xi, t) = (-i)^m \left(\frac{\overline{L(\xi)}}{|L(\xi)|}\right)^{1/2} \tilde{u}(\xi, t) \tag{5}$$

then we obtain the following equation

$$\frac{\partial^2 \tilde{v}}{\partial t^2} + |L(\xi)|\, \overline{\tilde{v}(-\xi, t)} = \tilde{f}_1(\xi, t)\,, \tag{6}$$

where

$$\tilde{f}_1(\xi, t) = (-i)^m \left(\frac{(\overline{L(\xi)}}{|L(\xi)|}\right)^{1/2} \tilde{f}(\xi, t)\,. \tag{7}$$

Because the factor to $\tilde{u}(\xi, t)$ in (5) is a homogeneous function of the order 0, so it is the symbol $\sigma_{A_0}(\xi)$ of a certain singular integral operator and the transformation (5) is equivalent to

$$v(x, t) = A_0 u(x, t)\,.$$

Analogously to (7) we have $f_1 = A_0 f$. The equation (6) is equivalent to the following integral equation

$$\tilde{v}(\xi, t) + |L(\xi)| \int_0^t (t - s)\, \overline{\tilde{v}(-\xi, s)}\, ds = \tilde{v}(\xi, 0) + t\tilde{v}_t(\xi, 0) + \int_0^t (t - s)\, \tilde{f}_1(\xi, s)\, ds \tag{8}$$

which will be solved by successive approximation

$$\begin{aligned} \tilde{v}_0(\xi, t) &= \tilde{v}(\xi, 0) + t\tilde{v}_t(\xi, 0) + \int_0^t (t - s)\, \tilde{f}_1(\xi, s)\, ds\,, \\ \tilde{v}_k(\xi, t) &= -\,|L(\xi)| \int_0^t (t - s)\, \overline{\tilde{v}_{k-1}(-\xi, s)}\, ds\,. \end{aligned} \tag{9}$$

We obtain the following representation for $\tilde{v}(\xi, t)$:

$$\tilde{v}(\xi, t) = \frac{\cosh\sqrt{|L(\xi)|}\,t + \cos\sqrt{|L(\xi)|}\,t}{2}\tilde{v}(\xi, 0) -$$

$$-\frac{\cosh\sqrt{|L(\xi)|}\,t - \cos\sqrt{|L(\xi)|}\,t}{2}\overline{\tilde{v}(-\xi, 0)} +$$

$$+\frac{1}{\sqrt{|L(\xi)|}}\left(\frac{\sinh\sqrt{|L(\xi)|}\,t + \sin\sqrt{|L(\xi)|}\,t}{2}\tilde{v}_t(\xi, 0) -\right.$$

$$\left.-\frac{\sinh\sqrt{|L(\xi)|}\,t - \sin\sqrt{|L(\xi)|}\,t}{2}\overline{\tilde{v}_t(-\xi, 0)}\right) +$$

$$+\frac{1}{\sqrt{|L(\xi)|}}\int_0^t\left(\frac{\sinh\sqrt{|L(\xi)|}\,(t-s) + \sin\sqrt{|L(\xi)|}\,(t-s)}{2}\tilde{f}_1(\xi, s) -\right.$$

$$\left.-\frac{\sinh\sqrt{|L(\xi)|}\,(t-s) - \sin\sqrt{|L(\xi)|}\,(t-s)}{2}\overline{\tilde{f}_1(-\xi, s)}\right)ds\,. \tag{10}$$

Because (10) includes parts which grow exponentially when $t \to \infty$ $\tilde{v}(\xi, t)$ and therefore $\tilde{u}(\xi, t)$ do not grow faster than powers of t when $t \to \infty$ if and only if

$$\tilde{v}(\xi, 0) - \overline{\tilde{v}(-\xi, 0)} + \frac{1}{\sqrt{|L(\xi)|}}\left(\tilde{v}_t(\xi, 0) - \overline{\tilde{v}_t(-\xi, 0)}\right) +$$

$$+\frac{1}{\sqrt{|L(\xi)|}}\int_0^\infty e^{-\sqrt{|L(\xi)|}s}\left(\tilde{f}_1(\xi, s) - \overline{\tilde{f}_1(-\xi, s)}\right)ds = 0\,. \tag{11}$$

The relation (11) means that the values of the solution $u(x, t)$ of (3) from the class K and the differentiation u_t can not be given independently of each other and of the right-hand side on the hyperplane $t = 0$.

Because the hyperplane $t = 0$ is not characteristic for the equation (3) so the relation only produces the property of the partial ellipticity of equation (1). Because of equation (11) we obtain from the representation (10) for $\tilde{v}(\xi, t)$ either the representation

$$\tilde{v}(\xi, t) = \frac{e^{-\sqrt{|L(\xi)|}t}}{2}\left(\tilde{v}(\xi, 0) - \overline{\tilde{v}(-\xi, 0)}\right) +$$

$$+\frac{1}{4\sqrt{|L(\xi)|}}\int_0^\infty e^{-\sqrt{|L|(\xi)|}(t+s)}\left(\tilde{f}_1(\xi, s) - \overline{\tilde{f}_1(-\xi, s)}\right)ds -$$

$$-\frac{1}{4}\int_0^\infty e^{-\sqrt{|L(\xi)|}s}\left(\tilde{f}_1(\xi, s+t) - \overline{\tilde{f}_1(-\xi, s+t)}\right)ds -$$

$$-\frac{1}{\sqrt{|L(\xi)|}}\int_0^t e^{-\sqrt{|L(\xi)|}(t-s)}\left(\tilde{f}_1(\xi, s) - \overline{\tilde{f}_1(-\xi, s)}\right)ds +$$

$$+\frac{\cos\sqrt{|L(\xi)|}\,t}{2}\left(\tilde{v}(\xi, 0) + \overline{\tilde{v}(-\xi, 0)}\right) +$$

$$+\frac{\sin\sqrt{|L(\xi)|}\,t}{2\sqrt{|L(\xi)|}}\left(\tilde{v}_t(\xi,0)+\overline{\tilde{v}_t(-\xi,0)}\right)+$$

$$+\int_0^t\frac{\sin\sqrt{|L(\xi)|}\,(t-s)}{2\sqrt{|L(\xi)|}}\left(\tilde{f}_1(\xi,s)+\overline{\tilde{f}_1(-\xi,s)}\right)\mathrm{d}s \tag{12}$$

or the following representation

$$\tilde{v}(\xi,t)=\frac{-\mathrm{e}^{-\sqrt{|L(\xi)|}t}}{2\sqrt{|L(\xi)|}}\left(\tilde{v}_t(\xi,0)-\overline{\tilde{v}(-\xi,0)}\right)-$$

$$-\frac{1}{4\sqrt{|L(\xi)|}}\int_0^\infty \mathrm{e}^{-\sqrt{|L(\xi)|}(t+s)}\left(\tilde{f}_1(\xi,s)-\overline{\tilde{f}_1(-\xi,s)}\right)\mathrm{d}s-$$

$$-\frac{1}{4}\int_0^\infty \mathrm{e}^{-\sqrt{|L(\xi)|}s}\left(\tilde{f}_1(\xi,s+t)-\overline{\tilde{f}_1(-\xi,s+t)}\right)\mathrm{d}s-$$

$$-\frac{1}{\sqrt{|L(\xi)|}}\int_0^t \mathrm{e}^{-\sqrt{|L(\xi)|}(t-s)}\left(\tilde{f}_1(\xi,s)-\overline{\tilde{f}_1(-\xi,s)}\right)\mathrm{d}s+$$

$$\frac{\cos\sqrt{|L(\xi)|}\,t}{2}\left(\tilde{v}(\xi,0)+\overline{\tilde{v}(-\xi,0)}\right)+\frac{\sin\sqrt{|L(\xi)|}\,t}{2\sqrt{|L(\xi)|}}\left(\tilde{v}_t(\xi,0)+\overline{\tilde{v}_t(-\xi,0)}\right)+$$

$$+\int_0^t\frac{\sin\sqrt{|L(\xi)|}\,(t-s)}{2\sqrt{|L(\xi)|}}\left(\tilde{f}_1(\xi,s)+\overline{\tilde{f}_1(-\xi,s)}\right)\mathrm{d}s\,. \tag{13}$$

We remark that we obtain the same equation (6) with $\tilde{f}_1(\xi,t)=(-\mathrm{i})^m\sqrt{|L(\xi)|}\,\tilde{f}(\xi,t)$, if we introduce the new function $\tilde{v}(\xi,t)$ by the formula

$$\tilde{v}(\xi,t)=(-\mathrm{i})^m\sqrt{|L(\xi)|}\,\tilde{u}(\xi,t) \tag{5'}$$

instead of (5).

Because the factor in (3) is a homogeneous function of the order m and $L(\xi)\neq 0$ for $\xi\neq 0$ so it is the symbol $\sigma_{A_1}(\xi)$ of a certain elliptic pseudodifferential operator A_1 and (5') is equivalent to $v=A_1u$.

Let the following condition be given on the hyperplane:

$$u(x,0)=\varphi(x)\,,\qquad \operatorname{Re}\left(Au_t(x,0)\right)=\psi(x) \tag{14}$$

where A is either the operator A_0 or the operator A_1 and $\varphi(x),\psi(x)\in\mathscr{H}(\mathbb{R}^n)$ are given functions.

In (14) let us go to the Fourier's transformation, so we have

$$\tilde{u}(\xi,0)=\tilde{\varphi}(\xi)\,,\qquad \sigma_A(\xi)\,\tilde{u}_t(\xi,0)+\overline{\sigma_A(-\xi)\,\tilde{u}_t(-\xi,0)}=2\tilde{\psi}(\xi) \tag{15}$$

where $\sigma_A(\xi)$ is either $\sigma_{A_0}(\xi)$ or $\sigma_{A_1}(\xi)$.

Because (5), (5') condition (15) means that

$$\tilde{v}(\xi,0)=\sigma_A(\xi)\,\tilde{\varphi}(\xi)\,,\qquad \tilde{v}_t(\xi,0)+\overline{\tilde{v}_t(-\xi,0)}=2\tilde{\psi}(\xi)\,.$$

Therefore with the help of (12) the function $\tilde{v}(\xi, t)$ is uniquely defined by the given functions φ, ψ:

$$\tilde{v}(\xi, t) = e^{-\sqrt{|L(\xi)|}t} \left(\sigma_A(\xi)\,\tilde{\varphi}(\xi) - \overline{\sigma_A(-\xi)}\,\overline{\tilde{\varphi}(-\xi)}\right) +$$

$$+ \frac{1}{4\sqrt{|L(\xi)|}} \int_0^\infty e^{-\sqrt{|L(\xi)|}(t+s)} \left(\sigma_A(\xi)\,\tilde{f}(\xi, s) - \overline{\sigma_A(-\xi)}\,\overline{\tilde{f}(-\xi, s)}\right) ds -$$

$$- \frac{1}{4} \int_0^\infty e^{-\sqrt{|L(\xi)|}s} \left(\sigma_A(\xi)\,\tilde{f}(\xi, s+t) - \overline{\sigma_A(-\xi)}\,\overline{\tilde{f}(-\xi, s+t)}\right) ds -$$

$$- \frac{1}{\sqrt{|L(\xi)|}} \int_0^t e^{-\sqrt{|L(\xi)|}(t-s)} \left(\sigma_A(\xi)\,\tilde{f}(\xi, s) - \overline{\sigma_A(-\xi)}\,\overline{\tilde{f}(-\xi, s)}\right) ds +$$

$$+ \frac{\cos\sqrt{|L(\xi)|}\,t}{2} \left(\sigma_A(\xi)\,\tilde{\varphi}(\xi) - \overline{\sigma_A(-\xi)}\,\overline{\tilde{\varphi}(-\xi)}\right) +$$

$$+ \frac{\sin\sqrt{|L(\xi)|}\,t}{\sqrt{|L(\xi)|}} \left(\tilde{\psi}(\xi) + \overline{\tilde{\psi}(-\xi)}\right) +$$

$$+ \int_0^t \frac{\sin\sqrt{|L(\xi)|}\,(t-s)}{2\sqrt{|L(\xi)|}} \left(\sigma_A(\xi)\,\tilde{f}(\xi, s) + \overline{\sigma_A(-\xi)}\,\overline{\tilde{f}(-\xi, s)}\right) ds\,. \tag{16}$$

From this formula it follows that

$$\begin{aligned}\tilde{v}(\xi, t) + \tilde{v}(-\xi, t) = {} & \cos\sqrt{|L(\xi)|}\,t\left(\sigma_A(\xi)\,\tilde{\varphi}(\xi) + \overline{\sigma_A(-\xi)}\,\overline{\tilde{\varphi}(-\xi)}\right) + \\ & + 2\sin\sqrt{|L(\xi)|}\,t\left(\tilde{\psi}(\xi) + \overline{\tilde{\psi}(-\xi)}\right) + \\ & + \int_0^t \frac{\sin\sqrt{|L(\xi)|}\,(t-s)}{\sqrt{|L(\xi)|}} \left(\sigma_A(\xi)\,\tilde{f}(\xi, s) + \sigma_A(-\xi)\tilde{f}(-\xi, s)\right) ds\end{aligned} \tag{17}$$

$$\begin{aligned}\tilde{v}(\xi, t) - \overline{\tilde{v}(-\xi, t)} = {} & 2e^{-\sqrt{|L(\xi)|}t} \left(\sigma_A(\xi)\,\tilde{\varphi}(\xi) - \overline{\sigma_A(-\xi)}\,\overline{\tilde{\varphi}(-\xi)}\right) + \\ & + \frac{1}{2\sqrt{|L(\xi)|}} \int_0^\infty e^{-\sqrt{|L(\xi)|}\,(t+s)} \left(\sigma_A(\xi)\,\tilde{f}(\xi, s) - \right. \\ & \left. - \overline{\sigma_A(-\xi)}\,\overline{\tilde{f}(-\xi, s)}\right) ds - \frac{1}{2}\int_0^\infty e^{-\sqrt{|L(\xi)|}s} \left(\sigma_A(\xi)\,\tilde{f}(\xi, s+t) - \right. \\ & \left. - \overline{\sigma_A(-\xi)}\,\overline{\tilde{f}(-\xi, s+t)}\right) ds - \\ & - \frac{2}{\sqrt{|L(\xi)|}} \int_0^t e^{-\sqrt{|L(\xi)|}(t-s)} \left(\sigma_A(\xi)\,\tilde{f}(\xi, s) - \overline{\sigma_A(-\xi)}\,\overline{\tilde{f}(-\xi, s)}\right) ds\end{aligned} \tag{18}$$

is valid. Hence we obtain under the use of Parseval's equation

$$\int_{R^n} |\operatorname{Re} Au(x, T)|^2 \, dx \leqq \text{const} \Big(\int_{R^n} |\operatorname{Re} A\varphi(x)|^2 \, dx + \int_{R^n} |\psi(x)|^2 \, dx + \\ + \mu(T) \int_0^T dt \int_{R^n} |\operatorname{Re} f(x, t)|^2 \, dx \Big) \qquad (0 < T < \infty) \tag{17'}$$

$$\sum_{|\alpha| \leqq 2m} \int_0^\infty dt \int_{R^n} |D^\alpha \operatorname{Im} Au(x, t)|^2 \, dx \leqq \text{const} \Big(\int_0^\infty dt \int_{R^n} |\operatorname{Im} f(x, t)|^2 \, dx + \\ + \int_{R^n} |\operatorname{Im} \overline{A\varphi(x)}|^2 (1 + |\xi|)^{2m-1/2} \, d\xi \Big) . \tag{18'}$$

Now we consider the following hyperbolic equation of the first order

$$\frac{\partial u}{\partial t} + L_1(D) \, \bar{u} = f(x, t) , \tag{19}$$

where $L_1(D)$ is a linear elliptic operator of the first order in R^n with complex coefficients.

We obtain the following representation for the Fourier transformation $\tilde{u}(\xi, t)$ of the wanted function:

$$\tilde{u}(\xi, t) = \cos |L_1(\xi)| \, t\tilde{u}(\xi, 0) + \frac{iL_1(\xi)}{|L_1(\xi)|} \sin |L_1(\xi)| \, t\overline{\tilde{u}(-\xi, 0)} + \\ + \int_0^t \cos |L_1(\xi)| \, (t - s) \tilde{f}(\xi, s) \, ds + \\ + \frac{iL_1(\xi)}{|L_1(\xi)|} \int_0^t \sin |L_1(\xi)| \, (t - s) \overline{\tilde{f}(-\xi, s)} \, ds . \tag{20}$$

This representation gives the solution of the Cauchy problem for the equation (19) with the start condition $u(x, 0) = \varphi(x)$. Among other things we remark if $n = 2$ and $L_1 = \partial x + i \, \partial y$ (the Cauchy-Riemann operator) then we obtain with the help of (20) a formula, which gives the solution of the Cauchy problem

$$u(z, t) = \frac{\partial}{\partial t} \Big(\frac{1}{2\pi} \int_{|\zeta - z| < t} \frac{\varphi(\zeta) \, d\xi \, d\eta}{\sqrt{t^2 - |\zeta - z|^2}} \Big) - \frac{\partial}{\partial \bar{z}} \Big(\frac{1}{\pi} \int_{|\zeta - z| < t} \frac{\overline{\varphi(\zeta)} \, d\xi \, d\eta}{\sqrt{t^2 - |\zeta - z|^2}} \Big) + \\ + \int_0^t \frac{\partial}{\partial t} \Big(\frac{1}{2\pi} \int_{|\zeta - z| < t - s} \frac{f(\zeta, s) \, d\xi \, d\eta}{\sqrt{(t - s)^2 - |\zeta - z|^2}} \Big) ds - \\ - \int_0^t \frac{\partial}{\partial \bar{z}} \Big(\frac{1}{\pi} \int_{|\zeta - z| < t - s} \frac{\overline{f(\zeta, s)} \, d\xi \, d\eta}{\sqrt{(t - s)^2 - |\zeta - z|^2}} \Big) ds . \tag{21}$$

In the general case it follows from the representation (20) with the help of Parseval's equation that the following inequality is right:

$$\int_{R^n} |u(x, T)|^2 \, dx = \int_{R^n} |\tilde{u}(\xi, T)|^2 \, d\xi \\ \leqq 2 \int_{R^n} |\tilde{\varphi}(\xi)|^2 \, d\xi + 2 \int_0^T dt \int_{R^n} |\tilde{f}(\xi, t)|^2 \, d\xi \\ = 2 \Big(\int_{R^n} |\varphi(x)|^2 \, dx + \int_0^T dt \int_{R^n} |f(x, t)|^2 \, dx \Big) . \tag{22}$$

The inequality (22) can be used for the proof of the existence of a unique solution of the Cauchy problem for the generalized hyperbolic equation

$$\frac{\partial u}{\partial t} + L_1(D)\,\bar{u} = a(x, t)\,u + b(x, t)\,\bar{u} + g(x, t)\,. \tag{23}$$

This is made by the same scheme, which is given in BERS, JOHN, and SCHECHTER [1, p. 87]. We can also obtain equations of a more general nature than (1)

$$\frac{\partial^2 u}{\partial t^2} + P(D)\,u + Q(D)\,\bar{u} = f\,. \tag{24}$$

The assumption of the ellipticity of the operator $P(D)\,u + Q(D)\,\bar{u}$ means that $|P(\xi)|^2 - |Q(\xi)|^2 \neq 0$ for $\xi \neq 0$.

If $|P(\xi)| < |Q(\xi)|$ then the characteristic form has the following nature

$$\big(\tau^2 - \tau_1(\xi)\big)\,\big(\tau^2 + \tau_2(\xi)\big)$$

$$\tau_1(\xi) = -\operatorname{Re} P(\xi) + \sqrt{|Q(\xi)|^2 - \big(\operatorname{Im} P(\xi)\big)^2}\,,$$

where

$$\tau_2(\xi) = \operatorname{Re} P(\xi) + \sqrt{|Q(\xi)|^2 - \big(\operatorname{Im} P(\xi)\big)^2} > 0\,.$$

Therefore system (24) is of the composed type. With some modifications of the above construction we can formulate the correct problem in $\mathbb{R}^{n+1}_+$ for this case.

13.3. Mixed problems

We regard a system of two real equations in a cylindrical domain

$$G = \{\big((x, y); t\big) : (x, y) \in D \subset \mathbb{R}^2, t > 0\}$$

$$\frac{\partial^2 U}{\partial t^2} + L(U) = F\,, \tag{25}$$

where $L(U)$ is an elliptic operator of second order in D:

$$L(U) \equiv AU_{xx} + BU_{xy} + CU_{yy}\,.$$

We assume that the elements of the quadratic matrices of second order A, B, C are given smooth functions of the variables x, y with $\det A(z) \neq 0$, $z = x + iy \in D$. Then the condition of the ellipticity of the operator $L(U)$ means, that all roots of the equation

$$\det\big(A\lambda^2 + B\lambda + C\big) = \det\big(a_{ij}(\lambda)\big) = 0 \tag{26}$$

are complex in D. We multiply the second equation of (25) by i and add it to the first one, so we obtain the following complex equation

$$\frac{\partial^2 U}{\partial t^2} + L_1 U + L_2 \bar{U} = f(z, t) \tag{27}$$

where

$$2L_k = \alpha_k \frac{\partial^2}{\partial x^2} + \beta_k \frac{\partial^2}{\partial x\,\partial y} + \gamma_k \frac{\partial^2}{\partial y^2}\,, \qquad k = 1, 2\,,$$

and

$$\alpha_1\lambda^2 + \beta_1\lambda + \gamma_1 = a_{11}(\lambda) + a_{22}(\lambda) + i\big(a_{21}(\lambda) - a_{12}(\lambda)\big)\,,$$

$$\alpha_2\lambda^2 + \beta_2\lambda + \gamma_2 = a_{11}(\lambda) - a_{22}(\lambda) + i\big(a_{21}(\lambda) + a_{12}(\lambda)\big)\,.$$

The condition of the ellipticity of the operator $L_1U + L_2\bar{u}$ means that the equation $|L_1(\lambda)|^2 - |L_2(\lambda)|^2 = 0$ has only complex roots λ.

We suppose that the equation $\alpha_k\lambda^2 + \beta_k\lambda + \gamma_k$ has at least one common root $\lambda = \lambda_1(z)$ which is a root of the equation $\det\left(a_{ij}(\lambda)\right) = 0$.

We transform the domain D with the help of the homeomorphic solution $\zeta = \zeta(z)$ of the Beltrami equation $\zeta_x - \lambda\zeta_y = 0$ with $\operatorname{Im}\lambda \neq 0$ into another domain D' (for instance on the union circle with the centre in the origin of the system of coordinats) (see I. N. VEKUA [5]). Then we obtain the following relations for the operators L_k:

$$L_k \equiv \left(\alpha_k|\zeta_x|^2 + \beta_k \operatorname{Re}(\lambda\zeta_x\bar{\zeta}_y) + \gamma_k|\zeta_y|^2\right)\frac{\partial^2}{\partial\zeta\,\partial\bar{\zeta}}$$

$$+ \frac{1}{2}\left(\alpha_k\zeta_x^2 + \beta_k\zeta_x\zeta_y + \gamma_k\zeta_y^2\right)\frac{\partial^2}{\partial\zeta^2} + \frac{1}{2}\left(\alpha_k\bar{\zeta}_x^2 + \beta_k\bar{\zeta}_x\bar{\zeta}_y + \gamma_k\bar{\zeta}_y^2\right)\frac{\partial^2}{\partial\bar{\zeta}^2}$$

$$= (\alpha_k|\lambda|^2 + \beta_k\operatorname{Re}\lambda + \gamma_k)|\zeta|^2\frac{\partial^2}{\partial\zeta\,\partial\bar{\zeta}} + \frac{1}{2}(\alpha_k\lambda^2 + \beta_k\lambda + \gamma_k)\zeta_y^2\frac{\partial^2}{\partial\zeta^2}$$

$$+ \frac{1}{2}(\alpha_k\bar{\lambda}^2 + \beta_k\bar{\lambda} + \gamma_k)\bar{\zeta}_y^2\frac{\partial^2}{\partial\bar{\zeta}^2},$$

where $2\partial_{\bar{\zeta}} = \partial_\xi + \mathrm{i}\,\partial\eta$, $2\partial_\zeta = \partial_\xi - \mathrm{i}\,\partial\eta$. If the roots of the equation $\alpha_1\lambda^2 + \beta_1\lambda + \gamma_1$ are denoted by λ_1, λ_2 and the roots of the equation $\alpha_2\lambda^2 + \beta_2\lambda + \gamma_2$ by λ_3, λ_4 then we obtain apparently

$$L_1 = \alpha_1\left(|\lambda|^2 - \frac{\lambda_1 + \lambda_2}{2}(\lambda + \bar{\lambda}) + \lambda_1\lambda_2\right)|\zeta_y|^2\frac{\partial^2}{\partial\zeta\,\partial\bar{\zeta}} + \frac{\alpha_1}{2}(\lambda - \lambda_1)(\lambda - \lambda_2)\zeta_y^2\frac{\partial^2}{\partial\zeta^2}$$

$$+ \frac{\alpha_1}{2}(\bar{\lambda} - \lambda_1)(\bar{\lambda} - \lambda_2)\bar{\zeta}_y^2\frac{\partial^2}{\partial\bar{\zeta}^2},$$

$$L_2 = \alpha_2\left(|\lambda|^2 - \frac{\lambda_3 + \lambda_4}{2}(\lambda + \bar{\lambda}) + \lambda_3\lambda_4\right)|\zeta_y|^2\frac{\partial^2}{\partial\zeta\,\partial\bar{\zeta}} + \frac{\alpha_2}{2}(\lambda - \lambda_3)(\lambda - \lambda_4)\zeta_y^2\frac{\partial^2}{\partial\zeta^2}$$

$$+ \frac{\alpha^2}{2}(\bar{\lambda} - \lambda_3)(\bar{\lambda} - \lambda_4)\bar{\zeta}_y^2\frac{\partial^2}{\partial\bar{\zeta}^2}.$$

We suppose $\lambda = \lambda_1$ and we obtain

$$L_1 \equiv \frac{\alpha_1}{2}(\bar{\lambda}_1 - \lambda_1)\left((\lambda_1 - \lambda_2)|\zeta_y|^2\frac{\partial^2}{\partial\zeta\,\partial\bar{\zeta}} + (\bar{\lambda}_1 - \lambda_2)\bar{\zeta}_y^2\frac{\partial^2}{\partial\bar{\zeta}^2}\right).$$

Besides if $\lambda_3 = \lambda_1$ then we obtain

$$L_2 \equiv \frac{\alpha_2}{2}(\bar{\lambda}_1 - \lambda_1)\left((\lambda_1 - \lambda_4)|\zeta_y|^2\frac{\partial^2}{\partial\zeta\,\partial\bar{\zeta}} + (\bar{\lambda}_1 - \lambda_4)\bar{\zeta}_y^2\frac{\partial^2}{\partial\bar{\zeta}^2}\right)$$

and if $\lambda_1 = \lambda_4$ then we obtain

$$L_2 \equiv \frac{\alpha_2}{2}(\bar{\lambda}_1 - \lambda_1)\left((\lambda_1 - \lambda_3)|\zeta_y|^2\frac{\partial^2}{\partial\zeta\,\partial\bar{\zeta}} + (\bar{\lambda}_1 - \lambda_3)\bar{\zeta}_y^2\frac{\partial^2}{\partial\bar{\zeta}^2}\right).$$

Because we have apparently $|\lambda_1 - \lambda_2|^2 = |\bar{\lambda}_1 - \lambda_2|^2 - \operatorname{Im}\lambda_1\operatorname{Im}\lambda_2$, $|\lambda_1 - \lambda_3|^2 = |\bar{\lambda}_1 - \lambda_3|^2 - \operatorname{Im}\lambda_1\operatorname{Im}\lambda_3$, $|\lambda_1 - \lambda_4|^2 = |\bar{\lambda}_1 - \lambda_4|^2 - \operatorname{Im}\lambda_1\operatorname{Im}\lambda_4$, so if $\operatorname{Im}\lambda_1, \operatorname{Im}\lambda_2$ have a various sign the operator L_1 has the form

$$L_1 \equiv k_1(\zeta)\left(\frac{\partial^2}{\partial\zeta\,\partial\bar\zeta} - \mu\frac{\partial^2}{\partial\bar\zeta^2}\right) \tag{28}$$

with $|\mu(\zeta)| < 1$, and if $\operatorname{Im}\lambda_1$, $\operatorname{Im}\lambda_2$ have the same sign the operator L_1 has the form

$$L_1 \equiv k_2(\zeta)\left(\frac{\partial^2}{\partial\bar\zeta^2} - \nu\frac{\partial^2}{\partial\zeta\,\partial\bar\zeta}\right) \tag{29}$$

with $|\nu(\zeta)| < 1$. It is analogously for the operator L_2. The boundary value problem for the equation (27) essentially depends on the kind of the operators (28) or (29). We consider the equation

$$\frac{\partial^2 u}{\partial t^2} + 4\frac{\partial^2\bar u}{\partial\bar z^2} = 0\,. \tag{30}$$

Mixed Problem. A solution of the equation (30) has to be found in a cylinder $G = \{(z, t) : z \in D, t > 0\}$ with the conditions:

$$u(z, 0) = \varphi(z)\,,$$
$$\operatorname{Im} u_{zt}(z, 0) = \psi(z)\,,\quad z \in D;\qquad \operatorname{Re}\,[a(\zeta)\,u_t(\zeta, 0)] = h(\zeta)\,,\qquad \zeta \in \Gamma = \partial D\,,$$

on the base and the homogeneous condition

$$u_z(\zeta, t) = 0\,,\qquad \zeta \in \Gamma\,,\qquad t \geqq 0\,.$$

on the mantle S of the cylinder.

We suppose that the given functions are smooth enough, for instance we suppose that $a(\zeta)$ and $h(\zeta)$ are continuous in the sense of Hölder on $\Gamma = \partial D$ where $a(\zeta) \neq 0$, $\zeta \in \Gamma$. Because the function $v(z, t) = u_z(z, t)$ accomplishes the equation $v_{tt} + \Delta\bar v = 0$ in the interior of the cylinder G apparently and $v(z, 0) = \varphi_z$, $\operatorname{Im} v_t(z, 0) = \psi(z)$, $z \in D$, on the base of the cylinder and $v = 0$ on the mantle S so it is as a solution of the Dirichlet and the Cauchy problem uniquely defined.

Then the function $u(z, t)$ has the representation

$$u(z, t) = \frac{2}{\pi i}\int_0^t ds \int_\Gamma \frac{t - s}{\bar\zeta - \bar z}\,\bar v_{\bar\zeta}(\zeta, s)\,d\bar\zeta - \frac{1}{2\pi i}\int_\Gamma \frac{\varphi(\zeta)}{\bar\zeta - \bar z}\,d\bar\zeta$$
$$- \frac{1}{\pi}\iint_D \frac{v(\zeta, t)}{\bar\zeta - \bar z}\,dD_\zeta + \overline{\gamma(z)}\,t \tag{31}$$

where $\gamma(z)$ is a holomorphic function in D. To find the function $\gamma(z)$ we use the assumption $\operatorname{Re}\,[a(\zeta)\,u_t(\zeta, 0)] = h(\zeta)$. We regard the following Hilbert boundary value problem:

$$\operatorname{Re}\,[\overline{a(\zeta)}\,\gamma(\zeta)] = h^*(\zeta)\,,\qquad \zeta \in \Gamma\,, \tag{32}$$

where

$$h^*(\zeta) = h(\zeta) + \operatorname{Re}\frac{\overline{a(\zeta)}}{\pi}\iint_D \frac{\overline{v_t(z, 0)}}{z - \zeta}\,dD_z\,. \tag{32'}$$

Let $\varkappa = (2\pi)^{-1}\,\Delta_\Gamma \arg a(\zeta)$, where $\Delta_\Gamma \arg a(\zeta)$ is the growth of the function $\arg a(\zeta)$ when we cover the curve Γ one time counter-clockwise.

We know (I. N. Vekua [5]) if $\varkappa \geqq 0$ then the problem (32) has always a solution and its general solution consists of the sum of a particular solution which is produced from the right part and a linear combination with real coefficients of the $2\varkappa + 1$ linearly independent solutions of the homogeneous problem (the right-hand side is equal to zero). However if $\varkappa < 0$ then the problem (32) can only have a solution if the assumption

$$\int_\Gamma h^*(\zeta)\, a(\zeta)\, \sigma_j(\zeta)\, \mathrm{d}\zeta = 0\,, \qquad j = 1, 2, \dots, -2\varkappa - 1\,, \tag{33}$$

is valid where $\sigma_j(z)$ is a solution of the homogeneous problem conjugated to (32)

$$\operatorname{Re}\,[\mathrm{i}\zeta'(s)\, a(\zeta)\, \sigma_j(\zeta)] = 0\,.$$

The relation (33) can be written:

$$\int_\Gamma h(\zeta)\, \sigma_j(\zeta)\, \mathrm{d}\zeta + 2\operatorname{Im}\iint_D \sigma_j(z)\, \overline{v_t(z, 0)}\, \mathrm{d}D_z = 0\,. \tag{34}$$

Because the function $v(z, t)$ can be represented by the functions $\varphi(z)$, $\psi(z)$ uniquely so in the case $\varkappa < 0$ the equation (34) is a condition for the right-hand side $h(\zeta)$, $\varphi(z)$, $\psi(z)$ of our problem, which is necessary and sufficient for its solubility.

So we have the following statement: If

$$\varphi \in \mathscr{C}^3(D) \cap \mathscr{C}^2(\overline{D})\,, \qquad \psi(z) \in \mathscr{C}^2(D) \cap \mathscr{C}^1(\overline{D})\,,$$

$$\varphi_z|_\Gamma = 0\,, \qquad \Delta\varphi_z,\, \Delta\psi \in L_2(D)\,, \qquad a(\zeta) \neq 0,\ \zeta \in \Gamma$$

and $a(\zeta)$, $h(\zeta)$ are continuous in the Hölder sense. If $\varkappa \geqq 0$ then the homogeneous problem (for instance if $h \equiv \varphi \equiv \psi \equiv 0$) has $2\varkappa + 1$ linearly independent solutions and the inhomogeneous problem is always solvable, however if $\varkappa < 0$ then it is necessary and sufficient for the solubility of the mixed problem that $-2\varkappa - 1$ conditions for the right-hand side are realized.

Now we consider instead of equation (30) the more generalized equation

$$\frac{\partial^2 u}{\partial t^2} + \alpha\, \Delta u + 4\beta \frac{\partial^2 \bar{u}}{\partial \bar{z}^2} = 0 \tag{35}$$

where α, β are complex variables with

$$|\beta| > |\alpha|\,. \tag{35'}$$

In this case the characteristic form of the equation (35) is

$$(\tau^2 - \tau_1)\,(\tau^2 + \tau_2)\,,$$

$$\tau_1 = -\operatorname{Re}\alpha + \sqrt{|\beta|^2 - (\operatorname{Im}\alpha)^2}\,, \qquad \tau_2 = \operatorname{Re}\alpha + \sqrt{|\beta|^2 - (\operatorname{Im}\alpha)^2} > 0\,.$$

The function $v(z, t) = u_z(z, t)$ is a solution of the equation

$$\frac{\partial^2 v}{\partial t^2} + \alpha\, \Delta v + \beta\, \Delta \bar{v} = 0\,. \tag{36}$$

We introduce the following real functions

$$\begin{aligned} v_1(z, t) &= -\operatorname{Im}(\alpha + \beta)\operatorname{Re} v(z, t) + \left(\sqrt{|\beta|^2 - (\operatorname{Im}\alpha)^2} + \operatorname{Re}\beta\right)\operatorname{Im} v(z, t)\,,\\ v_2(z, t) &= -\left(\sqrt{|\beta|^2 - (\operatorname{Im}\alpha)^2} + \operatorname{Re}\beta\right)\operatorname{Re} v(z, t) + \operatorname{Im}(\alpha - \beta)\operatorname{Im} v(z, t) \end{aligned} \tag{37}$$

and for these we see the conditions

$$\frac{\partial^2 v_k}{\partial t^2} - \lambda_k\, \Delta v_k = 0\,, \qquad k = 1, 2\,,$$

$$\lambda_1 = -\operatorname{Re}\alpha + \sqrt{|\beta|^2 - (\operatorname{Im}\alpha)^2} > 0\,, \qquad \lambda_2 = -\operatorname{Re}\alpha - \sqrt{|\beta|^2 - (\operatorname{Im}\alpha)^2} < 0$$

are right. Therefore the following mixed problem is properly posed for the equation (35).

A solution of the equation (35) in a cylinder $G = \{(z, t) : z \in D, t > 0\}$ with the conditions

$$u(z, 0) = \varphi(z) ,$$
$$\operatorname{Im} [(\lambda_1 + \bar{\alpha} + \bar{\beta})\, u_{zt}(z, 0)] = \psi(z) , \qquad z \in D ,$$
$$\operatorname{Re} [a(\zeta)\, u_t(\zeta, 0)] = h(\zeta) , \qquad \zeta \in \Gamma = \partial D ,$$

on the base and the homogeneous condition

$$u_z(\zeta, t) = 0 , \qquad \zeta \in \Gamma , \quad t \geqq 0 ,$$

on the mantle S of the cylinder must be found.

In this case the function $u(z, t)$ is uniquely defined by the function v below the assumptions of the problem:

$$u(z, t) = \frac{8}{\pi \mathrm{i}} \int_0^t \mathrm{d}s \int_\Gamma \frac{t - s}{\bar{\zeta} - \bar{z}} \left(\alpha v_{\bar{\zeta}}(\zeta, s) + \beta \bar{v}_{\bar{\zeta}}(\zeta, s)\right) \mathrm{d}s$$
$$- \frac{1}{2\pi \mathrm{i}} \int_\Gamma \frac{\varphi(\zeta)}{\bar{\zeta} - \bar{z}} \mathrm{d}\bar{\zeta} - \frac{1}{\pi} \iint_D \frac{v(\zeta, t)}{\bar{\zeta} - \bar{z}} \mathrm{d}D_\zeta + \overline{\gamma(z)}\, t .$$

In this case it is the same as in the case of the equation (30), the character of the solubility of the mixed problem depends on the sign of the number $\varkappa = (2\pi)^{-1} \Delta_\Gamma \arg a(\zeta)$.

13.4. Systems of first order

In the following we will show that certain first order systems of composed type can be reduced to one of the cases considered above.

We consider for instance the following first order system with reference to complex-valued functions $u(z, t)$, $v(z, t)$:

$$\frac{\partial u}{\partial t} + L_1 v = f , \qquad \frac{\partial v}{\partial t} + L_2 \bar{u} = g \tag{38}$$

where L_k is a homogeneous elliptic operator of first order with complex coefficients in $\mathbb{R}^n$.

Let $u(z, t)$, $v(z, t)$ be a solution of the system (38) in a certain domain G. Then the function

$$\omega(z, t) = \int_0^t v(z, s)\, \mathrm{d}s + w(z) , \tag{39}$$

where $w(z)$ is a solution of the inhomogeneous elliptic equation $L_1 w = -u(z, 0)$, obviously realizes the relation

$$\omega_t = v(z, t) , \qquad L_1 \omega = -u(z, t) + \int_0^t f(z, s)\, \mathrm{d}s$$

in G and with the help of (38) we obtain that it is a solution of the second order equation

$$\frac{\partial^2 \omega}{\partial t^2} + L_2 \bar{L}_1 \bar{\omega} = g(z, t) + \int_0^t L_2 \overline{f(z, s)}\, \mathrm{d}s . \tag{40}$$

The function

$$v(z, t) = \int_0^t u(z, s)\, ds + \chi(z)\,, \tag{41}$$

where $\chi(z)$ is a solution of the inhomogeneous elliptic equation $\bar{L}_2\chi = -\overline{v(z, 0)}$ obviously realizes the relation

$$v_t = u(z, t)\,, \qquad L_2\bar{v} = -v(z, t) + \int_0^t g(z, s)\, ds \tag{42}$$

and with the help of (36) we obtain that it is a solution of the second order equation

$$\frac{\partial^2 v}{\partial t^2} - L_1 L_2 \bar{v} = f(z, t) - \int_0^t L_1 g(z, s)\, ds\,. \tag{43}$$

Because the operators L_k are elliptic so the operators L_1L_2 and $L_2\bar{L}_1$ are also elliptic and therefore the equations (40) and (43) are of composed type of the kind (1). The knowledge of the correct problem for equation (40) and (42) permits to formulate and to study the correct problem for the system (38). Especially for instance the mixed problem in the cylinder $G = \{(z, t) : z \in D \subset \mathbb{R}^2, t \geqq 0\}$ for the system (38) with the operators $L_1 = -2\partial_z$, $L_2 = 2\partial_{\bar{z}}$ with the conditions

$$u(z, 0) = \varphi(z)\,,$$

$$\operatorname{Im}\,[v_z(z, 0)] = \psi(z)\,, \quad z \in D\,, \qquad \operatorname{Re}\,[a(\zeta)\, v(\zeta, 0)] = h(\zeta)\,, \quad \zeta \in \Gamma = \partial D\,,$$

on the base and the homogeneous condition $u(\zeta, t) = 0$ on the mantle S of the cylinder leads to the mixed problem for equation (30) considered above.

14. Differential Operators in the Theory of Elliptic Equations and Boundary Value Problems[1])

H. Florian and R. Heersink

14.1. Introduction

The starting point of this chapter is the homogeneous elliptic equation of second order in two independent real variables in its canonical form

$$\Delta u + A(x, y)\, u_x + B(x, y)\, u_y + C(x, y)\, u = 0\,, \qquad u = u(x, y)\,, \tag{1}$$

where the coefficients A, B and C are assumed to be analytic functions in some domain $D \subset \mathbb{R}^2$. A function $u(x, y)$, defined in D, is called a solution of (1) in D, if $u \in C^2(D)$ and u satisfies (1) in D. (According to a well-known theorem of Picard every solution of (1) is analytic in D.)

If we introduce the complex variable $z = x + iy$, equation (1) can be written in the form ($\bar{z} := x - iy$ denotes the conjugate complex variable)

$$w_{z\bar{z}} + a(z, \bar{z})\, w_z + b(z, \bar{z})\, w_{\bar{z}} + c(z, \bar{z})\, w = 0\,, \tag{2}$$

where $w(z, \bar{z}) := u\big((z + \bar{z})/2, (z - \bar{z})/2\mathrm{i}\big)$, $a := (A + \mathrm{i}B)/4$, $b := (A - \mathrm{i}B)/4$, $c := C/4$, $\partial/\partial z := (\partial/\partial x - \mathrm{i}\, \partial/\partial y)/2$, $\partial/\partial \bar{z} := (\partial/\partial x + \mathrm{i}\, \partial/\partial y)/2$.

For all further investigations it proves to be very useful to perform the analytic continuation of the coefficients A, B, and C in (1) to complex values of x and y. These in this way constructed new functions — for the sake of simplicity they are again denoted by the symbols $A(x, y)$, $B(x, y)$ and $C(x, y)$ — are holomorphic in some domain $G \subset \mathbb{C}^2$, $D \subset G$. In general, nothing is known about the size of G — the analytic continuation exists only "in the small".

Let us now introduce two new *independent* complex variables z and ζ in $\mathbb{C}^2$ by the relations $z := x + \mathrm{i}y$, $\zeta := x - \mathrm{i}y$ ($\zeta = \bar{z}$ if and only if $(x, y) \in \mathbb{R}^2$); for the considered domain D let us presume that the functions $a(z, \zeta)$, $b(z, \zeta)$, $c(z, \zeta)$, generated by the above mentioned analytic continuation and the following definitions

$$\left.\begin{aligned} a(z, \zeta) &:= [A\big((z + \zeta)/2, (z - \zeta)/2\mathrm{i}\big) + \mathrm{i}B\big((z + \zeta)/2, (z - \zeta)/2\mathrm{i}\big)]/4\,, \\ b(z, \zeta) &:= [A\big((z + \zeta)/2, (z - \zeta)/2\mathrm{i}\big) - \mathrm{i}B\big((z + \zeta)/2, (z - \zeta)/2\mathrm{i}\big)]/4\,, \\ c(z, \zeta) &:= [C\big((z + \zeta)/2, (z - \zeta)/2\mathrm{i}\big)]/4 \end{aligned}\right\} \tag{3}$$

are *defined and holomorphic in* $D \times \bar{D}$, where $\bar{D} := \{z : \bar{z} \in D\}$. Simply-connected domains D with the just formulated property will be called *fundamental domains* of equation (1) (see I. N. Vekua [3, p. 8]).

By using the functions a, b, and c, defined by (3), we can associate with equation (1) the following equation (4) in two independent complex variables z, ζ, which is defined in $D \times \bar{D}$:

$$w_{z\zeta} + a(z, \zeta)\, w_z + b(z, \zeta)\, w_\zeta + c(z, \zeta)\, w = 0\,, \qquad w = w(z, \zeta)\,. \tag{4}$$

[1]) Dedicated to Professor Wolfgang Hahn on his 70th birthday.

For $\zeta = \bar{z}$ equations (2) and (4) are obviously identical. Thus by substituting $\zeta = \bar{z}$ every solution $w(z, \zeta)$ of (4), holomorphic in $D \times \bar{D}$, gives a solution $u(x, y)$, analytic in D.

The following theorem gives the inversion of this statement and shows that for fundamental domains solving problem (1) is equivalent to solving (4) (see I. N. VEKUA [3, p. 35]).

Theorem 14.1.1 (I. N. VEKUA). *Let D be a fundamental domain of equation* (1) *and let $u(x, y)$ be a solution of* (1), *defined in D; if we denote the analytic continuation of $u(x, y)$ to complex values of x and y by $U(x, y)$, the function $w(z, \zeta) := U\big((z + \zeta)/2, (z - \zeta)/2\mathrm{i}\big)$ is defined and holomorphic in $D \times \bar{D}$ and represents a solution of equation* (4).

In his book [3] I. N. VEKUA gives representations of the solutions of equation (4) (and thus for (1)) by means of integral operators, which include the complex holomorphic Riemann function of the formally hyperbolic equation (4). BERGMAN uses a somewhat different representation by definite complex integrals in [1].

For certain — especially function theoretic — investigations, representations by means of differential operators for some important classes of equations have become the object of increasing interest in recent years. These representation operators will be the object of the following considerations.

First of all let us demonstrate in one special example of equation (1) how it is possible for certain cases to get a representation by means of differential operators, using Vekua's integral representation as a starting point. For this task we need the following theorem (see I. N. VEKUA [3, p. 23]); H_D, $H_{\bar{D}}$ and $H_{D\times\bar{D}}$, respectively, denote the sets of functions holomorphic in D, $\bar{D}$ and $D \times \bar{D}$, respectively, $R(z, \zeta, t, \tau)$ denotes the Riemann function of equation (4), $z_0 \in D$, $\zeta_0 \in \bar{D}$ denote arbitrary, but fixed points:

Theorem 14.1.2 (I. N. VEKUA). *Let D be a fundamental domain of equation* (1).

a$_1$) *For every solution $w \in H_{D\times\bar{D}}$ of equation* (4) *there exist two functions $f \in H_D$ and $f^* \in H_{\bar{D}}$ and a constant $\alpha \in \mathbb{C}$, such that w can be represented by*

$$w(z, \zeta) = \alpha R(z_0, \zeta_0, z, \zeta) + \int_{z_0}^{z} R(t, \zeta_0, z, \zeta) f(t)\, dt + \int_{\zeta_0}^{\zeta} R(z_0, \tau, z, \zeta) f^*(\tau)\, d\tau\,. \quad (5)$$

a$_2$) *Conversely, for arbitrary $\alpha \in \mathbb{C}, f \in H_D, f^* \in H_{\bar{D}}$,* (5) *represents solutions $w \in H_{D\times\bar{D}}$ of equation* (4).

a$_3$) *For a given solution w of* (4), *$\alpha \in \mathbb{C}$, $f \in H_D$ and $f^* \in H_{\bar{D}}$ in* (5) *are uniquely determined by*

$$\alpha = w(z_0, \zeta_0); \qquad f(z) = w_z(z, \zeta_0) + b(z, \zeta_0)\, w(z, \zeta_0);$$
$$f^*(\zeta) = w_\zeta(z_0, \zeta) + a(z_0, \zeta)\, w(z_0, \zeta)\,. \quad (6)$$

b$_1$) *For every solution $u(x, y)$ of* (1), *defined in D, there exist two functions $f \in H_D$ and $f^* \in H_{\bar{D}}$ and a constant $\alpha \in \mathbb{C}$, such that*

$$u(x, y) = \alpha R(z_0, \zeta_0, z, \bar{z}) + \int_{z_0}^{z} R(t, \zeta_0, z, \bar{z}) f(t)\, dt + \int_{\zeta_0}^{\bar{z}} R(z_0, \tau, z, \bar{z}) f^*(\tau)\, d\tau\,. \quad (7)$$

b$_2$) *Conversely, for arbitrary $\alpha \in \mathbb{C}, f \in H_D, f^* \in H_{\bar{D}}$,* (7) *represents solutions u of* (1), *defined in D.*

b$_3$) *For a given solution u of* (1), *$\alpha \in \mathbb{C}, f \in H_D$ and $f^* \in H_{\bar{D}}$ in* (7) *are uniquely determined by* (6), *where $w(z, \zeta) := U\big((z + \zeta)/2, (z - \zeta)/2\mathrm{i}\big)$ and $U(x, y)$ is the analytic continuation of $u(x, y)$ to complex values of x and y.*

Let us now consider the special equation (1)

$$\left.\begin{aligned}&\Delta u + C(x, y)\, u = 0\,,\\ &C(x, y) = \frac{4N(N+1)}{(1+x^2+y^2)^2} - \frac{4M(M+1)}{(1-x^2-y^2)^2}, \quad N, M \in \mathbb{N}_0, \quad \mathbb{N}_0 := \mathbb{N} \cup \{0\}.\end{aligned}\right\} \tag{8}$$

Thus $C(x, y)$ is a function analytic in $K_1 := \{(x, y) \in \mathbb{R}^2 : x^2 + y^2 < 1\}$. The complex equation (4), associated with (8) is now given by

$$w_{z\zeta} + c(z, \zeta)\, w = 0\,, \qquad c(z, \zeta) = \frac{N(N+1)}{(1+z\zeta)^2} - \frac{M(M+1)}{(1-z\zeta)^2}\,. \tag{9}$$

It is easy to see that K_1 is the largest fundamental disc of (8), containing the origin (0, 0).

The Riemann function of (9) is given by (see LANCKAU [1, p. 18—23]):

$$\left.\begin{aligned}&R(z, \zeta, t, \tau) = \sum_{k=0}^{N} \sum_{l=0}^{M} \alpha_{kl} U^k V^l\,, \qquad \alpha_{kl} = \frac{(-1)^{k+l}\,(N+k)!\,(M+l)!}{(N-k)!\,(M-l)!\,k!\,l!\,(k+l)!}\,,\\ &U(z, \zeta, t, \tau) = \frac{(z-t)(\zeta-\tau)}{(1+z\zeta)(1+t\tau)}\,, \qquad V(z, \zeta, t, \tau) = -\frac{(z-t)(\zeta-\tau)}{(1-z\zeta)(1-t\tau)}\,.\end{aligned}\right\} \tag{10}$$

If we put $z_0 = \zeta_0 = 0$, we get the following theorem by inserting (10) into Vekua's representation formula (5) and integrating by parts (see HEERSINK [2, p. 12—15]); H_{K_1} denotes the set of functions holomorphic in K_1; $H_{K_1}(j, z_0)$, $j \in \mathbb{N}$, denotes the set of functions $g(z) \in H_{K_1}$ with the property $g(z_0) = g'(z_0) = \ldots = g^{(j-1)}(z_0) = 0$:

Theorem 14.1.3. a) *For every solution $w \in H_{K_1 \times K_1}$ of (9) there exist two functions $g \in H_{K_1}(N + M, 0)$ and $g^* \in H_{K_1}(N + M, 0)$, such that w can be represented by*

$$w(z, \zeta) = Hg + H^*g^*\,, \tag{11}$$

where

$$\left.\begin{aligned}&(Hg)(z, \zeta) := \sum_{k=0}^{N} \sum_{l=0}^{M} A_{kl}^{NM} \frac{\zeta^{N+M-k-l}}{(1+z\zeta)^{N-k}(1-z\zeta)^{M-l}}\, g^{(k+l)}(z)\,,\\ &(H^*g^*)(z, \zeta) := \sum_{k=0}^{N} \sum_{l=0}^{M} A_{kl}^{NM} \frac{z^{N+M-k-l}}{(1+z\zeta)^{N-k}(1-z\zeta)^{M-l}}\, g^{*(k+l)}(\zeta)\,,\\ &A_{kl}^{NM} = (-1)^{N-k} \frac{(2N-k)!\,(2M-l)!}{k!\,l!\,(N-k)!\,(M-l)!}\,.\end{aligned}\right\} \tag{12}$$

b) *Conversely, for arbitrary $g, g^* \in H_{K_1}(N + M, 0)$, (11) represents solutions $w \in H_{K_1 \times K_1}$ of (9).*

c) *For a given solution w of (9), the functions g and g^* in the representation (11) are uniquely determined by*

$$g(z) = \int_0^z (z-t)^{N+M} f(t)\, dt + \frac{\alpha}{2} z^{N+M}\,,$$

$$g^*(\zeta) = \int_0^\zeta (\zeta-\tau)^{N+M} f^*(\tau)\, d\tau + \frac{\alpha}{2} \zeta^{N+M}\,,$$

where $\alpha = w(0, 0)$, $f(z) = w_z(z, 0)$, $f^*(\zeta) = w_\zeta(0, \zeta)$, if we require $g^{(N+M)}(0) = g^{*(N+M)}(0)$.

Remark 15.1.1. $Hg + H^*g^*$ represent solutions $w \in H_{K_1 \times K_1}$ of (9) even for all $g, g^* \in H_{K_1}$ (see HEERSINK [2, p. 15]).

The differential operators H and H^*, respectively, in the representation (11) are formally of the form $H = \sum_{i=0}^{N+M} a_i(z, \zeta)\, \partial^i/\partial z^i$, $H^* = \sum_{i=0}^{N+M} b_i(z, \zeta)\, \partial^i/\partial \zeta^i$, where $a_i, b_i \in H_{K_1 \times K_1}$, $i = 0, 1, \ldots, N + M$. Operators of this type for the representation of the solutions for equations that are different from (8) are quoted e.g. in BAUER [3–5], HEERSINK [1], JANK [1], and BAUER and FLORIAN [1]. In the second section of this chapter we will give a more detailed characterization of these differential operators.

14.2. Relations between integral and differential operators

We now consider the general equation (4) in D, where D always denotes a fundamental domain of equation (1). $z_0 \in D$, $\zeta_0 \in \overline{D}$ denote arbitrary, but fixed points, H_D, $H_{\overline{D}}$ and $H_{D \times \overline{D}}$ denote the sets of functions holomorphic in D, $\overline{D}$ and $D \times \overline{D}$, respectively, $H_D(j, z_0)$ and $H_{\overline{D}}(j, \zeta_0)$, $j \in \mathbb{N}$, denote the sets of functions $g \in H_D$ and $g^* \in H_{\overline{D}}$, respectively, with the properties $g(z_0) = g'(z_0) = \ldots = g^{(j-1)}(z_0) = 0$ and $g^*(\zeta_0) = g^{*\prime}(\zeta_0) = \ldots = g^{*(j-1)}(\zeta_0) = 0$, $R(z, \zeta, t, \tau)$ denotes the Riemann function of equation (4).

In order to get relations between Vekua's integral operators (introduced in theorem 14.1.2) and differential operators as defined in theorem 14.1.3 (for $\zeta = \bar{z}$ these differential operators have mainly been introduced by K. W. BAUER) we will start by giving the following

Definition 14.2.1. a) Integral operators I and I^* of the form

$$\left.\begin{aligned} (If)(z, \zeta) &:= \int_{z_0}^{z} R(t, \zeta_0, z, \zeta) f(t)\, dt, \qquad f \in H_D, \quad \text{and} \\ (I^*f^*)(z, \zeta) &:= \int_{\zeta_0}^{\zeta} R(z_0, \tau, z, \zeta) f^*(\tau)\, d\tau, \qquad f^* \in H_{\overline{D}} \end{aligned}\right\} \tag{13}$$

will be termed *I-operator* and *I^*-operator*, respectively, of equation (4).

b) Differential operators D_n and D_m^* of the form

$$\left.\begin{aligned} &D_n := \sum_{k=0}^{n} a_k(z, \zeta) \frac{\partial^k}{\partial z^k}, \quad n \in \mathbb{N}, \quad a_k \in H_{D \times \overline{D}} \ \text{ for } \ k = 0, 1, \ldots, n, \quad a_n \not\equiv 0, \\ &\text{and} \\ &D_m^* := \sum_{k=0}^{m} b_k(z, \zeta) \frac{\partial^k}{\partial \zeta^k}, \quad m \in \mathbb{N}, \quad b_k \in H_{D \times \overline{D}} \ \text{ for } \ k = 0, 1, \ldots, m, \quad b_m \not\equiv 0 \end{aligned}\right\} \tag{14}$$

will be termed *D_n-operator* and *D_m^*-operator*, respectively, of equation (4), if $D_n g$ and $D_m^* g^*$ for every $g \in H_D$ and $g^* \in H_{\overline{D}}$, respectively, represent solutions $w \in H_{D \times \overline{D}}$ of (4).

Since all solutions of equation (4) can be represented with the aid of I- and I^*-operators (theorem 14.1.2), the main question will be: What are the conditions on the D_n and D_m^*-operators for all solutions of the considered equation to be representable by $w = D_n g + D_m^* g^*$? In order to answer this question, it is advantageous to find relations between I- and I^*-operators and D_n- and D_m^*-operators. By way of example we prove

Lemma 14.2.1. *Let* $D_n = \sum_{k=0}^{n} a_k(z, \zeta)\, \partial^k/\partial z^k$ *be a given* D_n*-operator of equation* (4) *such that* $a_n(z, \zeta_0) \neq 0$ *for all* $z \in D$. *Then the following statements hold:*

a) *For every given function* $g \in H_D(n+1, z_0)$ *there exists a uniquely determined function* $f \in H_D$, *such that* $D_n g = If$. *This function* $f(z)$ *is given by*

$$f(z) = \sum_{k=0}^{n+1} g^{(k)}(z)\, [a_{k-1}(z, \zeta_0) + a_{k,z}(z, \zeta_0) + b(z, \zeta_0)\, a_k(z, \zeta_0)]\,, \tag{15}$$

where $a_{-1} := 0$, $a_{n+1} := 0$.

b) *Conversely, for every given function* $f \in H_D$ *there exists a uniquely determined function* $g \in H_D(n+1, z_0)$, *such that* $If = D_n g$. *This function* $g(z)$ *is the uniquely determined solution of the ordinary differential equation* (15) *with the initial conditions*

$$g(z_0) = g'(z_0) = \dots = g^{(n)}(z_0) = 0\,. \tag{16}$$

Proof. a) Every solution $w = D_n g$, $g \in H_D(n+1, z_0)$ of equation (4) can be represented (see theorem 14.1.2) by $w = \alpha R(z_0, \zeta_0, z, \zeta) + If + I^* f^*$, where α, f and f^* are given by (6). Since z_0 is a zero of order (at least) $n+1$ of g, it follows that $\alpha = 0$, $f^* \equiv 0$ and that $f(z)$ is given by (15), where the coefficients a_{-1} and a_{n+1} are defined by $a_{-1} := 0$ and $a_{n+1} := 0$, all other coefficients a_k, $k = 0, 1, \dots, n$, are determined by the form of the given D_n-operator, b is given by equation (4). Since $g \in H_D(n+1, z_0) \subset H_D$, $b(z, \zeta_0) \in H_D$ and $a_k(z, \zeta_0) \in H_D$, $k = 0, 1, \dots, n$, it is obvious that $f(z) \in H_D$.

b) It only remains to show that there exists one and only one solution $g \in H_D(n+1, z_0)$ of (15) for every $f \in H_D$. Since $a_n(z, \zeta_0) \neq 0$ in D, this results from the theory of initial value problems for ordinary differential equations with complex variables (see BIEBERBACH [1, p. 20]).

Analogous lemmas can be proved for the relations between the I^*-operator and the D_m^*-operator as well as for the relations between the representations $w = \alpha R(z_0, \zeta_0, z, \zeta)$ and $w = D_n g + D_m^* g^*$ (see HEERSINK [2, p. 19–21]). As a consequence we get the following general representation theorem for D_n-, D_m^*-operators (see HEERSINK [2, p. 22]):

Theorem 14.2.1. *Let* $D_n = \sum_{k=0}^{n} a_k(z, \zeta)\, \partial^k/\partial z^k$ *and* $D_m^* = \sum_{k=0}^{m} b_k(z, \zeta)\, \partial^k/\partial \zeta^k$ *be given* D_n*- and* D_m^**-operators, respectively, of equation* (4), *such that* $a_n(z, \zeta_0) \neq 0$ *for all* $z \in D$ *and* $b_m(z_0, \zeta) \neq 0$ *for all* $\zeta \in \overline{D}$. *Then, for every solution* $w \in H_{D \times \overline{D}}$ *of equation* (4) *there exists a uniquely determined pair of functions* $g \in H_D(n, z_0)$, $g^* \in H_{\overline{D}}(m, \zeta_0)$ *with the property* $a_n(z_0, \zeta_0)\, g^{(n)}(z_0) = b_m(z_0, \zeta_0)\, g^{*(m)}(\zeta_0)$, *such that*

$$w(z, \zeta) = D_n g + D_m^* g^*\,. \tag{17}$$

Remark 14.2.1. For a given solution w of (4), the pair of functions in (17) can be explicitly constructed. For the construction-algorithm see HEERSINK [2, p. 22].

Example 14.2.1. BAUER and FLORIAN [1] gave a representation of solutions of a certain differential equation, Theorem 14.2.1 enables us to state that *all* solutions of this equation can be represented in the given form. In BAUER and FLORIAN [1] the following equation is considered:

$$[\varphi(z) + \psi(\zeta)]^2 S_1 S_2 w + (n - n^*)\, [\varphi(z) + \psi(\zeta)]\, S_2 w - n(n^* + 1) w = 0\,,$$
$$n, n^* \in \mathbb{N}_0\,,$$
$$S_1 := \frac{1}{\varphi'(z)} \frac{\partial}{\partial z}, \quad S_2 := \frac{1}{\psi'(\zeta)} \frac{\partial}{\partial \zeta}, \qquad (\varphi + \varphi)\, \varphi' \psi' \neq 0 \quad \text{in } D \times \overline{D}\,.$$

It is shown that for all $g \in H_D$, $g^* \in H_{\bar{D}}$,

$$w = \sum_{k=0}^{n} \frac{(-1)^k (n + n^* - k)!}{k!(n-k)! [\varphi(z) + \psi(\zeta)]^{n-k}} S_k^1 g + \sum_{k=0}^{n^*} \frac{(-1)^k (n + n^* - k)!}{k!(n^* - k)! [\varphi(z) + \psi(\zeta)]^{n-k}} S_2^k g^*$$

represents solutions of this equation, holomorphic in $D \times \bar{D}$. If we arrange the right-hand side of this representation as the sum of two series

$$\sum_{k=0}^{n} a_k(z, \zeta)\, g^{(k)}(z) \quad \text{and} \quad \sum_{k=0}^{n^*} b_k(z, \zeta)\, g^{*(k)}(\zeta)\,,$$

we see that this is of the form $D_n g + D_{n^*}^* g^*$. The coefficients a_n and b_{n^*} of D_n and $D_{n^*}^*$, respectively, are of the form

$$a_n = \frac{(-1)^n n^*!}{n![\varphi'(z)]^n}, \quad b_n^* = \frac{(-1)^{n^*} n!}{n^*![\varphi(z) + \psi(\zeta)]^{n-n^*} [\psi'(\zeta)]^{n^*}},$$

so that $a_n(z, \zeta_0) \neq 0$ in D and $b_{n^*}(z_0, \zeta) \neq 0$ in $\bar{D}$. From theorem 14.2.1 it follows that the above representation gives all solutions of the equation, holomorphic in $D \times \bar{D}$, if $g \in H_D(n, z_0)$ and $g^* \in H_{\bar{D}}(n^*, \zeta_0)$.

In section 14.3 we will need the following theorem, concerning real solutions of equation (1). In the proof (see HEERSINK [2, p. 24, 25]) again relations between Vekua's I-operator and the D_n-operator are used.

Theorem 14.2.2. *Let D be a fundamental domain of* (1) *and let $C(x, y)$ be real in D and $A = B \equiv 0$ in D. Let $D_n = \sum_{k=0}^{n} a_k(z, \zeta)\, \partial^k/\partial z^k$ be a D_n-operator of the associated equation* (4), *such that $a_n(z, \zeta_0) \neq 0$ for all $z \in D$. Then the following statements hold:*

a) *For every solution $u(x, y)$ of equation* (1), *defined and real in D, there exists a function $g \in H_D(n, z_0)$, such that*

$$u(x, y) = \operatorname{Re}\{D_n g|_{\zeta=\bar{z}}\} = \operatorname{Re}\{\sum_{k=0}^{n} a_k(z, \bar{z})\, g^{(k)}(z)\}\,. \tag{18}$$

b) *For a given solution $u(x, y)$ of this kind of equation* (1), *the function g in* (18) *is uniquely determined by $u(x, y)$, if we require*

$$a_n(z_0, \bar{z}_0)\, g^{(n)}(z_0) = \overline{a_n(z_0, \bar{z}_0)\, g^{(n)}(z_0)}\,. \tag{19}$$

Remark 14.2.2. The problem of finding necessary and sufficient conditions for the existence of D_n- and D_m^*-operators for a given equation (4) is still unsolved. Only if the order n and m of the wanted operators D_n and D_m^*, respectively, is fixed, an answer to this question could be given in KRACHT [2] by using Bergman operators with polynomials as kernel functions. The following section will for the most part deal with a special case of equation (8) and (9), respectively, where the D_n- and D_m^*-operators are explicitly known.

14.3. Dirichlet problems for certain selfadjoint equations

In this section we will first consider certain selfadjoint equations (1) in fundamental domains D:

$$\Delta u + C(x, y)\, u = 0\,, \tag{20}$$

with real coefficients C, such that there exist D_n-operators $D_n = \sum_{k=0}^{n} a_k(z, \zeta)\, \partial^k/\partial z^k$ (with the property $a_n(z, \zeta_0) \neq 0$ for all $z \in D$ ($\zeta_0 \in \bar{D}$ fixed)) for the associated equation (4)

$$w_{z\zeta} + c(z, \zeta)\, w = 0\,. \tag{21}$$

$K_R := \{(x, y) \in \mathbb{R}^2 : x^2 + y^2 < R,\ R \in \mathbb{R}^+\}$, $K_R \subset D$, denotes the open disc of radius R and centre $(0, 0)$. $\bar{K}_R$ and ∂K_R denote the closed disc and the circle of radius R, respectively. H_{K_R} denotes the set of functions, holomorphic in K_R; in the same way as up to now we use the symbols $H_{K_R}(j, z_0)$, $j \in \mathbb{N}$, $z_0 \in K_R$, as well as $H_{K_R \times K_R}$.

$C^H(\partial K_R)$ denotes the set of functions, defined and Hölder-continuous on ∂K_R; $C^{j,H}(\partial K_R)$, $j \in \mathbb{N}$, denotes the set of functions $f \in C^H(\partial K_R)$ with Hölder-continuous derivatives up to order j.

In the present section we shall be concerned with solving the following Dirichlet problem:

Problem (D). Find a function $u(x, y) \in C^2(K_R) \cap C(\bar{K}_R)$, defined and real in $\bar{K}_R$, which is a solution of equation (20) in K_R and satisfies the boundary condition $u^+(\xi) := \lim_{z=x+iy\to\xi} u(x, y) = f(\xi)$, $\xi \in \partial K_R$, where $f \in C^H(\partial K_R)$ is a given Hölder-continuous real function on ∂K_R.

If we presume the existence of a solution of problem (D), this solution u is representable by

$$u(x, y) = \operatorname{Re}\{D_n g|_{\zeta=\bar{z}}\} = \operatorname{Re}\{\sum_{k=0}^{n} a_k(z, \bar{z})\, g^{(k)}(z)\}\,, \tag{22}$$

where the *generating function* $g \in H_{K_R}(n, 0)$ with the property $a_n(0, 0)\, g^{(n)}(0) = \overline{a_n(0, 0)\, g^{(n)}(0)}$ is uniquely determined by $u(x, y)$ (see theorem 14.2.2 with $z_0 = 0$, i.e. $z_0 \in K_R \subset D$).

So the task of solving problem (D) is reduced to the problem of determining this *generating function* $g(z)$ in (22).

By using the following facts:

a) $[g^{(k)}(\xi)]^+ \in C^{n-k,H}(\partial K_R)$ for $k = 0, 1, \dots, n$ $\quad (C^{0,H}(\partial K_R) := C^H(\partial K_R))$, $\xi \in \partial K_R$;

b) $[g^+(\xi)]^{(k)} = [g^{(k)}(\xi)]^+$ for $k = 1, 2, \dots, n$, $\quad \xi \in \partial K_R$,

where here and in the following we always use the notation $g^+(\xi) := \lim_{z\to\xi\in\partial K_R} g(z)$, we first get with (22) (for proofs see Heersink [2, p. 41–43]):

$$u^+(\xi) = f(\xi) = \operatorname{Re}\{\sum_{k=0}^{n} a_k(\xi)\, [g^{(k)}(\xi)]^+\}\,, \quad \text{where } a_k(\xi) := \lim_{z\to\xi\in\partial K_R} a_k(z, \bar{z})\,. \tag{23}$$

(23) is a so-called *generalized Riemann-Hilbert-Poincaré problem* (see Mushelišvili [2, p. 257]) for the unknown function $g(z)$. This problem is solved in general by using the concept of singular integral equations (see Gahov [3, p. 304], Mushelišvili [2, p. 257]). For special cases however, we will reduce this problem to a system of ordinary differential equations with periodic coefficients (which might be treated by the Floquet theory).

Therefore we write (23) in the form

$$f(\xi) = \operatorname{Re}\{\sum_{k=0}^{n} a_k(\xi)\, [g^+(\xi)]^{(k)}\}\,. \tag{24}$$

Since $g \in H_{K_R}(n, 0)$, we can represent $g(z)$ by a Cauchy integral of the form

$$g(z) = \frac{1}{2\pi i} \oint_{K_R} \frac{\tau^n \Phi(\tau)}{\tau - z} d\tau , \tag{25}$$

where $g^+(\xi) = \xi^n \Phi(\xi)$ and hence $\Phi(\xi) \in C^{n,H}(\partial K_R)$. So we get from (24)

$$f(\xi) = \operatorname{Re} \{ \sum_{j=0}^{n} \Phi^{(j)}(\xi)\, \xi^j S_j(\xi) \} , \tag{26}$$

where

$$S_j(\xi) := \sum_{k=j}^{n} \xi^{n-k} a_k(\xi) \binom{k}{j} \frac{n!}{(n-k+j)!} . \tag{27}$$

By using the parametric representation $\xi = R\, e^{it}$, $t \in [0, 2\pi]$, (and denoting $f(\xi(t))$ again by the symbol $f(t)$) as well as

$$\Phi(t) =: \Phi_1(t) + i\Phi_2(t) , \tag{28}$$

we finally get from (26):

$$f(t) = \sum_{k=0}^{n} [A_k(t)\, \Phi_1^{(k)}(t) + B_k(t)\, \Phi_2^{(k)}(t)] , \tag{29}$$

where

$$\left.\begin{aligned} &A_k(t) := \operatorname{Re} \hat{A}_k(t) , \qquad B_k(t) := -\operatorname{Im} \hat{A}_k(t) , \\ &\hat{A}_0(t) := S_0(t) \quad \text{and} \quad \hat{A}_j(t) := \sum_{k=j}^{n} \alpha_{k,j} S_k(t) \quad \text{for} \quad j = 1, 2, \dots, n , \\ &\alpha_{k,l} := (-1)^k (i)^l a_{k,l} , \quad \text{where} \quad a_{0,0} := 1,\ a_{0,1} := 0 \text{ and } a_{k,l} := a_{k-1,l-1} \\ &\text{for } k = 1, 2, \dots, n; \quad l = 1, 2, \dots, n . \end{aligned}\right\} \tag{30}$$

For the following considerations we will need

Definition 14.3.1. $\psi_1(\xi) \in C^H(\partial K_R)$ will be termed the *conjugate function* of $\psi_2(\xi) \in C^H(\partial K_R)$ and for this we will use the notation: $\psi_1 \overset{k}{\sim} \psi_2$, if ψ_1 and ψ_2 are the boundary values of two functions, conjugate harmonic in K_R, i.e. if there exists a function $G(z) \in H_{K_R}$, such that $G^+(\xi) \in C^H(\partial K_R)$ and $\psi_1(\xi) = [\operatorname{Re} G(\xi)]^+$ as well as $\psi_2(\xi) = [\operatorname{Im} G(\xi)]^+$.

This enables us to formulate

Theorem 14.3.1. a) *Let* $(\Phi_1, \Phi_2) \in C^{n,H}(\partial K_R) \times C^{n,H}(\partial K_R)$ *be a solution of*

$$\left.\begin{aligned} &f(t) = \sum_{k=0}^{n} [A_k(t)\, \Phi_1^{(k)}(t) + B_k(t)\, \Phi_2^{(k)}(t)] , \\ &\Phi_1 \overset{k}{\sim} \Phi_2 , \end{aligned}\right\} \tag{31}$$

then (22) *represents a solution of problem* (D), *if* $g(z)$ *in* (22) *is determined by* (25) *and* $\Phi(\xi)$ *in* (25) *is given by* (28).

b) *Conversely, let* $u(x, y)$ *be a solution of problem* (D) *with the representation* (22), *then* Φ_1 *and* Φ_2 (*in* (25), (28)) *are necessarily solutions of* (31).

Thus (31) and problem (D) are equivalent.

For our further investigations we shall put forward the following hypothesis:

Hypothesis 14.3.1. *Let the given equation* (20) *and the associated* D_n*-operator be of such a form that all coefficients* A_k, B_k, $k = 0, 1, \dots, n$, *in* (30) *and* (31), *respectively, do not depend on the variable* t, *i.e. all* A_k *and* B_k *are constants.*

Let us remark that hypothesis 14.3.1 is fulfilled for equation (20) with $C(x, y)$ given in (8).

By using the following obvious statements

$$\left.\begin{aligned}&(\Phi_1 \overset{k}{\sim} \Phi_2) \Leftrightarrow (c\Phi_1 \overset{k}{\sim} c\Phi_2)\,, \qquad c \in \mathbb{C};\\ &(\Phi_1 \overset{k}{\sim} \Phi_2) \Leftrightarrow (\Phi_2 \overset{k}{\sim} -\Phi_1);\\ &(\Phi_1 \overset{k}{\sim} \Phi_2) \wedge \Phi_1 \in C^{n,H}(\partial K)\,, \quad n \in \mathbb{N} \Leftrightarrow \Phi_2 \in C^{n,H}(\partial K_R) \wedge \left(\Phi_1^{(j)}(t) \overset{k}{\sim} \Phi_2^{(j)}(t)\right),\\ &j = 1, 2, \ldots, n;\end{aligned}\right\} \tag{31a}$$

we deduce from theorem 14.3.1:

Theorem 14.3.2. *Let hypothesis* 14.3.1 *be fulfilled. Then problem* (31) *and thus problem* (D) *are equivalent to problem* (32):

$$\left.\begin{aligned}&f(t) = \sum_{k=0}^{n} [A_k\Phi_1^{(k)}(t) + B_k\Phi_2^{(k)}(t)]\,,\\ &\tilde{f}(t) = \sum_{k=0}^{n} [A_k\Phi_2^{(k)}(t) - B_k\Phi_1^{(k)}(t)]\,,\\ &\textit{where } \Phi_1 \overset{k}{\sim} \Phi_2 \quad \textit{and } \tilde{f} \textit{ is defined by } f \overset{k}{\sim} \tilde{f}.\end{aligned}\right\} \tag{32}$$

Remark 14.3.1. $\tilde{f}$ in (32) is determined by f only up to a real constant. This constant enters in the generating function $g(z)$, but will be fixed by condition (19).

Let us now consider equation (20) with $C(x, y)$ given by (8), i.e.

$$\Delta u + \left[\frac{4N(N+1)}{(1 + x^2 + y^2)^2} - \frac{4M(M+1)}{(1 - x^2 - y^2)^2}\right] u = 0\,, \qquad N, M \in \mathbb{N}\,. \tag{33}$$

As already mentioned, the unit disc K_1 is the largest fundamental disc, containing the origin (0, 0). Thus problem (D) can only be formulated for discs K_R with $R < 1$. As it is well-known, for $M \geqq N$ (i.e. $C(x, y) \leqq 0$ in K_R) problem (D) is uniquely solvable for every boundary function $f \in C^H(\partial K_R)$, so that we deduce

Theorem 14.3.3. *Let the radius R of the disc K_R be $R < 1$ and let $M \geqq N$ in equation* (33). *The D_n-operator of order $n = N + M$ in the representation* (22) *is given by $D_n := H$, H given in* (12):

a) *Problem* (D) *for equation* (33) *is uniquely solvable for every $f \in C^H(\partial K_R)$.*

b) *Problem* (D) *for equation* (33) *is equivalent to problem* (32), *where the coefficients A_k, B_k are constants given by* (30).

c) *Let (Φ_1, Φ_2) be the solution of* (32); *then the solution $u(x, y)$ of problem* (D) *is given by* (22) (*with $g(z)$ given by* (25) *and $\Phi(t)$ given by* (28)).

Theorem 14.3.3 is particularly true for $N = 0$ in (33). In the case $M = 0$, $N \neq 0$ in (33) however, the situation is different; in general neither the existence nor the uniqueness of solutions of problem (D) is ensured. This case was investigated in Heersink [2, p. 50–59]. Further information was given in Bauer [2] and Bauer and Ruscheweyh [1, p. 205–210], where partly more elegant methods were used. The reason why we nevertheless consider this problem in the way as pointed out is the fact that these methods (using (32)) will allow some interesting statements, concerning the boundary behaviour of the solutions of the equations

$$\Delta u + \frac{4n(n+1)}{(1 + x^2 + y^2)^2} u = 0\,, \qquad n \in \mathbb{N}\,, \tag{34}$$

and the equivalent complex form

$$w_{z\bar{z}} + \frac{n(n+1)}{(1+z\bar{z})^2} w = 0\,, \tag{35}$$

respectively. These investigations will be described in 14.4.

For the sake of completeness let us give here some more information on the solution of problem (D) for equation (34). The representation of the solutions (22) has now the form

$$u(x, y) = \operatorname{Re}\{X_n g|_{\zeta=\bar{z}}\} = \operatorname{Re}\{\hat{X}_n g\}\,, \tag{36}$$

where

$$\hat{X}_n := \sum_{k=0}^{n} A_k^n \left(\frac{\bar{z}}{1+z\bar{z}}\right)^{n-k} \frac{\partial^k}{\partial z^k}, \qquad A_k^n := (-1)^{n-k} \frac{(2n-k)!}{k!(n-k)!}. \tag{37}$$

(We get this result, if we put $N = n$ and $M = 0$ in the D_n-operator H in (12).)

Since hypothesis 14.3.1 is fulfilled for equation (34), theorem 14.3.2 remains true for the associated problem (D). At this point it must be stated explicitly that the representation (36) for equation (34) is not only valid in fundamental domains (i.e. in subsets of K_1), but in *arbitrary subdomains of the whole complex plane* $\mathbb{C}$ (see BAUER [1,] HEERSINK [2, p. 52–55]); particularly all statements concerning problem (D) for equation (34) are now true for *arbitrary radii* $R \in \mathbb{R}^+$.

In BAUER and RUSCHEWEYH [1, p. 206] it is proved that problem (D) for equation (34) is uniquely solvable, if the radius R of the disc K_R satisfies the condition:

$$P_n^{(k)}\left(\frac{1-R^2}{1+R^2}\right) \neq 0 \quad \text{for} \quad k = 0, 1, \ldots, n-1 \tag{38}$$

($P_n(x) :=$ Legendre polynomial of degree n). Thus for such radii R (32) is uniquely solvable and the solution $u(x, y)$ of the problem (D) for equation (34) might be constructed in the same way as it was formulated in theorem 14.3.3 for equation (33). If R is a so-called *exceptional radius* (i.e. if (38) is not satisfied), problem (D) for equation (34) and the system (32), respectively, are either unsolvable or the solution is not unique. Let us conclude by quoting a theorem of BAUER and RUSCHEWEYH [1, p. 208], the statements of which are more smooth than the corresponding ones in HEERSINK [2, p. 55–58]:

Theorem 14.3.4 (RUSCHEWEYH). a) *Let the radius R of the disc K_R satisfy condition* (38). *Then for every continuous boundary function $f(\xi) \in C(\partial K_R)$ the uniquely determined solution $u(x, y)$ of problem* (D) *for equation* (34) *is given by*

$$\left.\begin{aligned} &u(x, y) = \frac{1}{2\pi} \int_0^{2\pi} f(R\,e^{it})\, k_R(\sigma - t, r)\, dt\,, \qquad x + iy = r\,e^{i\sigma}\,, \\ &\textit{where} \\ &k_R(t, r) := \hat{a}_0(r) + 2 \sum_{k=1}^{\infty} \hat{a}_k(r) \cos(kt)\,, \end{aligned}\right\} \tag{39}$$

where $\hat{a}_k(r) := s_n(r^k)/s_n(R^k)$, $s_n(\varrho^k) := \left(\frac{\varrho}{1+\varrho^2}\right)^k \binom{k+n}{n} {}_2F_1\left(k+n+1, k-n; k+1; \frac{\varrho^2}{1+\varrho^2}\right)$. ($s_n(R^k) \neq 0$, *if* (38) *is true*!)

b) *Let* $R > 0$ *and for* $s \in \mathbb{N}$, $j = 1, 2, \dots, s$, *let us assume* $P_n^{(k_j)}(1 - R^2/1 + R^2) = 0$;

b₁) *Let* $f \sim \operatorname{Re} \sum_{k=0}^{\infty} \alpha_k e^{ikt}$, $\alpha_0 \in \mathbb{R}$ *be the Fourier expansion of the boundary function* f *and let us assume*

$$\alpha_{k_j} = 0 \quad \text{for} \quad j = 1, 2, \dots, s \,. \tag{40}$$

Then the function $h(z) := \operatorname{Re} \sum_{\substack{k=0\\k \neq k_j}}^{\infty} (\alpha_k / s_n(R^k))\, z^k$ *is harmonic in* K_R *and the general solution* u *of problem* (D) *of equation* (34) *is given by*

$$u = T_n h + \sum_{j=1}^{s} \operatorname{Re} \{c_j T_n(z^{k_j})\}\,, \qquad c_j \in \mathbb{C}\,,$$

where (41)

$$(T_n h)(z) := \frac{(-1)^n}{(n!)^2} (1 + z\bar{z})^{n+1} \Delta^n \left(\frac{h(z)}{1 + z\bar{z}} \right).$$

b₂) *If condition* (40) *is not fulfilled, problem* (D) *of equation* (34) *is not solvable.*

14.4. On the boundary behaviour of a certain class of generalized holomorphic functions

Let us now perform some more detailed investigations, concerning the behaviour of the solutions $w = \hat{X}_n g$ of equation (35) on the boundary ∂K_R of the disc K_R, $R \in \mathbb{R}^+$. These solutions $w = \hat{X}_n g$ ($g \in H_{K_R}$ or $g \in H_{K_R}(n, 0)$, $\hat{X}_n$ given by (37)) are termed e.g. in Ruscheweyh [1, 2] *generalized holomorphic functions* (g.h. functions), since these and holomorphic functions have many properties in common (see also Bauer [1], Ruscheweyh [3], Heersink [2]. We therefore give the following

Definition 14.4.1. GH_{K_R} and $GH_{K_R}(l, 0)$, $l \in \mathbb{N}$, denote the sets of functions, defined by

$$GH_{K_R} := \{w(z, \bar{z}) : w = \hat{X}_n g, \text{ where } g \in H_{K_R}\}\,,$$

$$GH_{K_R}(l, 0) := \{w(z, \bar{z}) : w = \hat{X}_n g, \text{ where } g \in H_{K_R}(l, 0), l \in \mathbb{N}\}\,.$$

By using the results of 14.3., it can be proved (see Heersink [2, p. 60, 61]) that the following two lemmas hold (the first of them is a generalization of a theorem of Privalov); for a solution $w(z, \bar{z})$ of (35), defined in K_R, we use the notation $w^+(\xi, \bar{\xi}) := \lim_{z \to \xi \in \partial K_R} w(z, \bar{z})$:

Lemma 14.4.1. *Let* $w(z, \bar{z}) \in GH_{K_R}$ *be a g.h. function, such that* $[\operatorname{Re} w(\xi, \bar{\xi})]^+ \in C^H(\partial K_R)$. *Then* $[\operatorname{Im} w(\xi, \bar{\xi})]^+ \in C^H(\partial K_R)$.

Lemma 14.4.2. *Let the radius* R *of the disc* K_R *satisfy condition* (38); *let* $w(z, \bar{z})$ *be a solution of equation* (35), *such that*

$$[\operatorname{Re} w(\xi, \bar{\xi})]^+ = f(\xi) \in C^H(\partial K_R) \quad \text{and} \quad [\operatorname{Im} w(\xi, \bar{\xi})]^+ = F(\xi) \in C^H(\partial K_R)\,. \tag{42}$$

Then there does not exist any further solution of (35) *in* K_R *with the property* (42).

Now we are in the position to prove

Theorem 14.4.1. a) *Let* $w(z, \bar{z}) \in GH_{K_R}(n, 0)$ *be a given g.h. function, such that* $[\operatorname{Re} w(\xi, \bar{\xi})]^+ = f(\xi) \in C^H(\partial K_R)$ *and* $[\operatorname{Im} w(\xi, \bar{\xi})]^+ = F(\xi) \in C^H(\partial K_R)$. *Then there exists* *a uniquely determined function* $G(z) \in H_{K_R}$, *such that* $[\operatorname{Re} G(\xi)]^+ = f(\xi)$ *and* $[\operatorname{Im} G(\xi)]^+ = F(\xi)$.

b) *Let the radius* R *of the disc* K_R *satisfy condition* (38). *Then the inversion of statement* a) *is true, i.e.:*

Let $G(z) \in H_{K_R}$ *be a given holomorphic function, such that* $[\operatorname{Re} G(\xi)]^+ = f(\xi) \in C^H(\partial K_R)$ *and* $[\operatorname{Im} G(\xi)]^+ = F(\xi) \in C^H(\partial K_R)$; *then there exists a uniquely determined g.h. function* $w(z, \bar z) \in GH_{K_R}$, *such that* $[\operatorname{Re} w(\xi, \bar\xi)]^+ = f(\xi)$ *and* $[\operatorname{Im} w(\xi, \bar\xi)]^+ = F(\xi)$. *This g.h. function belongs to* $GH_{K_R}(n, 0)$, *i.e.* $w(z, \bar z) \in GH_{K_R}(n, 0)$.

Proof. a) From $f(\xi) = [\operatorname{Re} w(\xi, \bar\xi)]^+$ we get by using definition 14.4.1 and the considerations of 14.3. (see (22)—(30))

$$f(t) = \sum_{k=0}^{n} [A_k \Phi_1^{(k)}(t) + B_k \Phi_2^{(k)}(t)]\,, \tag{43}$$

where A_k, B_k are constants, given by (30) (and the operator $\hat X_n$ in (37)); Φ_1 and Φ_2 with the property $\Phi_1 \overset{k}{\sim} \Phi_2$ are also defined as in 14.3. ($\xi = R\,e^{it}$).

In an analogous manner it follows from $F(\xi) = [\operatorname{Im} w(\xi, \bar\xi)]^+$:

$$F(t) = \sum_{k=0}^{n} [A_k \Phi_2^{(k)}(t) - B_k \Phi_1^{(k)}(t)]\,. \tag{44}$$

The relations (31a) now show the interesting fact that the right-hand sides in (43) and (44), respectively, are conjugated functions in the sense of definition 14.3.1, i.e. $f \overset{k}{\sim} F$.

b) According to the theorems 14.3.2 and 14.3.4, ((43) and (44)) with $\Phi_1 \overset{k}{\sim} \Phi_2$ and $f \overset{k}{\sim} F$ has a uniquely determined solution $\Phi_1, \Phi_2 \in C^{n,H}(\partial K_R)$; if we construct a g.h. function $w \in GH_{K_R}(n, 0)$ by $w(z, \bar z) = \hat X_n g$, where g is given by (25) (and $\Phi = \Phi_1 + i\Phi_2$), we know for this function from 14.3. that $[\operatorname{Re} w(\xi, \bar\xi)]^+ = f(\xi)$. On the other hand it follows again $[\operatorname{Im} w(\xi, \bar\xi)]^+ = \sum_{k=0}^{n} [A_k \Phi_2^{(k)}(t) - B_k \Phi_1^{(k)}(t)]$; this proves part b) of the theorem, if we observe lemma 14.4.2.

Summarizing the statements of theorem 14.4.1 we can say that for every g.h. function $w \in GH_{K_R}(n, 0)$ there exists one and only one holomorphic function $G \in H_{K_R}$, which has the same boundary values as w, and vice versa.

This enables us to deal with boundary value problems for g.h. functions simply by solving the corresponding problems for holomorphic functions. (We just have to solve the boundary value problem for holomorphic functions and after that we evaluate the associated *density function* $\Phi = \Phi_1 + i\Phi_2$ in the *generating function* $g(z)$ by solving (32).)

This principle was used (see Heersink [2, p. 65—68]) in connection with Riemann boundary value problems for g.h. functions and for the corresponding singular integral equations (see Heersink [2, p. 69—74]). In Florian and Heersink [1] we considered the Hilbert boundary value problem for these g.h. functions.

14.5. Some applications in relation to Dirichlet problems and variational problems

This final section deals with relations between Dirichlet problems for certain elliptic equations and associated variational problems.

In 14.5.1. we define the considered problems in their general form in $\mathbb{R}^n$. Part 14.5.2 is devoted to a classical example of J. HADAMARD for $\mathbb{R}^2$: The Dirichlet problem for the unit disc K_1 and $\Delta_2 u = 0$ for a given continuous (but not Hölder-continuous) boundary function; the (classical C^∞-)solution is explicitely known, but cannot be constructed with the help of the associated variational problem. This is a consequence of the fact that for the given solution u, $u_x, u_y \in C^\infty(K_1)$, but $u_x, u_y \notin \mathcal{L}^2(K_1)$, so that the so-called *energy integral* does not converge. From this we can derive the further conclusion that those boundary value problems which can be solved by means of Sobolev space methods do not include all classical problems. An elastic membrane allows a "physical interpretation" of this result.

Part 14.5.3. generalizes the conclusions of 14.5.2. on to a more general elliptic equation in $\mathbb{R}^2$, namely the equation (34). The proof essentially depends on the function theoretic methods of the preceding sections.

Part 14,5.4. is devoted to the equation $\Delta_3 u = 0$ for a hemisphere. Again we choose a continuous, but not Hölder-continuous boundary function. Unlike the considered problems in $\mathbb{R}^2$ the variational formulation now shows to be applicable to the construction of the solution. Finally we are giving a "physical interpretation" by means of an electrostatic field.

14.5.1. Many problems of mathematical physics may be described by partial differential equations of second order. Stationary problems frequently are related to boundary value problems for elliptic equations; one of the basic problems is the Dirichlet problem (again denoted by problem (D)):

Problem (D). (Let $D \subset \mathbb{R}^n$ be a domain; ∂D and $\bar{D} := D \cup \partial D$ denote the boundary and the closure of D, respectively.)

Find a function $u \in C^2(D) \cap C(\bar{D})$, defined and real on $\bar{D}$, which is a solution of $Eu = f$ in D (E elliptic differential operator, $f: D \to \mathbb{R}$) and satisfies the boundary condition: $u|_{\partial D} = g$, where $g: \partial D \to \mathbb{R}$, $g \in C(\partial D)$.

Remark 14.5.1. It is well-known that this classical Dirichlet problem is uniquely solvable for great classes of domains (for so-called *regular* domains) and operators E (particularly for Δ_2 and Δ_3).

Thus mathematicians and even physicists sometimes forget that this mathematical formulation (problem (D)) of a physical problem generally is deduced from a more original problem. Since very often $Eu = f$ in problem (D) is the so-called Euler equation of a minimum problem, namely the variational problem (problem (V)), which follows from the Hamiltonian principle:

Problem (V). Find a function $u \in V_g$, $V_g := \{v: \bar{D} \to \mathbb{R} : v \in C^2(D) \cap C(\bar{D}), v|_{\partial D} = g\}$ — g defined as in problem (D) —, which minimizes the Dirichlet integral $DI[v] := \int_D F(v, \operatorname{grad} v, x)\,dx$, $x \in D$ (i.e. find a function $u \in V_g$ such that $DI[u] \in \mathbb{R}$ and $DI[u] = \inf_{v \in V_g} DI[v] \overset{!}{=} \min_{v \in V_g} DI[v]$; here F denotes the so-called Lagrange density with the associated Euler equation $Eu = f$.)

14.5.2

Remark 14.5.2. Under certain conditions problem (D) and problem (V) are equivalent, but in general this is not true; since the Euler equation is a necessary, but generally not sufficient condition for an extremum of the functional $DI[v]$.

Even if problem (D) is uniquely solvable, the solution does not necessarily solve problem (V). One of the best-known counterexamples was given by J. HADAMARD:

Problem (D_{Δ_2}). (Let $K_1 := \{(\xi, \eta) \in \mathbb{R}^2 : \xi^2 + \eta^2 < 1\}$ be the open unit disc, ∂K_1 and $\bar{K}_1$ denote the boundary and the closure of K_1, respectively, (ϱ, θ) denote polar coordinates: $\xi = \varrho \cos \theta$, $\eta = \varrho \sin \theta$.)

Find a function $u \in C^2(K_1) \cap C(\bar{K}_1)$, defined and real on $\bar{K}_1$, which is a solution of $\Delta_2 u = u_{\xi\xi} + u_{\eta\eta} = 0$ in K_1 and satisfies the boundary condition $u|_{\partial K_1} = g$, where

$$g \in C([0, 2\pi]) , \qquad g(\theta) = \sum_{n=1}^{\infty} \frac{1}{n^2} \cos (n!\theta) . \tag{45}$$

$\Delta_2 u = 0$ is the Euler equation of the functional $\iint\limits_{K_1} (v_\xi^2 + v_\eta^2) \, d\xi \, d\eta$, i.e. the associated variational problem reads:

Problem (V_{Δ_2}). Find a function $u \in V_g$, $V_g := \{v : \bar{K}_1 \to \mathbb{R} : v \in C^2(K_1) \cap C(\bar{K}_1), v|_{\partial K_1} = g\}$ — g defined as in problem (D_{Δ_2}) —, which minimizes the Dirichlet integral

$$ID_{\Delta_2}[v] := \iint\limits_{K_1} (v_\xi^2 + v_\eta^2) \, d\xi \, d\eta.$$

The function u in problems (D_{Δ_2}) and (V_{Δ_2}) may be physically interpreted as the displacement of an elastic membrane; $DI_{\Delta_2}[u]$ then represents the energy of the membrane and $DI_{\Delta_2}[v]$ therefore often is denoted as *energy integral*. According to the statements of the Hamiltonian principle, the given boundary function g will displace the membrane in such a way that the energy integral becomes a minimum. Since every solution of problem (V_{Δ_2}) is a solution of (D_{Δ_2}) (remark 14.5.2) and since problem (D_{Δ_2}) is uniquely solvable (remark 14.5.1), we can conclude:

a) Either the solution of (D_{Δ_2}) is the (only) solution of (V_{Δ_2}) and represents the displacement of the membrane or

b) the solution of (D_{Δ_2}) does not solve (V_{Δ_2}); this would imply that (V_{Δ_2}) is unsolvable and that the solution of (D_{Δ_2}) cannot be physically realized by some "displaced membrane".

It was shown by J. HADAMARD that b) is true:

Theorem 14.5.1 (HADAMARD). a) *The solution of problem* (D_{Δ_2}) *is given by*

$$u(\varrho, \theta) = \sum_{n=1}^{\infty} \frac{1}{n^2} \cos (n!\theta) \, \varrho^{n!} \qquad (\xi = \varrho \cos \theta, \eta = \varrho \sin \theta) . \tag{46}$$

b) *For this function u the Dirichlet integral $DI_{\Delta_2}[u]$ does not converge.*

Remark 14.5.1. Thus continuity of the boundary function does not guarantee the existence of the energy integral $DI_{\Delta_2}[u]$ (where u denotes the solution of problem (D_{Δ_2})).

Moreover it can be shown that $\hat{V}_g := \{v \in V_g : DI_{\Delta_2}[v] < \infty\} = \emptyset$, i.e. the set of competitive functions in problem (V_{Δ_2}) is actually empty (see RAGGER [1, p. 10]). The problem of finding significant conditions on the boundary function seems to be unsolved. A necessary condition for $\hat{V}_g$ not to be the empty set is e.g. $g \in C^1(\partial K_1)$. The function g in (45) does not satisfy this condition and is not even an element of $C^H(\partial K_1)$, the set of Hölder-continuous functions.

14.5.3. We now generalize Hadamard's example on to equation (34), the equation of generalized holomorphic functions (g.h. functions):

Problem ($\mathrm{D_{gh}}$). (K_1, ∂K_1 and $\bar{K}_1$ as well as (ϱ, θ) are defined as in problem D_{Δ_2}.)

Find a function $w \in C^2(K_1) \cap C(\bar{K}_1)$, defined and real on $\bar{K}_1$, which in K_1 is a solution of (34) with $n = 1$, i.e. of

$$w_{\xi\xi} + w_{\eta\eta} + \frac{8}{(1 + \xi^2 + \eta^2)^2} w = 0 \tag{47}$$

and satifies the boundary condition $w|_{\partial K_1} = h$, where

$$\left.\begin{aligned} h \in C([0,2\pi]), \quad h(\theta) = &- \cos\theta \sum_{n=1}^{\infty} \frac{1}{n^2(n!+1)} \cos[(n!+1)\theta] \\ &- \sin\theta \sum_{n=1}^{\infty} \frac{1}{n^2(n!+1)} \sin[(n!+1)\theta] \\ &+ \sum_{n=1}^{\infty} \frac{1}{n^2} \cos(n!\theta) . \end{aligned}\right\} \tag{48}$$

Equation (47) is the Euler equation of the functional $DI_{\mathrm{gh}}[v]$, given in the following

Problem ($\mathrm{V_{gh}}$). Find a function $w \in V_h$, $V_h := \{v : \bar{K}_1 \to \mathbb{R} : v \in C^2(K_1) \cap C(\bar{K}_1),$ $v|_{\partial K_1} = h\}$ — h defined as in problem ($\mathrm{D_{gh}}$) —, which minimizes the Dirichlet integral

$$DI_{\mathrm{gh}}[v] := \iint_{K_1} \left(v_\xi^2 + v_\eta^2 - \frac{8}{(1+\xi^2+\eta^2)^2} v^2 \right) \mathrm{d}\xi \, \mathrm{d}\eta .$$

Remark 14.5.4. The attentive reader will notice that problem ($\mathrm{D_{gh}}$) for the unit disc K_1 is not uniquely solvable ($R = 1$ is an "exceptional radius", i.e. $R = 1$ does not satisfy condition (38), see 14.3.). We could formulate problem ($\mathrm{D_{gh}}$) for a radius $R < 1$, where ($\mathrm{D_{gh}}$) always is uniquely solvable. But this would only complicate the following formulations and proofs without changing the main results.

The following theorem shows that the continuity of the boundary function is again not sufficient therefore that problem ($\mathrm{D_{gh}}$) might be solved via problem ($\mathrm{V_{gh}}$):

Theorem 14.5.2. a) *A solution of problem* ($\mathrm{D_{gh}}$) *is given by*

$$\left.\begin{aligned} w(\varrho, \theta) = &- \frac{2}{1+\varrho^2} \left\{ \varrho \cos\theta \sum_{n=1}^{\infty} \frac{1}{n^2(n!+1)} \cos[(n!+1)\theta] \varrho^{n!+1} \right. \\ &\left. + \varrho \sin\theta \sum_{n=1}^{\infty} \frac{1}{n^2(n!+1)} \sin[(n!+1)\theta] \varrho^{n!+1} \right\} \\ &+ \sum_{n=1}^{\infty} \frac{1}{n^2} \cos(n!\theta) \varrho^{n!} . \end{aligned}\right\} \tag{49}$$

b) *For this function the Dirichlet integral $DI_{\mathrm{gh}}[w]$ does not converge.*

Proof. a) We will use the function theoretic methods as described in the preceding sections: From (36) and (37) we know that

$$w(z, \bar{z}) := \operatorname{Re}\left\{ -2 \frac{\bar{z}}{1 + z\bar{z}} F(z) + F'(z) \right\}, \quad z := \xi + \mathrm{i}\eta, \quad \bar{z} := \xi - \mathrm{i}\eta, \tag{50}$$

for all holomorphic functions $F \in H_{K_1}$ represents real solutions of equation (47) in K_1. If we define functions U, V, u, v by $U(z) + \mathrm{i}V(z) := F(z)$, $u(z) + \mathrm{i}v(z) := f(z)$

$:= F'(z)$, (50) becomes

$$w(\xi, \eta) = -\frac{2}{1 + \xi^2 + \eta^2}(\xi U + \eta V) + u\,. \tag{51}$$

Let us choose for $f(z) = u(z) + iv(z)$ the function

$$f(z) = \sum_{n=1}^{\infty} \frac{1}{n^2} z^{n!} = \sum_{n=1}^{\infty} \frac{1}{n^2} (\cos n!\theta + \sin n!\theta)\, \varrho^{n!}\,, \qquad z = \varrho\, e^{i\theta};$$

(the radius of convergence of this series is $r = 1$). Thus we get in K_1

$$F(z) = \int f(z)\, dz = \sum_{n=1}^{\infty} \frac{1}{n^2(n!+1)} z^{n!+1}\,,$$

$$u(z) = u(\varrho, \theta) = \sum_{n=1}^{\infty} \frac{1}{n^2} \cos(n!\theta)\, \varrho^{n!}\,, \qquad v(\varrho, \theta) = \sum_{n=1}^{\infty} \frac{1}{n^2} \sin(n!\theta)\, \varrho^{n!};$$

$$U(\varrho, \theta) = \sum_{n=1}^{\infty} \frac{1}{n^2(n!+1)} \cos[(n!+1)\,\theta]\, \varrho^{n!+1}\,,$$

$$V(\varrho, \theta) = \sum_{n=1}^{\infty} \frac{1}{n^2(n!+1)} \sin[(n!+1)\,\theta]\, \varrho^{n!+1}\,.$$

All these series for u, v and thus for f as well as for U, V and thus for F do not only converge in K_1, but even converge uniformly on $\bar{K}_1$, so that u, v, U, V, f and F are continuous (extended) on $\bar{K}_1$. Note that $u(\varrho, \theta)$ for $\varrho = 1$ represents the boundary function g in Hadamard's problem (D_{Δ_2}).

If we put U, V and u into (51), we get the function w in (49) in theorem 14.5.2. By using the just mentioned continuity considerations, we see that this solution (51) of equation (47) takes the boundary values $h(\theta)$ on ∂K_1, as given in (48). This proves part a) of the theorem.

b) From (51) we derive $w_\xi = w_1(\xi, \eta) + u_\xi$, $w_\eta = w_2(\xi, \eta) + u_\eta$, where

$$w_1(\xi, \eta) := -\frac{4\xi}{(1 + \xi^2 + \eta^2)^2}(\xi U + \eta V) - \frac{2}{(1 + \xi^2 + \eta^2)}(U + \xi u + \eta v)\,,$$

$$w_2(\xi, \eta) := -\frac{4\eta}{(1 + \xi^2 + \eta^2)^2}(\xi U + \eta V) - \frac{2}{(1 + \xi^2 + \eta^2)}(V + \eta u - \xi v)\,.$$

Since u, v, U, V are continuous on $\bar{K}_1$, this also holds for w_1 and w_2; this implies $w_1, w_2 \in \mathcal{L}^2(K_1)$.

By using $u_\xi = u_\varrho \cos\theta - u_\theta \sin\theta/\varrho$,

$$u_\varrho = \sum_{n=1}^{\infty} \frac{n!}{n^2} \cos(n!\theta)\, \varrho^{n!-1} \quad \text{for} \quad 0 \leqq \varrho \leqq \varrho_1 < 1\,, \quad \theta \in [0, 2\pi]\,,$$

$$u_\theta = \sum_{n=1}^{\infty} \frac{n!}{n^2} \sin(n!\theta)\, \varrho^{n!} \quad \text{for} \quad 0 \leqq \varrho \leqq \varrho_1 < 1\,, \quad \theta \in [0, 2\pi]$$

(where the series are uniformly convergent), and changing the order of the integration and summation as well as applying the orthogonality of the trigonometric functions we get:

$$\iint_{K_1} u_\xi^2\, d\xi\, d\eta = \lim_{\varrho_1 \to 1} \int_0^{\varrho_1}\!\!\int_0^{2\pi} u_\xi^2(\varrho, \theta)\, \varrho\, d\varrho\, d\theta = \lim_{\varrho_1 \to 1} \left[\frac{\pi}{2} \sum_{n=1}^{\infty} \frac{n!}{n^2} \varrho_1^{2n!}\right]$$

$$= \infty\,, \quad \text{i.e} \quad u_\xi \notin \mathcal{L}^2(K_1)\,.$$

In an analogous way it follows that $u_\eta \notin \mathcal{L}^2(K_1)$. Thus $w_\xi \notin \mathcal{L}^2(K_1)$ and $w_\eta \notin \mathcal{L}^2(K_1)$.

Since $[8/(1 + \xi^2 + \eta^2)^2]\, w^2 \in C(\overline{K}_1)$, we have proved part b) of the theorem, i.e. $DI_{\text{gh}}[w] = \iint\limits_{K_1} (w_\xi^2 + w_\eta^2 - 8/(1 + \xi^2 + \eta^2)^2\, w^2)\, d\xi\, d\eta$ does not converge.

Remark 14.5.5. Theorem 14.5.2 shows that the Hölder discontinuity of the boundary function h in problem (D_{gh}) has the same effect as in problem (D_{Δ_2}): the boundary value problem cannot be solved via the associated variational problem. Whichever physical problem you might consider, it certainly will therefore never be connected with the considered Dirichlet integral $DI_{\text{gh}}[v]$ and the given boundary function h, since in this case $DI_{\text{gh}}[v]$ does not represent a "reasonable energy integral".

13.5.4. Let us finally consider a Dirichlet problem for the Laplace equation in $\mathbb{R}^3$, where the boundary function again is not Hölder-continuous:

Problem (D_{Δ_3}). (Let $HS := \{(x_1, x_2, x_3) \in \mathbb{R}^3 : x_1^2 + x_2^2 + x_3^2 = 1, x_3 \leqq 0\}$ be the "southern" hemisphere in $\mathbb{R}^3$, $UD := \{(x_1, x_2, x_3) \in \mathbb{R}^3 : x_1^2 + x_2^2 \leqq 1, x_3 = 0\}$ and $HB := \{(x_1, x_2, x_3) \in \mathbb{R}^3 : x_1^2 + x_2^2 + x_3^2 < 1, x_3 < 0\}$; let $\partial HB := UD \cup HS$ and $\overline{HB} := HB \cup \partial HB$ be the boundary and the closure of HB, respectively. Besides Cartesian coordinates (x_1, x_2, x_3) we will use spherical coordinates (r, φ, ϑ): $x_1 = r \cdot \cos\varphi \sin\vartheta$, $x_2 = r \cdot \sin\varphi \sin\vartheta$, $x_3 = r \cdot \cos\vartheta$.)

Find a function $W \in C^2(HB) \cap C(\overline{HB})$, defined and real on $\overline{HB}$, which is a solution of $\Delta_3 W = W_{x_1x_1} + W_{x_2x_2} + W_{x_3x_3} = 0$ in HB and satisfies the boundary condition $W/_{\partial HB} = G$, where $G(r, \varphi, \pi/2) = r \cdot \hat{w}(\varphi, \pi/2)$ on UD (i.e. for $0 \leqq r \leqq 1$, $0 \leqq \varphi \leqq 2\pi$, $\vartheta = \pi/2$), $G(1, \varphi, \vartheta) = \hat{w}(\varphi, \vartheta)$ on HS (i.e. for $r = 1$, $0 \leqq \varphi \leqq 2\pi$, $\pi/2 \leqq \vartheta \leqq \pi$); here $\hat{w}(\varphi, \vartheta)$ denotes the function that we get, if we replace θ by φ and ϱ by $\sin\vartheta/(1 - \cos\vartheta)$ in (49).

Remark 14.5.6. Thus one gets $\hat{w}(\varphi, \vartheta)$ by stereographic projection of the solution $w(\varrho, \theta)$ of problem (D_{gh}) from $\overline{K}_1$ on to HS. The boundary function G in problem (D_{Δ_3}) is therefore a continuous, but not Hölder-continuous function on HB. Problem (D_{Δ_3}) is uniquely solvable.

The associated variational problem is now given by

Problem (V_{Δ_3}). Find a function $W \in V_G$, $V_G := \{v : \overline{HB} \to \mathbb{R} : v \in C^2(HB) \cap C(\overline{HB})$ $v|_{\partial HB} = G\}$ — G defined as in problem (D_{Δ_3}) —, which minimizes the Dirichlet integral

$$DI_{\Delta_3}[v] := \iiint\limits_{HB} (v_{x_1}^2 + v_{x_2}^2 + v_{x_3}^2)\, dx_1\, dx_2\, dx_3 .$$

The following theorem shows that — in contrast to problems (D_{Δ_2}) and (D_{gh}) — the Hölder discontinuities of G in problem (D_{Δ_3}) are "too weak" to "prevent" the solubility via problem (V_{Δ_3}):

Theorem 14.5.3. a) *The solution of problem* (D_{Δ_3}) *is given by* $W(r, \varphi, \vartheta) = r\hat{w}(\varphi, \vartheta)$ (*where* $\hat{w}(\varphi, \vartheta) := w(\varrho, \theta)$, $\varrho = \sin\vartheta/(1 - \cos\vartheta)$, $\theta = \varphi$, *w given by* (49)).

b) *For this function W the Dirichlet integral $DI_{\Delta_3}[W]$ exists.*

c) *This function W is the uniquely determined solution of problem* (V_{Δ_3}).

For the proof of this theorem see Heersink [4, p. 287—290].

Remark 14.5.7. If we interprete $W(x_1, x_2, x_3)$ in problems (D_{Δ_3}) and (V_{Δ_3}) as the potential of an electrostatic field, the Dirichlet integral $DI_{\Delta_3}[W]$ represents the energy of this field. Since $DI_{\Delta_3}[W]$ now exists, it seems to be possible to interprete problems (D_{Δ_3}) and (V_{Δ_3}) as mathematical models of a physically realizable field.

Further references

Bauer and Ruscheweyh [1], Heersink [3, 4], Ruscheweyh [3].

15. Solving of Linear Partial Differential Equations by Using Complex Integral Transforms

E. Lanckau

15.1. Introduction

The method of integral transforms is well-known in the treatment of linear differential equations of mathematical physics. The usual way is as follows: Let a (partial or ordinary) differential equation for the function u be given

$$L_{a,b,c,\ldots}u = 0\,, \tag{1}$$

by $a, b, c, \ldots$ we denote coefficients or parameters in the linear differential operator L. Using an integral transform T_0, we get for the image $w = T_0 u$ a simpler equation (differential equation, algebraic equation)

$$l_{a',b',c',\ldots}w = 0\,. \tag{2}$$

Here the coefficients or parameters $a', b', c', \ldots$ come from those of the original equation (1). We find the original solution u by the inverse transform, $u = T_0^{-1}w$.

Applying this method — with a fixed, a priori choosen transform T_0 — the main problem is the study of the equation (2) for the images w.

Here we develop another type of integral transforms to get always (for all original equations (1) under consideration) the same fixed equation for the images w:

$$l_0 w = 0\,,$$

e.g. the two-dimensional Laplace potential equation

$$l_0 w = \Delta_2 w = 0\,. \tag{2'}$$

To ensure this our essential problem is to construct an integral transform which now depends on the coefficients or parameters in the original equation (1):

$$w = T_{a,b,c,\ldots}u$$

or, more interesting,

$$u = T^{-1}_{a,b,c,\ldots}w\,,$$

w being a solution of (2′).

For an equation (1) with three or more independent variables in this way we realize a *method of ascent* (a term of Gilbert [1], see also Rundell and Stecher [1]) by constructing the solution u with the aid of a "two-dimensional" function w. We attack this problem using complex integral operators. The construction and the study of the solutions of linear partial differential equations of elliptic type by complex integral operators have been originated by I. N. Vekua [3] and Bergman [1]. The basic idea of this method is to associate holomorphic functions of a complex variable and solutions of partial differential equations. Here the solution of the differential equation is found through an integral transform from the associated holomorphic function.

Recently these methods have found new applications, especially by the work of COLTON [5, 7] and GILBERT [1, 2]. We give a survey of these methods, but here we describe only the investigation of equations with three independent variables $x, y, \mathfrak{z}$, although the same methods may be used in higher dimensions too, see BHATNAGAR and GILBERT [1, 3], GILBERT and KUKRAL [1], KUKRAL [1]. Further we consider especially the methods for the explicit construction of the solutions which is the base of their detailed study.

As usually in complex methods — see the preceding section 14.1. — we introduce the new complex independent variables $z = x + iy$, $\xi = x - iy$ ($\mathfrak{z} = \mathfrak{z}$). (We mention $\Delta = \partial^2/\partial x^2 + \partial^2/\partial y^2 = 4\partial^2/\partial z\,\partial\xi$.) Let the function $u = u(z, \xi, \mathfrak{z})$ in the domain $D \times \bar{D} \times G$ ($0 \in D$) satisfy the differential equation

$$Lu = \frac{\partial^2}{\partial z\,\partial \xi} u + A_1\left[\frac{\partial u}{\partial z}\right] + A_2\left[\frac{\partial u}{\partial \xi}\right] + A_0 u\,, \tag{1'}$$

here ($i = 0, 1, 2$)

$$A_i u = a_{i1}(z, \xi, \mathfrak{z})\, u + a_{i2}(z, \xi, \mathfrak{z})\, S_i u$$

with analytical coefficients $a_{ij}(z, \xi, \mathfrak{z})$, and the S_i are linear operators, not depending on z, ξ, that is, $S_i h(\mathfrak{z})$ depends only on $\mathfrak{z}$ if $h(\mathfrak{z})$ depends only on $\mathfrak{z}$.

15.2. The Riemann transform

Let us develop a concept to solve equations (1'). We want to derive a general method to construct solutions to (1') with operators S_i as general as possible. Some well-known methods of GILBERT [1] and COLTON [7] for "three-dimensional" equations are special cases of our method. First we define a set of functions $f(z, \mathfrak{z})$ which are to be associated with the solutions of (1). The set depends on the operators S_i. Second we shall construct an integral operator which transforms these associated functions into solutions to (1).

Definition 15.1. $f(z, \mathfrak{z}) \in F$ if and only if

(i) $f(z, \mathfrak{z})$ is holomorphic with respect to z in D and continuous in the closure $\hat{D}$ of D.
(ii) Constants $A > 0$, $C > 0$, $r \geqq 0$ (integer) exist with

$$\left| \prod_i S_0^{a_i} S_1^{b_i} S_2^{c_i} [f(z, \mathfrak{z})] \right| < A C^m (m + r)!$$

for all integers $a_i, b_i, c_i \geqq 0$ with $m = \sum\limits_i (2a_i + b_i + c_i) \geqq m_0$ and for all $z \in D$, $\mathfrak{z} \in G_0 \subset G$.

Examples. 1. Let

$$S_i = \sum_{k=0}^{2} b_{ik}\, \partial^k/\partial^k \mathfrak{z} \qquad (i = 0, 1, 2) \tag{3}$$

or

$$S_0 = \mathfrak{z} \cdot \partial^2/\partial \mathfrak{z}^2\,, \tag{3'}$$

b_{ik} constant, $b_{12} = b_{22} = 0$.

$f \in F$ if $f(z, \mathfrak{z})$ is holomorphic with respect to $\mathfrak{z}$ in G, continuous in the closure $\hat{G}$ of G and $G_0 \subset G$ is given by $0 < \delta \leqq |\sigma - \mathfrak{z}| < 1$ for all $\sigma \in \partial G$, $\mathfrak{z} \in G_0$.

2. Let

$$S_i u = b_{i0} u + b_{i1} \int_0^{\mathfrak{z}} u(\dots, s)\,(\mathfrak{z} - s)^k \, ds \qquad (i = 0, 1, 2)\,, \tag{4}$$

b_{ij}, $k \geqq 0$ constant.

$f \in F$ if $f(z, \mathfrak{z})$ is bounded in $\hat{D} \times \hat{G}$ and integrable with respect to $\mathfrak{z}$ in $G_0 = G$, $0 \in G$ (also all $s \in G$).

Thus with (3), (3'), (4) we get partial differential equations (1) of elliptic, parabolic, hyperbolic, mixed, and composed type.

3. Let

$$S_i u = b_{i0} u + b_{i1} u(z, \xi, \mathfrak{z} + k) \qquad (i = 0, 1, 2)\,,$$

b_{ij}, k constant. (Thus (1') is an equation with transformed argument.) $f \in F$ if $f(z, \mathfrak{z})$ is in $\mathfrak{z}$ an entire function of exponential type, $G_0 = G$.

Proof see LANCKAU [7].

Definition 15.2. The transform $R[f(t, \mathfrak{z})] = h(z, \xi, t, \tau, \mathfrak{z})$ with $z, t \in D_0 \subset D$, $0 \in D_0$, $\xi, \tau \in \overline{D}_0$, $\mathfrak{z} \in G_0$ is called *Riemann transform* of (1) if and only if

(A) $\partial/\partial z\, R[f(t, \mathfrak{z})]$, $\partial/\partial \xi\, R[f(t, \mathfrak{z})]$, $\partial^2/\partial z\, \partial \xi\, R[f(t, \mathfrak{z})]$, $S_i R[f(t, \mathfrak{z})]$, $i = 0, 1, 2$, exist,

(B) $R[f(t, \mathfrak{z})]$ is integrable with respect to t in D_0,

(C) in $D_0 \times \overline{D}_0 \times G_0$ hold

$$LR[f(t, \mathfrak{z})] = 0 \quad \text{for all} \quad t \in D_0,\ \xi \in \overline{D}_0 \tag{5}$$

$$\partial/\partial \xi\, R[f(t, \mathfrak{z})] + A_1 R[f(t, \mathfrak{z})] = 0 \quad \text{for} \quad t = z \text{ and for all } \xi, \tau \in \overline{D}_0\,. \tag{6}$$

Example. Let $A_1 = A_2 = 0$, $a_{01} = p(z + \xi + 2a)^{-2}$, p be a complex constant, $a > \operatorname{diam} D$ a real constant, $a_{02} = 1$. Thus we have the (with respect to z, ξ) self-adjoint equation

$$\partial^2/\partial z\, \partial \xi\, u + p(z + \xi + 2a)^{-2}\, u + S_0 u = 0\,.$$

Here in a closed polydisc in $D_z \times \overline{D}_t \times D_\xi \times \overline{D}_\tau \cap \{(z, \xi, t, \tau) : |v| \leqq 1/C - \varepsilon,$ $|w| \leqq 1 - \varepsilon\}$

$$R[f(t, \mathfrak{z})] = \sum_{m=0}^{\infty} {}_2F_1(b, 1 - b, m + 1, w)\, \frac{v^m}{m!^2} \cdot S_0^m f(t, \mathfrak{z}) \tag{7}$$

with $v = (z - t)(\tau - \xi)$, $w = v(z + \xi + 2a)^{-1}(t + \tau + 2a)^{-1}$; C from condition (ii) in definition 15.1, $\varepsilon > 0$, $b(1 - b) = p$, ${}_2F_1$ is the Gauss hypergeometric function (and $f \in F$).

Proof. For $p = 0$ see LANCKAU [7], for $p \neq 0$ analogously.

Let now first $S_0 = \partial^2/\partial \mathfrak{z}^2$; thus we have an elliptic equation (1') of second order. From $f(t, \mathfrak{z}) = 1/2\pi i \oint_K f(t, s)\, ds/s - \mathfrak{z}$, $K\colon |s - \mathfrak{z}| = \delta > 0$, we have

$$R[f(t, \mathfrak{z})] = C_H f(t, \mathfrak{z}) = \frac{1}{2\pi i} \oint_K H(z, \xi, t, \tau, s, \mathfrak{z})\, f(t, s)\, ds\,, \tag{8}$$

the Riemann transform is a Cauchy integral with the kernel

$$H(z, \xi, t, \tau, s, \mathfrak{z}) = R\left[\frac{1}{s - \mathfrak{z}}\right]. \tag{9}$$

For $p = 0$ is (LANCKAU [3])

$$H(z, \xi, t, \tau, s, \mathfrak{z}) = \left((s - \mathfrak{z})^2 - 4v\right)^{-1/2} .$$

Now let second $S_0 = \int_0^{\mathfrak{z}} \cdot \, ds$; thus we have a pseudoparabolic equation (1′). From

$$S^m u = \frac{1}{m!} \frac{\partial}{\partial \mathfrak{z}} \int_0^{\mathfrak{z}} u(\dots, s)\, (\mathfrak{z} - s)^m \, ds \qquad (m \geqq 0)$$

we have

$$R[f(t, \mathfrak{z})] = D_H f(t, \mathfrak{z}) = \frac{\partial}{\partial \mathfrak{z}} \int_0^{\mathfrak{z}} H(z, \xi, t, \tau, \mathfrak{z} - s)\, f(t, s)\, ds , \tag{10}$$

the Riemann transform is a Duhamel product with the kernel

$$H(z, \xi, t, \tau, \mathfrak{z}) = R[1] . \tag{11}$$

For $p = 0$ (LANCKAU [8]) is $H = {}_0F_2(1, 1, \mathfrak{z}v)$ with the generalized hypergeometric function ${}_0F_2$ (with the notation of BATEMAN and ERDÉLYI [1]).

Remark. From (7) and (9) or (11) H may be found in many cases explicitly, e.g. for (3), (3′), (4), see LANCKAU [3—5].

Remark. With $a_{i2} = 0$ $(i = 0, 1, 2)$ we have a two-dimensional equation (1′); $\mathfrak{z}$ is only a parameter. Let $R^* = R^*(z, \xi, t, \tau)$ be the complex Riemann function of the adjoint equation (with respect to z, ξ) $L^*u = 0$. Here immediately by the definition 15.2.

$$R[f(t, \mathfrak{z})] = R^*(z, \xi, t, \tau)\, f(t, \mathfrak{z}) .$$

This is a version of the original result of I. N. VEKUA [3], see the preceding chapter 14, theorem 14.1.2.

Let N be a neighbourhood of $z = \xi = t = \tau = 0$.

Theorem 15.1. *Let the coefficients $a_{ij} = a_{ij}(z, \xi)$, $i = 0, 1, 2$, $j = 1, 2$, not depend on $\mathfrak{z}$. The Riemann transform $R[f]$ exists in $N \times G_0$ if $f \in F$.*

Proof. We only sketch the proof. A solution of the Volterra equation

$$h(z, \xi, t, \tau, \mathfrak{z}) + M[h(z, \xi, t, \tau, \mathfrak{z})] = f(t, \mathfrak{z})$$

with

$$M[h(z, \xi, t, \tau, \mathfrak{z})] = \int_\tau^\xi A_1[h(z, q, \dots)]\, dq + \int_t^z A_2[h(p, \dots)]\, dp + \iint_{t\tau}^{z\xi} (A - A_1' - A_2')\, [h(p, q, \dots)]\, dp\, dq ,$$

where

$$A_k'[h(p_1, p_2, \dots)] = \frac{\partial}{\partial p_k} a_{k1}(p_1, p_2)\, h(p_1, p_2, \dots) + \frac{\partial}{\partial p_k} a_{k2}(p_1, p_2)\, S_k[h(p_1, p_2, \dots)]$$

$(k = 1, 2)$, satisfies (5), (6); therefore it is a Riemann transform of $f(t, \mathfrak{z})$. From the analyticity of the a_{ij} we have in N a constant K with

$$|f(z, \mathfrak{z})| < K , \qquad |a_{ij}(z, \xi)| < K , \qquad \left| \frac{\partial}{\partial z} a_{1j}(z, \xi) \right| < K , \qquad \left| \frac{\partial}{\partial \xi} a_{2j}(z, \xi) \right| < K$$

$(i = 0, 1, 2, j = 1, 2)$.

Let $b^{(n)}_{m_1m_2}(z, \xi, \mathfrak{z})$ be a function found from $f(z, \mathfrak{z})$ by n multiplications with one of the 10 coefficients a_{ij}, $\partial/\partial z\, a_{1j}$, $\partial/\partial\xi\, a_{2j}$ (in an arbitrary order) and by m_1 integrations with respect to the first independent variable and by m_2 integrations with respect to the second independent variable. It is easily seen that $M^n[f(t, \mathfrak{z})]$, $n = 0, 1, \ldots$, consists of (not more than) 10^n terms as

$$B = \Pi\, S_0^{\lambda_i} S_1^{\mu_i} S_2^{\varkappa_i} b^{(n)}_{m_1m_2}(z, \xi, \mathfrak{z})\,,$$

here $\lambda_i, \mu_i, \varkappa_i \geqq 0$, $m_1 \geqq \sum\limits_i (\lambda_i + \varkappa_i) = \tilde{m}_1$, $m_2 \geqq \sum\limits_i (\lambda_i + \mu_i) = \tilde{m}_2$. Using condition (ii) of definition 15.1 we have in $N \times T_0$

$$|B| \leqq \frac{(2\delta)^{m_1+m_2}}{m_1!m_2!} K^{n+1} A C^{\tilde{m}_1+\tilde{m}_2} (\tilde{m}_1 + \tilde{m}_2 + p)!\,.$$

from this we find using a $C \geqq 1$

$$|B| \leqq (2C\delta)^{m_1+m_2} K^{n+1} A\, \frac{(m_1 + m_2 + p)!}{m_1!m_2!}\,.$$

The series $\sum\limits_{m_1=0}^{\infty} \sum\limits_{m_2=0}^{\infty} \frac{(m_1 + m_2 + p)!}{m_1!m_2!} s^{m_1} t^{m_2} = p!(1 - s - t)^{-p-1}$ converges absolutely for $s = t = 1/3$; therefore

$$\frac{(m_1 + m_2 + p)!}{m_1!m_2!} 3^{-m_1-m_2} \to 0\,,$$

that is $\frac{(m_1 + m_2 + p)!}{m_1!m_2!} \leqq K(p) \cdot 3^{m_1+m_2}$ for every $m_1, m_2 \geqq 0$ with a constant $K(p)$ depending only on p. Thus we have

$$|B| \leqq AK(p)\, (6C\delta)^{m_1+m_2} \cdot K^{n+1}\,.$$

Let now for the present $6C\delta \leqq 1$. From $n \leqq m_1 + m_2$ we have

$$|B| \leqq AKK(p)\, (6KC\delta)^n\,,$$

thus $M^n[f(t, \mathfrak{z})] \leqq AKK(p)\, (60KC\delta)^n$. Therefore the (Neumann) series

$$\sum_{n=0}^{\infty} (-1)^n M^n[f(t, \mathfrak{z})] \tag{*}$$

for a sufficiently small $\delta < 1/(60KC)$ $(\delta \leqq 1/6C)$ converges uniformly and absolutely, thus it is an analytic function of z, ξ, t, τ. But (*) is a solution of (5), (6).

Remark. The condition that all a_{ij} do not depend on $\mathfrak{z}$ is sufficient but not necessary for all special S_i, see 15.3.1.

Theorem 15.2. *Let $R[f]$ exist. Then*

$$u(z, \xi, \mathfrak{z}) = \int_0^z R[f(t, \mathfrak{z})]\, dt \tag{12}$$

solves (1').

Proof by insertion into (1') using definition 15.2.

Remark. The existence of $R[f]$ may be ensured by theorem 15.1 or otherwise! We remark again that theorem 14.1.2 gives the two-dimensional analogue.

Definition 15.2′. The transform $\tilde{R}[f(t, \mathfrak{z})] = h(z, \xi, t, \tau, \mathfrak{z})$ with $z, t \in D_0 \subset D$, $0 \in D_0$, $\xi, \tau \in \overline{D}_0 \subset \overline{D}$, $0 \in \overline{D}_0$, $\mathfrak{z} \in G_0$ is called a *proper Riemann transform of* (1′), if and only if

(A′) $\tilde{R}[f]$ is a Riemann transform

(B′) (additionally) in $D_0 \times \overline{D}_0 \times G_0$

$$\partial/\partial z\, \tilde{R}[f(z, \mathfrak{z})] + A_2 R[f(t, \mathfrak{z})] = 0 \quad \text{for} \quad \tau = \xi \text{ and for all } z, t \in D_0\,,$$

$$\overline{R}[f(t, \mathfrak{z})] = f(z, \mathfrak{z}) \quad \text{for} \quad t = z, \tau = \xi\,.$$

Remark. The transform $\tilde{R}[f]$ given in the above example is a proper Riemann transform.

Theorem 15.1′. *Let the conditions of theorem* 15.1 *hold. Then the proper Riemann transform* $\tilde{R}[f]$ *exists.*

Proof as for theorem 15.1. (The transform constructed there is the proper Riemann transform.)

Theorem 15.2′. *Let* $\tilde{R}[f]$ *exist. Then*

$$\tilde{u}(z, \xi, \mathfrak{z}) = \int_0^z \int_0^\xi \tilde{R}[G(t, \tau, \mathfrak{z})]\, dt\, d\tau$$

solves the equation

$$L\tilde{u} = \partial/\partial z\, G(z, \xi, \mathfrak{z})\,.$$

Proof by insertion using the definition 15.2′.

15.3. Riemann transforms as integral transforms

One of the main problems in the application of integral operators is to find a suitable representation of the Riemann transform, e.g. as integral transform. (In the two-dimensional case this is the problem to find the complex Riemann function or another kernel of an analogous transform, see e.g. Bauer [6], Kracht [3], Kracht and Kreyszig [1], Kreyszig [1], Püngel [1, 2].) We consider the problem of representation for differential equations (1′) with three independent variables $z, \xi, \mathfrak{z}$. (8) suggests:

Theorem 15.3. *Let the coefficients* $a_{ij} = a_{ij}(z, \xi, \mathfrak{z})$, $i = 0, 1, 2$, $j = 1, 2$, *and* $A = A(z, \xi, \mathfrak{z})$ *be analytical functions of* $z, \xi, \mathfrak{z}$, *and let*

$$S_1 = S_2 = \partial/\partial \mathfrak{z}\,,$$

$$S_0 = \frac{\partial}{\partial \mathfrak{z}}\left(A(z, \xi, \mathfrak{z}) \frac{\partial}{\partial \mathfrak{z}}\right).$$

1. *In* $N \times G_0 \times G_0$ *a function* $H = H(z, \xi, t, \tau, s, \mathfrak{z})$ *exists with*

$$LH = 0\,, \tag{13}$$

$$\partial/\partial\xi\, H + A_1 H = 0 \quad \textit{for} \quad t = z\,,$$

$$H = \frac{1}{s - \mathfrak{z}} \quad \textit{for} \quad t = z, \tau = \xi\,. \tag{14}$$

2. *With this* H *the Riemann transform of* (1′) *is given by* (8).

Proof. To 1. Modifying the proof of theorem 15.1 we find the existence of

$$H = R\left[\frac{1}{s - \mathfrak{z}}\right].$$

To 2. by insertion into (5), (6) using (13), (14).

Another proof follows the methods given by COLTON [2, 4] (construction of H with the aid of differential equations).

Remark. Theorem 15.3 is related with elliptic equations (1'). The same representation (8) is possible for parabolic equations, see HILL [1].

Now we consider pseudoparabolic equations. (10) suggests:

Theorem 15.4. *Let the coefficients* $a_{ij} = a_{ij}(z, \xi)$, $i = 0, 1, 2$, $j = 1, 2$, *be analytic functions of* z, ξ, $a_{12} \equiv 0$, *and let*

$$S_0 = S_2 = \int_0^{\mathfrak{z}} \cdot \, ds\,.$$

1. *In* $N \times G_0 \times G_0$ *a function* $H = H(z, \xi, t, \tau, s, \mathfrak{z})$ *exists with* (13), (14) *and with*

$$\left(\frac{\partial^2}{\partial z\, \partial \xi} + a_{11}\frac{\partial}{\partial z} + a_{21}\frac{\partial}{\partial \xi} + a_{01}\right) H = 0 \quad \text{for} \quad \mathfrak{z} = s\,,$$

$$H = 1 \quad \text{for} \quad t = z,\ \xi = \tau\,.$$

2. *With this* H *the Riemann transform of* (1') *is given by* (10).

Proof. To 1. Using the proof of theorem 15.1 we find the existence of $H = R[1]$. To 2. by insertion into (5), (6) using (13), (14).

Another proof follows the methods given by COLTON [3].

Remark. For more general equations the representation of the Riemann transform (and therefore the solution) by a single Cauchy integral or a single Duhamel product does not suffice. In general a partial differential equation (1') requires a Riemann transform which is a sum of a Cauchy integral and a Duhamel product. We give an example, for detailed results see LANCKAU [9].

Example. Consider the pseudoparabolic equation (of composed type)

$$\frac{\partial}{\partial \xi}\left[\frac{\partial^2}{\partial z\, \partial \mathfrak{z}} u - \xi\left(c_2 \frac{\partial^2}{\partial \mathfrak{z}^2} u + c_1 \frac{\partial}{\partial \mathfrak{z}} u + c_0 u\right)\right] = 0\,.$$

Here we have

$$R[f(t, \mathfrak{z})] = C_{H_1} f(t, \mathfrak{z}) + D_{H_{-1}} f(t, \mathfrak{z}) - H_0 f(t, \mathfrak{z}) - c_0 c_2 \oint_{K_0} H f(t, s) \frac{ds}{s}\,,$$

where the kernel H_1 is a hypergeometric function of two variables (BATEMAN and ERDÉLYI [1])

$$(s - \mathfrak{z})\, H_1 = e^{-c_1 v}\, \Xi_2(1, 1, 1; c_2 v(\mathfrak{z} - s)^{-1}, c_0 c_2 v^2)\,,$$

the kernel H_{-1} is a modified Bessel function

$$H_{-1} = e^{-c_1 v}\, I_0\left(2\sqrt{c_0 v(c_2 v + \mathfrak{z} - s)}\right)$$

and

$$H_0 = e^{-c_1 v}\, I_0(2\sqrt{c_0 c_2}\, v)$$

(all with $\tau = 0$), and $H = H(z, \xi, t, \tau, s)$ is a certain function, continuously depending on c_0 and c_2, $K = \{s : |s| < \delta\}$.

Proof by insertion (but with the requirement of difficult calculations!).

From this we find especially: For $c_0 = 0$ we have a parabolic equation, with $H_0 = H_{-1} = e^{-c_1 v}$ the Riemann transform becomes a single Cauchy integral with

$$H_1 = e^{-c_1 v} \frac{1}{s - \mathfrak{z} + c_2 v},$$

and we have the remarkable result

$$R[f(t, \mathfrak{z})] = e^{-c_1 v} f(t, \mathfrak{z} - c_2 v) .$$

For $c_2 = 0$ we have a pseudoparabolic equation, with $H_0 = (s - \mathfrak{z}) H_1 = e^{-c_1 v}$ the Riemann transform becomes a single Duhamel product with the kernel

$$H_{-1} = e^{-c_1 v} I_0\left(2 \sqrt{c_0 v(\mathfrak{z} - s)}\right) .$$

15.4. Higher order equations

A single equation of higher order may be written in the form of a system of equations of second order. All considerations of the above sections 15.2. and 15.3. may be done also for such systems.

In this generalizations in (1′) u is the vector of unknown functions, A_0, A_1, A_2 are matrix operators, and the integral transform (12) associates the vector u and a vector f of holomorphic (with respect to z) functions with the aid of a Riemann matrix operator R.

We do not give here the theory of such transforms (see Bhatnagar and Gilbert [2, 4], Brown [1], Brown and Gilbert [1], Gilbert and Hsiao [1], Gilbert and Roach [1], Kürcz [1], Lanckau [2]), but we consider an example.

Example Let $u(z, \xi, \mathfrak{z}) = \left(u_i(z, \xi, \mathfrak{z})\right)_{i=1,2,\ldots,N}$ the vector satisfying the equation

$$\frac{\partial^2}{\partial z\, \partial \xi} u + A[u] = 0 , \tag{15}$$

where A is a (quadratic) matrix operator, whose elements A_{ik} do not depend on z, ξ, and all elements $[A^n f(t, \mathfrak{z})]_{i,k}$ of $A^n[f(t, \mathfrak{z})]$ satisfy

$$|[A^n f(t, \mathfrak{z})]_{i,k}| \leqq A C^n (2n + r)!$$

with the constants $A > 0$, $C > 0$, $r \geqq 0$; and $f(t, \mathfrak{z}) = \left(f_i(t, \mathfrak{z})\right)_{i=1,2,\ldots,N}$ is a vector of holomorphic functions. (12) with

$$R = \sum_{n=0}^{\infty} \frac{1}{n!^2} (z - t)^n (\tau - \xi)^n A^n \tag{16}$$

solves (15).

Proof by insertion.

Example (Brown, Gilbert, and Hsiao [1]). Let $N = 2$, $A = \begin{pmatrix} 0 & -1 \\ a + cS & bS \end{pmatrix}$, a, b, c constant. With $S = \partial/\partial\mathfrak{z}$ (15) gives a "metaparabolic" equation of 4th order for u_1, with $S = \int_0^{\mathfrak{z}} \cdot\, ds$ (15) gives a "pseudoparabolic" equation of 5th order for u_1.

For both cases the knowledge of

$$A^n = \begin{pmatrix} -h_n & -h_{n-1} \\ h_{n+1} & h_n \end{pmatrix} \begin{pmatrix} -1 & 0 \\ bS & 1 \end{pmatrix}$$

with the polynomial $h_n = \sum_{k=0}^{n} \binom{n-k}{k} (-1)^k (bS)^{n-2k} (a + cS)^k$ gives an explicit representation of (16) as integral transforms similarly to (8) and (10) (now H is a matrix of functions $H_{ik}(z, \xi, t, \tau, s, \mathfrak{z})$ which are hypergeometric and related functions).

15.5. Other transforms

The above concept of a Riemann transform has been found by a generalization of I. N. VEKUA's method, which uses a complex Riemann function. Other operator methods (see the survey of KRACHT [1]), especially the method of S. BERGMAN, may in the same way be a starting point for the construction of integral operator methods in higher dimensions, see e.g. COLTON [2], GILBERT and KUKRAL [1], KUKRAL [1], GILBERT and ROACH [1]. Special cases very often allow much simpler transforms. We give (generalizing a result of H. FLORIAN [1]) an example.

Example. Let $S_1 S_2 = S_2 S_1$, S_1, S_2 be as above. The equation ($|a| > \operatorname{diam} D$, a constant)

$$(z + \xi + 2a)^2 \frac{\partial^2}{\partial z\, \partial \xi} u + \left((z + \xi + 2a) S_1 + 2 + S_2\right) \frac{\partial}{\partial \xi} u + S_2 u = 0$$

has in $D \times \overline{D} \times G_0$ the solution

$$u(z, \xi, \mathfrak{z}) = f(z, \mathfrak{z}) + (z + \xi + 2a)^{-1} \int_0^z S_2[\mathrm{e}^{(t-z)S_1}] [f(t, \mathfrak{z})] \, \mathrm{d}t$$

with $f \in F$.

Proof by insertion.

Examples. For $S_1 = a_0 + a_1\, \partial/\partial\mathfrak{z}$ (a_0, a_1 constants) we have

$$\mathrm{e}^{(t-z)S_1} [f(t, \mathfrak{z})] = \mathrm{e}^{a_0(t-z)} f\left(t, \mathfrak{z} + a_1(t - z)\right);$$

for $S_1 = a_0 + a_1 \int_0^{\mathfrak{z}} \cdot \, \mathrm{d}s$, we have

$$\mathrm{e}^{(t-z)S_1} [f(t, \mathfrak{z})] = \mathrm{e}^{a_0(t-z)} \frac{\partial}{\partial \mathfrak{z}} \int_0^{\mathfrak{z}} I_0\left(2 \sqrt{(t - z)(\mathfrak{z} - s)}\right) f(t, s) \, \mathrm{d}s$$

with the modified Bessel function I_0 of order 0.

15.6. Application: Boundary value problems

The method of integral operators has found many applications in mathematical physics, see e.g. the survey of v. KRZYWOBLOCKI [1, 2] and BERGMAN [2], CLEMENTS and ROGERS [1], COLTON [8, 12], COLTON and WENDLAND [1], IL'INSKIĬ et al. [1], KASIMOV [1], NAZAROV [1], VOL'MAN [1].

The applications of the method of complex integral operators have the following mathematical sources:

a) The regularity properties of the solution of (1′) (location and kind of singularities) are closely related to those of its associated holomorphic function (MITCHELL [1, 2]).

(b) By (12) it is possible to construct complete sets of particular solutions of (1′) in a simple manner (see the examples of Colton [6, 9, 10]). This gives the possibility to solve boundary value problems of (1′) by superposition of particular solutions, minimizing a suitable norm (Colton [13], Eisenstat [1], Gilbert and Linz [1], Hoefer [1], Reinartz [1], Schryer [1] ,Rundell and Stecher [2]).

Example. Consider the equation

$$\frac{\partial^2 u}{\partial z\, \partial \xi} + \mathfrak{z}^{k-2} \frac{\partial^2 u}{\partial \mathfrak{z}^2} = 0$$

(for integer odd k of mixed type). Using (7) and (12) we get for $k \neq 0$ the solutions

$$u_{mn}(z, \xi, \mathfrak{z}) = f(z, \mathfrak{z})\, {}_2F_1\left(\frac{m}{k}, \frac{m+1}{k};\ n+1;\ k^2 z(\tau - \xi)\, \mathfrak{z}^{-k}\right)$$

and for $k = 0$ the solutions

$$u_{mn}(z, \xi, \mathfrak{z}) = f(z, \mathfrak{z})\, \Gamma(n+1)\, \big(m(m+1)\, z(\tau - \xi)\big)^{-n/2} \, I_n\big(2\sqrt{m(m+1)\, z(\tau - \xi)}\big)$$

from the associated function $f(z, \mathfrak{z}) = z^n \mathfrak{z}^{-m}$.

c) Boundary value problems for solutions of (1′) may be reformulated as boundary value problems for holomorphic functions $f(z, \mathfrak{z})$ (see Bischoff [1]).

d) By the use of an integral representation of $f(t, \mathfrak{z})$ through real values on the boundaries we get singular (in special cases weakly singular) integral equations, whose solution solves the boundary value problem (Heersink [2]).

16. Function Theory in Algebras

K. HABETHA

16.1. Introduction

Many mathematicians have tried to generalize the powerful methods of function theory of one complex variable in $\mathbb{R}^2$ to higher dimensions. More than thousand papers appeared since the usually assumed starting point, a paper by SCHEFFERS in 1894 [1]. In a short survey as this it is impossible to touch all features of the subject. The author apologizes in advance for not mentioning many important papers and results. Other generalizations of function theory are dealt with in other parts of this book, you find some hints later on.

To fix the notation, let $\mathfrak{V}$, $\mathfrak{W}$ be real finite-dimensional vector spaces. The topology may be the usual Euclidean one, G is a domain in $\mathfrak{V}$. Then we discuss mappings

$$f: G \to \mathfrak{W} .$$

Further we assume the existence of an algebra $\mathfrak{A}$, which is a finite-dimensional real vector space with an (only) distributive multiplication and the rule

$$\alpha(ab) = (\alpha a)\, b = a(\alpha b)$$

for all $\alpha \in \mathbb{R}$, $a, b \in \mathfrak{A}$, so that always

$$\mathfrak{V}, \mathfrak{W} \subset \mathfrak{A} .$$

The multiplication in $\mathfrak{A}$ is governed by the multiplication constants ε_{ijk}: let

$$\{e_k \mid k \in K \subset \mathbb{N}_0\} , \qquad \mathbb{N}_0 = \{0, 1, 2, \dots\} ,$$

be a base of $\mathfrak{A}$, then we have

$$e_i e_j = \sum_{k \in K} \varepsilon_{ijk} e_k .$$

It is clear that the ε_{ijk} generally depend on the base, but we hold the base fixed and assume moreover, that $\{e_i \mid i \in I \subset K\}$ and $\{e_j \mid j \in J \subset K\}$, respectively, are bases of $\mathfrak{V}$ and $\mathfrak{W}$, respectively, always $I = \{0, 1, \dots, n\}$ or $= \{1, \dots, n\}$ and $\dim \mathfrak{W} = m$. The independent variable in $\mathfrak{V}$ is written

$$z = \sum_{i \in I} x_i e_i ,$$

$|z| := (\sum_{i \in I} x_i^2)^{1/2}$ is the Euclidean norm.

16.2. Differentiation and power series

In this section we assume $\mathfrak{W} = \mathfrak{A}$, so that $\mathfrak{V}$ is a subvectorspace of the algebra $\mathfrak{W}$, for convenience $\mathfrak{W}$ may be associative.

16.2.1. Definition. (i) Let $L(\mathfrak{V}, \mathfrak{W})$ be the space of linear transformations from $\mathfrak{V}$ to $\mathfrak{W}$ and let $D(\mathfrak{V}, \mathfrak{W})$ be the subspace generated by the special linear transformations azb with $a, b \in \mathfrak{W}$.

(ii) A function $f: G \to \mathfrak{W}$ is called $\mathfrak{V}$-*differentiable* or *Hausdorff-differentiable* in $z_0 \in G$ if there is an $f'(z_0) \in D(\mathfrak{V}, \mathfrak{W})$, so that for $z \to z_0$

$$f(z) = f(z_0) + f'(z_0)(z - z_0) + o(|z - z_0|).$$

One may say that $D(\mathfrak{V}, \mathfrak{W})$ contains those elements of $L(\mathfrak{V}, \mathfrak{W})$ which are acting on the variable z *as a whole*. $f'(z_0)$ has the not uniquely determined form

$$f'(z_0)(\zeta) = \sum_j a_j \zeta b_j$$

with at most m^2 terms, $m = \dim \mathfrak{W}$. Sometimes the uniquely defined number

$$\sum_j a_j b_j$$

is called $f'(z_0)$ (Portmann [1]).

If $\mathfrak{W}$ is not associative one has to take care of the brackets in azb, therefore we neglect this case here. There are the usual differentiation rules which will not be written down here.

16.2.2. Definition (i) Let $L_j(\mathfrak{V}, \mathfrak{W})$ be the space of j-linear mappings from $\mathfrak{V}$ to $\mathfrak{W}$ and let $D_j(\mathfrak{V}, \mathfrak{W})$ be the subspace generated by the transformations

$$a_0 z_{i_1} a_1 z_{i_2} a_2 \dots a_{j-1} z_{i_j} a_j$$

with $a_0, \dots, a_j \in \mathfrak{W}$, $(i_1, \dots, i_j)$ a permutation of $(1, \dots, j)$.

(ii) $P_j(\mathfrak{V}, \mathfrak{W})$ is defined to be the *linear space of all homogeneous polynomials* of degree j generated by $D_j(\mathfrak{V}, \mathfrak{W})$, that is: For every $p_j(z) \in P_j(\mathfrak{V}, \mathfrak{W})$ exists a $d \in D_j(\mathfrak{V}, \mathfrak{W})$ with

$$p_j(z) = d(z, \dots, z)$$

(all variables in d equal to z).

(iii) A series $\sum p_j(z)$, $p_j \in P_j(\mathfrak{V}, \mathfrak{W})$, is called a $\mathfrak{V}$-*power series* in z, a function $f: G \to \mathfrak{W}$ is $\mathfrak{V}$-*analytic* in $z_0 \in G$, if it is given by a $\mathfrak{V}$-power series in $z - z_0$ converging in a neighbourhood of z_0.

The idea behind definition 16.2.2 is the same as in 16.2.1, we have also the variable z *as a whole*. This additional assumption makes possible in some cases a theory which is more than the usual theory of differentiable functions. It depends on the algebra wether one is getting really all differentiable functions by definition 16.2.1 (ii) or not. But it should be said that this theory is not so satisfactory as the classical theory of power series in $\mathbb{C}$, as general integral theorems fail to exist.

But some remarks are necessary: As usual a polynomial $p_j \in P_j(\mathfrak{V}, \mathfrak{W})$ is generated by a uniquely determined symmetric element π_j of $D_j(\mathfrak{V}, \mathfrak{W})$, that is

$$\pi_j(z_1, \dots, z_j) = \pi_j(z_{\sigma(1)}, \dots, z_{\sigma(j)})$$

for all permutations σ. If $\|\cdot\|$ is a norm in $\mathfrak{W}$ we can define the usual operator norms $\|p_j\|$ and $\|\pi_j\|$. J. H. Grace — see Hörmander [1] — has proved, that these norms are equal and we have

$$\|p_j(z)\| \leqq \|p_j\| \, \|z\|^j.$$

Now we can define the convergence of $\sum p_j(z)$ in two ways: First the usual norm convergence has the set of convergence

$$K := \{z \mid \sum_{j=0}^{\infty} \|p_j(z)\| < \infty\},$$

this does not depend on the norm, as all norms in $\mathfrak{W}$ are equivalent. Second we can demand

$$\sum_{j=0}^{\infty} \|p_j\| \, \|z\|^j < \infty,$$

this is a ball in the given norm with the classical radius of convergence. This ball may depend on the norm, but

$$K_p := \bigcup_{\|\cdot\|} \{z \mid \sum_{j=0}^{\infty} \|p_j\| \, \|z\|^j < \infty\}$$

is independent of it and contained in K. One can prove that a power series may be rearranged around a z_0 in the interior of K_p,

$$\sum_{j=0}^{\infty} p_j(z) = \sum_{j=0}^{\infty} \tilde{p}_j(z - z_0),$$

which is one of the important properties of analytic functions. The proof uses the binomial formula

$$p_j(z + \zeta) = \sum_{k=0}^{j} \binom{j}{k} \pi_j(z, \dots, z, \zeta, \dots, \zeta)$$

(k times z, $(j - k)$ times ζ). So a $\mathfrak{B}$-analytic function in $z = 0$ is at least $\mathfrak{B}$-analytic in the interior of K_p. An interesting open question is wether this is also true in the interior of K.

16.2.3. Historical remarks. Scheffers in the already mentioned paper [1] was the first who tried to generalize the complex differentiability to the case of algebras of higher dimensions, Hausdorff [1] had a definition similar to 16.2.1 (ii). Later on many authors had sometimes quite different formulations for the same thing, I mention Ward [1] who used matrix representations of $\mathfrak{W}$. The book by Roşculeţ [1] and the dissertations by Eichhorn [1] and Edenhofer [1] contain extensive lists of relevant literature. At last one remark to the terminology: Most authors use the word monogenic for functions which posses a derivative not depending on the direction of differentiation, so they use it in the case of Hausdorff-differentiable functions. But in the last years the term monogenic has also been used for functions which are looked at in the next sections.

16.3. Cauchy-Riemann equations and Cauchy integral theorems

In analogy to the classical Cauchy-Riemann operator

$$\frac{\partial}{\partial x} + i \frac{\partial}{\partial y}$$

we have the

16.3.1. Definition. (i) $D := \sum_{j \in I} e_j \, \partial/\partial x_j$.

(ii) A function $f\colon G \to \mathfrak{W}$, G domain in $\mathfrak{V}$, is called *left-regular* (*right-regular*) in G, if there

$$Df = 0 \qquad (fD = 0)\,.$$

It is called *biregular*, if it is left- and right-regular. $Df = 0$ is called the *Cauchy-Riemann system* with respect to $\mathfrak{V}$, $\mathfrak{W}$ and $\mathfrak{A}$.

It is clear that there are many possibilities of generalizing the Cauchy-Riemann equations of classical function theory. The above way seems to be very fruitful. Introducing the multiplication constants, we get with $f = \sum_{j\in J} f_j e_j$

$$Df = \sum_{i\in I}\sum_{j\in J} e_i e_j \frac{\partial f_j}{\partial x_i} = \sum_{i\in I}\sum_{j\in J}\sum_{k\in K} \varepsilon_{ijk} e_k \frac{\partial f_j}{\partial x_i}\,.$$

The equation $Df = 0$ is therefore equivalent to the real system of first order partial differential equations

$$\sum_{i\in I}\sum_{j\in J} \varepsilon_{ijk} \frac{\partial f_j}{\partial x_i} = 0\,, \qquad k \in K\,.$$

There are $N = \dim \mathfrak{A}$ equations, $m = \dim \mathfrak{W}$ unknown functions and n or $n + 1$ independent variables. If on the other hand such a system is given, the dimensions of $\mathfrak{V}$, $\mathfrak{W}$ and $\mathfrak{A}$ are defined and also the multiplication constants in $\mathfrak{A}$, the latter at least partially and the remaining may be defined conveniently.

So every homogeneous system with constant coefficients can be looked at as a Cauchy-Riemann system for suitable algebras and the above concept of a Cauchy-Riemann system turns out to be fairly general. But therefore one cannot expect to get an interesting function theory in all cases.

Now we shall prepare for the Cauchy integral theorem and by the way we shall get a slight generalization of the concept of Cauchy-Riemann system:

16.3.2. Definition and Lemma. (i) *Let*

$$\mathrm{d}z = \sum_{i\in I} e_i\, \mathrm{d}x_i$$

and let for every differential form Ω the Hodge operator Ω^* be defined by linearity and

$$(\mathrm{d}x_{i_1} \wedge \ldots \wedge \mathrm{d}x_{i_p})^* := \operatorname{sgn}(i_1, \ldots, i_n)\, \mathrm{d}x_{i_{p+1}} \wedge \ldots \wedge \mathrm{d}x_{i_n}$$

(*if* $\dim \mathfrak{V} = n + 1$, *we have to start with* i_0).

(ii) *If f is once differentiable in the sense of Fréchet the following three statements are equivalent*: α) $Df = 0$, β) $\mathrm{d}z^* \wedge \mathrm{d}f = 0$, γ) $\mathrm{d}z^* f$ *is closed.*

(iii) *If Ω is a closed differential form in a domain $G \subset \mathfrak{V}$ with values in $\mathfrak{A}$, a function $G \to \mathfrak{W}$ is called left-Ω-regular (right-Ω-regular) if in G*

$$\Omega \wedge \mathrm{d}f = 0 \Leftrightarrow \mathrm{d}(\Omega f) = 0$$

$$(\mathrm{d}f \wedge \Omega = 0 \Leftrightarrow \mathrm{d}(f\Omega) = 0)\,.$$

$\Omega \wedge \mathrm{d}f = 0$ *is called the* (*left*) *Cauchy-Riemann system of* Ω.

(ii) is a simple calculation and (iii) is the announced generalization of the notion of a Cauchy-Riemann system. With suitable forms Ω it is possible to get many interesting systems, within functions in algebras mostly the two cases $\Omega = \mathrm{d}z$ or $\Omega = \mathrm{d}z^*$ have been dealt with.

An application of Stoke's theorem is giving the

16.3.3. Cauchy integral theorem. (i) *If $f\colon H \to \mathfrak{W}$ is continuously differentiable and left-regular in a domain $H \subset \mathfrak{V}$ and if $G, \overline{G} \subset H$, is a domain with sufficiently smooth boundary, then*

$$\int_{\partial G} \mathrm{d}z^* f = 0\,.$$

(ii) *If f is continuously differentiable in a domain $H \subset \mathfrak{V}$, if Ω is a continuously differentiable closed differential form of degree p in H, if M_{p+1} is a differentiable manifold of dimension $p + 1$ in H with sufficiently smooth boundary, and if f is left-Ω-regular, then*

$$\int_{\partial M_{p+1}} \Omega f = 0\,.$$

For right-regular functions f and $\mathrm{d}z^*$ (or Ω) have to be changed. It is possible to prove the theorem by the method of Goursat if f is only Frechet-differentiable, or for even weaker assumptions. The inverse direction of the theorem — a theorem of Morera type — can be proved, provided f is continuously differentiable. Without this last assumption the inverse theorem would be false in most cases as we have not specified Ω or the algebras.

Theorem 16.3.3. (i) is valid for $(n + 1)$- or n-dimensional domains G, but in the latter case there are integral theorems in lower dimensional manifolds.

16.3.4. Lemma. (i) *Let $\omega_{i_1\ldots i_p} := (\mathrm{d}x_{1_p} \wedge \ldots \wedge \mathrm{d}x_{i_p} \wedge \mathrm{d}z)^*$ for $\mathfrak{V} = \mathbb{R}^n$ and $1 \leqq p \leqq n - 1$, let f be left-regular and continuously differentiable in the domain $G \subset \mathfrak{V}$, and M_{n-p} be a differentiable manifold in G of dimension $n - p$ with sufficiently smooth boundary. Then for all $(i_1, \ldots, i_p)$*

$$\int_{\partial M_{n-p}} \omega_{i_1\ldots i_p} f = \int_{M_{n-p}} \sum_{j=1}^{p} (-1)^{n+j+p} \omega_{i_1\ldots i_{j-1} i_{j+1}\ldots i_p} \frac{\partial f}{\partial x_{i_j}}$$

(ii) *If f is moreover biregular we have*

$$\int_{\partial M_{n-1}} \sum_{j=1}^{n} \omega_j f e_j = 0\,.$$

The proof follows from the formula

$$d(\omega_{i_1\ldots i_p} f) = (-1)^{n-1} (\mathrm{d}x_{i_1} \wedge \ldots \wedge \mathrm{d}x_{i_p})^* Df + \sum_{j=1}^{p} (-1)^{n+p+j}\, \omega_{i_1\ldots i_{j-1} i_{j+1}\ldots i_p} \frac{\partial f}{\partial x_{i_j}}$$

which can easily be verified.

16.3.5. Historical remarks. Nearly all papers in function theory in algebras are dealing with one or another form of the Cauchy-Riemann system and the Cauchy integral theorem. Therefore it is impossible to give here only the important citations. I mention a paper by Kriszten [1] because of his look on forms Ω in 16.3.2 (iii) with variable coefficients, including the Beltrami system, further Dentoni and Rizza [1] as they work with an Ω of general degree p. Something will be said in 16.6. on the case $\Omega = dg$ with some function g, which was introduced by Fedorov [2]. Other work is mostly investigating $\Omega = \mathrm{d}z$ or $\Omega = \mathrm{d}z^*$. Not much has been done for non-associative algebras, see Dentoni and Sce [1].

A quite other way to define Cauchy-Riemann systems has been given in an interesting paper by STEIN and WEISS [1]. If $\mathfrak{V} = \mathbb{R}^n$ and $\mathfrak{W} = \mathbb{R}^m$ they look for those classes of functions which are invariant under transformations of the form $R_\varrho^{-1} \circ f \circ \varrho$, where $\varrho \in SO(n)$ is an orthogonal transformation in $\mathbb{R}^n$ and R_ϱ is some representation of $SO(n)$ in $SO(m)$. In the plane this leads to classical function theory and in higher dimensional space they get also most of the interesting cases, giving so an important property of these Cauchy-Riemann systems.

16.4. The Cauchy integral formula

The question of getting a representation formula like Cauchy's for a regular function is much more difficult than the considerations in the last section. Therefore such a formula seems to be the decisive step into a function theory. On this way we need first a generalization of the Cauchy integral theorem, which is called generally

16.4.1. Green's formula. (i) *Let $f, g: G \to \mathfrak{W}$ be continuously differentiable, Ω a closed continuously differentiable differential form in G with values in $\mathfrak{A}$ and of degree p, let M_{p+1} be a differentiable manifold of dimension $p+1$ in G with sufficiently smooth boundary, then*

$$\int_{\partial M_{p+1}} g\Omega f = \int_{M_{p+1}} (\mathrm{d}g \wedge \Omega f + (-1)^p g\Omega \wedge \mathrm{d}f)\,.$$

(ii) *If moreover g is right- and f left-Ω-regular then*

$$\int_{\partial M_{p+1}} g\Omega f = 0\,.$$

The proof follows from Stoke's theorem. This formula is the first one for regular functions in which nonassociativity plays a role, in this case one has two such expressoins which differ in the brackets.

For the purpose of a Cauchy formula for a left-Ω-regular f one has to look for a suitable right-Ω-regular function g with the correct singularity at a point z_0 and this is just the difficulty.

If Ω is of degree $p < n-1$, a manifold ∂M_{p+1} has no interior points, so the information of the values of f on ∂M_{p+1} is not very great relatively to a point z_0. But nevertheless there are Cauchy formulas also for the case $\Omega = \mathrm{d}z$, then ∂M_{p+1} has to be a circle round a zero divisor line of the algebra, see EDENHOFER [1].

If Ω is of degree $n-1$, especially $\Omega = \mathrm{d}z^*$, ∂M_n is the boundary of a domain M_n and one can hope for more information in a point $z_0 \in M_n$. At least for elliptic systems which are normal determined (N equations for N unknown functions, i.e. $\mathfrak{W} = \mathfrak{A}$) there is a Cauchy integral formula in vectorial form (cf. JOHN [1]):

$$f_j(z_0) = -\int_{\partial G} \sum_{i \in I} \sum_{p,q \in K} \varepsilon_{ipq} k_{pj}(z_0, \zeta) f_q(\zeta)\, \mathrm{d}\xi_i^*$$

for z_0 in the domain G, k_{pq} are the elements of the so-called fundamental matrix. Ellipticity of the system means $\det\left(Q_{pq}(\xi)\right) \neq 0$ for all real vectors $(\xi_i)_{i \in I} \neq 0$ with

$$Q_{pq}(\xi) = \sum_{i \in I} \varepsilon_{ipq} \xi_i\,.$$

The restriction to the elliptic type seems natural, but it is an open question for which algebras the above Cauchy formula can be written in the algebra. A partial answer to this question has been given on other lines by the author [5].

The important case is that of quaternions and Clifford algebra which will be dealt with in the next section. Also the historical remarks will be postponed to that section.

16.5. Quaternions and Clifford algebra

The function theory in the quaternions and in the Clifford algebra is in this connection the most developed theory which has already been applied in many cases, therefore a short introduction to this theory is given here.

A Clifford algebra $\mathfrak{C}(\mathfrak{V}, \varphi)$ over a vector space $\mathfrak{V}$ with respect to a given symmetric bilinear form φ over $\mathfrak{V}$ is defined so that there are square roots of $\varphi(v, v)$ in $\mathfrak{C}(\mathfrak{V}, \varphi)$. For example this leads to a solution of the Dirac problem in quantum mechanics. For simplicity we give the following

16.5.1. Definition. Let $e_1, \ldots, e_n$ be an orthonormal base of $\mathbb{R}^n$, then the Clifford algebra $\mathfrak{C}_n(\mathbb{R})$ over $\mathbb{R}^n$ (with respect to the usual scalar product) has the base $e_0 = 1, e_1, \ldots, e_n$, and for $2 \leqq p \leqq n$ the products $e_{i_1}e_{i_2} \ldots e_{i_p}$, $i_1 < \ldots < i_p$. Together with the relations

$$e_i e_j + e_j e_i = 0 \quad \text{for} \quad i \neq j,\ 1 \leqq i, j \leqq n$$

$$e_i^2 = -1 \quad \text{for} \quad i = 1, \ldots, n$$

all products in $\mathfrak{C}_n(\mathbb{R})$ are defined. If Q_n is the power set of $\{1, \ldots, n\}$, an element of $\mathfrak{C}_n(\mathbb{R})$ can be written

$$w = \sum_{B \in Q_n} w_B e_B \,.$$

For $B = \emptyset$ we write always $B = 0$ and for $B = \{i\}$ only $B = i$. The dimension of $\mathfrak{C}_n(\mathbb{R})$ is clearly 2^n. There are other possibilities for the defining relations, for example $e_i^2 = +1$ for some i.

For the following we assume

$$\mathfrak{V} = \mathbb{R}^{n+1} = \{z \mid z = \sum_{i=0}^{n} x_i e_i\} \subset \mathfrak{W}$$

and

$$\mathfrak{W} = \mathfrak{A} = \mathfrak{C}_n(\mathbb{R})$$

with one exception: It has to be remarked that $\mathfrak{C}_1(\mathbb{R}) = \mathbb{C}$ and $\mathfrak{C}_2(\mathbb{R}) = \mathbb{H}$, the quaternion algebra. For $n = 1$ we choose $\mathfrak{V} = \mathfrak{W} = \mathbb{C}$, i.e. we have the classical function theory, for $n = 2$ we admit also the choice

$$\mathfrak{V} = \mathbb{R}^4 = \mathfrak{C}_2(\mathbb{R}) = \mathbb{H} = \mathfrak{W} = \mathfrak{A}\,,$$

then we have the function theory in the quaternions. This possibility rests on the fact that $\mathbb{H}$ is a skew field, which is false for $\mathfrak{C}_n(\mathbb{R})$ with $n > 2$. There will be no formal difference for the two cases $\mathfrak{V} = \mathbb{R}^3$ or $\mathfrak{V} = \mathbb{R}^4$ in the following.

The importance of the concept of the Clifford algebra lies in the fact that all special algebras of physics are equal to, or subalgebras of, or ideals in some convenient $\mathfrak{C}_n(\mathbb{R})$, see Porteous [1] and Lounesto [1] with further references, for the physical relevance see e.g. Hestenes [1, 2], and Gürsey and Tze [1]. This connection to physics is far from being exhausted.

For example two of these possible restrictions should be mentioned: For $\mathfrak{V} = \mathfrak{W} = \mathbb{R}^n$ the Cauchy-Riemann system is the Riesz system $\operatorname{rot} \mathfrak{v} = 0$, $\operatorname{div} \mathfrak{v} = 0$ for a vector-valued function $\mathfrak{v}$; $\mathfrak{V} = \mathbb{R}^n$ and $\mathfrak{W}$ = subspace of $\mathfrak{C}_n(\mathbb{R})$, spanned by the basic elements $e_{i_1} \dots e_{i_p}$ of order p gives the Hodge system $d\omega = 0$, $\delta\omega = 0$ for a differential form ω of degree p.

First we will formulate some properties of the Clifford algebra $\mathfrak{C}_n(\mathbb{R})$:

16.5.2. Definition and Lemma. (i) *The multiplication in* $\mathfrak{C}_n(\mathbb{R})$ *is associative.*

(ii) *There is an involution on* $\mathfrak{C}_n(\mathbb{R})$ *defined by*

$$\overline{e_{i_1} \dots e_{i_p}} := (-1)\, p(p+1)/2 \;\; e_{i_1} \dots e_{i_p}$$

and linear continuation, such that

$$\bar{e}_B e_B = e_B \bar{e}_B = 1 \quad \textit{for all} \quad B \in Q_n \,.$$

(iii) *For all* $z = x_0 + \sum_{i=1}^{n} x_i e_i =: x_0 + \mathfrak{x} \in \mathfrak{B}$ *we have* $z\bar{z} = \bar{z}z = |z|^2$ *and therefore, provided* $z \neq 0$,

$$z^{-1} =: \frac{1}{z} = \frac{\bar{z}}{|z|^2} \,.$$

(iv) *For* $a, b \in \mathfrak{C}_n(\mathbb{R})$ *we have the real part*

$$\operatorname{Re} a = a_0 \quad \textit{if} \quad a = \sum_{B \in Q_n} a_B e_B \,,$$

$\operatorname{Re}(a\bar{b})$ *is the usual scalar product and* $\operatorname{Re}(a\bar{a}) = |a|^2$ *the Euclidean norm.*

The proofs are easy. Besides 16.5.2 (iv) we have an algebra norm $|a|_0 := 2^n\,|a|$ in $\mathfrak{C}_n(\mathbb{R})$ with $|ab|_0 \leqq |a|_0\,|b|_0$. The Cauchy-Riemann system for left-regular (right-regular) functions is

$$Df = 0\,, \qquad (fD = 0)\,, \qquad D = \sum_{i=0}^{n} e_i \frac{\partial}{\partial x_i} \,.$$

The Cauchy integral theorem is given in 16.3.4, and Green's formula in 16.4.1. We have now a very simple Cauchy kernel (in $\mathbb{H}$ is $n = 3$)

$$g(z, \zeta) = \frac{\overline{\zeta - z}}{|\zeta - z|^{n+1}} \,,$$

for $n = 1$ this is just the classical kernel in $\mathbb{C}$. The usual proof gives the

16.5.3. Cauchy integral formula. *If f is left-regular in a domain $H \subset \mathfrak{B}$ and if G is a domain with $\bar{G} \subset H$ and sufficiently smooth boundary then we have for all $z \in G$*

$$f(z) = \frac{1}{\omega_{n+1}} \int_{\partial G} \frac{\overline{\zeta - z}}{|\zeta - z|^{n+1}}\, d\zeta^* f(\zeta) \,.$$

ω_{n+1} *is the surface of the unit ball in* $\mathbb{R}^{n+1}$.

For right-regular functions $f(\zeta)$ and $(\overline{\zeta - z})/|\zeta - z|^{n+1}$ have to change their places. If we restrict the consideration to $\mathfrak{B} = \mathbb{R}^n$ (i.e. $x_0 = 0$), the exponent has to be n, and in the quaternions $n = 3$. Some of the usual consequences of the Cauchy integral formula may be mentioned:

16.5.4. Corollary. (i) *A left-(or right-)regular function is real analytic, its components are harmonic functions, so we have Poissons and the mean value formula.*

(ii) **Liouville's theorem.** *If f is left-regular and bounded in $\mathfrak{B}$ it is constant.*

(iii) **Maximum principle.** $f =$ const *in G or*

$$|f(z)| < M = \sup_{\zeta \in G} |f(\zeta)| \quad \textit{for all} \quad z \in G \,.$$

(i) can be proved by developing $1/|z-\zeta|^{n+1}$ in some neighbourhood of $z_0 \in G$ in a multiple power series and integrating the series term by term. $\overline{D}D = \Delta$ gives the harmonicity and then the real Poisson formulas for the components can be added up.

(ii) is as usual proved by differentiating the Cauchy formula for a ball of radius R and then $R \to \infty$. Also the stronger form can be shown, that f is a polynomial of degree at most s, if $|f(z)| \leqq K\,|z|^s$.

The first step to the maximum principle follows from the mean value formula:

$$|f(z)| \leqq \frac{1}{\omega_{n+1}\varrho^{n+1}} \int\limits_{|\zeta - z_0| = \varrho} |f(\zeta)|\, do(\zeta)$$

gives as usual $|f(z)| < M$ in G or $|f(z)| = M$ in G. To prove $f = \text{const}$ for the last case one can proceed as follows: $|f(z)| = M$, then

$$0 = D\,|f(z)|^2 = 2 \sum_{B \in Q_n} f_B D f_B$$

and because of $\Delta f = 0$

$$0 = \overline{D}D\,|f(z)|^2 = 2 \sum_{B \in Q_n} (\overline{D} f_B)\,(D f_B) + 2 \sum_B f_B\,\Delta f_B$$

$$= 2 \sum_B (\overline{D} f_B)\,(D f_B) = 2 \sum_B |D f_B|^2\,,$$

this means $D f_B = 0$, $f_B = \text{const}$ for all $B \in Q_n$.

Next we wish to give the Taylor and Laurent expansion.

16.5.5. Definition. (i) For $j = 1, \ldots, n$ let $z_j := x_j - x_0 e_j$, let $\alpha = (\alpha_1, \ldots, \alpha_n)$ be a multiindex with

$$|\alpha| = \alpha_1 + \ldots + \alpha_n\,, \qquad \alpha! = \alpha_1! \ldots \alpha_n!$$

and with $\nabla := \left(\frac{\partial}{\partial x_1}, \ldots, \frac{\partial}{\partial x_n}\right)$

$$\nabla^\alpha = \frac{\partial_{|\alpha|}}{\partial x_1^{\alpha_1} \ldots \partial x_n^{\alpha_n}}$$

(in the quaternions n has to be 3).

(ii) For given α let in $(i_1, \ldots, i_{|\alpha|})$ the first α_1 numbers be equal to 1, the next α_2 numbers equal to 2 and so on. Then the Fueter polynomials are defined to be

$$P_\alpha(z) := \frac{1}{|\alpha|!} \sum_{\sigma \in S_{|\alpha|}} z_{i_{\sigma(1)}} \ldots z_{i_{\sigma(|\alpha|)}} \quad \text{for} \quad |\alpha| > 0\,,$$

$$P_\alpha(z) := 1 \quad \text{for} \quad |\alpha| = 0\,,$$

where $S_{|\alpha|}$ is the symmetric group of $|\alpha|$ elements. For convenience $P_\alpha = 0$ if α has a negative entry.

The P_α are homogeneous polynomials of degree $|\alpha|$. For $n = 1$ we have $z_1 = x_1 - x_0 e_1$ $= -e_1(x_0 + e_1 x_1) = -iz$ in the usual notation, $\alpha = (\alpha_1)$ and $P_\alpha(z) = z_1^{\alpha_1} = (-iz)^{\alpha_1}$. It is possible to use $\tilde{z}_j := x_0 + e_j z_j$ instead of z_j, and $\tilde{z}_j$ reduces really to z in the case $n = 1$, but then the formulas are much more complicated.

For the following properties of P_α it is important to know that many authors are dealing with $|\alpha|!\,P_\alpha(z)/\alpha!$ and the formulas are slightly different then. This normalization makes the results a bit more similar to the usual ones of real and complex analysis.

16.5.6. Lemma. (i) *Let ε_j be the multiindex with a 1 at place j and zeroes otherwise, then*

$$|\alpha|\, P_\alpha(z) = \sum_{j=1}^{n} \alpha_j z_j P_{\alpha-\varepsilon_j} = \sum_{j=1}^{n} \alpha_j P_{\alpha-\varepsilon_j} z_j\,.$$

(ii) $\partial P_\alpha/\partial x_k = \alpha_k P_{\alpha-\varepsilon_k}$, $k = 1, \ldots, n$, *and therefore for* $\beta \leqq \alpha$

$$\nabla^\beta P_\alpha = \frac{\alpha!}{(\alpha-\beta)!}\, P_{\alpha-\beta}$$

and for $|\alpha| = |\beta|$ *with the Kronecker symbol*

$$\nabla^\beta P = \alpha!\,\delta_{\alpha\beta}\,.$$

(iii) P_α *is left- and right-regular.*

(iv) *Every left-regular homogeneous polynomial P of degree k possesses the uniquely determined representation*

$$P = \sum_{|\alpha|=k} P_\alpha(z)\, c_\alpha \quad \textit{with} \quad c_\alpha = \frac{1}{\alpha!}\nabla^\alpha P(0)\,.$$

For right-regular P the c_α are standing left from P_α.

(v) *Binomial formula*:

$$P_\alpha(\zeta + z) = \sum_{\beta+\gamma=\alpha} \frac{\alpha!}{\beta!\gamma!} P_\beta(z)\, P_\gamma(\zeta) = \sum_{\beta+\gamma=\alpha} \frac{\alpha!}{\beta!\gamma!} P_\beta(\zeta)\, P_\gamma(z)\,.$$

Some hints for the proof: For (i) one has to group the terms in P_α according to the first (or last) factor z_j, the rest is just $P_{\alpha-\varepsilon_j}$, but as one cannot distinguish the α_j different z_j, there has to be the factor α_j. (ii) can be proved by induction over $|\alpha|$ with the aid of (i). By Euler's formula for homogeneous functions and by (ii) one has

$$|\alpha|\, P_\alpha(z) = \sum_{j=0}^{n} x_j \frac{\partial P_\alpha}{\partial x_0}(z) = \sum_{j=1}^{n} \alpha_j z_j P_{\alpha-\varepsilon_j}(z) + x_0 D P_\alpha$$

which gives $DP_\alpha = 0$ by (i), thus proving (iii). In (iv) P is uniquely determined by $\nabla^\alpha P$ with $|\alpha| = k$, as the derivatives with respect to x_0 can be calculated via $DP = 0$. From (ii)

$$\nabla^\beta \Big(P(z) - \sum_{|\alpha|=k} P_\alpha(z)\, c_\alpha\Big) = \nabla^\beta P(z) - \beta! c_\beta = 0$$

follows for $|\beta| = k$ which gives the desired result. (v) is proved by induction over $|\alpha|$, for the step from $|\alpha| - 1$ to $|\alpha|$ one uses the fact that

$$F(z, \zeta) := P_\alpha(z + \zeta) - \sum_{\beta+\gamma=\alpha} \frac{\alpha!}{\beta!\gamma!} P_\beta(z)\, P_\gamma(\zeta)$$

is a left-regular polynomial in z. Differentiating with respect to x_k, $k = 1, \ldots, n$, one gets $\partial F/\partial x_k = 0$ and from $DF = 0$ also $\partial F/\partial x_0 = 0$. $F(0, \zeta) = 0$ shows the assertion.

The next step towards Taylor and Laurent expansion is a suitable series development of the Cauchy kernel. For this purpose we use the known development

$$\frac{1}{(1 - 2st + t^2)^\gamma} = \sum_{k=0}^{\infty} C_{k,\gamma}(s)\, t^k$$

with the Gegenbauer polynomials $C_{k,\gamma}$ (sometimes they are not normalized as here). With the ordinary scalar product

$$\langle z, \zeta\rangle := \sum_{j=0}^{n} x_j \xi_j$$

we have (once again: for the quaternions is $n = 3$)

$$\frac{1}{|\zeta - z|^{n-1}} = \frac{1}{\left||z|^2 + |\zeta|^2 - 2\langle z, \zeta\rangle\right|^{(n-1)/2}}$$
$$= \frac{1}{|\zeta|^{n-1}} \sum_{k=0}^{\infty} C_{k,\frac{n-1}{2}}\left(\left\langle \frac{z}{|z|}, \frac{\zeta}{|\zeta|}\right\rangle\right)\left|\frac{z}{\zeta}\right|^k .$$

The single term of this series is harmonic in z and in ζ. The series and the term by term differentiated series converge locally uniformly in $|z| < |\zeta| < R$. Defining

$$\pi_k(z, \zeta) := C_{k,\frac{n-1}{2}}\left(\left\langle \frac{z}{|z|}, \frac{\zeta}{|\zeta|}\right\rangle\right)\frac{|z|^k}{|\zeta|^{k+n-1}} \tag{1}$$

we get (the index at D indicates the differentiated variable)

$$\frac{1}{n-1}\,\overline{D}_z \frac{1}{|\zeta - z|^{n-1}} = \frac{\overline{\zeta - z}}{|\zeta - z|^{n+1}} = \sum_{k=0}^{\infty} \frac{1}{n-1}\,\overline{D}_z \pi_k(z, \zeta)\,. \tag{2}$$

The $\overline{D}_z\pi_k(z, \zeta)/(n-1)$ are homogeneous polynomials in z of degree $k-1$. As the series in (2) is biregular, these polynomials are also biregular. By 16.5.6 (iv) we have the following

16.5.7. Definition and Lemma. *The functions $Q_\alpha(\zeta)$ are defined by* (1) *and*

$$\sum_{|\alpha|=k} P_\alpha(z)\, Q_\alpha(\zeta) = \sum_{|\alpha|=k} Q_\alpha(\zeta)\, P_\alpha(z) = \frac{1}{n-1}\,\overline{D}_z \pi_{k+1}(z, \zeta)\,.$$

We have

$$Q_\alpha(\zeta) = \frac{1}{n-1}\,\nabla_z^\alpha \overline{D}_z \pi_{k+1}(z, \zeta)\,.$$

Because of $\overline{D}_z \frac{1}{|\zeta - z|^{n-1}} = -\overline{D}_\zeta \frac{1}{|\zeta - z|^{n-1}}$ and the $\pi_k(z, \zeta)$ being homogeneous functions of ζ of degree $-k-n+1$ the $Q_\alpha(\zeta)$ are also biregular functions. The $Q_\alpha(\zeta)$ are the analogues of the negative powers of z in classical function theory as the $P_\alpha(z)$ are the analogues of the positive powers. As the involved series (2) converges uniformly for $|z/\zeta| \leqq \varrho < 1$, we can put it into the Cauchy formula and integrate term by term getting

16.5.8. Theorem. (i) *Taylor series: If f is left-regular for $|z| < R$ we have there the locally uniformly converging Taylor series*

$$f(z) = \sum_{k=0}^{\infty} \sum_{|\alpha|=k} P_\alpha(z)\, c_\alpha$$

with $(0 < \varrho < R)$

$$c_\alpha = \frac{1}{\omega_{n+1}} \int_{|\zeta|=\varrho} Q_\alpha(\zeta)\, \mathrm{d}\zeta^* f(\zeta) = \frac{1}{\alpha!}\nabla^\alpha f(0)\,.$$

(ii) *Laurent series: If f is left-regular for $R_1 < |z| < R_2$ we have there the locally uniformly converging Laurent series*

$$f(z) = \sum_{k=0}^{\infty} \sum_{|\alpha|=k} P_\alpha(z)\, c_\alpha + \sum_{k=0}^{\infty} \sum_{|\alpha|=k} Q_\alpha(z)\, d_\alpha$$

with $(R_1 < \varrho < R_2)$

$$c_\alpha = \frac{1}{\omega_{n+1}} \int_{|\zeta|=\varrho} Q_\alpha(\zeta)\, \mathrm{d}\zeta^* f(\zeta)\,, \qquad d_\alpha = \frac{1}{\omega_{n+1}} \int_{|\zeta|=\delta} P_\alpha(\zeta)\, \mathrm{d}\zeta^* f(\zeta)\,.$$

For the **proof** of (ii) one has to apply the Cauchy formula as usual in the annular region $R_1 < \varrho_1 < |z| < \varrho_2 < R_2$, the integral over the outer ball gives the Taylor part, that over the inner ball has to be handled in an analogous manner, ζ and z changing places.

We may stop here introducing function theory in Clifford algebras. There are numerous further developments: Residue theory, theorems of Mittag-Leffler and Runge, subharmonicity of $|f|^\mu$ for $\mu \geqq (n-2)/(n-1)$ and consequences, growth problems, representation of distributions in $\mathbb{R}^n$ as boundary values of regular functions in $\mathbb{R}^{m+1}$, generalized Fourier, Laplace, and Borel transformations, H^p-theory. The best reference for all these things is the book by BRACKX, DELANGHE, and SOMMEN [1], also the lecture notes by FUETER [1, 3].

16.5.9. Historical remarks. The starting point for the quaternion theory are papers by MOISIL and THÉODORESCU [1] and by FUETER [2]. The latter published a long series of papers and the above mentioned two lecture notes [1, 3]. The Fueter school also began to look on Clifford algebras, see HAEFELI [1], the first paper in this direction being written by MOISIL [1]. Later on there must be mentioned IFTIMIE [1], DELANHGE, e.g. [1], DELANGHE and BRACKX [1], see the book BRACKX, DELANGHE, and SOMMEN [1]. A major contribution from the last time are the thesis by SOMMEN [1].

16.6. Further investigations

This section will be even more incomplete than the other ones. But I shall at least try to give some hints and references to developments which cannot be dealt with in detail here.

There are interesting investigations using infinite-dimensional algebras, I refer to the book by SNYDER [2] and a paper by LAMMEL [1].

A second point is the Douglis theory, often called hyperanalytic function theory. DOUGLIS [1] started this dealing with general elliptic systems in the plane, he got a normal form which can be described in our notation by $\mathfrak{B} = \mathbb{C}$, $\mathfrak{W} = \mathbb{C}^p$ with the base $1, e, \ldots, e^{p-1}$ and $e^p = 0$. He used $D = \partial/\partial\bar{z} + a(z)\, e\, \partial/\partial z$, so systems with variable coefficients can be investigated. This theory was developed during the last years by GILBERT and his school, see the book [2] and chapter 12 in this book.

Another development was begun by FEDOROV [2], who defined the derivative of a function f with respect to some other function g, both having values in a given algebra $\mathfrak{A}$. This can be described by the assumption $\mathrm{d}f = \lambda\, \mathrm{d}g$, and in a commutative algebra by $\mathrm{d}f \wedge \mathrm{d}g = 0$, so that we can speak of dg-regularity. Here also variable coefficients of the underlying Cauchy-Riemann system are possible. For the later state of this theory see MOREV [1].

Further investigation has been done for solutions of $D^p f = 0$, $p \in \mathbb{N}$, $p > 1$. For classical function theory these functions are called polyanalytic (bianalytic or areolar monogenic for $p = 2$), this theory started in the early thirthies with N. THÉODORESCU now there is a school around M. B. BALK, see chapter 3 in this book. For functions in other algebras than $\mathbb{C}$ (especially quaternion and Clifford theory) see the publications by COROI-NEDELCU, most of them in matrix notation e.g. [1], and BRACKX [2].

There are very interesting papers in matrix or vector notation which without doubt have to be put in this context, and not only because of the fact that quadratic matrices form algebras. I cite Coroi-Nedelcu [1], Hile [2], and Obolašvili [3], who give also more references.

Another branch are the solutions of equations $Df = Lf$ with a right side linear in f. For classical function theory the solutions are the pseudoanalytic or generalized functions in the sense of Bers and Vekua, these are investigated by some other authors in this book. For the case of algebras I refer to Obolašvili [3] (in vector notation) and Goldschmidt [4].

There are many papers trying to deal with higher order differential equations by means of function theories in suitable algebras. This is generally done by decomposing the given differential operator K in a product D_1D of two or more operators with a D in the sense of this article, ideal is $K = D^p$. A procedure like this was leading to the Dirac algebra of quantum mechanics, what is closely related to the number theoretical question of writing the product of two quadratic forms as the square of a third one. For work with respect to higher order equations I cite Sobrero [1] and Eichhorn [2].

At last I should like to mention geometric properties of functions between algebras. Generalization of the notion of conformal transformations in the plane has led to the important theory of quasiconformal mappings. Very few work has been done for function theory in algebras, I cite Faber [1], Haefeli [2], and Rizza [2].

At the end the reader may allow an attempt of a small vocabulary on these many different sorts of analytic functions:

monogenic: differentiable within the algebra, Hausdorff-differentiable, sometimes also used instead of regular.

regular: (also holomorphic, but seldom used, much worse is analytic alone because of the so many sorts of analytic functions) functions which satisfy a Cauchy-Riemann system and integral theorems, often a prefix is used for better characterization: Ω-regular, hyperregular, D-regular, ...

pseudo or generalized: solutions of $Df = Lf$ with a right side linear in f, it follows the nearer characterization of $Df = 0$, e.g. pseudohyperanalytic.

poly: prefix for solutions of $D^pf = 0$ with $p > 1$.

17. Analytic Continuation of Solutions to Partial Differential Equations with Applications to Problems in Scattering Theory

D. L. Colton

17.1. Introduction

In this chapter we shall indicate some of the current directions in the study of the analytic continuation of solutions to partial differential equations. Our intended audience is not only those mathematicians working in the area of partial differential equations, but also complex analysists interested in seeing how a major topic in the theory of analytic functions can be extended to include certain classes of linear partial differential equations. In order for such a generalization to be worthwhile it seems to us that ideally two conditions should be met: The extension should disclose new phenomena that are peculiar to partial differential equations and not characteristic of analytic functions, and the investigation of such new phenomena should illuminate areas of physical applications having as their mathematical model the partial differential equations being studied. Our specific aim in this paper is to therefore provide a survey of the general area of the analytic continuation of solutions to partial differential equations and to show how such a study elegantly unites both analytic function theory and the theory of partial differential equations and their applications. For the sake of brevity we shall concentrate on three main topics: the unique continuation of solutions to partial differential equations, reflection principles for equations of elliptic and parabolic type, and, as an example of an area of application, the inverse scattering problem for acoustic waves.

17.2. Unique continuation

First we recall that a family of functions is said to have the *unique continuation property* with respect to a surface S if any two functions of this family which are defined in a neighbourhood containing a portion of S in its interior, and agree on one side of the surface S, are in fact equal to each other in the entire neighbourhood. Of particular interest in the theory of partial differential equations is the case when the family of functions is the class of all solutions to a given partial differential equation defined in a domain containing the surface S. The unique continuation property in this case clearly depends on both the equation and the surface, and for a linear homogeneous partial differential equation is equivalent to the uniqueness of the solution to Cauchy's problem with data prescribed on S. The classical result in this direction is Holmgren's uniqueness theorem which states that classical solutions of linear partial differential equations with analytic coefficients possess the unique continuation property with respect to any smooth noncharacteristic surface (cf. Bers, John, and Schechter [1]). The case of partial differential equations with non-analytic coefficients is only partly solved (cf. CALDERÓN [1] and the references contained in BERS, JOHN, and SCHECHTER [1]).

Now assume that the linear partial differential equation under consideration is uniformly elliptic with smooth coefficients, and write it as

$$L[u] = 0 . \tag{1}$$

A remarkable result established by LAX [1], MALGRANGE [1], and BROWDER [1] is that the unique continuation property for solutions of $L[u] = 0$ is equivalent to a Runge approximation property for the adjoint equation:

Runge Approximation Property. Solutions of $L[u] = 0$ are said to have the *Runge approximation property* if, whenever D_1 and D_2 are two bounded simply connected domains, D_1 a subset of D_2, any solution in D_1 can be approximated uniformly on compact subsets of D_1 by a sequence of solutions which can be extended as solutions to D_2.

The theorem of Lax, Malgrange and Browder can now be formulated as follows:

Solutions of $L[u] = 0$ have the Runge approximation property if and only if solutions of the adjoint equation have the unique continuation property with respect to any smooth surface S.

As a simple consequence of this result we have, using Holmgren's uniqueness theorem, that the harmonic polynomials are complete (with respect to the maximum norm over compact subsets) in the space of solutions of Laplace's equation defined in a bounded simply connected domain. This follows by choosing D_2 to be a sphere in the definition of the Runge approximation property.

It would be desirable to extend the result of Lax, Malgrange, and Browder to the case of parabolic equations defined in domains with moving boundaries. Some partial results in this direction can be found in COLTON [13]. For example, let $D = \{(x, t) : s_1(t) < x < s_2(t), 0 \leqq t \leqq t_0\}$ where t_0 is a positive constant and $s_1(t)$, $s_2(t)$ are analytic functions of t for $0 \leqq t \leqq t_0$. Then it can be shown that the polynomials

$$h_n(x, t) = \sum_{k=0}^{[n/2]} \frac{x^{n-2k} t^k}{(n - 2k)!\, k!} \tag{2}$$

are complete in the space of solutions to the heat equation

$$u_{xx} = u_t \tag{3}$$

defined in D.

17.3. Reflection principles

Given the possibility of unique continuation, the problem which naturally arises is how to carry out this continuation in specific cases. Occasionally one has an explicit integral representation of a solution in terms of analytic functions with known singularity manifolds, and in this case the continuation can often be performed by using classical results from analytic function theory (cf. GILBERT [1], MILLAR [2]). However in general one is given a solution of a partial differential equation defined in a domain D and satisfying known boundary data f on a portion σ of the boundary of D, and from a knowledge only of D and f it is desired to continue the solution across σ. In order to solve this problem it is necessary to develop reflection principles for partial differential equations. In particular assume that a preliminary change of variables has been made such that σ is a portion of a hyperplane. Then the partial differential equation is said to have a *reflection principle* with respect to σ if, under suitable conditions on the boundary data f, it is possible to continue the solution u across σ into the domain D^*, where D^* is the mirror image of D with respect to σ. The classical example of such behaviour is the Schwarz reflection principle for solutions of Laplace's equation that vanish on a portion of a hyperplane, and in this case an explicit reflection rule is provided by the formula

$$u(-x_1, x_2, \dots, x_n) = -u(x_1, x_2, \dots, x_n) \tag{4}$$

where we have taken σ to be $x_1 = 0$. For example of other classes of partial differential equations with constant coefficients admitting an explicit reflection rule for solutions having vanishing Dirichlet data on a hyperplane see John [2].

The case of partial differential equations with constant coefficients and solutions having vanishing Dirichlet data on a portion of the boundary is rather limited and hence efforts have been made to extend such results to linear equations with analytic coefficients and nonhomogeneous boundary data. The somewhat surprising result is that in general this is only possible for partial differential equations in two independent variables. To state the results in the case of two independent variables we assume, according to our convention, that σ is a portion of one of the coordinate axis. For second order elliptic equations in two independent variables with analytic coefficients a reflection principle was established by Lewy [1] under the condition that the boundary data f prescribed on σ is the restriction to σ of an analytic function $f(z)$ for $z \in D \cup \sigma \cup D^*$. The case for second order parabolic equations in two independent variables with analytic coefficients has been studied by Colton [13] with the result that a reflection principle is valid provided f is analytic on σ, where now we must insist that σ is the t-axis so that it is not characteristic. Both of these results have been extended to the case where the solution satisfies a first order linear boundary condition on σ (Colton [13], Lewy [1]).

The problem of when a reflection principle is valid for partial differential equations in more than two independent variables is considerably more complicated than the case of equations in two independent variables. In the case of equations with constant coefficients and σ a hyperplane we refer the reader to the previously cited work of John. Reflection principles are also possible for solutions of the Laplace or Helmholtz equation defined in a ball and vanishing on a portion of the spherical boundary (the case of Laplace's equation is due to Schwarz and is classical, whereas the reflection principle for the Helmholtz equation has only recently been obtained by Colton [13]). On the other hand, Lewy has given an example of a solution of Laplace's equation satisfying a first order linear homogeneous boundary condition with constant coefficients along a hyperplane such that the solution cannot be continued across this hyperplane into the mirror image of its original domain of definition (Lewy [1]). The problem of when solutions of higher-dimensional elliptic equations can be reflected across analytic boundaries was taken up by Garabedian [1] who showed that the domain of dependence associated with a solution of an n-dimensional elliptic equation at a point on one side of an analytic surface is in general a whole n-dimensional ball on the other side. Only in exceptional circumstances does some kind of degeneracy occur which causes the domain of dependence to collapse onto a lower-dimensional subset, thus allowing a continuation into a larger region than that afforded in general. Such is the situation for example in the case of the Schwarz reflection principle for harmonic functions defined in a ball or half space (where the domain of dependence is a point) and Colton's reflection principle for solutions of the Helmholtz equation defined in a ball (where the domain of dependence is a one-dimensional line segment). Such a collapsing of the domain of dependence can be viewed as a *Huygen's principle for reflection*, analogous to the classical Huygen's principle for hyperbolic equations.

The analysis discussed above still leaves open the possibility of obtaining a reflection principle for parabolic equations in two space variables with analytic coefficients. In the case of *analytic* solutions of such equations having analytic Dirichlet data on a portion of a plane boundary, a reflection principle has been established by Colton [14] and the possibility of obtaining explicit reflection formulas has been investigated by Hill [1]. In this case the domain of dependence is a one-dimensional line segment, in conformity with Huygen's principle for reflection. An open problem is to consider the case of first-order boundary conditions and to remove the assumption of analyticity of the solutions. This last problem is intimately connected with the regularity of solutions to initial value problems for equations having the operator

$$L[u] = u_{xy} - u_t \tag{5}$$

as principal part (Colton [14]).

17.4. The inverse scattering problem

Recent work in the area of analytic continuation of solutions to partial differential equations has often been divorced from practical applications and as a result the desired interplay between "pure" and "applied" mathematics which has been characteristic of the history of this field has in recent years been muted. It is the purpose of this paper to try and rectify this imbalance by considering problems in analytic continuation which arise in the mathematical treatment of the scattering of acoustic waves by an obstacle. Although there have been many interesting developments connected with the analytic dependence of the far field pattern on the wave number, leading to the theory of Regge poles and resonant states, we shall restrict our attention here to the more narrowly defined problem of the analytic continuation of solutions to the Helmholtz equation. This problem is basic to the mathematical development of scattering theory, appearing for example in the proof of the uniqueness of the solution to the direct scattering problem (cf. Hellwig [2], I. N. Vekua [3]). However the full appreciation of the role of analytic continuation methods in scattering theory only becomes apparent when one considers the inverse scattering problem, i.e. the problem of determining the shape of the scattering obstacle from a knowledge of the far field pattern, and it is this problem that we shall be interested in here.

In recent years there has been a rapidly growing interest in various classes of *inverse problems*, in particular those in which from measured data it is desired to either determine the domain of definition of the solution or, given the domain, to determine the boundary conditions satisfied by the solution. Typical examples of this type of problem are the backwards heat equation, the inverse Stefan problem, and the inverse scattering problem (cf. Colton [13]). A broad class of these problems (in particular the ones just listed) can be characterized as problems in the analytic continuation of solutions to partial differential equations, but where now the solutions are not known exactly but only in some approximate sense. An interesting characteristic of such problems is that they are in general improperly posed, i.e. the solution either does not exist, is not unique, or does not depend continuously on the initial data. Due to what now appears to be an inexplicable bias, the area of improperly posed problems in applied mathematics was left almost untouched by mathematicians until the late 1950's when the beginning of a theory was initiated. Thus in 1961 Courant wrote (cf. COURANT and HILBERT [1]): "The stipulations about existence, uniqueness and stability of solutions dominate classical mathematical physics. They are deeply inherent in the ideal of a unique, complete and stable determination of physical events by appropriate conditions at the boundaries, at infinity, at time $t = 0$, or in the past. Laplace's vision of the possibility of calculating the whole future of the physical world from complete data of the present state is an extreme expression of this attitude. However, this rational ideal of casual-mathematical determination was gradually eroded by confrontation with physical reality. Nonlinear phenomena, quantum theory, and the advent of powerful numerical methods have shown that "properly posed" problems are by far not the only ones which appropriately reflect real phenomena. So far, unfortunately, little mathematical progress has been made in the important task of solving or even identifying and formulating such problems which are not "properly posed" but still are important and motivated by realistic situations".

Since the time of these observations by COURANT, there has been a virtual explosion of interest in improperly posed problems, and we refer the reader to the recent conferences by ANGER [1] and NASHED [1] for current developments in this area. However, as the interest has risen, the problems have multiplied, and many of the most important improperly posed problems arising in applications are still waiting for a solution that is suitable for practitioners.

In order to investigate a problem that is improperly posed, we must answer two basic questions: (1) What do we mean by a solution? and (2) How do we construct this solution? The answers to these questions are by no means trivial. For example, as initially posed a solution may not even exist in the classical sense, or if it does exist, may not be defined in a large enough domain to be of practical use (this is the place where analytic continuation often comes into play). A classic example of such a situation arises in the study of the inverse scattering problem, a subject which has already been the topic of numerous research papers and monographs. Here we restrict ourselves to the inverse scattering problem for acoustic waves, i.e. we consider the scattering of a time harmonic plane wave by a rigid bounded obstacle, and from a knowledge of the asymptotic behaviour of the scattered wave we want to determine either the location of the so-called *equivalent sources* generating the field, or the shape of the scattering obstacle.

Two distinct problems present themselves, one in which the far field pattern is known exactly and the other when it is only known to within a given experimental error. The first problem is mainly theoretical, and addresses itself to the problem of characterizing the analytic nature of the far field pattern and the relation of this analytic behaviour to the location of the singularities of the scattered field. This problem has been the object of attention of numerous investigations (BATES and WALL [1], COLTON [1], MILLAR [1], MÜLLER [1], SLEEMAN [1]) and will be the subject matter of 17.5. The second problem is of more direct practical interest and is basically the problem of analytically continuing solutions of the Helmholtz equation where the solution is not known exactly in its original domain of definition. This can most easily be seen as follows. If we assume that the scattering obstacle D is contained in a ball B centered at the origin, then an approximate knowledge of the far field pattern means that we approximately know the scattered wave u_s outside B. The aim is to use this knowledge to determine D, and if this can be done then we know u_s outside D, i.e. we have continued u_s from the exterior of B into the exterior of D. This problem is clearly improperly posed and hence in order to arrive at a "solution" we must carefully define what this means and describe a stable procedure for constructing it. This problem will be discussed in 17.6. In the author's opinion this is an exceptionally promising area of future research, and one of the main purposes of this paper is to kindle interest in this subject in the hopes new approaches can be found to solve the many outstanding problems which still remain.

17.5. The inverse scattering problem: exact data

We first define more precisely what we mean by the inverse scattering problem for acoustic waves. Let $D \subset \mathbb{R}^n$ be a bounded simply connected acoustically "soft" obstacle with smooth boundary ∂D. Suppose an incoming time harmonic plane wave is scattered by D. Then the *direct* scattering problem is to find a solution u of

$$\Delta_n u + k^2 u = 0 \quad \text{in} \quad \mathbb{R}^n \setminus \bar{D}, \tag{6a}$$

$$u = 0 \quad \text{on } \partial D, \tag{6b}$$

$$u = u_i + u_s, \tag{6c}$$

$$\lim_{r \to \infty} r^{(n-1)/2} \left(\frac{\partial u_s}{\partial r} - iku_s \right) = 0 \tag{6d}$$

where $r = |x|$, $k > 0$, is the wave number, u_i denotes the (known) incident wave and u_s the (unknown) scattered wave. It is well known (cf. I. N. Vekua [3]) that a solution to (6) exists, is unique, and that at infinity u_s has the asymptotic behaviour

$$u_s \sim e^{ikr} r^{-(n-1)/2} f(\hat{x}; k); \qquad r \to \infty \tag{7}$$

where $\hat{x} = x/|x|$ and f is known as the *far field pattern*. The *inverse* scattering problem is to determine D from a knowledge of f.

The fact that solutions of the Helmholtz equation (6a) are analytic in $\mathbb{R}^n \setminus \bar{D}$ allows us to establish uniqueness of the solution to the inverse scattering problem (Lax and Phillips [1]):

Theorem 17:1. *D is uniquely determined by a knowledge of f for all angles $\hat{x}$ and k on any interval of the positive real axis.*

Proof. Suppose there existed two obstacles D_1 and D_2 having the same far field pattern f. Consider first the case when $\bar{D}_1$ and $\bar{D}_2$ are disjoint. Then since f uniquely determines u_s outside a ball containing D_1 and D_2 in its interior we can conclude by analytic continuation that u_s is an entire solution of (6a) satisfying the radiation condition (6d). But this implies $u_s \equiv 0$ (cf. Hellwig [2]), i.e. from (6b) $u_i = 0$ on ∂D_1. But this is a contradiction since u_i is a plane wave. Now consider the case when $D = D_1 \cap D_2$ has a nonempty interior. Then by the above arguments u is a solution of (6a) in $D_1 \setminus \bar{D}$ or $D_2 \setminus \bar{D}$ satisfying $u = 0$ on the boundary, i.e. k^2 is an eigenvalue of the Laplacian in this region. But since the set of eigenvalues is discrete, we again arrive at a contradiction. The case when D_1 and D_2 are tangent can be easily handled by simply considering D_1 instead of $D_1 \setminus \bar{D}$ or $D_2 \setminus \bar{D}$.

We are now interested in characterizing the class of radiation patterns f and determining the relationship between the analytic behaviour of f and the location of the singularities of u_s. To fix our ideas we only consider axially symmetric solutions of the three-dimensional Helmholtz equation regular in the exterior of a (unknown) bounded domain D and without loss of generality assume that the wave number k is equal to one. Then in cylindrical coordinates (r, z, φ) such solutions satisfy

$$\frac{\partial^2 u_s}{\partial z^2} + \frac{\partial^2 u_s}{\partial r^2} + \frac{1}{r}\frac{\partial u_s}{\partial r} + u_s = 0 \tag{8}$$

where we have assumed the axis of symmetry to be $r = 0$. If we further assume that u_s satisfies the radiation condition

$$\lim_{R=\infty} R\left(\frac{\partial u_s}{\partial R} - iu_s\right) = 0; \qquad R = \sqrt{r^2 + z^2} \tag{9}$$

then u_s may be regarded as being generated by a set of equivalent sources all of which are contained in $\bar{D}$. The equivalent sources are not in general unique, although they are confined to some finite region of space, and one of the fundamental problems in scattering theory is to determine the extent and location of this region from a knowledge of the behaviour of u_s at infinity. To this end we note that it follows from (9) that u_s behaves asymptotically like

$$u_s \sim \frac{e^{iR}}{R} f(\cos\theta); \qquad R \to \infty \tag{10}$$

and hence our problem is to determine the location of the sources which generate u_s from a knowledge of f. But before we can do this we must first characterize the class of functions f which can be far field patterns. It turns out that this class is neither the class of continuous functions nor the class of analytic functions, but can best be characterized by introducing a certain class of harmonic functions (MÜLLER [1]). Recall that if g is an entire function of a complex variable and

$$M_g(r) = \max_{|z|=r} |g(z)| \tag{11}$$

then the *order* ϱ of g is defined by

$$\varrho = \overline{\lim_{r\to\infty}} \frac{\log\log M_g(r)}{\log r} \tag{12}$$

and the *type* σ is given by

$$\sigma = \overline{\lim_{r\to\infty}} \frac{\log M_g(r)}{r^\varrho}. \tag{13}$$

We now have the following theorem due to MÜLLER [1]:

Theorem 17.2. *A necessary and sufficient condition for a function f to be a far field pattern of a solution to* (8), (9) *is that there exists an axially symmetric harmonic function $h(z, r) = \tilde{h}(R, \theta)$ which is regular in the entire space and satisfies $\tilde{h}(1, \theta) = f(\cos\theta)$ such that*

$$\int_0^\pi |\tilde{h}(R, \theta)|^2 \sin\theta \, d\theta$$

is an entire function of R of order one and finite type C. In particular when this condition holds there exists a solution u_s of (8) *that is regular for $R > C$ having f as its far field pattern.*

Suppose now that f is a far field pattern corresponding to a solution u_s of (8) which is regular for $R \geqq C$. We want to determine the maximal domain of regularity of u_s, i.e. we want to continue u_s across the circle $R = C$. We shall determine such a region in terms of the entire harmonic function h defined in Müller's theorem. In particular we consider the function $h(2iz, 0)$ and note that from Müller's theorem it follows that this is an entire function of order 1 and type C. Let $H(z)$ be the Laplace transform of $h(2iz, 0)$ and define

$$K(\theta) = \overline{\lim_{R\to\infty}}\, R^{-1} \log |h(2iRe^{i\theta}, 0)| \, . \tag{14}$$

Then it can be shown (cf. LEVIN [1]) that K is the supporting function of a convex set G, called the indicator diagram of $h(2iz, 0)$, and a result of Polya (see LEVIN [1]) says that H is analytic in the exterior of the conjugate of G which we denote by G^*. Note that since $h(2iz, 0)$ is of order 1 and type C, G and G^* are contained in the disk $|z| \leqq C$. The following theorem is due to COLTON ([13, 1]) and is based on an application of Gilbert's envelope method (COLTON [13], GILBERT [1]) and a reflection principle for solutions of the three-dimensional Helmholtz equation which vanish on a portion of a sphere (COLTON [13, 11]):

Theorem 17.3. *u_s is regular in the exterior of $G \cup G^*$.*

For a proof of this theorem the reader is referred to COLTON [13].

17.6. The inverse scattering problem: inexact data

We now want to consider the problem of determining the shape of the scattering obstacle from *measurements* of the far field pattern. We note that the analysis of the previous section is not applicable to this problem since it is not possible to detect from experimental data what the growth properties of $h(2iz, 0)$ are. Furthermore, the problem is clearly improperly posed since if the function chosen to approximate the far field pattern is not entire of order one and finite type, no solution to the problem can exist, and even if a solution does exist for $R \geqq C$ there is no guarantee that this solution agrees with $-u_i$ on a simple closed curve ∂D as it must if it is to be a solution of the inverse scattering problem. In what follows we shall show how these problems can be overcome by reformulating our inverse scattering problem as a problem in optimal control and then assuming "a priori" restraints on the nature of the scattering obstacle in order to assure the existence of a solution to this optimal control problem. For the sake of simplicity we shall restrict ourselves to the case of two-dimensional domains (i.e. the scattering obstacle is an infinite cylinder with cylindrical cross section D). For further details of the analysis to follow the reader is referred to COLTON and KIRSCH [1] and COLTON and KLEINMAN [1].

Consider the inverse scattering problem associated with the direct problem (6) for $n = 2$, and in order to avoid unimportant constant factors in the sequel, let

$$u_s \sim \frac{1}{4} e^{i(kr+\pi/4)} \sqrt{\frac{2}{\pi k r}}\, F(\theta; k) \tag{15}$$

where F is the (normalized) far field pattern and (r, θ) are polar coordinates. If we assume the incoming wave u_i is $u_i = e^{ikx_1}$ where $x = (x_1, x_2)$ and expand F in a Fourier series

$$F(\theta; k) = \sum_{n-\infty}^{\infty} a_n(k) e^{in\theta} \tag{16}$$

then it is possible to show by using the method of integral equations that

$$a_n(k) = \frac{i^{-n}k^n}{2^n n! \log k} \int_{\partial D} \frac{\partial}{\partial \nu} \log |g^{-1}(\varrho e^{i\varphi})| \varrho^n e^{-in\varphi} ds + O\left(\frac{|k|^n}{|\log k|^2}\right) \tag{17}$$

where (ϱ, φ) are the polar coordinates of a point on ∂D, ν is the unit outward normal to ∂D, and $w = g^{-1}(z)$, $z = x_1 + ix_2$, is the unique analytic function that conformally maps the exterior of D onto the exterior of the unit disk Ω such that at infinity g^{-1} has the Laurent expansion

$$g^{-1}(z) = \frac{z}{a} + b + \frac{c}{z} + \frac{d}{z^2} + \dots, \qquad a > 0\,. \tag{18}$$

We note that a can be identified as the *mapping radius* or *transfinite diameter* of D (Pommerenke [1]) and our first aim is to show how to determine this number from a knowledge of F. To this end we first define μ_n by

$$\mu_n = (-i)^n 2^n n! \lim_{k\to 0} \frac{\overline{a_n(k)} \log k}{k^n} \tag{19}$$

and note that from the Cauchy-Riemann equations and an integration by parts we can rewrite (17) as

$$\mu_n = i \int_{|w|=1} \frac{1}{w} [g(w)]^n dw\,. \tag{20}$$

Since g has a Laurent expansion at infinity of the form

$$g(w) = aw + b_0 + \frac{b_1}{w} + \frac{b_2}{w^2} + \dots \tag{21}$$

an elementary residue calculation now shows that

$$\mu_0 = -2\pi\,, \qquad \mu_1 = -2\pi b_0\,, \qquad \mu_2 = -2\pi[b_0^2 + 2ab_1]. \tag{22}$$

We now begin all over again and consider our inverse scattering problem with $u_i = e^{-ikx_1}$ (i.e. we transmit a second plane wave from the opposite direction) and combine this with our original problem to arrive at problem (6) with $u_i = \sin kx_1$. If we denote the resulting combined far field pattern by

$$F^*(\theta; k) = \sum_{n-\infty}^{\infty} a_n^*(k) e^{in\theta} \tag{23}$$

and define μ_n^* by (19) with $a_n(k)$ replaced by $a_n^*(k)$ we find by arguing as in the above that

$$\mu_n^* = a \int_{|w|=1} \left[1 + \frac{1}{w^2}\right] [g(w)]^n dw\,. \tag{24}$$

Computing the residue in (24) now yields

$$\mu_1^* = 2\pi a i[b_1 + a] \tag{25}$$

and hence from (22), (25) we can determine the transfinite diameter a in terms of the measured data μ_1, μ_2 and μ_1^*. We can now assume that a is known, although if F is only known approximately we of course only know a approximately.

Our aim now is to view our inverse scattering problem as the problem of determining a conformal mapping $g\colon \mathbb{R}^2 \setminus \Omega \to \mathbb{R}^2 \setminus D$ such that D has given transfinite diameter a and such that the far field pattern corresponding to D is a best approximation in some sense to the measured far field pattern F. We further require that ∂D be smooth in order to guarantee that the calculations made up to now are valid. To this end we first define a compact family Σ of univalent functions $g\colon \mathbb{R}^2 \setminus \Omega \to \mathbb{R}^2 \setminus D$ (where $D = D(g)$) as follows: $g \in \Sigma$ if

1. there exists a positive constant b independent of g such that D is contained in the disk $|z| \leqq b$;
2. D has transfinite diameter $a > 0$;
3. there exists a positive constant M_1 independent of g such that

$$\max_{0 \leqq \varphi \leqq 2\pi} \left| \frac{\mathrm{d}}{\mathrm{d}\varphi} g(\mathrm{e}^{\mathrm{i}\varphi}) \right| \leqq M_1 \,.$$

4. There exists a positive constant M_2 independent of g such that

$$\left| \frac{\mathrm{d}}{\mathrm{d}\varphi} g(\mathrm{e}^{\mathrm{i}\varphi_1}) - \frac{\mathrm{d}}{\mathrm{d}\varphi} g(\mathrm{e}^{\mathrm{i}\varphi_2}) \right| \leqq M_2 |\varphi_1 - \varphi_2| \quad \text{for} \quad \varphi_1, \varphi_2 \in [0, 2\pi] \,.$$

Under the above assumptions it can be shown (Colton and Kirsch [1]) that Σ is compact with respect to the topology of uniform convergence on compact subsets of $\mathbb{R}^2 \setminus D$ and if $g \in \Sigma$, $g\colon \partial\Omega \to \partial D$, then ∂D is smooth. At the risk of a slight abuse of notation we shall henceforth identify a given boundary ∂D by its associated conformal mapping in the set Σ and denote this by either $\partial D \in \Sigma$ or $g \in \Sigma$. By using the method of integral equations it is easy to show that the mapping $\partial D \to F$ is a continuous mapping from Σ into $C[0, 2\pi]$.

We are now in a position to reformulate our inverse scattering problem as a problem in optimal control and show that a solution to this optimal control problem exists. Let F be a measured far field pattern and let F_0 be a far field pattern associated with a domain D and conformal mapping g such that $g \in \Sigma$. Then we define our optimal control problem $P(F)$ as follows: Minimize (for fixed k)

$$C_F(g) = \int_0^{2\pi} |F_0(\theta; k) - F(\theta; k)|^2 \, \mathrm{d}\theta \tag{26}$$

subject to the control that F_0 is the far field pattern of a solution to (6) (for $n = 2$) such that ∂D is in the control set Σ. Note that if $\partial D^* \in \Sigma$ is a solution of the inverse scattering problem corresponding to the far field pattern F then $g^*\colon \partial\Omega \to \partial D^*$ is a solution of $P(F)$ since $C_F(g^*) = 0$. Conversely, if g^* is a solution of $P(F)$ and $C_F(g^*) = 0$ then ∂D^* is a solution of the inverse scattering problem. If $C_F(g^*) > 0$ then the inverse scattering problem is not solvable for $\partial D \in \Sigma$, but ∂D^* is a best approximation in Σ in the sense that $C_F(g^*)$ is minimal. Finally, note that we could also consider the problem where F is only measured over a sector $0 < \theta_0 < \theta < \theta_1 < 2\pi$ by replacing the interval of integration $(0, 2\pi)$ in (26) by the interval (θ_0, θ_1).

The compactness of Σ and the continuity of C_F immediately imply the following existence theorem (Colton and Kirsch [1]).

Theorem 17.4. *There exists a solution to* $P(F)$.

In general the solution of $P(F)$ is not unique. However let $\Phi^*(F)$ be the set of all solutions g^* of $P(F)$. Then we have the following result on the continuous dependence of $\partial D \in \Sigma$ on the measured far field data F (Colton and Kirsch [1]):

Theorem 17.5. *The set* $\Phi^*(F)$ *is graph compact, i.e. if* $F_n \to F$ *in* $C[0, 2\pi]$, $g_n^* \in \Phi^*(F_n)$, *then there exists a convergent subsequence of* $\{g_n^*\}$ *and every limit point lies in* $\Phi^*(F)$.

Note that it follows from the theorem proved at the beginning of 17.1. that even if F is a far field pattern for some $g \in \Sigma$ we cannot conclude that the solution of $P(F)$ is unique. However if we look at the set $\Phi^*(F)$ for an interval of k values, $k_0 < k < k_1$, then there exists a unique $g^* \in \bigcap_{k_0<k<k_1} \Phi^*(F)$. Therefore in practice one should compute the solution of $P(F)$ for sufficiently many values of k such that there is only one g^* lying in the intersection of all the solution sets. This g^* will map $\partial\Omega$ onto ∂D^* where ∂D^* is the solution of the inverse scattering problem.

A major area that remains to be investigated is to develop efficient computational methods for constructing the solution to $P(F)$. It is hoped that investigations in this and related directions will be forthcoming in the not too distant future.

18. Numerical Methods for Boundary Value Problems of Elliptic Systems in the Plane

W. L. Wendland

18.1. Introduction

Here we consider some numerical methods for solving Riemann-Hilbert boundary value problems for generalized analytic vectors. The formulation of these boundary value problems will be given in 18.2. They appear in many different applications. In 18.3. we present some of them as plane elasticity problems, the Maxwell equations and compressible as well as viscous flows.

Nowadays the finite element methods are the most popular and efficient numerical methods. In 18.4. we shall use finite elements for the first order systems by utilizing the Ritz method, i.e. the least squares method. We obtain a numerical scheme which combines the features of variational methods with the treatment of general boundary conditions.

For first order systems one cannot apply Galerkin's method with the same trial and test functions directly. Therefore Nitsche's remark on least squares methods (Nitsche [3]) does not apply to our situation. Our method is optimal for first order systems and it corresponds to the mixed methods if the first order system is equivalent to equations of higher order. Since the trial functions do not satisfy the boundary conditions, our method is not conforming. We enforce the boundary conditions approximately.

Our presentation is mainly based on Wendland [4, chapter 8] but here we consider also nonlinear problems. For their treatment we use the Newton imbedding as in Wendland [3] and find a Newton imbedding for the discrete problems. If the necessary integrations in the stiffness matrix are executed with appropriate numerical quadratures then we eventually find finite difference schemes (Strang [1]). In this way, certain finite difference approximations are nothing but special finite element approximations. Therefore we shall not consider finite difference approximations here. (For generalized analytic functions one finds direct difference approximations in Wendland [4, chapter 7]).

More recently the integral equation methods are becoming also desirable for numerical computations and we give a brief introduction in 18.5. In many engineering applications, the integral equations are implemented on computing machines in the form of the so-called boundary integral methods or boundary element methods (Brebbia [1]). For their execution one needs the fundamental solution explicitly which restricts its applicability. On the other hand, these methods have also many advantages as the direct computation of desired boundary data and smoothing effects.

Here we present only the most popular standard procedures. In addition, there are several other effective methods as e.g. the numerical methods based on the Bergman-Vekua operators. Those can be found in chapter 15 by LANCKAU.

18.2. Regular boundary value problems for first order systems in the plane

Here we characterize the class of equations that we shall approximate with numerical methods. Let us begin with linear regular boundary value problems for real first order systems for $2n$ unknown real-valued functions which are elliptic in the Petrovski sense.

These linear systems can always be transformed into the quasinormal form (see WENDLAND [4, p. 121]):

$$\boldsymbol{D}\boldsymbol{w} := \boldsymbol{w}_{\bar{z}} + \boldsymbol{Q}\boldsymbol{w}_z + \boldsymbol{A}\boldsymbol{w} + \boldsymbol{B}\overline{\boldsymbol{w}} = \boldsymbol{f}(x, y)\,. \tag{1}$$

Here $\boldsymbol{w} = (w_1, \dots, w_n)\,(x, y)$ is now an unknown n-dimensional complex vector-valued function whose real and imaginary parts define the original unknowns. $\boldsymbol{Q}$, $\boldsymbol{A}$ and $\boldsymbol{B}$ are given complex-valued $n\times n$-matrix functions depending in the linear case only on x, y. The ellipticity for the original system is equivalent to the property $|q_j| \leqq q_0 < 1$ for all the eigenvalues $q_1, \dots, q_n$ of $\boldsymbol{Q}$. $\boldsymbol{f}$ in (1) is a given n-dimensional complex-valued vector function.

In many particular cases, $\boldsymbol{Q}$ in (1) can be found having special form and (1) then corresponds to one of the normal forms (see WENDLAND [4, chapter 3]).

For (1) we consider boundary value problems, i.e. we require (1) in a domain G in the x-y-plane. For simplicity G is bounded with the boundary curve $\dot{G}$. The boundary is assumed to consist of a simply closed exterior curve $\dot{G}_0$ and m respective simply closed and mutually disjoint inner curves $\dot{G}_1, \dots, \dot{G}_m$. For simplicity, however, we shall frequently consider only the simply connected case.

The boundary conditions are given in the form

$$\operatorname{Re} \boldsymbol{P}\boldsymbol{w}|_{\dot{G}} = \boldsymbol{\psi}\,, \tag{2}$$

where $\boldsymbol{P}$ is a given complex-valued $n\times n$-matrix function on $\dot{G}$ satisfying the Lopatinskiĭ condition (WENDLAND [4, p. 81]):

$$\det \boldsymbol{P} \neq 0\,. \tag{3}$$

$\boldsymbol{\psi}$ is a given real-valued n-vector function on $\dot{G}$. As everybody knows, (1), (2) in general is not uniquely solvable but defines a suitable Fredholm mapping between appropriate function spaces. For instance it is a continuous Fredholm mapping from the space of Hölder-continuously differentiable functions $\boldsymbol{w} \in \mathscr{C}^{1+\varepsilon}(\overline{G})$, $\varepsilon > 0$, into the product space $(\boldsymbol{f}, \boldsymbol{\psi}) \in \mathscr{C}^{\varepsilon}(\overline{G}) \times \mathscr{C}^{1+\varepsilon}(\dot{G})$. It also defines a continuous Fredholm mapping from the Sobolev space $\mathscr{H}^t(G)$ into $\mathscr{H}^{t-1}(G) \times \mathscr{H}^{t-1/2}(\dot{G})$ for $t \geqq 1$, (AGMON, DOUGLIS, and NIRENBERG [1], WENDLAND [4]). Here $\mathscr{H}^t = \mathscr{W}_2^t$ denotes the usual Sobolev space for integer t, $\mathscr{H}^0 = \mathscr{L}_2$ and for arbitrary $t \in \mathbb{R}$, $\mathscr{H}^t$ denotes the corresponding dual and interpolation spaces (TRIEBEL [1]). Because of the form (1), (2) of our boundary value problem all the above spaces are supposed to be vector spaces over the field of reals $\mathbb{R}$, otherwise (1), (2) will not be linear anymore.

In both the above cases the Fredholm index is given by the Bojarski-Vekua formula

$$N - N^* = n - mn - \frac{1}{\pi} \oint_{\dot{G}} \mathrm{d} \arg \det \boldsymbol{P} \,, \tag{4}$$

where m denotes the number of inner components of the boundary curve $\dot{G}$, N denotes the dimension of the eigenspace of (1), (2) and N^* is the codimension of the range. In case of a simply connected domain we have $m = 0$ in (4) (cf. WENDLAND [4, f. (3.39)]). Thus, the solution of (1), (2) becomes unique if we augment the problem by N side conditions,

$$\Lambda_l \boldsymbol{w} = \gamma_l \in \mathbb{R} \,, \qquad l = 1, \dots, N \,, \tag{5}$$

where $\gamma_1, \dots, \gamma_N$ are given real numbers and $\Lambda_1, \dots, \Lambda_N$ are suitable *real* linear functionals on the space of solutions.

In order to modify (1), (2) to be a surjective mapping, we also augment the desired unknowns $\boldsymbol{w}$ by N^* unknown Lagrange multipliers $\alpha_1, \dots, \alpha_{N^*} \in \mathbb{R}$. The modified problem then reads as

$$D\boldsymbol{w} = \boldsymbol{w}_{\bar{z}} + \boldsymbol{Q}\boldsymbol{w}_z + \boldsymbol{A}\boldsymbol{w} + \boldsymbol{B}\bar{\boldsymbol{w}} = \boldsymbol{f} + \sum_{p=1}^{N^*} \alpha_p \boldsymbol{g}_p \quad \text{in } G \,, \tag{6}$$

$$\operatorname{Re} \boldsymbol{P}\boldsymbol{w}|_{\dot{G}} = \boldsymbol{\psi} + \sum_{p=1}^{N^*} \alpha_p \boldsymbol{\varphi}_p \quad \text{on } \dot{G} \,, \tag{7}$$

$$\Lambda_l \boldsymbol{w} = \gamma_l \,, \qquad l = 1, \dots, N \,, \tag{8}$$

where $\boldsymbol{g}_p$ and $\boldsymbol{\varphi}_p$ are suitably given additional functions. For the modified problem (6)–(8), the functions $\boldsymbol{f}, \boldsymbol{\psi}$ and the constants $\gamma_1, \dots, \gamma_N$ are given and $\boldsymbol{w}$ and $\alpha = (\alpha_1, \dots, \alpha_{N^*})$ are the desired unknowns. Thus, now the mapping A defined by (6), (7), (8) with

$$A : \mathscr{C}^{1+\varepsilon}(\bar{G}) \times \mathbb{R}^{N^*} \to \mathscr{C}^{\varepsilon}(\bar{G}) \times \mathscr{C}^{1+\varepsilon}(\dot{G}) \times \mathbb{R}^N$$

is a continuous bijection. The corresponding bijectivity holds for

$$A : \mathscr{H}^t(G) \times \mathbb{R}^{N^*} \to \mathscr{H}^{t-1}(G) \times \mathscr{H}^{t-1/2}(\dot{G}) \times \mathbb{R}^N$$

for $t \geqq 1$ due to the well-known results by AGMON, DOUGLIS, and NIRENBERG [1]. Hence, we have the a priori estimates

$$(\|\boldsymbol{w}\|_{\mathscr{C}^{1+\varepsilon}} + |\alpha|) \leqq c_1 \{ \|D\boldsymbol{w} - \sum_{p=1}^{N^*} \alpha_p \boldsymbol{g}_p\|_{\mathscr{C}^\varepsilon} + \|\operatorname{Re} \boldsymbol{P}\boldsymbol{w} - \sum_{p=1}^{N^*} \alpha_p \boldsymbol{\varphi}_p\|_{\mathscr{C}^{1+\varepsilon}(\dot{G})} + \sum_{l=1}^{N} |\Lambda_l \boldsymbol{w}| \} \tag{9}$$

$\varepsilon > 0$, ε is non integer

and

$$(\|\boldsymbol{w}\|_{\mathscr{H}^t} + |\alpha|) \leqq c_2 \{ \|D\boldsymbol{w} - \sum_{p=1}^{N^*} \alpha_p \boldsymbol{g}_p\|_{\mathscr{H}^{t-1}} + \|\operatorname{Re} \boldsymbol{P}\boldsymbol{w} - \sum_{p=1}^{N^*} \alpha_p \boldsymbol{\varphi}_p\|_{\mathscr{H}^{t-1/2}(\dot{G})} + \sum_{l=1}^{N} |\Lambda_l \boldsymbol{w}| \} \,, \qquad t \geqq 1 \tag{10}$$

provided Λ_l are continuous on $\mathscr{C}^{1+\varepsilon}$ or $\mathscr{H}^t$, correspondingly and where here and in the following $c_1, c_2, c, \dots$ denote generic constants of different size and meaning at different places. In order to apply duality arguments and the Aubin-Nitsche lemma for finite element approximations we need (10) also for $t \geqq 0$. For $0 \leqq t \leqq \frac{1}{2}$ however, (10) is not valid anymore in general. But (10) remains valid for sufficiently *smooth* solutions, i.e. for $w \in \mathscr{H}^1$ according to DIKANSKIĬ's results [1]. In this case the functionals Λ_l should also be continuous on $\mathscr{L}_2$.

For practical purposes, the functionals Λ_l and the pairs (g_p, φ_p) need to be chosen first appropriately depending on the special problem but we omit this detail here due to the more general purpose of this chapter. For many boundary value problems as e.g. for those involving generalized hyperanalytic functions with triangular matrix coefficients A, B, one has for $N - N^* > 0$ and for $m = 0$ the property $N^* = 0$ and then the Λ_l can be given explicitly in the form of point conditions or total flux conditions (see WENDLAND [2]).

Besides the above linear problems we shall also consider some nonlinear problems. To this end we shall always assume that the nonlinear problem can be imbedded into a family of problems involving the parameter t of imbedding:

$$\text{differential equations} \quad \Phi(t; w_{\bar{z}}, w_z, w, \alpha, x, y) = 0 \quad \text{in } G, \tag{11}$$

$$\text{boundary conditions} \quad R(t; w, \alpha, s) = 0 \quad \text{on } \dot{G}, \tag{12}$$

$$\text{side conditions} \quad \Lambda(t; w) = 0. \tag{13}$$

For the dependence on t we assume arbitrary smoothness, e.g. analyticity. For the dependence on $w_{\bar{z}}, w_z, w$ we do *not* assume holomorphy but smooth dependence on the corresponding real and imaginary parts, in particular let us require continuity of the second Fréchet derivatives of (11)–(13).

Since we are interested in the numerical solution of (11)–(13) we assume that the corresponding mapping properties as existence, regularity and uniqueness are known and that the following assumptions (i)–(iii) are satisfied.

(i) For all $t \in [0, T]$ we have an a priori estimate $\|w\| + |\alpha| \leqq M$ where $\|\cdot\|$ denotes a suitable norm, e.g. one of the foregoing Hölder norms or Sobolev norms.

(ii) For all t and $w(t)$ satisfying (i) we assume for the Davidenko differential equation

$$\begin{aligned}\Phi_t(t; w_{\bar{z}}, w_z, w, \alpha, x, y) + \Phi_{w_{\bar{z}}}\dot{w}_{\bar{z}} + \Phi_{\overline{w_{\bar{z}}}}\overline{\dot{w}_{\bar{z}}} + \Phi_{w_z}\dot{w}_z + \Phi_{\overline{w_z}}\overline{\dot{w}_z} \\ + \Phi_w\dot{w} + \Phi_{\bar{w}}\overline{\dot{w}} + \Phi_\alpha\dot{\alpha} = 0 \quad \text{in } G,\end{aligned} \tag{14}$$

$$R_t(t; w, \alpha, s) + R_w\dot{w} + R_{\bar{w}}\overline{\dot{w}} + R_\alpha\dot{\alpha} = 0 \quad \text{on } \dot{G}, \tag{15}$$

$$\Lambda_t(t; w) + \Lambda_w\dot{w} + \Lambda_{\bar{w}}\overline{\dot{w}} = 0 \tag{16}$$

i.e. that the linear first order system for $\dot{w}$ and $\dot{\alpha}$ is equivalent to (6)–(8) and that it is uniquely solvable with respect to $\dot{w}$ and $\dot{\alpha}$.

(iii) For $t = 0$ the problem (11)–(13) takes the form (6)–(8) and is uniquely solvable.

For more general nonlinear elliptic systems we refer to BERGER and SCHECHTER [1] and NIRENBERG [1].

With the rather strong assumptions (i)–(iii) we find exactly one family of solutions $w = w(t)$ of (11)–(13) depending continuously differentiable on $t \in [0, T]$. (For fixed $t = t_0$, however, there might exist more than one solution but the other solutions will *not* belong to the above family $w(t)$.)

For the computation of $w(T)$ belonging to the above family we may use the method of *Newton imbedding* (see WACKER [1]).

For utilizing this method we execute Newton's procedure for a fixed value where $t = t_j$ in the form

$$\begin{aligned}
&\Phi(t_j; w_{\bar{z}}^{(m)}, w_z^{(m)}, w^{(m)}, \alpha^{(m)}, x, y) \\
&+ \Phi_{w_{\bar{z}}}(t_j; w_{\bar{z}}^{(m)}, w_z^{(m)}, w^{(m)}, \alpha^{(m)}, x, y)\, (w_{\bar{z}}^{(m+1)} - w_{\bar{z}}^{(m)}) \\
&+ \Phi_{\overline{w_{\bar{z}}}}(t_j; w_{\bar{z}}^{(m)}, w_z^{(m)}, w^{(m)}, \alpha^{(m)}, x, y)\, \overline{(w_{\bar{z}}^{(m+1)} - w_{\bar{z}}^{(m)})} \\
&+ \Phi_{w_z}^{(m)}(w_z^{(m+1)} - w_z^{(m)}) + \Phi_{\overline{w_z}}^{(m)}\overline{(w_z^{(m+1)} - w_z^{(m)})} + \Phi_w^{(m)} \cdot (w^{(m+1)} - w^{(m)}) \\
&+ \Phi_{\overline{w}}^{(m)}\overline{(w^{(m+1)} - w^{(m)})} + \Phi_\alpha^{(m)}(\alpha^{(m+1)} - \alpha^{(m)}) = 0 \quad \text{in } G, \qquad (17)
\end{aligned}$$

$$\begin{aligned}
&R(t_j; w^{(m)}, \alpha^{(m)}, s) + R_w^{(m)}(w^{(m+1)} - w^{(m)}) \\
&+ R_{\overline{w}}^{(m)}\overline{(w^{(m+1)} - w^{(m)})} + R_\alpha^{(m)}(\alpha^{(m+1)} - \alpha^{(m)}) = 0 \quad \text{on } \dot{G}, \qquad (18)
\end{aligned}$$

$$\begin{aligned}
&\Lambda(t_j; w^{(m)}) + \Lambda_w^{(m)}(w^{(m+1)} - w^{(m)}) + \Lambda_{\overline{w}}^{(m)}\overline{(w^{(m+1)} - w^{(m)})} = 0, \\
&\qquad m = 0, 1, 2, \ldots, \qquad (19)
\end{aligned}$$

using the initial solution

$$w^{(0)}(t_j) = w^{(\infty)}(t_{j-1}) \qquad (20)$$

or correspondingly

$$w^{(0)}(t_j) = w^{(m')}(t_{j-1}) \qquad (21)$$

with m' sufficiently large.

Assuming (i)—(iii) one obtains with the a priori estimate and smoothness and boundedness of the Fréchet derivatives the following theorem (cf. WENDLAND [3]):

Theorem 18.2.1. *There exists a finite partition* $0 = t_0 < t_1 < \ldots < t_M = T$ *such that the Newton imbedding* (17)—(20) *converges with*

$$w(T) = \lim_{m \to \infty} w^{(m)}(T). \qquad (22)$$

For the proof one first shows uniform a priori estimates in a suitable neighbourhood of the family $w(t)$ defining a path in $\mathscr{C}^{1+\varepsilon} \times \mathbb{R}^{N^*}$. We omit the details. (For semilinear problems see BEGEHR and HSIAO [1, 2] and WENDLAND [3]). Here one needs in addition the Tychonov implicit function theorem. Thus it is possible to follow a path of regular solutions by using the Newton imbedding. In case of bifurcations the above method naturally fails and one needs the Ljapunov-Schmidt procedure.

18.3. Some examples of applications of first order systems

Let us collect here a few examples of the foregoing boundary value problems following WENDLAND [4].

1. It is well-known that the already classical Riemann-Hilbert problems for *generalized analytic functions* belong to the class considered in 18.2. The mathematical theory can be found in I. N. VEKUA [5] and for $m = 0$ also in HAACK and WENDLAND [1]. In this case we have $n = 1$ and (after using an appropriate Beltrami homeomorphism) without loss of generality $Q = 0$. For a simply connected domain, i.e. $m = 0$, there are only the two cases $\{N > 0, N^* = 0\}$ or $\{N = 0, N^* > 0\}$. Let

us consider first the case $N > 0$:

$$N = 1 - \frac{1}{\pi} \oint_{\dot{G}} \mathrm{d} \arg \boldsymbol{P} = 1 - 2\nu > 0 \quad \text{and} \quad N^* = 0 \,. \tag{23}$$

Here the most popular side conditions are of the form

$$\begin{aligned} \Lambda_l w &= \operatorname{Re} w(z_l) = \gamma_l \,, && l = 1, \dots, |\nu| \,, \\ \Lambda_l w &= \operatorname{Im} w(z_{l-|\nu|}) = \gamma_l \,, && l = |\nu| + 1, \dots, 2\,|\nu| \,, \\ \Lambda_N w &= \oint_{\dot{G}} \sigma \operatorname{Im} \boldsymbol{P} w \,\mathrm{d}s = \gamma_N && \text{with} \quad \sigma > 0 \,. \end{aligned} \tag{24}$$

The other case is

$$N^* = \frac{1}{\pi} \oint_{\dot{G}} \mathrm{d} \arg \boldsymbol{P} - 1 > 0 \quad \text{and} \quad N = 0 \,. \tag{25}$$

Here the functions g_p and φ_r are to be chosen in such a way that the matrix

$$\langle (g_r, \varphi_r), v_q \rangle = -2 \operatorname{Im} \iint_G g_r v_q \,\mathrm{d}x \,\mathrm{d}y - \operatorname{Re} \oint_{\dot{G}} \varphi_r \left(\frac{1}{\boldsymbol{P}} v_q \right) \mathrm{d}z \,,$$

$$r, q = 1, \dots, N^* \,,$$

is regular where $v_1, \dots, v_{N^*}$ is a basis of the eigenspace of the adjoint boundary value problem.

Applications of these Riemann-Hilbert problems can be found for analytic functions in incompressible ideal flows (Mises and Friedrichs [1]), infiltration flows in porous media (Scheidegger [1]) and in electrostatics (Wendt [1]) and for generalized analytic functions in the momentum free shell theory (I. N. Vekua [5]).

2. As a second example let us consider *Maxwell's equations* for time harmonic plane electromagnetic fields (cf. Borgnis and Papas [1]):

$$E_{2x} - E_{1y} - \mathrm{i}\omega\mu H = J \,, \tag{26}$$

$$\begin{pmatrix} H_y \\ -H_x \end{pmatrix} + \mathrm{i}\omega\boldsymbol{\eta} \begin{pmatrix} E_1 \\ E_2 \end{pmatrix} = \begin{pmatrix} K_1 \\ K_2 \end{pmatrix} \text{in} \quad G, \omega \neq 0 \,, \tag{27}$$

$$\dot{y} E_2 + \dot{x} E_1 = 0 \quad \text{on } \dot{G} \text{ where} \quad \dot{x} = \left.\frac{\mathrm{d}x}{\mathrm{d}s}\right|_{\dot{G}} , \quad \dot{y} = \left.\frac{\mathrm{d}y}{\mathrm{d}s}\right|_{\dot{G}} \tag{28}$$

with s the arc length on $\dot{G}$. Here E_1, E_2 denote the components of the electrical field and H denotes the magnetic field. J, K_1, K_2 and μ, ω are given as well as the positive definite matrix function $\boldsymbol{\eta}$. For $\omega \neq 0$, the system (26), (27) is elliptic in the sense of Agmon-Douglis-Nirenberg (see Hörmander [1, p. 268], with $s_1 = 1$, $s_2 = s_3 = t_1 = t_2 = 2$, $t_3 = 3$.). For our purposes we reformulate (26), (27) as an equivalent first order properly elliptic system

$$\{(\eta_{11} E_1 + \eta_{12} E_2)_x + (\eta_{21} E_1 + \eta_{22} E_2)_y\} = -\frac{\mathrm{i}}{\omega} (K_{1x} + K_{2y}) \,, \tag{29}$$

$$E_{1y} - E_{2x} + \mathrm{i}\omega\mu H = -J \,,$$

$$\begin{aligned} H_{1x} - H_{2y} - \mathrm{i}(1, 1)\, \omega\boldsymbol{\eta} \begin{pmatrix} E_1 \\ E_2 \end{pmatrix} &= -(K_1 + K_2) \,, \\ H_{1y} + H_{2x} + \mathrm{i}(1, -1)\, \omega\boldsymbol{\eta} \begin{pmatrix} E_1 \\ E_2 \end{pmatrix} &= K_1 - K_2 \end{aligned} \tag{30}$$

together with the boundary conditions (28) and

$$H_1 - H_2 = 0 \quad \text{on } \dot{G} . \tag{31}$$

In case of *isotropic* media we have $\eta_{jk} = \eta(x, y)\, \delta_{jk}$. With

$$w_1 = E_2 + \mathrm{i}E_1 \quad \text{and} \quad w_2 = H_1 + \mathrm{i}H_2 = (1 + \mathrm{i})\, H \tag{32}$$

the equations (29), (30) take Pascali's normal form:

$$\begin{aligned} &w_{1\bar z} - \frac{\mathrm{i}}{4}\,\omega\mu(w_2 + \overline{w}_2) + \frac{\mathrm{i}}{4\eta}\big(\mathrm{i}\eta_x(\overline{w}_1 - w_1) + \eta_y(w_1 + \overline{w}_1)\big) \\ &= \frac{1}{2}\left(J + \frac{1}{2\omega}(K_{1x} + K_{2y})\right), \\ &w_{2\bar z} + \tfrac{1}{2}\,\omega\eta(\overline{w}_1 - \mathrm{i}\overline{w}_1) = \tfrac{1}{2}\,\big(\mathrm{i}(K_1 - K_2) - (K_1 + K_2)\big) , \end{aligned} \tag{33}$$

$$\operatorname{Re}\begin{pmatrix} -\mathrm{i}\dot{z} & 0 \\ 0 & (1 + \mathrm{i}) \end{pmatrix}\begin{pmatrix} w_1 \\ w_2 \end{pmatrix} = 0 \quad \text{on } \dot{G} . \tag{34}$$

For a simply connected domain G we find $m = N^* = 0$. Hence here (33), (34) is already of the form (6)—(8) with $\Lambda_l = 0$.

3. One finds a similar Pascali system for two-dimensional *Oseen flows* (cf. BERKER [1, f. (78.11) and (23)]):

$$w_{1\bar z} = 0 , \qquad w_{2\bar z} - \frac{\mathrm{i}\lambda}{2}(w_2 - \overline{w}_2) = 0 , \qquad w_{3\bar z} - \overline{w}_2 = 0 \tag{35}$$

together with the boundary conditions

$$\operatorname{Re}\begin{pmatrix} 1 & (2\lambda)^{-1} & 0 \\ -1 & i(2\lambda)^{-1} & 0 \\ 0 & 0 & i \end{pmatrix}\begin{pmatrix} w_1 \\ w_2 \\ w_3 \end{pmatrix} = \begin{pmatrix} \psi_1 \\ \psi_2 \\ 0 \end{pmatrix} \text{on } \dot{G} . \tag{36}$$

For $m = 0$ we find from (4): $N - N^* = 3$. Here (35), (36) have the three linearly independent eigensolutions $(\mathrm{i}, 0, 0)^{\mathsf T}$, $(w_1, (1 + \mathrm{i})\, \mathrm{e}^{\lambda(y-x)}, w_3)^{\mathsf T}$, $(0, 0, 1)^{\mathsf T}$. Hence we may choose the side conditions

$$\begin{aligned} &\Lambda_1 \boldsymbol{w} = \oint_{\dot G} \operatorname{Im} w_1 \, \mathrm{d}s = \gamma_1 , \qquad \Lambda_2 \boldsymbol{w} = \oint_{\dot G} \operatorname{Re} w_2 \, \mathrm{e}^{\lambda(x-y)}\, \mathrm{d}s = \gamma_2 , \\ &\Lambda_3 \boldsymbol{w} = \oint_{\dot G} \operatorname{Re} w_3 \, \mathrm{d}s = \gamma_3 \end{aligned} \tag{37}$$

and $N = 0$.

4. The fundamental problems of *plane elasticity* (WENDLAND [4, p. 386]) with the complex components of the stress tensor,

$$\begin{aligned} w_1 &= (2\mu + \lambda)\,(u_{1x} + u_{2y}) + \mathrm{i}\mu(u_{2x} - u_{1y}) , \\ w_2 &= 2\mu\, \frac{2\mu + \lambda}{\mu + \lambda}\{u_{2y} - u_{1x} + \mathrm{i}(u_{1y} + u_{2x})\} , \end{aligned} \tag{38}$$

where (u_1, u_2) denotes the vector of displacement and λ, μ are the Lammé coefficients, yield a system in the Douglis normal form

$$\begin{aligned} &w_{1\bar z} - A_1 \boldsymbol{w} - B_1 \overline{\boldsymbol{w}} = f_1 , \\ &w_{2\bar z} - w_{1z} - A_2 \boldsymbol{w} - B_2 \overline{\boldsymbol{w}} = f_2 . \end{aligned} \tag{39}$$

If the Lammé coefficients λ, μ are constant then $A_j = B_j = 0, j = 1, 2$.

In the case of the *first fundamental problem*, the displacement (u_1, u_2) is given on $\dot{G}$ yielding the boundary conditions

$$\operatorname{Re}\begin{pmatrix} 2\dot{x}(2\mu+\lambda)^{-1} + 2i\dot{y}\mu^{-1}, & -(\lambda+\mu)(2\mu+\lambda)^{-1}(\dot{x}+i\dot{y}\mu^{-1}) \\ 2\dot{y}(2\mu+\lambda)^{-1} - 2i\dot{x}\mu^{-1}, & \mu^{-1}(2\mu+\lambda)^{-1}(\dot{y} - i\dot{x}(\lambda+\mu)) \end{pmatrix} \boldsymbol{w} = 4\begin{pmatrix} \dot{u}_1 \\ \dot{u}_2 \end{pmatrix} \tag{40}$$

where $\dot{u}_j = du_j/ds$. Here $N = 0$, $N^* = 2$ and because of the solvability conditions

$$\oint_{\dot{G}} \dot{u}_1\, ds = \oint_{\dot{G}} \dot{u}_2\, ds = 0 \quad \text{for} \quad f_1 = f_2 = 0$$

we can choose for (6), (7) the formulations

$$\boldsymbol{\varphi}_1 = \begin{pmatrix} 1 \\ 0 \end{pmatrix}, \quad \boldsymbol{\varphi}_2 = \begin{pmatrix} 0 \\ 1 \end{pmatrix} \quad \text{and} \quad \boldsymbol{g}_r = 0\,. \tag{41}$$

In the case of the *second fundamental problem* on $\dot{G}$ the boundary stress (τ_1, τ_2) is given, i.e.

$$\operatorname{Re}\begin{pmatrix} \dot{y}(2\mu+\lambda) - i\dot{x}\mu, & -(\lambda+\mu)(\dot{x}+i\dot{y}) \\ -\dot{x}\lambda + i\dot{y}\mu, & -i(\lambda+\mu)(\dot{x}+i\dot{y}) \end{pmatrix} \boldsymbol{w} = (2\mu+\lambda)\begin{pmatrix} \tau_1 \\ \tau_2 \end{pmatrix}. \tag{42}$$

Again, (41) can be used for defining (6), (7).

5. The fundamental problems of *plate bending* with the normal displacement u can also be formulated as a first order system in the Douglis normal form (WENDLAND [4, p. 389]). In case of the clamped plate, i.e. if the *boundary displacement*

$$u\Big|_{\dot{G}} = \psi_1\,, \quad \frac{\partial u}{\partial n}\Big|_{\dot{G}} = \psi_2$$

is given then one finds with $u_{\bar{z}} = \frac{1}{2}(w_1 + \overline{w}_2)$ for w_1, w_2 the Douglis system

$$w_{1\bar{z}} = 0\,, \qquad w_{2\bar{z}} - w_{1z} = 0 \tag{43}$$

and the boundary conditions

$$\operatorname{Re}\begin{pmatrix} \dot{\bar{z}} & \dot{z} \\ i\dot{\bar{z}} & -i\dot{z} \end{pmatrix} \boldsymbol{w} = \begin{pmatrix} \dot{\psi}_1 \\ \psi_2 \end{pmatrix}. \tag{44}$$

Here we find $N = 3$ and $N^* = 1$. Since the vector functions $(-1, 1)^{\mathsf{T}}$, $i(1, 1)^{\mathsf{T}}$ and $i(z, \bar{z})^{\mathsf{T}}$ are eigensolutions and since we have the compatibility condition $\oint_{\dot{G}} \dot{\psi}_1\, ds = 0$ as a side condition available we can choose in this case

$$\Lambda_1\boldsymbol{w} = \operatorname{Re} w_1(0) = \gamma_1\,, \qquad \Lambda_2\boldsymbol{w} = \operatorname{Im} w_1(0) = \gamma_2\,, \qquad \Lambda_3\boldsymbol{w} = \operatorname{Re}\oint_{\dot{G}} w_2\, dx\,,$$

$$\boldsymbol{g}_1 = \boldsymbol{0}\,, \quad \boldsymbol{\varphi}_1 = \begin{pmatrix} 1 \\ 0 \end{pmatrix}. \tag{45}$$

If the transversal forces ψ_1 and the bending moments ψ_2 are given then one finds with

$$w_2 = 2u_{zz} \tag{46}$$

again the Douglis system (43), this time with the boundary conditions

$$\operatorname{Re}\begin{pmatrix} (1+\nu)/2 & -(1-\nu)\,\dot{z}^2 \\ -i/2 & -i(1-\nu)\,\dot{z}^2 \end{pmatrix} \boldsymbol{w} = \begin{pmatrix} \psi_1 \\ \int_0^s \psi_2\, d\sigma \end{pmatrix}. \tag{47}$$

Here we have $N = 0$, $N^* = 2$ and we choose for (6), (7) the functions $\boldsymbol{g}_r = \boldsymbol{0}$, $\varphi_1 = \begin{pmatrix}1\\0\end{pmatrix}$, $\varphi_2 = \begin{pmatrix}0\\1\end{pmatrix}$.

6. The *Stokes flow* models the incompressible slow viscous flow. With the vector of velocity $\boldsymbol{u} = (u_1, u_2)$ and the pressure p the Stokes problem is given (cf. BERKER [1]) by

$$\Delta \boldsymbol{u} - R \operatorname{grad} p = R\boldsymbol{f} \quad \text{and} \quad \operatorname{div} \boldsymbol{u} = 0 \qquad \text{in } G\,, \tag{48}$$

where $\boldsymbol{u}_{|\dot{G}} = (v_1, v_2)$ is given on the boundary. R denotes the Reynolds number. If we introduce $\boldsymbol{u} = (\varphi_y, -\varphi_x)$ with the potential φ and if we apply the curl operator to the first equation of (48) then we find the bilaplacian applied to φ, and $\boldsymbol{u}$ can be obtained from solving (43), (44) with

$$\dot{\psi}_1 = v_1\dot{y} - v_2\dot{x}\,, \qquad \psi_2 = -v_1\dot{x} - v_2\dot{y} \qquad \text{on } \dot{G} \tag{49}$$

and

$$\mathrm{i}(u_1 + \mathrm{i}u_2) = w_1 + \overline{w}_2 \qquad \text{in } G\,. \tag{50}$$

For (6)—(8) we again choose (45).

7. The *compressible subsonic flow* in two dimensions is governed by the quasilinear system

$$w_{\bar{z}} - \mu(w)\, w_z - \overline{\mu(w)}\, \overline{w_z} = 0 \qquad \text{in } G \tag{51}$$

with

$$\mu(w) = -\frac{M^2}{w\overline{w}} \frac{\overline{w}^2}{2(2 - M^2)} \tag{52}$$

where $w = (u_1 - \mathrm{i}u_2)$ denotes the complex velocity corresponding to the velocity vector $\boldsymbol{u} = (u_1, u_2)$ and where M denotes the Mach number (cf. BERS [2] and BOJARSKI [6]. There, however, one finds the often more important exterior problems.). The Mach number is given by the equation of state of the corresponding gas as

$$M^2 = \frac{2w\overline{w}}{2 - (\gamma - 1)\, w\overline{w}}\,, \qquad \text{with } \gamma > 1\text{, for isentropic adiabatic flows} \tag{53}$$

or

$$M^2 = \frac{2\nu w\overline{w}}{1 + w\overline{w}}\,, \qquad 0 < \nu \leqq \frac{1}{2}\,, \qquad \text{for a generalized Chaplygin gas.} \tag{54}$$

In order to guarantee uniform ellipticity, i.e. subsonic flows we require $|w|^2 \leqq c_0^2 < 2/(1 - \gamma)$ for (53) and $\nu < \frac{1}{2}$ for (54). For the boundary conditions we use e.g.

$$\operatorname{Re} \mathrm{i}\dot{z}w = -u_1\dot{y} + u_2\dot{x} = \boldsymbol{u}\boldsymbol{n} = 0 \qquad \text{on } \dot{G} \tag{55}$$

($\boldsymbol{n}$ the exterior normal vector to $\dot{G}$). In the case $M = 0$ the boundary value problem (51), (55) becomes linear and here we have $N - N^* = m - 1$. E.g. for the flow in a ring domain we have $m = 1$ and $N = N^* = 1$. Here we choose

$$\Lambda_1 w := \oint_{\dot{G}} \operatorname{Re} \dot{z}w\, ds = \gamma_1\,, \qquad \varphi_1 = 1\,, \qquad g_1 = 0\,. \tag{56}$$

Then the modified imbedded boundary value problem defined by

$$w_{\bar{z}} - t\mu(w)\, w_z - \overline{t\mu(w)}\, \overline{w_z} = 0 \qquad \text{in } G,\ t \in [0, 1]\,, \tag{57}$$

$$\operatorname{Re} \mathrm{i}\dot{z}w|_{\dot{G}} = \alpha\,, \qquad \int_{\dot{G}} \operatorname{Re} \dot{z}w\, ds = \gamma \tag{58}$$

is uniquely solvable for any $t \in [0, 1]$. In particular we may apply the Newton imbedding (17)—(20) i.e. we solve the sequence of *linear* problems

$$\begin{aligned} & w_{\bar z}^{(k+1)} - t_j\{\mu(w^{(k)})\, w_z^{(k+1)} + \overline{\mu(w^{(k)})}\,\overline{w_z^{(k+1)}}\} \\ & \quad - t_j \left\{\frac{\partial\mu}{\partial w}(w^{(k)})\, w_z^{(k)} + \overline{\frac{\partial\mu}{\partial \bar w}(w^{(k)})}\,\overline{w_z^{(k)}}\right\}(w^{(k+1)} - w^{(k)}) \\ & \quad - t_j \left\{\frac{\partial\mu}{\partial \bar w}(w^{(k)})\, w_z^{(k)} + \overline{\frac{\partial\mu}{\partial w}(w^{(k)})}\,\overline{w_z^{(k)}}\right\}\overline{(w^{(k+1)} - w^{(k)})} = 0 \end{aligned} \tag{59}$$

$$\operatorname{Re} \mathrm{i}\dot z w^{(k+1)}|_{\dot G} = \alpha\,, \qquad \int_{\dot G} \operatorname{Re} \dot z w^{(k+1)}\, ds = \gamma\,, \qquad k = 0, 1, 2, \dots,$$

$$w^{(0)}(z, t_j) = w^{(\infty)}(z, t_{j-1})\,, \quad j = 0, 1, \dots, M\,, \quad 0 = t_0 < t_1 < \dots < t_M = 1\,.$$

Each of these boundary value problems (59) is uniquely solvable due to the principle of argument as long as in case (53) the condition $|w|^2 \leqq c_0^2 < 2/(1-\gamma)$ is satisfied.

8. The incompressible viscous flow is modeled by the *Navier-Stokes equations* for the velocity vector $\boldsymbol{u} = (u, v)$ and the pressure p (TEMAM [1]), i.e.

$$-\Delta \boldsymbol{u} + t(uu_x + vu_y,\, uv_x + vv_y) + t \operatorname{grad} p = t\boldsymbol{f}\,, \tag{60}$$

$$\operatorname{div} \boldsymbol{u} = 0 \quad \text{in } G\,, \qquad \boldsymbol{u} = \boldsymbol{g} \quad \text{on } \dot G\,.$$

Here t denotes the Reynolds number. In order to rewrite (60) into an equivalent boundary value problem with a first order system we introduce some new unknown functions and eliminate the pressure p since p can only be computed from (60) for $t > 0$. We introduce

$$w_1 = u - \mathrm{i}v\,, \qquad \operatorname{div} \boldsymbol{u} = u_x + v_y = 0\,, \qquad u_y - v_x = \omega\,. \tag{61}$$

Then (60) yields

$$\Delta\omega = tF \quad \text{with} \quad F = (uu_x + vu_y)_y - (uv_x + vv_y)_x - f_{1y} + f_{2x}\,. \tag{62}$$

Substituting (61) into the nonlinear right-hand side F we find

$$F = u\omega_x + v\omega_y - f_{1y} + f_{2x}\,. \tag{63}$$

With these substitutions the equations (61), (62) can easily be formulated as a first order system by introducing

$$w_2 := \omega\,, \qquad w_3 := \omega_y + \mathrm{i}\omega_x\,. \tag{64}$$

Then the right-hand side takes the form

$$F = -\frac{\mathrm{i}}{2}(\bar w_1 w_3 - w_1 \bar w_3) + \frac{\mathrm{i}}{2}(f_{2x} - f_{1y}) \tag{65}$$

and we find the system

$$\begin{aligned} & w_{1\bar z} = \frac{\mathrm{i}}{4}(w_2 + \bar w_2)\,, \qquad w_{2\bar z} = \frac{\mathrm{i}}{2}\bar w_3\,, \\ & w_{3\bar z} = t\frac{\mathrm{i}}{2}F = t\frac{1}{4}(\bar w_1 w_3 - w_1 \bar w_3) + t\frac{\mathrm{i}}{2}(f_{2x} - f_{1y})\,. \end{aligned} \tag{66}$$

Besides (66) we also need the boundary conditions corresponding to $\boldsymbol{u} = \boldsymbol{g}$ on $\dot G$. But here it turns out that one needs besides the special derivatives of $\boldsymbol{u}$ in

$$w_2 = u_y - v_x \quad \text{and} \quad w_3 = (u_y - v_x)_y + \mathrm{i}(u_y - v_x)_x$$

also all other derivatives such that all first and second derivatives of u and v are available. Therefore we need to introduce the further complex-valued functions

$$\widetilde{w}_4 := u_y + \mathrm{i} u_x\,, \qquad \widetilde{w}_5 := \widetilde{w}_{4z} = u_{xy} + \frac{\mathrm{i}}{2}(u_{xx} - u_{yy})\,,$$
$$w_6 := v_y + \mathrm{i} v_x\,, \qquad \widetilde{w}_7 := w_{6z} = v_{xy} + \frac{\mathrm{i}}{2}(v_{xx} - v_{yy})\,. \tag{67}$$

Now the integrability conditions between $w_1, \dots, w_6$ together with (66) yield the system

$$\widetilde{w}_{4\bar z} = \frac{\mathrm{i}}{4}(w_3 + \overline{w}_3)\,, \qquad \widetilde{w}_{5\bar z} - \frac{\mathrm{i}}{4} w_{3z} = -\frac{t}{8} F\,,$$
$$w_{6\bar z} = -\frac{1}{4}(w_3 - \overline{w}_3)\,, \qquad \widetilde{w}_{7\bar z} + \frac{1}{4} w_{3z} = t\frac{\mathrm{i}}{8} F\,. \tag{68}$$

Now the boundary conditions together with side conditions can be formulated as follows (we omit the elementary computations):

$$\operatorname{Re} w_1 = g_1 \quad \text{on } \dot G\,, \qquad \Lambda_1 \boldsymbol{w} = \oint_{\dot G} \operatorname{Im} w_1 \,\mathrm{d}s = -\oint_{\dot G} g_2 \,\mathrm{d}s\,, \tag{69}$$

$$\operatorname{Re}\{w_2 - (\dot x^2 + \mathrm{i}\dot x\dot y)\,\widetilde{w}_4 + (\dot y^2 + \mathrm{i}\dot x\dot y)\,w_6\} = \dot y\dot g_1 - \dot x\dot g_2\,, \tag{70}$$

$$\Lambda_2 \boldsymbol{w} = \oint_{\dot G} \operatorname{Im} w_2 \,\mathrm{d}s = 0\,,$$

$$\operatorname{Re}\{-\mathrm{i}\dot z w_3 - (2\dot x\ddot x + \mathrm{i}(\ddot x\dot y + \dot x\ddot y)\,\widetilde{w}_4 + (\dot x^3 - \dot x\dot y^2 + 2\mathrm{i}\dot x\dot y)\,w_5$$
$$- (\ddot x\dot y + \dot x\ddot y + 2\dot y\ddot y)\,w_6 + (\dot x^2\dot y - \dot y^3 + 2\mathrm{i}\dot x\dot y^2)\,w_7\} = \dot y\ddot g_1 - \dot x\ddot g_2\,, \tag{71}$$

with one solvability condition on each $\dot G_\nu$: $\int_{\dot G_\nu} \operatorname{Re} \mathrm{i}\dot z w_3 \,\mathrm{d}s = 0$, $\nu = 0, \dots, m$,

$$\operatorname{Re}\{-\mathrm{i}\dot z\widetilde{w}_4\} = \dot g_1\,, \tag{72}$$

with one solvability condition on each $\dot G_\nu$: $\oint_{\dot G_\nu} \dot g_1 \,\mathrm{d}s = 0$,

$$\operatorname{Re}\{-\mathrm{i}\dot z^2\widetilde{w}_5 + \tfrac{1}{2} w_3 - \mathrm{i}\ddot z\widetilde{w}_4\} = \ddot g_1\,, \tag{73}$$

with three solvability conditions on each $\dot G_\nu$, two of them are $\oint_{\dot G_\nu} \ddot g_1 \,\mathrm{d}s = 0$, $\oint s\ddot g_1 \,\mathrm{d}s - L\dot g_1(0) = 0$,

$$\operatorname{Re}\{-\mathrm{i}\dot z w_6\} = \dot g_2\,, \tag{74}$$

with one solvability condition on each $\dot G_\nu$: $\oint_{\dot G_\nu} g_2 \,\mathrm{d}s = 0$,

$$\operatorname{Re}\left\{-\mathrm{i}\dot z^2\widetilde{w}_7 + \frac{\mathrm{i}}{2} w_3 - \mathrm{i}\ddot z w_6\right\} = \ddot g_2\,, \tag{75}$$

with three solvability conditions on each $\dot G_\nu$, two of them are and $\oint_{\dot G_\nu} \ddot g_2 \,\mathrm{d}s = 0$, $\oint s\ddot g_2 \,\mathrm{d}s - L\dot g_2(0) = 0$.

The system (66), (68) has almost the Bojarski normal form (cf. 18.5., p. 265). If we introduce the new unknowns

$$w_4 = \tfrac{1}{2}(\widetilde{w}_5 - \mathrm{i}\widetilde{w}_7),\ w_5 = \widetilde{w}_4 \ \text{ and } \ w_7 = \tfrac{1}{2}(\widetilde{w}_5 + \mathrm{i}\widetilde{w}_7)\,, \quad \text{i.e.}$$
$$\widetilde{w}_5 = w_4 + w_7,\ \widetilde{w}_4 = w \ \text{ and } \ \widetilde{w}_7 = \mathrm{i}(w_4 - w_7) \tag{76}$$

then (68) is replaced by

$$\begin{aligned} &w_{4\bar z} - \frac{i}{4} w_{3z} = 0\,, \qquad w_{5\bar z} = \frac{i}{4}(w_3 + \bar w_3)\,, \\ &w_{6\bar z} = -\frac{1}{4}(w_3 - \bar w_3)\,, \qquad w_{7\bar z} = -\frac{t}{8} F\,. \end{aligned} \tag{77}$$

Correspondingly, the boundary conditions (73), (75) are changed due to (76).

The system (66), (77) is now a semilinear system in Bojarski's normal form with constant coefficients in the principal part. The corresponding boundary conditions (69)–(75) are linear. For the corresponding boundary matrix $\boldsymbol{P}$ we obtain

$$\det \boldsymbol{P} = -\dot{z}^7 \tag{78}$$

and, hence, from (4) the Fredholm index

$$N - N^* = 7(1 - m) - 14(1 - m) = 7(m - 1) \tag{79}$$

for the linear system (66), (77) in the case $t = 0$.

If G is simply connected, i.e. $m = 0$ then to (69), (70) there correspond the two functionals Λ_1, Λ_2 and to (71)–(75) there belong nine solvability conditions. Here $N = 2$ and $N^* = 9$. Thus, we augment (6), (7) by nine additional constants $\alpha_1, \dots, \alpha_9$. For $(\boldsymbol{g}_r, \boldsymbol{\varphi}_r)$ we choose $\boldsymbol{g}_r = \boldsymbol{0}$ and

$$\begin{aligned} &\boldsymbol{\varphi}_q(s) = (\delta_{r,q})\,, \qquad q = 1, \dots, 5 \\ &\text{and} \\ &\boldsymbol{\varphi}_6 = (x\delta_{r,5})\,, \quad \boldsymbol{\varphi}_7 = (y\delta_{r,5})\,, \quad \boldsymbol{\varphi}_8 = (x\delta_{r,7})\,, \quad \boldsymbol{\varphi}_9 = (y\delta_{r,7}) \end{aligned} \tag{80}$$

where δ_{jk} denotes the Kronecker symbol.

Thus we have obtained a semilinear boundary value problem of the form (11), (12), (13) with linear boundary conditions. To this system we may apply the results from WENDLAND [2, 3] and BEGEHR and HSIAO [1] obtaining the existence of the solution of the Navier-Stokes equations for small Reynolds numbers t recovering a well-known result (LADYŽENSKAJA [1], LIONS [1] and TEMAM [1]). On the other hand, the results from WENDLAND [3] imply much more, namely the convergence of the Newton imbedding method in connection with finite element approximations via the least squares method in 18.4.

18.4. The finite element approximation of first order systems

The usual Galerkin procedure with the same space of trial and test functions cannot be used for first-order systems including the function theoretic boundary value problems anymore since the corresponding bilinear forms are not coercive. Hence, in these cases one either is forced to use different test and trial functions or one uses the least squares method. Here we shall formulate the latter method which yields numerical procedures combining the advantages of the usual finite element methods for even order equations with the manipulation of most general boundary conditions. Here we follow essentially the presentation in WENDLAND [4, chapter 8].

In the sequel we denote by $\tilde H_h$ a family of regular finite element spaces with $\tilde H_h \subset \mathscr{H}^m(G) \cap \mathscr{C}^{m-1}(\bar G)$, $m \geqq 1$. One of the popular choices is a family of piecewise polynomials of degree $\tilde m$ on regular triangulations. To this end let the region $\bar G$ be covered by a family of triangular nets having the following properties:

Let $h > 0$ denote the maximal side length of the triangles of one of these nets. Then we require that every triangle of the particular net contains a disk of radius ch where c is a constant independent of h, $0 < c < \frac{1}{2}$. We further require for those triangles T_k containing points of the boundary curve $\dot{G}$ that the intersection $T'_k := \overline{T}_k \cap G$ contains also a disk of radius ch.

With every triangulation we then associate the space $\widetilde{H}_h$ of piecewise polynomials of degree $\widetilde{m}$, where $\widetilde{m}$ is independent of h. Let us here omit the details of these choices, we refer the reader to BABUŠKA and AZIZ [1] and FIX and STRANG [1, p. 85]. However, we further require the existence of a basis $\{\chi_j\}$ of the trial functions satisfying the "uniformity condition" of order m (FIX and STRANG [1, p. 137]). If the finite element functions are chosen in the above manner then the family $\widetilde{H}_h$ provides the following fundamental properties:

Approximation property (A). *Let $k \leqq r$, $-m-1 \leqq k \leqq m$, $-m \leqq r \leqq m+1$. Then to any $v \in \mathscr{H}^r(G)$ we find an approximating finite element function $V \in \widetilde{H}_h$ inpependently of r and k such that the estimate*

$$\|V - v\|_{\mathscr{H}^k(G)} \leqq c_{rk} h^{r-k} \|v\|_{\mathscr{H}^r(G)} \tag{81}$$

holds with a constant c_{rk} independent of V, v and h. (See BRAMBLE and SCHATZ [1, p. 659], FIX and STRANG [1, p. 145], BABUŠKA and AZIZ [1, Theorem 4.1.5.].)

For $v \in \mathscr{H}^r(G) \cap \mathscr{C}^0(\overline{G})$ with $1 \leqq r \leqq m+1$ and V in (81) we further have

$$|V(z) - v(z)| \leqq c\,|\log h|^{1/2}\, h^{r-1}\, \|v\|_{\mathscr{H}^r(G)} \tag{82}$$

with a constant c independent of V, v, h. (See WENDLAND [4, p. 308].)

Besides the approximation property (A) often the following property is very useful if we have a regular triangulation family:

Inverse assumption (S) (NITSCHE [4]). *Let $-m \leqq k \leqq r \leqq m$. Then there exist constants $M_{rk} > 0$ independently of h and $\chi \in \widetilde{H}_h$ such that*

$$\|\chi\|_{\mathscr{H}^r(G)} \leqq M_{rk} h^{k-r} \|\chi\|_{\mathscr{H}^k(G)} \quad \textit{for all} \quad \chi \in \widetilde{H}_h\,. \tag{83}$$

Moreover, there is a constant M independent of h and χ such that

$$|\chi(z)| \leqq M\,|\log h|^{1/2}\, \|\chi\|_{\mathscr{H}^1(G)} \quad \textit{for all} \quad \chi \in \widetilde{H}_h \quad \textit{and} \quad z \in \overline{G}\,. \tag{84}$$

(See WENDLAND [4, p. 351].)

Since the solutions $\boldsymbol{w}(z)$ are complex n-vector-valued functions forming a vector space over $\mathbb{R}$, we choose for their finite element approximation, subject to the basis $\{\chi_j\}$ of $\widetilde{H}_h$ via $\boldsymbol{e}_l\chi_j$ and $i\boldsymbol{e}_l\chi_j$ with $\boldsymbol{e}_l = (\delta_{lp})$, a basis $\boldsymbol{\chi}_1, \dots, \boldsymbol{\chi}_M$. Replacing $|\cdot|$ by the Euclidean length and $\|\cdot\|_{\mathscr{H}^k(G)}$ by the corresponding product space norms we have the approximation property (A) and the inverse assumption (S) also on the product spaces. We shall in the sequel also denote by $\widetilde{H}_h$ the space spanned by $\{\boldsymbol{\chi}_j\}$.

The least squares for the approximate solution of the linear boundary value problem (6)—(8) now reads as follows. For the approximate solution we choose the ansatz

$$\boldsymbol{W} = \sum_{j=1}^{M} \beta_j \boldsymbol{\chi}_j \quad \text{with} \quad \beta_j \in \mathbb{R} \tag{85}$$

and we determine the coefficients $\beta_1, \dots, \beta_M$ and the approximate constants $\tilde{\alpha}_1, \dots, \tilde{\alpha}_{N^*}$ in such a way that the defects subject to (6)—(8) are minimal, i.e.

$$\varrho_h^2[\boldsymbol{W}, \tilde{\alpha}] := \min_{\chi \in \tilde{H}_h, \tilde{\nu} \in R^{N^*}} \varrho_h^2[\chi, \tilde{\nu}] := \min_{\chi \in \tilde{H}_h, \tilde{\nu} \in R^{N^*}} \{\iint_G |D\chi - \boldsymbol{f} - \sum_{p=1}^{N^*} \tilde{\nu}_p \boldsymbol{g}_p|^2 \, dx \, dy$$

$$+ \frac{1}{h} \oint_{\dot{G}} |\mathrm{Re}\, \boldsymbol{P}\chi - \psi - \sum_{p=1}^{N^*} \tilde{\nu}_p \varphi_p|^2 \, ds + \sum_{l=1}^{N} |\Lambda_l \chi - \gamma_l|^2\} \,. \tag{86}$$

Note that the defect of the boundary conditions carries the penalty factor $1/h$. Without the penalty, in general we can not obtain quasioptimal asymptotic error estimates (cf. WENDLAND [4, p. 351]). For even order equations the corresponding least squares method goes back to BRAMBLE and SCHATZ [1]. STEPHAN and WENDLAND extended these results in [1] to a rather large class of elliptic pseudodifferential problems. Here we follow WENDLAND [4].

Since (86) is a quadratic functional, one finds in the usual manner the Gaussian normal equations or Ritz equations

$$\mathrm{Re} \iint_G (\sum_{j=1}^{M} \beta_j D\chi_j - \boldsymbol{f} - \sum_{p=1}^{N^*} \tilde{\alpha}_p \boldsymbol{g}_p) \overline{D\chi_k} \, dx \, dy$$

$$+ \frac{1}{h} \oint_{\dot{G}} (\mathrm{Re} \sum_{j=1}^{M} \beta_j \boldsymbol{P}\chi_j - \psi - \sum_{p=1}^{N^*} \tilde{\alpha}_p \varphi_p) \, \mathrm{Re}\, \boldsymbol{P}\chi_k \, ds$$

$$+ \sum_{l=1}^{N} (\Lambda_l \boldsymbol{W} - \gamma_l) \Lambda_l \chi_k = 0 \,, \qquad k = 1, \dots, M \tag{87}$$

and furthermore

$$\mathrm{Re} \iint_G (\sum_{j=1}^{M} \beta_j D\chi_j - \boldsymbol{f} - \sum_{p=1}^{N^*} \tilde{\alpha}_p \boldsymbol{g}_p) \overline{\boldsymbol{g}}_q \, dx \, dy$$

$$+ \frac{1}{h} \oint_{\dot{G}} (\mathrm{Re} \sum_{j=1}^{M} \beta_j \boldsymbol{P}\chi_j - \psi - \sum_{p=1}^{N^*} \tilde{\alpha}_p \varphi_p) \; \varphi_q \, ds = 0 \,, \qquad q = 1, \dots, N^* \,. \tag{88}$$

These equations form a quadratic system of $M + N^*$ linear equations for the unknowns $\beta_1, \dots, \beta_M$ and $\tilde{\alpha}_1, \dots, \tilde{\alpha}_{N^*}$. Since the original problem (6)—(8) was supposed to be uniquely solvable, also the system (7), (8) *has always a unique solution.*

Since to the finite elements one can always find a basis formed by functions having a small support (FIX and STRANG [1, p. 101 ff.]), the equations (87) have a sparse matrix of stiffness coefficients. Only the last terms with $\Lambda_l \chi_k$ yield N fully distributed rows as well as the $\boldsymbol{g}_p$ yield N^* fully distributed columns. Correspondingly, the equations (88) yield a fully distributed $(M + N^*) \times N^*$ matrix.

Hence we combine in (87), (88) the advantages of finite element methods for variational problems with the treatment of most general boundary conditions (7). For (87), (88) we have the following asymptotic error estimates:

Theorem 18.4.1. *Let* $0 \leqq t \leqq s \leqq m + 1$, $1 \leqq s$, $t \leqq m$. *If the functionals* Λ_l *are continuous on* $\mathcal{H}^0(G) = \mathcal{L}_2(G)$ *then we have the asymptotic error estimate*

$$|\tilde{\alpha} - \alpha| + \|\boldsymbol{W} - \boldsymbol{w}\|_{\mathcal{H}^t(G)} \leqq c h^{s-t} \|\boldsymbol{w}\|_{\mathcal{H}^s(G)} \tag{89}$$

with a constant c *independent of* $\boldsymbol{W}$, $\boldsymbol{w}$ *and* h.

If the functionals Λ_l are only continuous on $\mathscr{C}^0(\overline{G})$ as e.g. point conditions and if $w \in \mathscr{H}^s(\dot{G}) \cap \mathscr{C}^\varepsilon(\overline{G})$ with any $\varepsilon > 0$ then we have the asymptotic error estimate

$$||\boldsymbol{W} - \boldsymbol{w}||_{\mathscr{C}^0} + ||\boldsymbol{W} - \boldsymbol{w}||_{\mathscr{H}^t(G)} + |\tilde{\alpha} - \alpha| \leqq c\,|\log h|\, h^{s-t}\,||\boldsymbol{w}||_{\mathscr{H}^s(G)} \tag{90}$$

provided $1 \leqq t \leqq s \leqq m + 1$, $t \leqq m$.

The proof of (89) can essentially be found in WENDLAND [4, § 8.4.]. Correspondingly, the proof of (90) can be obtained with a simple modification from WENDLAND [3, p. 310 to 320]. The fundamental tools are Dikanskiĭ's a priori estimate (10) and the Agmon-Douglis-Nirenberg estimates (9), (10) and eventually Nitsche's trick for the least squares method (STEPHAN and WENDLAND [1]).

For the numerical implementation of the Gauss equations (87), (88), all the integrations must be replaced by numerical quadratures and this procedure introduces new errors. The effects of the quadrature error was studied extensively by CIARLET and RAVIART [1]. Here, however, we present only some cruder estimates from WENDLAND [4, § 8.5.]. There it is shown that one should use differently accurate quadratures for the different integrals of the stiffness matrix, i.e. the weights on one hand,

$$\begin{aligned}
&\operatorname{Re} \iint\limits_{G \cap \operatorname{supp}\chi_j \cap \operatorname{supp}\chi_k} D\chi_j \overline{D\chi_k}\, dx\, dy + \frac{1}{h} \int\limits_{\dot{G} \cap \operatorname{supp}\chi_j \cap \operatorname{supp}\chi_k} \operatorname{Re} \boldsymbol{P}\chi_j \operatorname{Re} \boldsymbol{P}\chi_k\, ds + \sum_{l=1}^{N} \Lambda_l\chi_j\Lambda_l\chi_k\,,\\
&\operatorname{Re} \iint\limits_{G \cap \operatorname{supp}\chi_k} g_p \overline{D\chi_k}\, dx\, dy + \frac{1}{h} \int\limits_{\dot{G} \cap \operatorname{supp}\chi_k} \varphi_p \operatorname{Re} \boldsymbol{P}\chi_k\, ds\,,\\
&\operatorname{Re} \iint\limits_{G} g_p \overline{g}_q\, dx\, dy + \frac{1}{h} \oint\limits_{\dot{G}} \varphi_p\varphi_q\, ds\,,
\end{aligned} \tag{91}$$

and for the right-hand sides, i.e.

$$\begin{aligned}
L_1(\chi_k) &:= \operatorname{Re} \iint\limits_{G \cap \operatorname{supp}\chi_k} f\overline{D\chi_k}\, dx\, dy + \frac{1}{h} \oint\limits_{\dot{G} \cap \operatorname{supp}\chi_k} \psi \operatorname{Re} \boldsymbol{P}\chi_k\, ds + \sum_{l=1}^{N} \gamma_l\Lambda_l\chi_k\,,\\
L_2(g_q, \varphi_q) &:= \operatorname{Re} \iint\limits_{G} f\overline{g}_q\, dx\, dy + \frac{1}{h} \oint\limits_{\dot{G}} \psi\varphi_q\, ds
\end{aligned} \tag{92}$$

on the other hand. In order to obtain still quasi optimal asymptotic error estimates we have the following theorem:

Theorem 18.4.2. *For the surface integrals in* (91) *use quadrature formulas in* $T_k' = T_k \cap \overline{G}$ *which are exact for polynomials in* x, y *up to the degree* $m + 2\tilde{m} - 2$. *For the boundary integrals in* (91) *use quadrature formulas on* $\dot{G} \cap T_k$, *which are exact for polynomials in* s *up to the degree* $m + 2\tilde{m}$. *For the right-hand sides* (92) *we require numerical integrations with error estimates*

$$\begin{aligned}
&\textit{integration error of} \quad L_1(\chi_k) = O(h^{s+3/2})\,,\\
&\textit{integration error of} \quad L_2(g_q, \varphi_q) = O(h^{s+1/2})\,, \qquad 1 \leqq s \leqq m + 1\,.
\end{aligned} \tag{93}$$

Let $0 \leqq t \leqq s \leqq m + 1$, $1 \leqq s$, $t \leqq m$. *If* Λ_l *are continuous on* $\mathscr{H}^0(G) = \mathscr{L}_2(G)$ *then the solutions* $\boldsymbol{W}^*$, α^* *of the numerical equations also satisfy the asymptotic error estimate*

$$|\alpha^* - \alpha| + ||\boldsymbol{W}^* - \boldsymbol{w}||_{\mathscr{H}^t(G)} = O(h^{s-t})\,. \tag{94}$$

If the Λ_l are continuous only on $\mathcal{C}^0(\overline{G})$ then we have

$$|\alpha^* - \alpha| + ||\boldsymbol{W}^* - \boldsymbol{w}||_{\mathcal{C}^0} + ||\boldsymbol{W}^* - \boldsymbol{w}||_{\mathcal{H}^t(G)} = 0(|\log h|\, h^{s-t})\,. \tag{95}$$

provided the solution $\boldsymbol{w} \in \mathcal{H}^s(G) \cap \mathcal{C}^\varepsilon(\overline{G})$ with $\varepsilon > 0$.

The proof follows as in Wendland [4, p. 365—366] by the use of (93) in connection with the Schwarz inequality.

After replacing the weights (91), (92) by the above quadrature formulas we find that the discrete system is equivalent to a modified finite difference scheme. In this scheme the expressions coming from the boundary integrals define finite difference approximations of the boundary conditions (7).

Since the systems of linear equations (87), (88) have symmetric and positive definite coefficient matrices they can be solved with accordingly developed solvers. This property is completely transformed to the fully discretized systems.

If we apply our least squares method to the special case of Cauchy-Riemann systems, i.e. (1) with $\boldsymbol{Q} = \boldsymbol{A} = \boldsymbol{B} = \boldsymbol{0}$ then we find methods corresponding to the mixed method in Fix, Gunzburger, and Nicolaides [1]. In Neittaanmäki and Saranen [1] one also finds the least squares method for the Cauchy-Riemann equations and $\boldsymbol{P} \equiv 1$. There the boundary condition (2) is required only in suitable grid points. This method was recently generalized to our general problems (1), (2) by Saranen [2].

If we apply (87) to the Maxwell equations (26)—(28) in the form (33), (34) then our method is rather closely related to the approximations in Saranen [1] and Adam, Gourdin-Servenier, Nedelec, and Raviart [1]. In Saranen [1], however, the boundary conditions are satisfied pointwise on $\dot{G}$ whereas here we approximate (28) by (86). In Adam, Gourdin-Servenier, Nedelec, and Raviart [1] the boundary condition (28) is utilized and approximated as a natural boundary condition whereas (26) is satisfied identically by the especially designed finite element functions.

For the two-dimensional Oseen flows our scheme (86) for (35)—(37) corresponds to mixed methods similarly as for the plane elasticity problems (39)—(41) or (42) (Oden and Lee [1]), the plate problem (Ciarlet [1, § 7.1]) and Stokes flows (cf. Girault [1]).

The nonlinear problems (11)—(13) require for the Newton imbedding (17)—(21) for each iteration step the solution of a *linear* problem (6)—(8). If we approximate any of these problems via the least squares method (86) we would use the same finite element functions for all values of t, i.e. a Newton imbedding method for the corresponding discrete equations. Then there arise several questions: whether the parameter steps $0 < t_1 < t_2 < \dots < t_M$ can be chosen independently of the meshwidth h; whether the eventually obtained approximation $\boldsymbol{W}^{(\infty)}(T)$ converges to $\boldsymbol{w}(T)$ if $h \to 0$ independently of the choice of t_j; whether the convergence $\boldsymbol{W}^{(m)}(t_j) \to \boldsymbol{W}^{(\infty)}(t_j)$ is uniform with respect to t_j and to h. If we assume (i)—(iii) then one finds uniform a priori estimates in a vicinity of $\boldsymbol{w}(t)$ not only in the spaces (9), (10) but also in the Sobolev spaces $\mathcal{W}_p^1$ with $p > 2$ and in appropriate Hölder spaces corresponding to Wendland [3, lemma 4.1]. Then we are able to generalize the proof in Wendland [3] to our more general situation (17)—(21) finding asymptotic convergence as in Wendland [3, theorems 4.3 and 4.4].

Theorem 18.4.3. *There exists $h_0 > 0$ and a finite partition $0 = t_0 < t_1 < \dots < t_M = T$ such that the Newton imbedding* (17)—(21) *in connection with* (87), (88) *for the linearized problems with $\boldsymbol{W}^{(m)} \in \tilde{H}_h$, $\boldsymbol{W}_{\bar{z}}^{(m)}, \dots$ instead of $\boldsymbol{w}^{(m)}, \boldsymbol{w}_{\bar{z}}^{(m)}, \dots$ in* (17)—(19) *converges with respect to the maximum norm uniformly for all $h \leq h_0$.*

Theorem 18.4.4. *For $1 \leqq r \leqq s \leqq m+1$, $r \leqq m$ and $\boldsymbol{w} \in C^s \cap \mathcal{H}^s$ there exists a constant c independently of h and $\boldsymbol{w}$ such that the asymptotic estimates*

$$|\tilde{\alpha} - \alpha| + ||\boldsymbol{W}^{(\infty)}(T) - \boldsymbol{w}(T)||_{\mathcal{H}^r(G)} \leqq ch^{s-r} \, ||\boldsymbol{w}(T)||_{\mathcal{H}^s(G)} , \tag{96}$$

$$|\tilde{\alpha} - \alpha| + ||\boldsymbol{W}^{(\infty)}(T) - \boldsymbol{w}(T)||_{\mathcal{C}^0(G)} \leqq c \, |\log h|^{1/2} \, h^{s-1} \, ||\boldsymbol{w}(T)||_{\mathcal{H}^s(G)} \tag{97}$$

hold provided $\Lambda_{\boldsymbol{w}}$, $\Lambda_{\overline{\boldsymbol{w}}}$ in (19) *are continuous on $\mathcal{H}^0(G)$. If $\Lambda_{\boldsymbol{w}}$, $\Lambda_{\overline{\boldsymbol{w}}}$ in* (19) *are continuous only on $\mathcal{C}^0(\overline{G})$ then we have*

$$|\tilde{\alpha} - \alpha| + ||\boldsymbol{W}^{(\infty)}(T) - \boldsymbol{w}(T)||_{\mathcal{H}^r(G)} \leqq c \, |\log h| \, h^{s-r} \, ||\boldsymbol{w}(T)||_{\mathcal{H}^s(G)} , \tag{98}$$

respectively

$$|\tilde{\alpha} - \alpha| + ||\boldsymbol{W}^{(\infty)}(T) - \boldsymbol{w}(T)||_{\mathcal{C}^0(G)} \leqq c \, |\log h| \, h^{s-1} \, ||\boldsymbol{w}(T)||_{\mathcal{H}^s(G)} . \tag{99}$$

Hence, if we apply our least squares finite element approximation in connection with Newton imbedding for the computation of compressible subsonic flows (51), (55); i.e. if we solve the linear problems in (58), (59) with (86) for every $k = 0, 1, \ldots$ then we obtain approximations which converge with optimal order if $h \to 0$.

The application of our method to the Navier-Stokes system (66), (69)—(77) corresponds to the mixed methods in Fix, Gunzburger, and Nicolaides [1] and Girault and Raviart [1]. In Girault and Raviart [1], however, the equation $\operatorname{div} \boldsymbol{u} = 0$ is enforced on $\tilde{\boldsymbol{H}}_h$ in a weak form.

Besides the computation of a finite element approximation, our Newton imbedding also allows us to compute the regular branch of solutions starting with Reynolds number $t = 0$. This method along the branch for larger Reynolds numbers (with appropriate modification of φ_p) works up to a vicinity of bifurcation points. There one needs to incorporate the Ljapunov-Schmidt method also into the approximating procedure correspondingly to the rather far going results in Brezzi, Rappaz, and Raviart [1]. (There one finds many corresponding references.) This combination probally yields also approximation results for the bifurcation. But all the latter is yet to be done for our first order systems.

18.5. Remarks on the boundary integral method

With the use of the fundamental solution to the linear system (1) or to its principal part (i.e. $\boldsymbol{A} = \boldsymbol{B} = \boldsymbol{0}$), the boundary value problem (1), (2) has been reduced to equivalent boundary integral equations by many authors as I. N. Vekua [5], Agmon [1], Bojarski [5], Fichera [1], Gilbert and Hile [2], Pascali [3], to name a few. (See also Wendland [4, chapters 5 and 6]). In fact, this method goes back to Gauss and Poisson.

In this section let us restrict to systems (1) in *Bojarski's normal form*. In this case, $\boldsymbol{Q}$ consists of lower triangular matrices whose shape is the same throughout the domain. Moreover, in each triangular submatrix, the diagonal elements, subdiagonal elements, and so forth, are the same, respectively, i.e.

$$Q_{j+1,k+1} = Q_{j,k} \quad \text{holds in any of the submatrices} , \tag{100}$$

$$\boldsymbol{Q}(x, y) = \begin{pmatrix} \triangle & & 0 \\ & \triangle & \\ 0 & & \triangle \end{pmatrix} . \tag{101}$$

For the principal part of a system in Bojarski's normal form it is possible to find the *generating matrix* $\Xi(z, \bar{z})$ of the same form as Q in (101) and satisfying the Beltrami system

$$\Xi_{\bar{z}} + Q\Xi_z = 0 \tag{102}$$

such that the diagonal elements of Ξ define appropriate Beltrami homeomorphisms. Following BOJARSKI [5], the generating function can be evaluated from the representation

$$\Xi = z\mathbf{1} + \frac{1}{\pi} \iint_G \frac{\omega(t)}{t - z} \, d\xi \, d\eta , \tag{103}$$

where $\mathbf{1}$ denotes the identity matrix, $z = x + iy$, $t = \xi + i\eta$ and where ω denotes a density matrix function which can be evaluated as the solution of the singular integral equation

$$\omega - Q \frac{1}{\pi} \iint_G \frac{\omega(t)}{(t - z)^2} \, d\xi \, d\eta = Q \tag{104}$$

by the method of successive approximation. For numerical computations, however, this method requires too much effort. Instead of solving (104) for finding Ξ and lateron the corresponding fundamental solution of the principal part, it would be as sensible to formulate a system of integral equations in G and on $\dot{G}$ for (1), (2) directly which also contains the Hilbert transformation (104).

For constant Q or for real analytic Q, however, there is a much simpler method for finding Ξ based on the analytic continuation of Q and (102) into $\mathbb{C}^2$. For the continued equation one may use the classical method of characteristics. This method was already used by Gauss, see WENDLAND [4, § 2.3]. The method of characteristics yields Volterra integral equations in $\mathbb{C}^2$ whose solutions give the elements ζ_{jk} of Ξ *implicitly*. E.g. for $\zeta_1 := \zeta_{11}$ we find the implicit equation

$$z(z^*, \zeta_1) = \zeta_1 - \int_0^{z^* - \zeta_1} Q_{11}\big(z(\sigma, \zeta_1), \sigma + \zeta_1\big) \, d\sigma , \tag{105}$$

where $Q_{11}(z, z^*)$ denotes the analytic continuation of $Q_{11}(x + iy, x - iy)$ into $\mathbb{C}^2$. For finding ζ_1, one has to solve (105) first with respect to $z(z^*, \zeta_1)$. Once z is known, ζ_1 can be found from (105) by using the implicit function theorem.

Correspondingly we find $\zeta_{21}(z, z^*)$ by eliminating s and τ from the system

$$\begin{aligned} z(s, \tau) &= \tau + \int_0^s Q_{11}\big(z(\sigma, \tau), \sigma + \tau\big) \, d\sigma , \\ z^* = \tau + s , \qquad \zeta_{21}(s, \tau) &= - \int_0^s Q_{21}\zeta_{1z}\big(z(\sigma, \tau), \sigma + \tau\big) \, d\sigma + \tau . \end{aligned} \tag{106}$$

Again, (106) first serves for finding $z(s, \tau)$ and $\zeta_{21}(s, \tau)$. Then the elimination of s, τ via the implicit function theorem yields $\zeta_{21}(z, z^*)$.

For constant Q, however, the generating matrix can be found very easily. For the examples (39) and (43) we have

$$\Xi = \begin{pmatrix} z, & 0 \\ \bar{z}, & z \end{pmatrix}$$

and for (66), (77)

$$\boldsymbol{\Xi} = \begin{pmatrix} z & & & & & & \\ & z & & & & 0 & \\ & & z & & & & \\ & & \frac{1}{4}\ \bar{z} & z & & & \\ & & 0 & 0 & z & & \\ & 0 & 0 & 0 & 0 & z & \\ & & 0 & 0 & 0 & 0 & z \end{pmatrix}.$$

In the following let us assume that the generating matrix $\boldsymbol{\Xi}$ is known explicitly. Let us denote

$$\boldsymbol{Z} := \boldsymbol{\Xi}(z, \bar{z}), \qquad \boldsymbol{T} := \boldsymbol{\Xi}(t, \bar{t}), \tag{107}$$

then Cauchy's integral formula yields with the Plemelj-Sohockiĭ jump relations the following system of integral equations:

$$\boldsymbol{w}(z, \bar{z}) = \frac{1}{2\pi \mathrm{i}} \oint_{\dot{G}} \frac{\boldsymbol{v}(t)\, \mathrm{d}\boldsymbol{T}}{\boldsymbol{T} - \boldsymbol{Z}} + \frac{1}{2\pi \mathrm{i}} \iint_{G} \frac{\boldsymbol{A}\boldsymbol{w} + \boldsymbol{B}\bar{\boldsymbol{w}} - \boldsymbol{f}(t)}{\boldsymbol{T} - \boldsymbol{Z}} [\mathrm{d}\boldsymbol{T} \wedge \mathrm{d}\bar{t}] + \mathrm{i}\boldsymbol{k} \quad \text{in } G, \tag{108}$$

$$\begin{aligned} 2\boldsymbol{\psi}(z) = {} & \frac{1}{2} (\boldsymbol{P} + \bar{\boldsymbol{P}})\, \boldsymbol{v} + \frac{1}{2\pi \mathrm{i}} (\boldsymbol{P} - \bar{\boldsymbol{P}}) \oint_{\dot{G}} \frac{\boldsymbol{v}(t)\, \mathrm{d}\boldsymbol{T}}{\boldsymbol{T} - \boldsymbol{Z}} \\ & + \frac{1}{\pi} \bar{\boldsymbol{P}} \oint_{\dot{G}} \boldsymbol{v}(t) \operatorname{Im} \frac{\mathrm{d}\boldsymbol{T}}{\boldsymbol{T} - \boldsymbol{Z}} + \mathrm{i}(\boldsymbol{P} - \bar{\boldsymbol{P}})\, \boldsymbol{k} \\ & + 2 \operatorname{Re} \frac{1}{2\pi \mathrm{i}} \boldsymbol{P}(z) \iint_{G} \frac{\boldsymbol{A}\boldsymbol{w} + \boldsymbol{B}\boldsymbol{w} - \boldsymbol{f}}{\boldsymbol{T} - \boldsymbol{Z}} [\mathrm{d}\boldsymbol{T} \wedge \mathrm{d}\bar{t}] \quad \text{on } \dot{G}. \end{aligned} \tag{109}$$

In this system the unknowns are the desired solution $\boldsymbol{w}$ in $\bar{G}$, the real-valued boundary charge $\boldsymbol{v}$ on $\dot{G}$ and the real constants $\boldsymbol{k}$. The expression

$$[\mathrm{d}\boldsymbol{T} \wedge \mathrm{d}\bar{t}] = \boldsymbol{T}_t[\mathrm{d}t \wedge \mathrm{d}\bar{t}] = -2\mathrm{i}\boldsymbol{T}_t\, \mathrm{d}\xi\, \mathrm{d}\eta$$

denotes the exterior product of the differential forms $\mathrm{d}\boldsymbol{T}$ and $\mathrm{d}\bar{t} = \mathrm{d}\xi - \mathrm{i}\, \mathrm{d}\eta$. (108) is a domain integral equation with Cauchy integrals defining coupling terms. (109) is a system of singular integral equations on the curve $\dot{G}$ having domain integrals as coupling terms. The singular integral equations (109) are of the normal type provided the Lopatinski condition (3) is satisfied (see Michlin and Prössdorf [1, chapter VII], Wendland [4, chapter 5]).

In the special case of $\boldsymbol{Q}$ being of the Douglis normal form, the generating matrix $\boldsymbol{\Xi}$ gives the basic tool for the development of the Douglis function theory of hypercomplex functions by Gilbert and Hile (Douglis [1], Gilbert [2], Gilbert and Hile [2]). There one also finds further more specific integral equations.

If $\boldsymbol{Q}$ belongs to the Pascali normal form then with $\zeta = \zeta_{11}$ the system (1) can be transformed into a system of the same form but with $\boldsymbol{Q} = \boldsymbol{0}$. In this case we find $\boldsymbol{Z} = z\boldsymbol{1}$ ($\boldsymbol{1}$ identity matrix) and (108), (109) can be simplified significantly.

In any case, our system (108), (109) of integral equations is just one specific system among many other possible choices associated to (1), (2). For numerical computations based on (108), (109) there are many different approximations available. Here let us discuss briefly the most common methods.

1. The domain integrals in (108) and (109) can be approximated by introducing polar coordinates about z and using numerical integration whilst approximating $\boldsymbol{w}$ at the corresponding nodal points. This approach yields collectively compact approximating families of operators in the sense of ANSELONE [1].

2. The Cauchy integrals and corresponding charges $\boldsymbol{v}$ can be approximated via Fourier series. Applying Fourier interpolation one obtains the collocation methods by PRÖSSDORF and SILBERMANN [3] which converge asymptotically of optimal order (see also PRÖSSDORF [1]).

3. If one approximates $\boldsymbol{w}$ by finite element functions in $\bar{G}$ and $\boldsymbol{v}$ by piecewise polynomials on $\dot{G}$ then one may either use Galerkin's method in the form of Galerkin collocation (HSIAO, KOPP, and WENDLAND [1]) or just simple collocation. For (109) one obtains a boundary element method. The collocation on $\bar{G}$, however, converges only if (109) provides additional properties as *strong ellipticity* (see PRÖSSDORF and SCHMIDT [1] and WENDLAND [5, 6]).

4. For many special cases of boundary conditions (7), the equations (109) take the form of Fredholm integral equations of the second kind with a smooth kernel (WENDLAND [4, p. 185]). For such equations there are many very effective numerical approximation methods available (ATKINSON [1] and NOBLE [1]).

5. FICHERA and RICCI (FICHERA [1], FICHERA and RICCI [1], RICCI [1]), HSIAO and MACCAMY [1] and also the author in WENDLAND [4, p. 185] have characterized boundary conditions $\boldsymbol{P}$ yielding Fredholm integral equations of the first kind with logarithmic kernel instead of (109). These integral equations have the form

$$-\oint_{\dot{G}} \log|z-t|\, \boldsymbol{v}(s)\, ds_z + \oint_{\dot{G}} L(z,t)\, \boldsymbol{v}(s)\, ds_z = \boldsymbol{\psi}(t) + \sum_{p=1}^{L} \alpha_p \boldsymbol{\varphi}_p(t) \quad \text{on } \dot{G}\,, \tag{110}$$

$$\oint_{\dot{G}} \boldsymbol{v}(s)\, \boldsymbol{\lambda}_p(s)\, ds = b_p\,, \qquad p = 1, \dots, L\,. \tag{111}$$

Here L is a smooth kernel, $\boldsymbol{\varphi}_p$ and $\boldsymbol{\lambda}_p$ are suitably given functions, $\boldsymbol{\psi}$ and $b_1, \dots, b_L \in \mathbb{R}$ are given right-hand sides and $\boldsymbol{v}(s)$ and $\alpha_1, \dots, \alpha_L$ are the desired unknowns. Again, the functions $\boldsymbol{\varphi}_p$ and the side conditions (111) are chosen in such a way that (110), (111) is uniquely solvable. These equations govern conformal mappings (GAIER [1], SYMM [1], WENDLAND [6]), viscous flows (HSIAO and MACCAMY [1]), problems of electrostatics and acoustics (CHRISTIANSEN and RASMUSSEN [1], LE ROUX [1], MARIN [1]) and elasticity problems (CHRISTIANSEN [1], FICHERA [1], HANSEN [1], HEIN [1], JAWSON and SYMM [1], NEDELEC [1]).

The basic property of (110), (111) is the coerciveness in form of Gårding's inequality,

$$-\oint_{\dot{G}}\oint_{\dot{G}} \log|z-t|\, \boldsymbol{v}(s)\cdot(\sigma)\, ds_z\, d\sigma_t + \oint_{\dot{G}}\oint_{\dot{G}} \boldsymbol{v}(\sigma)\cdot L(z,t)\, \boldsymbol{v}(s)\, ds_z\, d\sigma_t$$
$$\geqq \gamma\, \|\boldsymbol{v}\|^2_{\mathcal{H}^{-1/2}(\dot{G})} - \delta\, \|\boldsymbol{v}\|^2_{\mathcal{H}^{-1}(\dot{G})} \tag{112}$$

with constants $\gamma > 0$ and $\delta \geqq 0$. This inequality was proved in NEDELEC and PLANCHARD [1], LE ROUX [2], HSIAO and WENDLAND [1] and follows also from the more general approach given in STEPHAN and WENDLAND [1]. (112) implies that (110), (111) defines a Fredholm mapping with index zero from $\mathcal{H}^s(\dot{G}) \times \mathbb{R}^L$ into $\mathcal{H}^{s+1}(\dot{G}) \times \mathbb{R}^L$ for any $s \in \mathbb{R}$ if $\dot{G} \in \mathcal{C}^\infty$ or a suitable variety of real s corresponding to the smoothness of $\dot{G}$. Therefore the approximation of $\boldsymbol{v}$ by piecewise polynomials on $\dot{G}$ and the application of Galerkin's method prove to be very efficient. For this method we use a sufficiently smooth, regular parameter representation $z = z(t)$ of the boundary $\dot{G}$ with $t \in [0, 1]$ and $z(t)$, $\boldsymbol{v}(t)$ being 1-periodic functions. For $\tilde{H}_h$ we use 1-periodic piecewise polynomials of degree m in $\mathscr{C}^{m-1}(\dot{G})$ on a regular family of boundary grids. Then the approximation property (A), (81) and also the inverse assumption (S), (83) both hold on $\dot{G}$. If $\boldsymbol{v}_h$, $\boldsymbol{\alpha}_h$ denote the Galerkin solutions subject to the above approximations of (110), (111) then in HSIAO and WENDLAND [1, 2] the following error estimates are shown:

Theorem 18.5.1. *Let* $-m - 2 \leqq t \leqq m$, $t \leqq s \leqq m + 1$, $-\frac{1}{2} \leqq s$. *Then we have the asymptotic error estimates*

$$\begin{aligned} ||\boldsymbol{v}_h - \boldsymbol{v}||_{\mathscr{H}^t(\dot{G})} &\leqq c h^{s-t} \, ||\boldsymbol{v}||_{\mathscr{H}^s(\dot{G})} , \\ |\boldsymbol{\alpha}_h - \boldsymbol{\alpha}| &\leqq c h^{s+1} \, ||\boldsymbol{v}||_{\mathscr{H}^s(\dot{G})} \end{aligned} \tag{113}$$

where the constants c are independent of h, $\boldsymbol{v}$, $\boldsymbol{v}_h$, $\boldsymbol{\alpha}$ and $\boldsymbol{\alpha}_h$ but depend on s and t.

Remark. The estimate for the constants in (113) is not optimal. Recently it was shown that

$$|\boldsymbol{\alpha}_h - \boldsymbol{\alpha}| \leqq c h^{s+m+1} \, ||\boldsymbol{v}||_{\mathscr{H}^s(\dot{G})}$$

holds. This result will be published elsewhere.

The above method has been further developed to a fully discretized numerical scheme by the use of appropriate numerical integration formulas (HSIAO, KOPP, and WENDLAND [1], WENDLAND [6]). One finds a modified collocation method which also converges asymptotically with optimal orders as in (113).

6. In special cases as $\boldsymbol{Q} \equiv \boldsymbol{0}$ and $\boldsymbol{P} \equiv \boldsymbol{1}$ and for a simply connected domain G it is possible to find Fredholm integral equations of the second kind for $\boldsymbol{w}$ in the domain G which are equivalent to (1), (2). To this end one needs the conformal mapping of G onto the unit disk or its approximation, respectively. The approximation of these integral equations and corresponding asymptotic error estimates can be found in WENDLAND [4, chapter 6].

19. Optimal Control Problems with Elliptic Differential Equation Systems of First Order in the Plane

L. v. Wolfersdorf

19.1. Introduction

Optimal control problems for elliptic differential equation systems on the plane (written in complex form) have been dealt with only by a few authors yet. The investigation of such problems were started by Tutschke [6, 13], whose research has been continued by his pupil Klink [1—3], and independently of them by the author [2—4, 6, 7, 9] and by C. and Cl. Simionescu [1], Cl. Simionescu [1]. Tutschke and Klink consider problems of distributed control for general linear and nonlinear complex differential equation systems of m equations in some normal form. Either by means of a Dirichlet boundary condition with an additional condition in a boundary point or by utilizing a homeomorphism between the holomorphic functions and the solutions of the complex differential equation system they obtain a uniquely determined solution to the complex equation system for any admissible control function. In the author's papers various kinds of control problems with distributed and boundary control for semilinear, quasilinear and general nonlinear equations and with linear and nonlinear Riemann-Hilbert boundary conditions without and with additional conditions, respectively, are treated. Finally, C. and Cl. Simionescu deal with linear-quadratic control problems for analytic and also for generalized analytic functions. Their (unfortunately partly insufficient and incorrect) results are completed and improved in the author's paper [8].

Tutschke and Klink derive necessary optimality conditions for the optimal control functions only, whereas in the author's papers [7—9] and in the papers by C. and Cl. Simionescu also existence assertions are given. The derived optimality conditions throughout possess the form of a variational inequality in the control function, i.e. the form of a weak maximum principle in case of pointwise constraints to the control functions. A strong maximum principle analogous to Pontrjagin's principle seems to hold true only in some exceptional cases (if there is nonlinear dependence on the control function). Such a case might be the problem with distributed control for semilinear differential equation and suitable unique correspondence between control functions and solutions of the differential equation.

19.2. The papers of Tutschke and Klink

In the paper [6] Tutschke deals with a control problem with *distributed control* for the *semilinear differential equation system*

$$\frac{\partial w_j}{\partial \bar{z}} = f_j\,(z, w_1, \dots, w_m) + c_j(z)\, u(z)\,, \qquad j = 1, \dots, m\,, \tag{1}$$

into which the real-valued control function $u(z)$ enters linearly. The given functions f_j and c_j are continuous and the f_j are continuously differentiable with respect to the w_i. For simplicity the f_j are supposed to be holomorphic functions of $w_1, \ldots, w_m$. The continuous (or piecewise continuous) control functions $u(z)$ are subject to the constraint $|u(z)| \leq K$ with $K > 0$. Moreover, it is required that the solution $w = (w_1(z), \ldots, w_m(z))$ of the system (1) in a region G of the z-plane is homeomorph to a prescribed m-tupel $\Phi = (\Phi_1(z), \ldots, \Phi_m(z))$ of functions Φ_i holomorphic in G and continuous in $\bar{G}$. In this way the solution w of (1) for any fixed $u(z)$ is uniquely determined. Such a homeomorphism of the solutions w to (1) with the m-tupel Φ always exists if the region G has a sufficiently small diameter (cf. TUTSCHKE [5]). Aim of control is the minimization of a functional of the form

$$\operatorname{Re} \iint_G [\hat{f}_0(z, w_1(z), \ldots, w_m(z)) + \hat{c}_0(z)\, u(z)]\, dx\, dy\,, \tag{2}$$

where $\hat{f}_0$ and $\hat{c}_0$ are continuous functions of their arguments (and $\hat{f}_0$ is holomorphic in $w_1, \ldots, w_m$).

In analogy with the classical proof of Pontrjagin's maximum principle for ordinary differential equation systems the system (1) is supplemented by the equation

$$\frac{\partial w_0}{\partial \bar{z}} = f_0(z, w_1, \ldots, w_m) + c_0(z)\, u(z)\,, \tag{3}$$

where $f_0(z, w_1, \ldots, w_m) = -\pi(z - z_0)\, \hat{f}_0(z, w_1, \ldots, w_m)$, $c_0(z) = -\pi(z - z_0)\, \hat{c}_0(z)$ with an arbitrarily chosen point $z_0 \in G$. It is shown that the solution vector $(w_0^0(z), w_1^0(z), \ldots, w_m^0(z))$ of the augmented system (1), (3) corresponding to an optimal solution pair $(w^0(z), u^0(z))$ is a boundary point of the convex set of attainability $M(z_0)$ of the admissible variations δu of $u^0(z)$. The existence of a supporting hyperplane to $M(z_0)$ at this boundary point then implies the existence of a non-trivial solution $\lambda = (\lambda_0(z), \ldots, \lambda_m(z))$ of the adjoint system

$$\frac{\partial \lambda_j}{\partial \bar{z}} = -\sum_{i=0}^{m} \frac{\partial f_i}{\partial w_j}(z, w_1^0, \ldots, w_m^0)\, \lambda_i\,, \qquad j = 0, 1, \ldots, m\,, \tag{4}$$

possessing at most a pole of first order in z_0 such that in each point $z \in G$ the optimal control function $u^0(z)$ maximizes the real part of the complex Hamilton function

$$H(z, w, \lambda, u) = \sum_{j=0}^{m} f_j(z, w_1, \ldots, w_m)\, \lambda_j(z) + u \sum_{j=0}^{m} c_j(z)\, \lambda_j(z) \tag{5}$$

for $w = w^0(z)$ with respect to the admissible values $|u| \leq K$. An analogous assertion surely holds true in case the right-hand sides of the system (1) and the functional (2) depend on the control $u(z)$ in nonlinear manner.

In the special case of a *linear differential equation system*

$$\frac{\partial w_j}{\partial \bar{z}} = \sum_{i=1}^{m} A_{ji}(z)\, w_i(z) + d_j(z)\, u(z)\,, \qquad j = 1, \ldots, m\,, \tag{6}$$

with the *linear objective functional*

$$l(u) = \operatorname{Re} \iint_G [\sum_{i=1}^{m} A_{0i}(z)\, w_i(z) + d_0(z)\, u(z)]\, dx\, dy \tag{7}$$

TUTSCHKE [13] also considers a problem with boundary condition. The solution $w = (w_1(z), \ldots, w_m(z))$ of (6) is determined by the Dirichlet conditions

$$\operatorname{Re} w_j(t) = g_j(t) \quad \text{on } \partial G, \text{ the boundary of } G\,, \tag{8}$$

and the additional condition

$$\operatorname{Im} w_j(t_0) = 0 \tag{9}$$

in an arbitrarily chosen point $t_0 \in \partial G$. The coefficients A_{jl} and d_j in (6) lie in the space $\mathscr{C}^\alpha(\bar{G})$ of functions satisfying a Hölder condition with exponent α, $0 < \alpha < 1$, in $\bar{G}$, and the Hölder norms of the A_{jl} are sufficiently small. Also the control function $u(z)$ and the functions $w_j(z)$ shall lie in $\mathscr{C}^\alpha(\bar{G})$. Utilizing a suitable representation of the solution w to (6) with (8), (9) by means of a fundamental system and transforming the functional (7) in a suitable way TUTSCHKE shows that for an optimal control function $u^0(z)$ the inequality

$$l_1(\delta u) \geqq l_2(\delta u) \quad \text{with} \quad \delta u = u - u^0 \tag{10}$$

holds true for any admissible $u(z)$. Here l_1 is a linear functional constructed with the help of a solution of a Dirichlet problem for holomorphic functions and l_2 such one defined by means of suitable solutions of the homogeneous adjoint system to (6).

General problems with *distributed control* for the *nonlinear differential equation system in normal form*

$$\frac{\partial w_j}{\partial \bar{z}} = f_j\left(z, w(z), \frac{\partial w}{\partial z}(z), u(z)\right), \qquad j = 1, \dots, m, \tag{11}$$

are studied by KLINK in his thesis [1] (see also [2, 3]). The solutions $w = \big(w_1(z), \dots, w_m(z)\big)$ of (11) in a disk G shall fulfil the Dirichlet condition (8) with functions $g_j \in \mathscr{C}_1^\alpha(\partial G)$ and the (inhomogeneous) additional condition (9). The complex-valued control function $u(z)$ lies in $\mathscr{C}^\alpha(\bar{G})$ and takes up values from a convex control domain $U(z)$ depending on $z \in \bar{G}$. The functions $f_j(z, w, h, u)$ and their partial complex derivatives of first order with respect to the components of w, h, $\bar{w}$, $\bar{h}$ and u, $\bar{u}$ are Hölder-continuous functions of their arguments which moreover satisfy certain Lipschitz conditions of first and second order in the Hölder norms with respect to the variables w, h, u. If the Lipschitz constants in these conditions and the Hölder norms of the functions f_j and g_j are sufficiently small a uniquely determined state function $w(z) \in \mathscr{C}_1^\alpha(\bar{G})$ exists to any admissible control function $u(z)$ as can be proved by means of the Banach fixed point theorem.

The *cost functional* has the general form

$$F(u) = \operatorname{Re} \iint_G f_0\left(z, w(z), \frac{\partial w}{\partial z}(z), u(z)\right) dx\, dy, \tag{12}$$

where the function f_0 fulfils similar assumptions as the f_j.

The differential equation system (11) together with the boundary conditions (8), (9) is equivalent to the relations

$$\begin{aligned} w_j(z) &= \Phi_{j1}(z) + \Phi_{j2}(z) + Tf_j(z, w, h, u)\,[z], \\ h_j(z) &= \Phi_{j1}'(z) + \Phi_{j2}'(z) + \Pi f_j(z, w, h, u)\,[z] \qquad (j = 1, \dots, m) \end{aligned} \tag{13}$$

for w and $h = \partial w/\partial z$, where the functions Φ_{j1}, Φ_{j2} are holomorphic in G and Hölder-continuously differentiable in $\bar{G}$. Besides the well-known integral operators

$$T\varphi[z] = -\frac{1}{\pi}\iint_G \frac{\varphi(\zeta)}{\zeta - z}\, d\xi\, d\eta, \qquad \Pi\varphi[z] = -\frac{1}{\pi}\iint_G \frac{\varphi(\zeta)}{(\zeta - z)^2}\, d\xi\, d\eta$$

$$(\zeta = \xi + i\eta) \tag{14}$$

have been introduced in (13). The functions $\Phi_{j1}(z)$ satisfy the boundary conditions (8), (9), whereas the $\Phi_{j2}(z)$ equalize the corresponding values of the $Tf_j(z, w, h, u)$ on ∂G to zero. If on the other hand the functions $\Phi_{j2}(z)$ are chosen identically zero the above-mentioned homeomorphism of the solutions w of (11) to a fixed m-tupel of holomorphic functions dealt with by TUTSCHKE [6] is obtained as a special case.

For deriving necessary optimality conditions the *variational method* is used. The problem (13) and the functional (12) are in a suitable way linearized in a neighbourhood of the considered optimal control function $u^0(z)$ with a corresponding optimal state function $w^0(z)$. By means of suitable estimations for sufficiently small variations $\delta u = u - u^0$ and $\delta w = w - w^0$ of u^0 and w^0, respectively, it is shown that in any point $z \in \overline{G}$ the real part of the complex Hamilton function

$$H(z, w, a, b, u) = \sum_{j=0}^{m} \left[\frac{\partial f_j}{\partial u}(z)\, u + \frac{\partial f_j}{\partial \overline{u}}(z)\, \overline{u}\right] \frac{a_j(z)}{z - t_0} + \left[\frac{\partial \overline{f_j}}{\partial \overline{u}}(z)\, \overline{u} + \frac{\partial \overline{f_j}}{\partial u}(z)\, u\right] \frac{b_j(z)}{z - t_0} \tag{15}$$

is minimized with respect to the control set $U(z)$ by the optimal control function $u^0(z)$. The functions $a_j(z)$, $b_j(z)$ in (15) are the uniquely determined solutions of an associate operator problem to the equations (13) linearized in (w^0, u^0) and also the derivatives of the f_j, $\overline{f_j}$ in (15) have to be calculated in (w^0, u^0).

In some cases the *Hamilton function* can be stated in the simpler form (15) without the denominator $z - t_0$. This is the case for the above homeomorphism of the solutions w of (11) to a fixed m-tupel of holomorphic functions and also for the Dirichlet boundary condition (8) with (9) if the functions f_j do not depend on w. If in the first case the f_j are holomorphic functions of $w_1, \dots, w_m$ and u as in the above problem (1), (2) of TUTSCHKE [6] moreover the functions b_j in (15) are identically zero. It is remarkable that the Hamilton function constructed by KLINK in contrast with the Hamilton function (5) does not posses a pole in some point $z_0 \in G$. Recently KLINK (cf. [4]) also studies analogous control problems for linear differential equation systems with Riemann-Hilbert boundary conditions introducing suitable adjoint problems.

19.3. The papers of the author

In a series of papers [2—4, 6, 7, 9] the author derived necessary optimality conditions for various control problems with first order elliptic systems on the plane. These derivations are based on a *theory of optimality conditions* for abstract control problems with Noetherian operator equations in Banach spaces which was developed by M. GOEBEL and the author. Briefly this theory reads as follows (cf. v. WOLFERSDORF [2]).

Let X, Y, U be real Banach spaces, and let $V \subseteq U$ be a non-void convex subset of U. Further, $J(x, u)\colon X \times U \to \mathbb{R}$ shall be a real-valued functional and $T(x, u)\colon X \times U \to Y$ an operator into the space Y. The following optimal control problem P is considered:

$$\inf J(x, u) \quad \text{subject to} \quad T(x, u) = \theta\,, \quad u \in V\,, \tag{16}$$

where θ means the zero element of Y.

Let (x^0, u^0) be an optimal solution pair to problem P. The following assumptions should be satisfied.

(i) The functional J is Fréchet differentiable in (x^0, u^0), where the partial Fréchet derivatives $J_x^0 \equiv J_x(x^0, u^0) \in X^+$, $J_u^0 \equiv J_u(x^0, u^0) \in U^+$. X^+ and U^+ being so called conjugate spaces, that means total subspaces of the dual spaces X^* and U^* to X and U, respectively.

(ii) The operator T is continuously Fréchet differentiable in a neighbourhood of (x^0, u^0) and $T_x^0 \equiv T_x(x^0, u^0)$, $T_u^0 \equiv T_u(x^0, u^0)$ are bounded linear operators from X and U into Y, respectively.

(iii) With a suitable conjugate space Y^+ to Y the corresponding conjugate operators T_x^{0+}, T_u^{0+} are bounded linear operators from Y^+ into X^+ and U^+, respectively, and the operators $T_x^0: X \to Y$, $T_x^{0+}: Y^+ \to X^+$ form a Noetherian pair. The (finite-) dimensions of the kernels of T_x^0 and T_x^{0+} are denoted by $p = \dim N(T_x^0)$ and $q = \dim N(T_x^{0+})$. $\varkappa = p - q$ is the index of the problem.

(iv) If $q > 0$ there is $\dim T_u^{0+}N(T_x^{0+}) \doteq q$, and there exists a Banach space $U_1 \subseteqq U$ everywhere dense in U with the property, that the intersection $V_1 = V \cap U_1$ has a non-void interior V_0 with respect to U_1.

Then there holds the *orthogonality relation*

$$\langle J_x^0, z\rangle = 0 \qquad \forall z \in N(T_x^0) \tag{17}$$

and the *quasi-variational inequality*

$$\langle J_u^0 - T_u^{0+}\zeta^0, v - u^0\rangle \geqq 0 \qquad \forall v \in V_0 \tag{18}$$

subject to

$$\langle T_u^{0+}g, v - u^0\rangle = 0 \qquad \forall g \in N(T_x^{0+}), \tag{19}$$

where $\zeta^0 \in Y^+$ is an arbitrary solution to the conjugate equation

$$T_x^{0+}\zeta = J_x^0 \quad \text{in } X^+ . \tag{20}$$

For affin-linear operators $T(x, u)$ the assumption (iv) is superfluous, if $q = 0$ the inequality (18) holds true for all $v \in V$.

Advantages in applying this theory are the facts that the nonlinear operator equation $T(x, u) = \theta$ need not be investigated and that the linearized equation with the Fréchet derivative only with a suitable conjugate operator T_x^{0+} (not necessarily the adjoint operator T_x^{0*}) has to form a Noetherian pair. Especially this allows an immediate application of the well-known theories of boundary value problems for generalized analytic functions developed by I. N. Vekua, W. Haack and G. Hellwig, B. Bojarski, V. S. Vinogradov, etc. Only for general nonlinear problems one has to pose additional smoothness assumptions for setting up an adjoint boundary value problem to which these theories can be applied. Without such additional assumptions adjoint problems in case of general nonlinear control problems only exist for related integral equation or more general operator equation problems to which the problems may be reduced as in the work of Klink [1]. In the author's papers [2—4, 6, 7, 9] adjoint boundary value problems are always used. Moreover, only first order elliptic systems for two unknown real functions are dealt with, where owing to I. N. Vekua's [5] theory for the linearized problems the numbers p and q are explicitly known in dependence on the index $\varkappa$ of the problem. But in principle the above theory of optimality conditions may be applied also to general elliptic systems with more than two unknown functions if the related linearized problem possesses the Noetherian property.

In the paper V. Wolfersdorf [6] (see also I. N. Vekua [5]) a control problem with *distributed and boundary control* for a *semilinear complex differential equation* with general nonlinear boundary condition is treated. Let G be a bounded, simple connected region with boundary ∂G of class $\mathscr{C}^1_\mu$, $0 < \mu < 1$. The state function $W(z) = w(x, y) + iv(x, y) \in \mathscr{C}^\mu(\overline{G})$ with $\partial W/\partial \bar{z} \in L_\infty(G)$ satisfies the differential equation

$$\partial W/\partial \bar{z} = \Psi(z, \bar{z}, W, \overline{W}, U) \quad \text{in } G \tag{21}$$

and the boundary condition

$$\Phi\big(t, w(t), v(t), u(t)\big) = 0 \quad \text{on } \partial G\,. \tag{22}$$

The real-valued distributed control function $U(z)$, $z \in G$, is chosen from $\mathscr{L}_\infty(G)$, and the boundary control function $u(t)$, $t \in \partial G$, from $\mathscr{C}^\mu(\partial G)$. Both control functions take values from prescribed intervals of the real axis. The cost functional to minimize has the form

$$J = \iint_G F_1(z, \bar{z}, W, \overline{W}, U)\, dx\, dy + \int_{\partial G} F_2\big(t, w(t), v(t), u(t)\big)\, ds\,. \tag{23}$$

The given complex- and real-valued functions Ψ, Φ, and F_k, $k = 1, 2$, respectively, are sufficiently smooth such that the operator $T(W; U, u)$ defined by (21) and (22) and the functional $J(W; U, u)$ given by (23) are continuously Fréchet-differentiable in the spaces $X = \mathscr{C}^\mu(\overline{G})$, $Y = \mathscr{L}_\infty(G) \times \mathscr{C}^\mu(\partial G)$, and $U = \mathscr{L}_\infty(G) \times \mathscr{C}^\mu(\partial G)$.

Let $\omega^0 = \big(U^0(z), u^0(t)\big)$ be an optimal pair of controls with corresponding optimal state function $W^0(z)$. We suppose $\lambda(t) = \Phi^0_w(t) + i\Phi^0_v(t) \neq 0$ on ∂G. Then, for non-negative index $n = \operatorname{ind}\lambda = (1/2\pi)\,[\arg \lambda(t)]_{\partial G} \geqq 0$ the optimality conditions

$$\iint_G \left[\frac{\partial F_1^0}{\partial U}(z) - \operatorname{Re}\{\Psi^0_U(z)\, Z(z)\}\right][U(z) - U^0(z)]\, dx\, dy \geqq 0\,, \tag{24}$$

$$\int_{\partial G} \left[\frac{\partial F_2^0}{\partial u}(t) - \Phi^0_u(t)\, q(t)\right][u(t) - u^0(t)]\, ds \geqq 0 \tag{25}$$

hold for any pair $\omega = \big(U(z), u(t)\big)$ of admissible control functions. The adjoint state function $Z(z)$ is the uniquely determined solution to the adjoint boundary value problem

$$\partial Z/\partial \bar{z} + [\Psi^0_W Z + \overline{\Psi^0_{\overline{W}} Z}] = \overline{H(z)} \quad \text{in } G\,, \tag{26}$$

$$\operatorname{Re}[\lambda(t)\, t'(s) Z] = 2 \operatorname{Im}[\lambda(t)\, \overline{\sigma(t)}] \quad \text{on } \partial G\,, \tag{27}$$

where $H(z) = 2\partial F_1^0/\partial \overline{W}$, $\sigma(t) = \partial F_2^0/\partial w + i\, \partial F_2^0/\partial v$. The function $q(t)$ in (25) is given by

$$q(t) = \frac{1}{2\,|\lambda|^2}\{\operatorname{Im}[\lambda(t)\, t'(s)\, Z] + 2 \operatorname{Re}[\lambda(t)\, \overline{\sigma(t)}]\}\,. \tag{28}$$

Analogous optimality conditions hold true in case of negative index $n < 0$ and in case of complex-valued control function $U(z)$.

A similar control problem for the equation (21) with *linear boundary condition*

$$\operatorname{Re}[\overline{\lambda(t)}\, W] = f(t) + g(t)\, u(t) \quad \text{on } \partial G\,, \tag{29}$$

where the prescribed functions $f(t)$, $g(t) \in \mathscr{C}^\mu(\partial G)$ and $\lambda(t) \in \mathscr{C}^\mu(\partial G)$ with $\lambda(t) \neq 0$ on ∂G and $n = \operatorname{ind}\lambda \geqq 0$, is considered in the paper v. Wolfersdorf [4].

There the following *additional conditions* on the state function $W(z)$ are posed: $W(z_i) = c_i$, $i = 1, \ldots, n$, in n given fixed points $z_i \in G$ and either

$$\operatorname{Im}[\overline{\lambda(t_0)}\, W(t_0)] = c_0 \tag{30}$$

in a given point $t_0 \in \partial G$ or

$$\int_{\partial G} \operatorname{Im}\,[\overline{\lambda(t)}\,W]\,\varrho(t)\,ds = c_0 \tag{31}$$

with a prescribed positive function $\varrho(t) \in \mathscr{C}^{\mu}(\partial G)$. Again the optimality conditions (24), (25) (with $\Phi_u^0(t) = -g(t)$) hold true. But now the adjoint state function $Z(z)$ in (26), (27) may have poles of first order in the points z_i and in case of condition (30) a pole of first order of the kind

$$Z(z) \sim \frac{1}{\pi i}\,\frac{C_0\overline{\lambda(t_0)}}{z - t_0}\,, \qquad C_0 \text{ real parameter}\,, \tag{32}$$

in the point t_0, too, whereas in the case of condition (31) the boundary condition (31) has to be fulfilled apart from an additional term of the form $C_0\,|\lambda(t)|^2\,\varrho(t)$ with arbitrary real parameter C_0 only.

Finally, in the paper v. WOLFERSDORF [3] control problems for the *general nonlinear differential equation of first order*

$$\Psi(z, \bar z, W, \overline{W}, W_{\bar z}, W_z, \overline{W}_{\bar z}, \overline{W}_z, U) = 0 \quad \text{in } G \tag{33}$$

with boundary condition (22) and cost functional (23) are dealt with, where the function F_1 in (23) may also depend on the derivatives of W. The state functions $W(z) \in \mathscr{C}_1^{\mu}(\overline{G})$ are classical solutions to (33) and the control functions $U(z) \in \mathscr{C}^{\mu}(\overline{G})$, $u(t) \in \mathscr{C}_1^{\mu}(\partial G)$. Again $\lambda(t) = \Phi_w^0(t) + i\Phi_v^0(t) \neq 0$ on ∂G. Besides we assume $Q_k(z) \in \mathscr{C}_1^{\mu}(\overline{G})$, $k = 1, 2, 3$, for the functions $Q_k(z) = -A_k(z)/A_0(z)$, where

$$A_0(z) = \frac{\partial \Psi^0}{\partial W_{\bar z}}\,, \qquad A_1(z) = \frac{\partial \Psi^0}{\partial W_z}\,, \qquad A_2(z) = \frac{\partial \Psi^0}{\partial \overline{W}_{\bar z}}\,,$$
$$A_3(z) = \frac{\partial \Psi^0}{\partial \overline{W}_z}\,, \tag{34}$$

and the ellipticity condition

$$|Q_1(z)| + |Q_2(z)| + |Q_3(z)| \leqq q_0 < 1 \quad \text{in } G\,. \tag{35}$$

We further put $P_j(z) = B_j(z)/A_0(z)$, $j = 1, 2$, where

$$B_1(z) = \partial\Psi^0/\partial W\,, \qquad B_2(z) = \partial\Psi^0/\partial \overline{W}\,. \tag{36}$$

Moreover, *additional assumptions* for setting up an adjoint boundary value problem are necessary: $\partial F_1^0/\partial W \in \mathscr{C}^{\varepsilon}(\overline{G})$, $\sigma(t) \in \mathscr{C}_1^{\varkappa}(\partial G)$, and $G \in \mathscr{C}_2^{\nu}$, if the classical solution theory for the adjoint boundary value problem is used, or $\partial F_1^0/\partial W \in \mathscr{L}_q(G)$, $q > 2$, $\sigma(t) \in \mathscr{C}^{\varkappa}(\partial G)$, $\varkappa > 1/2$, and $G \in \mathscr{C}_1^{\nu}$, $\nu > 1/2$, if the theory of V. S. VINOGRADOV for elliptic differential equation systems of first order, which are not given in normal form, is utilized (cf. I. N. VEKUA [5]). Again optimality conditions of the form (24), (25) are obtained, where in (26) the adjoint differential operator

$$Z_{\bar z} - [Q_1 Z]_z - [\overline{Q}_2 \overline{Z}]_z - [\overline{Q}_3 \overline{Z}]_{\bar z} - P_1 Z - \overline{P}_2 \overline{Z} \tag{37}$$

and in (27) the boundary operator

$$\operatorname{Re}\,[\lambda(t)\,t'(s)\,R(t)\,Z] \tag{38}$$

with $R(t) = 1 + \overline{t'(s)}/t'(s)\,Q_1(t) + \overline{\lambda(t)}/\lambda(t)\,Q_2(t) - \overline{\lambda(t)}/\lambda(t)\,\overline{t'(s)}/t'(s)\,Q_3(t)$ occur.

The derivation of optimality conditions in control problems with the general equation (33) and the boundary condition (22) becomes difficult if more general classes of control functions are considered, particularly for boundary control. Using the above-mentioned theory of VINOGRADOV in the paper v. WOLFERSDORF [9] optimality conditions are derived only for problems with *quasi-linear equations*

$$A_0 W_{\bar{z}} - A_1 W_z - A_2 \overline{W}_{\bar{z}} - A_3 \overline{W}_z = B \quad \text{in } G\,, \tag{39}$$

where $A_k = A_k(z, W, U)$, $k = 0, \dots, 3$, $B = B(z, W, U)$, and *distributed control functions* $U(z) \in \mathcal{L}_\infty(G)$. In addition, a linear boundary condition of the form (29) without boundary control is prescribed. Here $W(z) \in \mathcal{W}_p^1(G)$, $p > 2$, and the assumptions $\lambda(t) \in \mathcal{C}^\mu(\partial G)$, $\mu > (p-1)/p$, and $G \in \mathcal{C}_1^\nu$, $\nu > 1/2$, are made.

The paper v. WOLFERSDORF [9] also contains *existence theorems* for optimal controls, where the state equation is assumed in the following normal form:

$$W_{\bar{z}} - \mu_1 W_z - \mu_2 \overline{W}_{\bar{z}} = d \quad \text{in } G, \text{ the unit disk}\,. \tag{40}$$

The coefficients $\mu_k(z, W, U)$, $k = 1, 2$, are Lipschitz continuous in W and satisfy the ellipticity condition

$$|\mu_1(z, W, U)| + |\mu_2(z, W, U)| \leqq \mu_0 < 1 \quad \text{in } G\,. \tag{41}$$

The function $d(z, W, U)$ possesses linear growth in W and fulfils a Lipschitz condition in W with Lipschitz coefficients from $\mathcal{L}_p(G)$, $p > 2$. For non-negative index $n = \operatorname{ind}\lambda \geqq 0$ the state function $W(z) \in \mathcal{W}_p^1(G)$ satisfies $2n + 1$ suitable real additional conditions, whereas for negative index $n < 0$ the homogeneous linear boundary condition (29) has to be fulfilled up to a certain linear combination of $2n - 1$ suitable real functions only. Then there exists exactly one state function $W(z)$ corresponding to any admissible control function $U(z)$. On suitable conditions to the functions F_k, $k = 1, 2$, in functional (23) existence of an optimal control function $U^0(z)$ is proved if either the admissible control set V is a non-void convex, bounded, and weak* closed subset of $\mathcal{L}_\infty(G)$ and the coefficients μ_k do not depend on U while the function d depends linearly on it or on the other hand V is a non-void compact subset of $\mathcal{L}_\infty(G)$ or $\mathcal{C}(\overline{G})$, respectively. In particular V can be a bounded subset of a Hölder space $\mathcal{C}^\tau(\overline{G})$, $0 < \tau \leqq 1$, which is closed in the norm of $\mathcal{C}(\overline{G})$.

In the special case of a *linear differential equation*

$$\partial W/\partial \bar{z} + a(z)\, W + b(z)\, \overline{W} = c(z) + d(z)\, U(z) \quad \text{in } G \tag{42}$$

with coefficients $a, b, c, d \in \mathcal{L}_p(G)$, $p > 2$, and the linear boundary condition (29) with non-negative index $n \geqq 0$ and corresponding additional conditions the existence of an optimal distributed control function $U^0(z)$ from a non-void convex, bounded, and weak* closed control set of $\mathcal{L}_\infty(G)$ and also the existence of an optimal boundary control function $u^0(t)$ from a non-void, bounded, and closed in the norm of $\mathcal{C}(\partial G)$ control set of $\mathcal{C}^\mu(\partial G)$, $0 < \mu < 1$, is shown in the paper v. WOLFERSDORF [7].

19.4. Linear-quadratic control problems

In the above-mentioned papers by TUTSCHKE, KLINK and the author the controls are taken as bounded measurable, Hölder continuous or even Hölder continuously differentiable functions. In case of the basic linear-quadratic control problems for generalized analytic functions the use of quadratic summable (distributed and boundary) control functions is natural. Important special cases of these control problems were dealt with by C. and CL. SIMIONESCU [1], CL. SIMIONESCU [1], who also discussed some interesting applications of these problems in hydrodynamics and theory of elasticity. Unfortunately they did not pay sufficient attention to the Noetherian character of the underlying Riemann-Hilbert problems and to the fact that the states are complex-valued functions. So, they obtained an incorrect form of the optimality conditions.

In the sequel the case of *distributed control* (with joint distributed and boundary observation) for non-negative index will be considered only. A full treatment of the linear-quadratic control problems for generalized analytic functions with distributed and boundary control and corresponding observation in case of non-negative and negative index of the underlying Riemann-Hilbert problem, respectively, is given in the author's paper [8], for which his results of [1] served as a base.

The state function $W(z)$ is defined in a bounded, simply connected region G with boundary ∂G of the class $\mathscr{C}_1^\varkappa$, $0 < \varkappa \leqq 1$, and satisfies the differential equation

$$\partial W/\partial \bar{z} + a(z)\, W + b(z)\, \overline{W} = f(z) + BU(z) \quad \text{in } G \tag{43}$$

with coefficients $a(z), b(z) \in \mathscr{L}_p(G)$, $p > 2$, and the boundary condition

$$\operatorname{Re}\,[\overline{\lambda(t)}\, W] = g(t) \quad \text{on } \partial G \tag{44}$$

with non-vanishing coefficient $\lambda(t) \in \mathscr{C}^\mu(\partial G)$, $0 < \mu < 1$, possessing non-negative index n. Further let be either

(α) $f(z) \in \mathscr{L}_p(G)$, $g(t) \in \mathscr{C}^\mu(\partial G)$ with $W(z) \in \mathscr{C}^\sigma(\overline{G})$, $\sigma = \min\,[(p-2)/p, \mu]$, or

(β) $f(z) \in \mathscr{L}_q(G)$, $q > 1$, $g(t) \in \mathscr{L}_\gamma(\partial G)$, $\gamma > 1$, with $W(z) \in \mathscr{L}_r(G)$ and boundary values $W(t) \in \mathscr{L}_\sigma(\partial G)$, where $r = 2\delta > 2$ and $\delta > 1$ depend on q and γ.

The (complex-valued) control functions $U(z)$ are elements of a non-void closed, convex subset U_{ad} of $\mathscr{L}_2(G)$, and B is a linear bounded operator from $\mathscr{L}_2(G)$ into $\mathscr{L}_p(G)$ or $\mathscr{L}_q(G)$, respectively.

The *cost functional* has the form

$$J = (NU, U)_{\mathscr{L}_2(G)} + \|MW(z) - c(z)\|^2_{\mathscr{L}_2(G)} + \|QW(t) - d(t)\|^2_{\mathscr{L}_2(\partial G)}\,, \tag{45}$$

where $N\colon \mathscr{L}_2(G) \to \mathscr{L}_2(G)$ is a linear bounded positive self-adjoint operator, $c(z) \in \mathscr{L}_2(G)$ and $d(t) \in \mathscr{L}_2(\partial G)$ are prescribed functions, and M and Q are linear bounded operators from $\mathscr{C}^\sigma(\overline{G})$ into $\mathscr{L}_2(G)$ and $\mathscr{L}_2(\partial G)$ or from $\mathscr{L}_r(G)$ and $\mathscr{L}_\sigma(\partial G)$ into $\mathscr{L}_2(G)$ and $\mathscr{L}_2(\partial G)$, respectively.

There exists a uniquely determined optimal control function $U^0(z) \in U_{ad}$ for the control problem. The corresponding optimal state function $W^0(z)$ is uniquely determined if for no non-trivial solution $W(z)$ of the homogeneous boundary value problem (43), (44) the relations $MW(z) = 0$ in G and $QW(t) = 0$ on ∂G hold simultaneously.

Let the operators M and Q have the form

$$MW(z) = M_1 W(z) + M_2 \overline{W}(z)\,, \qquad QW(t) = Q_1 W(t) + Q_2 \overline{W}(t)\,, \tag{46}$$

where M_k, Q_k, $k = 1, 2$, are sums of multiplication operators with certain functions and suitable integral operators. Then the following (necessary and sufficient) *optimality condition* holds:

$$\operatorname{Re} \iint_G [U(z) - U^0(z)]\, \overline{NU^0(z)}\, dx\, dy \tag{47}$$

$$-\operatorname{Re} \iint_G Z(z)\, B[U - U^0]\,(z)\, dx\, dy \geqq 0 \quad \forall\, U \in U_{ad}\,,$$

where the adjoint state $Z(z)$ is the solution to the adjoint boundary value problem

$$\partial Z/\partial \bar{z} - a(z)\, Z - \overline{b(z)}\, \bar{Z} = \overline{H(z)} \qquad \text{in } G\,, \tag{48}$$

$$\operatorname{Re}\,[\lambda(t)\, t'(s)\, Z] = 2 \operatorname{Im}\,[\lambda(t)\, \overline{\sigma(t)}] \qquad \text{on } \partial G\,. \tag{49}$$

The functions $H(z)$, $\sigma(t)$ are given by

$$H(z) = M_1^* \xi^0 + \overline{M_2^* \xi^0}\,, \qquad \sigma(t) = Q_1^* \eta^0 + \overline{Q_2^* \eta^0} \tag{50}$$

with $\xi^0 = MW^0 - c$, $\eta^0 = QW^0 - d$ and the adjoint operators M_k^*, Q_k^* to M_k, Q_k in the sense of (complex) spaces $\mathcal{L}_2(G)$ and $\mathcal{L}_2(\partial G)$, respectively.

In the *case without constraints* $U_{ad} = \mathcal{L}_2(G)$ we suppose assumptions (β) with operators $BU(z) = \beta_1(z)\, U(z) + \beta_2(z)\, \overline{U(z)}$, $N = I$ (identity operator), and $MW(z) = m_1(z) \times W(z) + m_2(z)\, \overline{W(z)}$, $QW(t) = q_1(t)\, W(t) + q_2(t)\, \overline{W(t)}$ with suitable functions $\beta_k(z)$, $m_k(z)$, $q_k(t)$, $k = 1, 2$. Then the optimal control function is

$$U^0(z) = \beta_2(z)\, Z(z) + \overline{\beta_1(z)}\, \overline{Z(z)}\,, \tag{51}$$

where $W = W^0(z)$, Z satisfy the *optimality system*

$$\partial W/\partial \bar{z} + aW + b\bar{W} - [2\beta_1\beta_2 Z + (|\beta_1|^2 + |\beta_2|^2)\, \bar{Z}] = f(z)\,, \tag{52}$$

$$\partial Z/\partial \bar{z} - aZ - \bar{b}\bar{Z} - [2m_1\bar{m}_2 W + (|m_1|^2 + |m_2|^2)\, \bar{W}] = -(\bar{c}m_1 + c\bar{m}_2) \tag{53}$$

with boundary conditions (44) and

$$\operatorname{Re}\,[(\lambda t' Z + 2\mathrm{i}\{2\lambda q_1 \bar{q}_2 - \bar{\lambda}(|q_1|^2 + |q_2|^2)\}\, W] = -2 \operatorname{Im}\,[\lambda(\bar{d}q_1 + d\bar{q}_2)]\,. \tag{54}$$

The boundary value problem (52)—(54) has the index zero. If the optimal state function $W^0(z)$ is uniquely determined this is the case also for the adjoint state function $Z(z)$ being the solution to problem (53), (54).

Further references

GOEBEL and v. WOLFERSDORF [1].

20. Methods of the Theory of Analytic Functions in the Theory of Elasticity

G. F. MANDŽAVIDZE

The theory of analytic functions has many applications in two-dimensional problems of mechanics of continua. The present chapter mainly aims at giving a survey on applications of analytic functions in elasticity, no applications in hydro-aeromechanics will be considered here (there is a vast literature available on the subject; see, for example, LAVRENT'EV and ŠABAT [1], SEDOV [1].

20.1. The differential equation of Airy's function

Let $Oxyz$ be some Cartesian coordinate system. The body is said to be *subjected to plane deformation* if the displacement component w (along the z-axis) is zero and the remaining components u, v are independent of z. In this case the deformation components

$$e_{yz} = \frac{1}{2}\left(\frac{\partial w}{\partial y} + \frac{\partial v}{\partial z}\right) = 0\,, \quad e_{zx} = \frac{1}{2}\left(\frac{\partial u}{\partial z} + \frac{\partial w}{\partial x}\right) = 0\,, \quad e_{zz} = \frac{\partial w}{\partial z} = 0$$

and the components

$$e_{xx} = \frac{\partial u}{\partial x}, \quad e_{yy} = \frac{\partial v}{\partial y}, \quad e_{xy} = \frac{1}{2}\left(\frac{\partial v}{\partial x} + \frac{\partial u}{\partial y}\right)$$

are independent of z.

From the six conditions of compatibility of Saint-Venant, five are automatically satisfied and one condition

$$\frac{\partial^2 e_{xx}}{\partial y^2} + \frac{\partial^2 e_{yy}}{\partial x^2} = 2\,\frac{\partial^2 e_{xy}}{\partial x\,\partial y} \tag{1}$$

will remain. If the body is elastic and has, at each point, a plane of elastic symmetry, which is parallel to the plane xy (LOVE [1]), then, in our case, on account of Hooke's law, we will have

$$\left.\begin{aligned} &e_{xx} = \beta_{11}X_x + \beta_{12}Y_y + \beta_{16}X_y\,, \qquad e_{yy} = \beta_{12}X_x + \beta_{22}Y_y + \beta_{26}X_y\,,\\ &2e_{xy} = \beta_{16}X_x + \beta_{26}Y_y + \beta_{66}X_y\,, \qquad Y_z = Z_x = 0\,,\\ &Z_z = \gamma_1 X_x + \gamma_2 Y_y + \gamma_3 X_y\,, \end{aligned}\right\} \tag{2}$$

where β_{ik}, γ_k are expressed through the elastic constants. In particular, when the body is isotropic, the relations (2) have the form

$$\left.\begin{aligned} e_{xx} &= \frac{1}{E}[(1-\sigma^2)X_x - \sigma(1+\sigma)Y_y],\\ e_{yy} &= \frac{1}{E}[(1-\sigma^2)Y_y - \sigma(1+\sigma)X_x]\\ e_{xy} &= \frac{1+\sigma}{E}X_y, \qquad Y_z = Z_x = 0, \qquad Z_z = \sigma(X_x + Y_y) \end{aligned}\right\} \tag{3}$$

E is Young's modulus and σ is Poisson's coefficient.

In the sequel we will consider homogeneous bodies. In view of (2) the equilibrium equations have the form

$$\frac{\partial X_x}{\partial x} + \frac{\partial X_y}{\partial y} + X = 0, \qquad \frac{\partial Y_x}{\partial x} + \frac{\partial Y_y}{\partial y} + Y = 0, \tag{4}$$

while the third equation becomes the equality $Z = 0$; therefore, for the plane deformation to take place it is necessary that the body force components $Z = 0$, while the components X and Y must be independent of z.

After constructing the particular solution of system (4), this system may be reduced to the system of homogeneous equations

$$\frac{\partial X_x}{\partial x} + \frac{\partial X_y}{\partial y} = 0, \qquad \frac{\partial X_y}{\partial x} + \frac{\partial Y_y}{\partial y} = 0. \tag{5}$$

System (5) is equivalent to the existence of the function $U(x, y)$ (the latter is called the *Airy function* or the *stress function*), by means of which the stress components are expressed by the formulas

$$X_x = \frac{\partial^2 U}{\partial y^2}, \qquad Y_y = \frac{\partial^2 U}{\partial x^2}, \qquad X_y = -\frac{\partial^2 U}{\partial x\,\partial y}. \tag{6}$$

From (6) it follows that

$$(X_n + \mathrm{i}Y_n)\,\mathrm{d}s = -\mathrm{i}d\left(\frac{\partial U}{\partial x} + \mathrm{i}\frac{\partial U}{\partial y}\right), \tag{6'}$$

where $\mathrm{d}s$ is an element of an arc of some contour drawn in the domain D (the cross section of the body) or on its boundary, $\boldsymbol{n}$ is the normal directed to the right; (X_n, Y_n) is the stress vector applied to $\mathrm{d}s$ on the right.

As it is to be seen from the compatibility condition (1), the function U must satisfy the equation

$$\beta_{22}\frac{\partial^4 U}{\partial x^4} - 2\beta_{26}\frac{\partial^4 U}{\partial x^3\,\partial y} + (2\beta_{12} + \beta_{66})\frac{\partial^4 U}{\partial x^2\,\partial y^2} - 2\beta_{16}\frac{\partial^4 U}{\partial x\,\partial y^4} + \beta_{11}\frac{\partial^4 U}{\partial y^4} = 0. \tag{7}$$

In the case of an isotropic body equation (7) does not contain the elastic constants and has the form

$$\Delta\Delta U \equiv \frac{\partial^4 U}{\partial x^4} + 2\frac{\partial^4 U}{\partial x^2\,\partial y^2} + \frac{\partial^4 U}{\partial y^4} = 0 \tag{8}$$

(the biharmonic equation).

The function $U(x, y)$ is single-valued if the cross-section of the body is a finite simply connected domain and is, in general, multi-valued in the case of a multiply connected or infinite domain. In the sequel, by the solutions (7) or (8) we will mean the functions $U(x, y)$ the derivatives of which are continuous up to 4th order, while the derivatives starting from the 2nd order are single-valued; the latter requirement is natural and follows from the formulas (6).

Thus there arises the question of constructing the general solution of a differential equation of the form

$$\mathcal{L}U \equiv a_0 \frac{\partial^n U}{\partial x^n} + a_1 \frac{\partial^n U}{\partial x^{n-1}\,\partial y} + \dots + a_n \frac{\partial^n U}{\partial y^n} = 0\,, \tag{9}$$

where $a_0, \dots, a_n$ are the given real constants, $U(x, y)$ is the unknown real function of two real variables x and y in the domain D.

We begin with the construction of complex solutions (the constants a_k may also be complex). Consider the characteristic equation, corresponding to equation (9)

$$a_0 + a_1 s + \dots + a_n s^n = 0\,. \tag{10}$$

Denote the roots of the characteristic equation by s_k (of multiplicity m_k), $k = 1, \dots, l$, $m_1 + \dots + m_l = n$ (without loss of generality it may be assumed that $a_n \neq 0$). The operator $\mathcal{L}$ is partitioned

$$\mathcal{L} = a_n \prod_{k=1}^{l} \left(\frac{\partial}{\partial y} - s_k \frac{\partial}{\partial x}\right)^{m_k}$$

and the integration of equation (9) is reduced to successively integrating equations of 1st order. The general (complex) solution of equation (9) will have the form

$$U(x, y) = \sum_{k=1}^{l} \sum_{\nu=1}^{m_k} y^{\nu-1} f_{k\nu}(x + s_k y) \tag{11}$$

where $f_{k\nu}$ are the differentiable functions of their arguments (if s_k is a complex number, then $f_{k\nu}$ are analytic functions).

For the biharmonic equation the roots of the characteristic equation are $s_1 = \mathrm{i}$, $s_2 = -\mathrm{i}$, $m_1 = m_2 = 2$; the general (complex) solution has the form

$$U(x, y) = f_{11}(x + \mathrm{i}y) + y f_{12}(x + \mathrm{i}y) + f_{21}(x - \mathrm{i}y) + y f_{22}(x - \mathrm{i}y)\,. \tag{12}$$

Introducing the complex variable $z = x + \mathrm{i}y$, formula (12) may be rewritten as

$$U(x, y) = F_{11}(z) + \bar{z} F_{12}(z) + \overline{F_{21}(z)} + z \overline{F_{22}(z)}\,, \tag{13}$$

where $F_{\alpha\beta}(z)$ ($\alpha, \beta = 1, 2$) are arbitrary analytic functions of z (the above line denotes a complex conjugate value).

It easily follows from formula (13) that the general real solution of the biharmonic equation has the form

$$U(x, y) = \tfrac{1}{2}\,[\bar{z}\varphi(z) + \chi(z) + z\overline{\varphi(z)} + \overline{\chi(z)}] = \operatorname{Re}\,[\bar{z}\varphi(z) + \chi(z)] \tag{14}$$

where $\varphi(z)$, $\chi(z)$ are arbitrary analytic functions of z (Goursat's formula).

For equation (7), too, the characteristic equation

$$\beta_{22} - 2\beta_{26} s + (2\beta_{12} + \beta_{66})\, s^2 - 2\beta_{16} s^3 + \beta_{11} s^4 = 0 \tag{15}$$

has only complex roots Lehnickiĭ ([1, 2]). If it has the multiple roots, then the general (real) solution of the equation will be represented in the form

$$2U(x, y) = \bar{z}\varphi(z) + \chi(z) + z\overline{\varphi(z)} + \overline{\chi(z)}\,, \tag{16}$$

just like in the case of an isotropic body, but here the complex variable has the form

$$z = x + sy = x + \alpha y + i\beta y \, .$$

By the affine transformation $x' = x + \alpha y$, $y' = \beta y$ we obtain the usual complex variable

$$z' = x' + iy'$$

changing in the domain D' which is obtained by the affine transformation from the original domain D.

Formula (16) and the expressions which this formula gives for the stress and displacement components show, that the case under consideration (i.e. the case of multiple roots) is almost a complete analogy to the case of an isotropic body and for this reason this case is not discussed in the literature.

If the characteristic equation (15) has no multiple roots (in the sequal this case will be refered to as the case (B), while the case of an isotropic body the case (A)), i.e. it has four different pairwise conjugate roots

$$s_1 = \alpha_1 + i\beta_1 \, , \qquad s_2 = \alpha_2 + i\beta_2 \, , \qquad s_3 = \bar{s}_1 \, , \qquad s_4 = \bar{s}_2 \, ,$$

the general real solution is represented in the form

$$U(x, y) = 2 \operatorname{Re} [F_1(z_1) + F_2(z_2)] \tag{17}$$

by means of two analytic functions of the variable

$$z_1 = x + s_1 y \, , \qquad z_2 = x + s_2 y$$

which may be replaced by the ordinary complex variables

$$x_1 + iy_1 \, , \qquad x_2 + iy_2$$

which changes within the domains D_1 and D_2, respectively, obtained from the domain D by the respective affine transformations.

Proceeding from the formulas (14), (17), we obtain the formulas for the stress and displacement components: In the case (A)

$$\left.\begin{aligned} X_x + Y_y &= 4 \operatorname{Re} \Phi(z) \, , \\ Y_y - X_x + 2iX_y &= 2[\bar{z}\Phi'(z) + \Psi(z)] \, , \\ 2\mu(u + iv) &= \varkappa\varphi(z) - \overline{z\varphi'(z)} - \overline{\psi(z)} \, , \end{aligned}\right\} \tag{18}$$

λ, μ are Lamé constants, $\varkappa = (\lambda + 3\mu)/(\lambda + \mu)$, $\Phi(z) = \varphi'(z)$, $\Psi(z) = \psi'(z)$, $\psi(z) = \chi'(z)$, in the case (B)

$$\left.\begin{aligned} &X_x = 2 \operatorname{Re} [s_1^2\Phi_1'(z_1) + s_2^2\Phi_2'(z_2)] \, , \qquad Y_y = 2 \operatorname{Re} [\Phi_1'(z_1) + \Phi_2'(z_2)] \, , \\ &X_y = -2 \operatorname{Re} [s_1\Phi_1'(z_1) + s_2\Phi_2'(z_2)] \, , \\ &u = 2 \operatorname{Re} [p_1\Phi_1(z) + p_2\Phi_2(z_2)] \, , \qquad v = 2 \operatorname{Re} [q_1\Phi_1(z_1) + q_2\Phi_2(z_2)] \, , \end{aligned}\right\} \tag{19}$$

the constants p_k, q_k $(k = 1, 2)$ are expressed through elastic constants, $\Phi_k(z_k) = \mathrm{d}/\mathrm{d}z_k \, F_k(z_k)$.

In the case of a finite simply connected domain the introduced analytic functions $\varphi(z)$, $\chi(z)$, $F_1(z_1)$, $F_2(z_2)$ will be single-valued (holomorphic) functions; in the case of a multiple or infinite domain these functions may be multi-valued; the multi-valuedness of these functions is described by logarithmic functions (see Mushelišvili [3], Lehnickiĭ [2]).

20.2. Boundary value problems for an isotropic body

In the sequel, for the sake of simplicity, the discussion will be limited to the case of the finite simply connected domain D, bounded by the simple smooth contour case of the finite simply connected domain D, bounded by the simple smooth contour L.

In the case (A) the boundary value problems of statics of an elastic body are reduced to finding two holomorphic functions $\varphi(z)$ and $\psi(z)$ from the following boundary conditions.

I. The first basic boundary value problem: On the boundary L external stresses are given; due to the formulas (6′) and (18) the boundary condition of this problem has

$$\varphi(t) + t\overline{\varphi'(t)} + \overline{\psi(t)} = f(t)\,, \qquad t \in L\,, \tag{20}$$

where $f(t)$ is a function given on L.

II. The second basic boundary value problem: Displacements are given on the boundary L. On the basis of the formulas (18) we obtain the boundary condition

$$\varkappa\varphi(t) - t\overline{\varphi'(t)} - \overline{\psi(t)} = g(t)\,, \qquad t \in L\,, \tag{21}$$

where $g(t)$ is a given function. Combining the conditions I and II, we write

$$k\varphi(t) + t\overline{\varphi'(t)} + \overline{\psi(t)} = f(t)\,, \qquad t \in L\,, \tag{21'}$$

where $k = 1$ (problem I) or $k = -\varkappa$ (problem II).

The function $f(t)$ in problem I is defined by the equality

$$f(t) = \mathrm{i} \int\limits_{t_0t} (X_n + \mathrm{i}Y_n)\,\mathrm{d}s\,, \tag{22}$$

where (X_n, Y_n) is the external stress vector, the integral in (22) is taken over some arc t_0t of the contour L; s is an arc abscissa. In problem II $f(t) = 2\mu\big(u(t) + \mathrm{i}v(t)\big)$, where $u(t)$, $v(t)$ are the boundary values of the displacement components given on L.

III. The basic mixed problem: Stresses are given on some part $L' = U a_k b_k$ of the boundary and displacements on the remainder $L'' = U b_k a_{k+1}$ of the boundary ($a_k b_k$, $k = 1, 2, \ldots, n$, denote the non-overlapping arcs of L, arranged in the same order in which they by-pass L in the positive direction). Here we have the boundary condition

$$k(t)\,\varphi(t) + t\overline{\varphi'(t)} + \overline{\psi(t)} = h(t) + C(t)\,, \qquad t \in L\,, \tag{23}$$

$$k(t) = \begin{cases} 1, & t \in L', \\ -\varkappa, & t \in L'', \end{cases} \qquad h(t) = \begin{cases} \mathrm{i} \int\limits_{t_0t} (X_n + \mathrm{i}Y_n)\,\mathrm{d}s, & t \in a_k b_k, \\ -2\mu\big(u(t) + \mathrm{i}v(t)\big), & t \in L'', \end{cases}$$

$$C(t) = \begin{cases} c_k, & t \in a_k b_k, \\ 0, & t \in L'', \end{cases}$$

the constants c_k (except one constant that may be fixed arbitrarily) are to be defined in the process of the problem solution.

The boundary condition of problem I may also be written in the form

$$\Phi(t) + \overline{\Phi(t)} - \mathrm{e}^{2\mathrm{i}\alpha}\,\{\bar{t}\Phi'(t) + \Psi(t)\} = F(t)\,,$$

where $F(t)$ is the given function, while α is an angle forming the normal to L with the x-axis.

Besides the above-mentioned problems, there are other boundary value problems, for example, the problem of contact with a rigid profile (MUSHELIŠVILI [3]).

Since for the given stresses and displacements the functions $\varphi(z)$, $\psi(z)$ are somewhat arbitrarily defined by the formulas (18), to obtain the complete definition of $\varphi(z)$, $\psi(z)$ some additional conditions are used. These additional conditions have the form (z_0 is an arbitrarily fixed point of the domain) in problem I

$$\operatorname{Im} \varphi'(z_0) = 0 , \qquad \varphi(z_0) = 0 \qquad \left(\text{or } \operatorname{Im} \varphi'(z_0) = 0, \psi(z_0) = 0\right) \tag{24}$$

in problem II

$$\varphi(z_0) = 0 \qquad \left(\text{or } \psi(z_0) = 0\right) . \tag{25}$$

In the basic mixed problem the additional conditions have the form of (25); however, if any of the constants c_k is fixed, then the additional conditions do not arise.

20.3. Boundary value problems for an anisotropic body

For an anisotropic body (case (B)) the boundary value problems are reduced to finding two analytic functions $\Phi_1(z_1)$ and $\Phi_2(z_2)$ in the domains D_1 and D_2 by the following boundary conditions:

I. The first basic problem

$$\left.\begin{aligned} &\operatorname{Re}\left[\Phi_1(t_1) + \Phi_2(t_2)\right] = f_1(t) , \\ &\operatorname{Re}\left[s_1\Phi_1(t_1) + s_2\Phi_2(t_2)\right] = f_2(t) , \end{aligned} \quad t \in L , \right\} \tag{26}$$

where $f_1(t)$, $f_2(t)$ are the given functions of the point $t = x + iy$ of the boundary L, $t_k = x + s_k y$ are the corresponding points on the boundaries of the domain D_k.

II. The second basic problem

$$\left.\begin{aligned} &\operatorname{Re}\left[p_1\Phi_1(t_1) + p_2\Phi_2(t_2)\right] = g_1(t) , \\ &\operatorname{Re}\left[q_1\Phi_1(t_1) + q_2\Phi_2(t_2)\right] = g_2(t) , \end{aligned} \quad t \in L , \right\} \tag{27}$$

where $g_1(t)$, $g_2(t)$ are the functions given on L.

In this case, too, the boundary conditions of the first and second problems are quite similar.

III. The basic mixed problem: Let a_k, b_k, L' and L'' denote the same as above; the boundary conditions have the form

$$\begin{aligned} &\operatorname{Re}\left[\Phi_1(t_1) + \Phi_2(t_2)\right] = f_1(t) + C_1(t) , \\ &\operatorname{Re}\left[s_1\Phi_1(t_1) + s_2\Phi_2(t_2)\right] = f_2(t) + C_2(t) , \end{aligned} \quad t \in L' ,$$
$$\begin{aligned} &\operatorname{Re}\left[p_1\Phi_1(t_1) + p_2\Phi_2(t_2)\right] = g_1(t) , \\ &\operatorname{Re}\left[q_1\Phi_1(t_1) + q_2\Phi_2(t_2)\right] = g_2(t) , \end{aligned} \quad t \in L'' , \tag{28}$$

$f_k(t)$, $g_k(t)$ $(k = 1, 2)$ are the given functions, $C_k(t) = c_k^\alpha$, $t \in a_k b_k$, $\alpha = 1, \dots, n$, $k = 1, 2$; the real constants c_k^α (except one pair $c_k^{\alpha_0}$ that can be fixed arbitrarily) are to be defined in the process of the problem solution.

The additional conditions are similar to those of the proceding section and have the form (z_k^0 is an arbitrarily fixed point in the domain D_k): In the problems I and II

$$\operatorname{Im} \Phi_k'(z_k^0) = 0 , \qquad \Phi_k(z_k^0) = 0 , \qquad k = 1 \text{ or } 2 , \tag{29}$$

in the basic mixed problem there are no additional conditions, if a pair of values of $c_k^{\alpha_0}$ is fixed.

It is important to note that the above boundary value problems of the theory of analytic functions (21′), (23), (26), (27), (28) can be used not only in the case of plane deformation, but they also enable us to solve the boundary value problems of elasticity in two other cases; of a generalized plane stressed state and the normal bending of thin plates.

20.4. Solution for an isotropic body

In the next section we will concentrate on some methods of the solution of boundary value problems for an isotropic body (case (A)).

20.4.1. Power series method (using conformal mappings). Let D be a finite simply connected domain bounded by the simple closed contour L of the class C^3. The problems I and II will be considered simultaneously.

It is required to find two functions $\varphi_1(z)$ and $\psi_1(z)$ holomorphic in D by the boundary condition

$$\mathfrak{k}\varphi_1(t) + t\overline{\varphi_1'(t)} + \overline{\psi_1(t)} = f_1(t)\,, \qquad t \in L\,, \tag{30}$$

where the constant $\mathfrak{k} = 1$ or $-\varkappa$; the given function $f_1(t)$ will be assumed to be sufficiently smooth and the so-called regular solution of the problem will be sought when $\varphi_1(z)$, $\varphi_1'(z)$, $\psi_1(z)$ are continuously extended to all points of the boundary L of the domain D.

Let $z = \omega(\zeta)$ conformally map the domain D on the unit circle Ω of the plane ζ. Denote by γ the circumference $\sigma = e^{i\theta}$, $0 \leqq \theta \leqq 2\pi$.

In our case when the contour is sufficiently smooth ($L \in C^3$), the functions $\omega(\zeta)$, $\omega'(\zeta)$, $\omega''(\zeta)$ will be continuously extended on γ; besides,

$$\omega'(\sigma) = \frac{d\omega(\sigma)}{d\sigma}\,, \qquad \omega'(\sigma) \neq 0\,, \qquad \omega''(\sigma) = \frac{d^2\omega(\sigma)}{d\sigma^2}\,.$$

Applying the conformal mapping $z = \omega(\zeta)$, instead of the boundary condition (30) we will have

$$\mathfrak{k}\varphi(\sigma) + \frac{\omega(\sigma)}{\overline{\omega'(\sigma)}}\overline{\varphi'(\sigma)} + \overline{\psi(\sigma)} = f(\sigma)\,, \qquad \sigma \in \gamma\,, \tag{31}$$

where $\varphi(\sigma)$, $\psi(\sigma)$ and $\varphi'(\sigma)$ are the boundary values on γ of the unknown holomorphic functions $\varphi(\zeta) = \varphi_1[\omega(\zeta)]$, $\psi(\zeta) = \psi_1[\omega(\zeta)]$ and $d\varphi(\zeta)/d\zeta$, $f(\sigma) = f^1 + if^2 = f_1(\omega(\sigma))$.

The functions $\varphi(\zeta)$, $\psi(\zeta)$ will be found in the form of their Taylor series

$$\varphi(\zeta) = \sum_{k=1}^{\infty} a_k\zeta^k\,, \qquad \psi(\zeta) = \sum_{k=0}^{\infty} a_k'\zeta^k\,. \tag{32}$$

We assume $a_0 = 0$, which agrees with the additional conditions (24), (25). Expanding the given functions $f(\sigma)$ and $\omega(\sigma)/\overline{\omega'(\sigma)}$ into the complex Fourier series

$$f(\sigma) = \sum_{k=-\infty}^{+\infty} A_k\sigma^k\,, \qquad \frac{\omega(\sigma)}{\overline{\omega'(\sigma)}} = \sum_{k=-\infty}^{+\infty} b_k\sigma^k$$

we obtain, from the boundary condition (31),

$$\mathfrak{k}a_m + \sum_{k=1}^{\infty} k\bar{a}_k b_{m+k-1} = A_m \qquad (m = 1, 2, \ldots)\,, \tag{33}$$

$$a_m' + \sum_{k=1}^{\infty} k a_k b_{-m+k-1} = \bar{A}_{-m} \qquad (m = 0, 1, \ldots)\,. \tag{34}$$

The equations (33) may be considered as an infinite system of linear algebraic equations with respect to the (real) unknowns $\alpha_m = \operatorname{Re} a_m$, $\beta_m = \operatorname{Im} a_m$ $(m = 1, 2, \ldots)$. If one succeeds in solving this system, the coefficients a_m' may be found from the equalities (34).

Let us now consider a special case when $\omega(\zeta)$ is a polynomial

$$\omega(\zeta) = \sum_{k=0}^{n} c_k \zeta^k , \qquad c_n \neq 0 .$$

Then

$$\frac{\omega(\sigma)}{\overline{\omega'(\sigma)}} = \sigma^{n-1} \frac{\sum\limits_{k=0}^{n} \dot{c}_k \sigma^k}{\sum\limits_{k=1}^{n} \overline{c_k} \sigma^{n-k}} ;$$

the right-hand side is a rational function of the variable σ which in the closed domain $|\sigma| \geqq 1$ has no poles except the pole of order n at the point at infinity. Therefore $b_k = 0$ for $k > n$ and the equations (33) are reduced to

$$\mathfrak{k} a_m = A_m \qquad (m \geqq n + 1) \tag{33'}$$

and

$$\mathfrak{k} a_m + \sum_{k=1}^{n-m+1} \overline{\mathfrak{k} a_k} b_{m+k-1} = A_m \qquad (m = 1, \ldots , n) \tag{33''}$$

while equality (34) has the form

$$a_m' + \sum_{k=1}^{m+n+1} k a_k \bar{b}_{-m+k-1} = \bar{A}_{-m} \qquad (m = 0, 1, \ldots) . \tag{34'}$$

Thus for defining a_m $(m = 1, \ldots , n)$ we also have the equations (33'') which form a system of $2n$ real equations for defining $2n$ real values α_m, β_m. If this system is solvable, then the other coefficients will be defined by the formulas (33'), (34') and the series for the functions $\varphi(\zeta)$, $\varphi'(\zeta)$, $\psi(\zeta)$ will absolutely and uniformly converge in the closed circle $|\zeta| \leqq 1$, if, for example, the given functions $f^1(\theta)$, $f^2(\theta)$ have the second derivatives with respect to θ which are functions of bounded variation. Indeed, in this case the coefficients A_m admit the estimate

$$|A_m| \leqq \frac{C}{|m|^3} .$$

Therefore

$$|m a_m| \leqq \frac{C}{m^2} \quad \text{for} \quad m \geqq n .$$

The last inequality proves the absolute and uniform convergence of the series for $\varphi(\zeta)$, $\varphi'(\zeta)$; it is also easy to investigate the convergence of the series for $\psi(\zeta)$.

20.4.2. Method of integral equations. We will proceed from the boundary condition (31). The function $f(\sigma)$ is assumed to have a derivative with respect to σ, which is continuous in the Hölder sense, while $\omega(\zeta)$ satisfies the previous assumptions.

Multiplying (32) by $\mathrm{d}\sigma/2\pi\mathrm{i}(\sigma - \zeta)$, $\zeta \in \Omega$ and assuming $\psi(0) = 0$, we obtain

$$\mathfrak{k}\varphi(\zeta) + \frac{1}{2\pi\mathrm{i}} \int_{\gamma} \frac{\omega(\sigma)}{\overline{\omega'(\sigma)}} \frac{\overline{\varphi'(\sigma)}\, \mathrm{d}\sigma}{\sigma - \zeta} = A(\zeta) , \qquad A(\zeta) = \frac{1}{2\pi\mathrm{i}} \int_{\gamma} \frac{f(\sigma)\, \mathrm{d}\sigma}{\sigma - \zeta} . \tag{35}$$

Differentiating (35) with respect to ζ and then considering the limit for ζ approaching an arbitrary point σ_0 of the contour γ, we have

$$\mathfrak{k}\varphi'(\sigma_0) + \int_\gamma K(\sigma_0, \sigma)\, \overline{\varphi'(\sigma)}\, d\sigma = A'(\sigma_0)\,, \tag{36}$$

where

$$K(\sigma_0, \sigma) = \frac{1}{2\pi i \overline{\omega'(\sigma)}} \left\{ \frac{\partial}{\partial \sigma_0} \left[\frac{\omega(\sigma) - \omega(\sigma_0)}{\sigma - \sigma_0} \right] + \frac{\omega'(\sigma_0)}{\sigma} \right\}.$$

The kernel $K(\sigma_0, \sigma)$ is continuous by virtue of the properties of the function $\omega(\zeta)$.

If in equation (36) we divide the real and imaginary parts, then we will obtain a system of two integral Fredholm equations; therefore the Fredholm theory holds for equation (36).

On account of the uniqueness theorem of boundary value problems of elasticity it may be proved that for problem II the integral equation (36) is always solvable, while for problem I it is solvable provided the resultant moment of external forces vanishes (the resultant vector vanishes automatically due to the continuity of the function $f(\sigma)$) and that the solution of equation (36) allows us to construct solutions of the problems under consideration, MUSHELIŠVILI [3].

If $\omega(\zeta)$ is a rational function, the kernel $K(\sigma_0, \sigma)$ is degenerate, i.e. it has the from

$$K(\sigma_0, \sigma) = \sum_{k=1}^{m} a_k(\sigma_0)\, b_k(\sigma)$$

and the solution of equation (36) and hence the solution of boundary value problems will be reduced to the solution of the system of linear algebraic equations. This yields MUSHELIŠVILI's result, according to which the solution of plane problems for the considered class of domains may be obtained in quadratures to an accuracy of the solution of the system of linear algebraic equations. The great practical importance of MUSHELIŠVILI's proposition becomes evident if we recall that for any simply connected domain the mapping function can always be approximated with any degree of accuracy by the rational function.

There are also other methods for reducing the boundary value problems to integral equations. MUSHELIŠVILI proposed a method which is also suitable for the case of multiply connected domains. The essence of the method will be explained by assuming, for the sake of simplicity, that the domain is finite and simply connected.

From the boundary condition (21′) we have

$$\psi(t) = \overline{f(t)} - \mathfrak{k}\overline{\varphi(t)} - \bar{t}\varphi'(t)$$

and therefore

$$\frac{1}{2\pi i} \int_L \frac{\overline{f(t)} - \mathfrak{k}\overline{\varphi(t)} - \bar{t}\varphi'(t)}{t - z}\, dt = 0$$

for all z lying outside $D \cup L$ or

$$\begin{gathered} \frac{\mathfrak{k}}{2\pi i} \int_L \frac{\overline{\varphi(t)}\, dt}{t - z} + \frac{1}{2\pi i} \int_L \frac{\bar{t}\varphi'(t)\, dt}{t - z} = A(z)\,, \\ A(z) = \frac{1}{2\pi i} \int_L \frac{\overline{f(t)}\, dt}{t - z}, \qquad z \,\bar{\in}\, D \cup L\,. \end{gathered} \tag{37}$$

The functional equation (37) may be reduced to the Fredholm equation in the following manner. Due to the Sohockiĭ-Plemelj formulas (see MUSHELIŠVILI [2] or 1.5.2.) (37) gives (it is assumed that the functions $\varphi(t)$, $\varphi'(t)$, $f(t)$ satisfy the Hölder condition)

$$-\frac{\mathfrak{k}}{2}\overline{\varphi(t_0)} + \frac{\mathfrak{k}}{2\pi i}\int_L \frac{\overline{\varphi(t)}\,dt}{t-t_0} - \frac{\bar{t}_0}{2}\varphi'(t_0) + \frac{1}{2\pi i}\int_L \frac{\bar{t}\varphi'(t)\,dt}{t-t_0} = A(t_0),$$

$$A(t_0) = -\frac{1}{2}f(t_0) + \frac{1}{2\pi i}\int_L \frac{\overline{f(t)}\,dt}{t-t_0}. \tag{38}$$

Adding, to the left-hand side of equation (38), the expression

$$\frac{\bar{t}_0}{2}\varphi'(t_0) - \frac{\bar{t}_0}{2\pi i}\int_L \frac{\varphi'(t)\,dt}{t-t_0} - \frac{\mathfrak{k}}{2}\overline{\varphi(t_0)} - \frac{\mathfrak{k}}{2\pi i}\int_L \frac{\overline{\varphi(t)}\,\overline{dt}}{\bar{t}-\bar{t}_0}$$

which is, as easily seen, equal to zero, we obtain

$$-\mathfrak{k}\overline{\varphi(t_0)} - \frac{\mathfrak{k}}{2\pi i}\int_L \overline{\varphi(t)}\,d\ln\frac{\bar{t}-\bar{t}_0}{t-t_0} - \frac{1}{2\pi i}\int_L \varphi(t)\,d\frac{\bar{t}-\bar{t}_0}{t-t_0} = A(t_0). \tag{39}$$

MUSHELIŠVILI's equation holds also if the contour L has angular points if the integrals contained in the equation are understood in some generalized sense (MAGNARADZE [1]).

ŠERMAN [1] proposed a method which consists in representing the functions $\varphi(z)$, $\psi(z)$ in the form of Cauchy type integrals, containing one unknown function of the point on the contour. Limiting our discussion, as before, to the case of a finite simply connected domain, we will derive the Šerman equation.

Consider the basic mixed problem and find the solution in the form

$$\varphi(z) = \frac{1}{2\pi i}\int_L \frac{\omega(t)\,dt}{t-z}, \qquad \psi(z) = -\frac{1}{2\pi i}\int_L \frac{\varkappa\overline{\omega(t)} + \bar{t}\omega'(t)}{t-z}\,dt, \tag{40}$$

where $\omega(t)$ is the unknown function.

Substituting the expressions (40) into the boundary condition (23), we obtain the singular integral equation with discontinuous coefficients

$$A(t_0)\,\omega(t_0) + \frac{B(t_0)}{2\pi i}\int_L \frac{\omega(t)\,dt}{t-t_0} + \int_L k_1(t_0,t)\,\omega(t)\,dt + \overline{\int_L k_2(t_0,t)\,\omega(t)\,dt}$$
$$= h(t_0) + C(t_0), \tag{41}$$

$$A(t) = \begin{cases} \dfrac{1-\varkappa}{2}, & t \in L', \\ -\varkappa, & t \in L'', \end{cases} \qquad B(t) = \begin{cases} \dfrac{1+\varkappa}{2}, & t \in L', \\ 0, & t \in L''. \end{cases}$$

If the contour L is sufficiently smooth, then the kernels $k_1(t_0, t)$, $k_2(t_0, t)$ are continuous. A detailed investigation of the equation is available in MUSHELIŠVILI [2].

20.4.3. Method of boundary value problems of linear conjugation. The reduction of problems of plane elasticity to problems of linear conjugation is one of the effective methods of the solution of these (especially mixed) problems. This method is due to MUSHELIŠVILI.

By the *boundary-value problem of linear conjugation* we mean the following: Find the function $F(z)$, which is holomorphic on the complex plane cut along the line L by the boundary condition (cf. 1.10.3. and 10.2.; concerning the cases of holomorphic vectors see chapter 5)

$$F^+(t) = a(t)\, F^-(t) + b(t)\,, \qquad t \in L\,, \tag{42}$$

where $a(t)$ and $b(t)$ are the functions given on L, $F^+(t)$ and $F^-(t)$ are the boundary values of the unknown function on L on the left- and right-hand sides with respect to the chosen direction along the line. It is assumed that these boundary values exist everywhere except, perhaps, the finite number of points $c_1, \ldots, c_m$ on the line L, in the neighbourhood of which $F(z)$ admits the estimate $|F(z)| \leqq A\,|z - c_k|^{-\alpha}$ (A and α are constants, $0 \leqq \alpha < 1$).

Sometimes solutions of the boundary value problem (42), admitting the pole at some point of the plane not lying on L, are sought; usually, this point is assumed to be the point at infinity.

Problem (42) will be considered under the following assumptions: L consists of the finite number of non-intersecting closed or open smooth contours; the functions $a(t)$ and $b(t)$ satisfy the Hölder condition on L except the finite number of points of the discontinuity of first kind, $a(t) \neq 0$. In this case problem (42) is solved explicitly (in quadratures). The solution (having the pole at infinity) has the form

$$F(z) = \frac{X(z)}{2\pi i} \int_L \frac{b(t)\, dt}{X^+(t)\,(t - z)} + X(z)\, P(z)\,, \tag{43}$$

where $P(z)$ is an arbitrary polynomial, $X(z)$ is the so-called *canonical solution* which is explicitly constructed (in quadratures).

For illustration we will give the solution of the basic mixed problem for a circle by reducing it to the problem of linear conjugation (the same method may be used to solve the problem in a much more general case when the domain under consideration is mapped on the circle by means of the rational function).

Let an isotropic body occupy the unit circle $|z| < 1$ denoted by D^+; D^- denotes the infinite domain $|z| > 1$, L the circumference $|z| = 1$.

We will proceed from the formulas of the general complex representation of stresses and displacements; in particular, we will apply the formulas

$$\left.\begin{aligned} \widehat{rr} + \widehat{\vartheta\vartheta} &= 2[\Phi(z) + \overline{\Phi(z)}]\,,\\ \widehat{rr} + i\widehat{r\vartheta} &= \Phi(z) + \overline{\Phi(z)} - \bar{z}\overline{\Phi'(z)} - z^{-1}\bar{z}\overline{\Psi(z)}\,,\\ 2\mu\left(\frac{\partial u}{\partial \vartheta} + i\,\frac{\partial v}{\partial \vartheta}\right) &= iz[\varkappa\Phi(z) - \overline{\Phi(z)} + \bar{z}\overline{\Phi'(z)} + z^{-1}\bar{z}\overline{\Psi(z)}\,,\end{aligned}\right\} \tag{44}$$

where $\widehat{rr}$, $\widehat{\vartheta\vartheta}$, $\widehat{r\vartheta}$ are the stress components in terms of polar coordinates.

Instead of two functions $\Phi(z)$ and $\Psi(z)$, holomorphic in D^+, we introduce one piecewise holomorphic function defined both in D^+ and D^-; in D^- it will be defined so that its values will analytically continue the values of $\Phi(z)$ in D^+ through the unloaded parts, if any.

Let us complete the definition of $\Phi(z)$ in D^- by

$$\Phi(z) = -\overline{\Phi(\bar{z}^{-1})} + z^{-1}\overline{\Phi'(\bar{z}^{-1})} + z^{-2}\overline{\Psi(\bar{z}^{-1})}\,. \tag{45}$$

The function $\Phi(z)$ defined by formula (45) is holomorphic in D^- including the point at infinity. For the function $\Psi(z)$, corresponding to $\Phi(z)$ by formula (45), to be holomorphic in D^+ it is necessary and sufficient that

$$A_0 + \bar{B}_0 = 0 , \qquad B_1 = 0 , \tag{46}$$

where A_k, B_k are the expansion coefficients

$$\Phi(z) = \sum_{k=0}^{\infty} A_k z^k \quad (|z| < 1) , \qquad \Phi(z) = \sum_{k=0}^{\infty} B_k z^{-k} \quad (|z| > 1) .$$

It will be assumed that the function $\Phi(z)$ is continuously extended, on L, from D^+ and D^- except, perhaps, the points a_k, b_k $(k = 1, \dots, n)$ in the neighbourhood of which

$$|\Phi(z)| \leqq A\,|z - c_k|^{-\alpha} , \qquad 0 \leqq \alpha < 1 , \qquad c_k = a_k \ \text{or}\ b_k$$

besides, it will be also assumed that for all points t on L except the points a_k, b_k;

$$\lim_{z \to t} (1 - r)\, \Phi'(z) = 0 , \qquad z = r\,e^{i\vartheta} .$$

For defining $\Phi(z)$ we thus obtain the boundary value problem of linear conjugation

$$\Phi^+(t) = a(t)\, \Phi^-(t) + b(t) , \qquad t \in L , \tag{47}$$

where

$$a(t) = \begin{cases} 1, & t \in L' , \\ -\dfrac{1}{\varkappa}, & t \in L'' , \end{cases}$$

while the function $b(t)$ is defined by the stresses and displacements given on L,

$$b(t) = \begin{cases} \widehat{rr} + i\widehat{r\vartheta} & \text{on} \quad L' , \\ \dfrac{2\mu}{\varkappa} \dfrac{dg(t)}{dt} & \text{on} \quad L'' , \end{cases} \qquad g(t) = u(t) + iv(t) .$$

Using formula (43) for the solution of conjugation problem (47), we obtain (see also theorem 10.1 in 10.2)

$$\Phi(z) = \frac{X(z)}{2\pi i} \int_L \frac{b(t)\, dt}{X^+(t)\,(t - z)} + X(z)\, P(z) , \tag{48}$$

where $X(z)$ (the canonical solution) has the form

$$X(z) = \prod_{k=1}^{n} (z - a_k)^{-1/2 - i\beta}\, (z - b_k)^{-1/2 + i\beta} \qquad (\beta = \ln \varkappa / 2\pi) .$$

By $X(z)$ we denote a branch such that in the neighbourhood of the point at infinity

$$X(z) = z^{-n} + O(z^{-n-1}) .$$

Since $\Phi(z)$ is bounded at infinityand the order of $X(z)$ at infinity is equal to $-n$, the degree of the polynomial $P(z)$ must not exceed n:

$$P(z) = C_0 z^n + \dots + C_n .$$

By reducing the problem under consideration to the problem of linear conjugation we differentiated the boundary condition along the arcs $b_k a_{k+1}$ $(k = 1, \dots, n$, $a_{n+1} = a_1)$, therefore we have so far succeded in satisfying the boundary conditions along $b_k a_{k+1}$ with an accuracy of some constant summands. If we can satisfy the boundary conditions on $b_k a_{k+1}$ to within a constant which is to be one and the same for all arcs $b_k a_{k+1}$, then we shall be able exactly to satisfy the boundary conditions by choosing the primitive function of $\Phi(z)$ in D^+. From the formula

$$2\mu[u(t) + iv(t)] = \int_{a_1 t} [\varkappa\Phi^+(t_0) + \Phi^-(t_0)]\, dt_0$$

it directly follows that the above mentioned requirement can be expressed by the equalities

$$\int_{b_k a_{k+1}} [\varkappa\Phi^+(t_0) + \Phi^-(t_0)]\, dt_0 = 2\mu[g(a_{k+1}) - g(b_k)]\,, \tag{49}$$

$$k = 1, \dots, n\,, \qquad a_{n+1} = a_1\,.$$

The equalities (49) represent a system of n linear equations with respect to $C_0, \dots, C_n$. It remains to satisfy the conditions (46). The condition

$$B_1 = 2\pi i \int_L \Phi^-(t_0)\, dt_0 = 0$$

is easily shown to be the consequence of the equalities (49) and the condition

$$A_0 + \bar{B}_0 = 0$$

can be written as

$$\bar{C}_0 + X(0)\, C_n + \frac{X(0)}{2\pi i} \int_L \frac{b(t)\, dt}{X^+(t)\, t} = 0\,. \tag{50}$$

Finally we have obtained $n + 1$ equations of (49) and (50); these equations may be considered as a system of $2n + 2$ real linear algebraic equations with respect to $\operatorname{Re} C_k$, $\operatorname{Im} C_k$ $(k = 0, \dots, n)$.

Further, it may be shown that the above obtained system is solvable. For this it is sufficient to show that the homogeneous system obtained for $b(t) \equiv 0$ has no solutions except $C_0 = \dots = C_n = 0$. This is a consequence of the uniqueness theorem of the solution for the mixed problem.

20.5. Solution for an anisotropic body

Now we proceed to the methods of solution of boundary value problems for an anisotropic body (case (B)). In this case a class of domains for which the boundary value problems can be efficiently solved (in quadratures) is considerably poorer in comparison with the case of an isotropic body.

One can, for example, solve in quadratures the boundary value problems for a half-plane, for an external domain of an ellipse.

We cite the solution of the first and the second boundary value problems for the nfinite plane with an elliptic hole given by Savin [1, 2].

Note that unlike the finite domain (recall that till now we have considered the boundary value problems for the finite simply connected domains only) in the case of an infinite domain (when the point at infinity belongs to the domain), besides boundary conditions some specific quantities are to be prescribed. In the first problem one must prescribe the limiting values of stress components and the rotation component in the point at infinity, while in the second and in the mixed problems the resultant vector of the external forces applied to the boundary of a domain is to be given as well.

Let D denote an infinite anisotropic plane with an elliptic hole and let L denote an ellipse, i.e. the boundary of a domain ($x = a\cos\theta$, $y = b\sin\theta$, $0 \leqq \theta \leqq 2\pi$ is the equation of L).

For simplicity we assume the resultant vector or external forces to be equal to zero; stress components and rotation components at infinity are also equal to zero.

The unknown functions $\Phi_k(z_k)$ will be holomorphic, including the point at infinity, in the domains D_k, $k = 1, 2$, which are obtained from D by the corresponding affine transformations.

To define the functions $\Phi_1(z_1)$, $\Phi_2(z_2)$ the boundary conditions have the form

$$\left.\begin{aligned} &\operatorname{Re}[a_{11}\Phi_1(t_1) + a_{12}\Phi_2(t_2)] = f_1(t)\,, \\ &\operatorname{Re}[a_{21}\Phi_1(t_1) + a_{22}\Phi_2(t_2)] = f_2(t)\,, \end{aligned}\quad t \in L\,, \right\} \tag{51}$$

where $f_1(t), f_2(t)$ are the given functions on L, a_{ik} are the given constants as, for example, in problem I, $a_{11} = a_{12} = 1$, $a_{21} = s_1$, $a_{22} = s_2$.

Having conformally mapped the domains D_k ($k = 1, 2$) onto the unit circle Ω of plane ζ by means of the functions

$$z_k = \omega_k(\zeta) = \frac{a + \mathrm{i}s_k b}{2}\zeta + \frac{a - \mathrm{i}s_k b}{2}\frac{1}{\zeta}\ (k = 1, 2) \tag{52}$$

we reduce the boundary value problem (51) to the definition of holomorphic functions $\varrho_k(\zeta) = \Phi_k(\omega_k(\zeta))$ in the circle Ω according to the boundary condition on $\gamma = Fr\Omega$

$$\left.\begin{aligned} &\operatorname{Re}[a_{11}\varrho_1(\sigma) + a_{12}\varrho_2(\sigma)] = h_1(\sigma)\,, \\ &\operatorname{Re}[a_{21}\varrho_1(\sigma) + a_{22}\varrho_2(\sigma)] = h_2(\sigma)\,, \end{aligned}\right\} \tag{53}$$

where $h_1(\sigma)$, $h_2(\sigma)$ are the given functions.

It should be noted that, by the mapping (52), one and the same point σ on the boundary γ of a unit circle Ω corresponds to the points t_1, t_2 of the boundaries of the domains D_1, D_2, which are in affine correspondence.

Using Schwarz' formula, from (53) we get

$$\left.\begin{aligned} a_{11}\varrho_1(\zeta) + a_{12}\varrho_2(\zeta) &= \frac{1}{2\pi\mathrm{i}}\int_\gamma h_1(\sigma)\frac{\sigma + \zeta}{\sigma - \zeta}\frac{d\sigma}{\sigma} + \mathrm{i}\delta_1\,, \\ a_{21}\varrho_1(\zeta) + a_{22}\varrho_2(\zeta) &= \frac{1}{2\pi\mathrm{i}}\int_\gamma h_2(\sigma)\frac{\sigma + \zeta}{\sigma - \zeta}\frac{d\sigma}{\sigma} + \mathrm{i}\delta_2\,. \end{aligned}\right\} \tag{54}$$

From the equations (54) we obtain $\varrho_k(\zeta)$ ($k = 1, 2$); returning to the variables z_1, z_2 we obtain the functions

$$\Phi_k(z_k) = \varrho_k\left(\frac{z_k - \sqrt{z_k^2 - (a^2 + s_k^2 b^2)}}{a + \mathrm{i}s_k b}\right).$$

Using the similar method for the mixed problem we obtain instead of (53) the boundary conditions

$$\left.\begin{aligned} &\operatorname{Re}[A_{11}(\sigma)\,\varrho_1(\sigma) + A_{12}(\sigma)\,\varrho_2(\sigma)] = f_1(\sigma) + C_1(\sigma)\,, \\ &\operatorname{Re}[A_{21}(\sigma)\,\varrho_1(\sigma) + A_{22}(\sigma)\,\varrho_2(\sigma)] = f_2(\sigma) + C_2(\sigma)\,, \qquad \sigma \in \gamma\,, \\ &A_{11}(\sigma) = \begin{cases} 1, & \sigma \in \gamma'\,, \\ p_1, & \sigma \in \gamma''\,, \end{cases} \quad A_{12}(\sigma) = \begin{cases} 1, & \sigma \in \gamma'\,, \\ p_2, & \sigma \in \gamma''\,, \end{cases} \quad A_{21}(\sigma) = \begin{cases} s_1, & \sigma \in \gamma'\,, \\ q_1, & \sigma \in \gamma''\,, \end{cases} \\ &A_{22}(\sigma) = \begin{cases} s_2, & \sigma \in \gamma'\,, \\ q_2, & \sigma \in \gamma''\,, \end{cases} \quad C_1(\sigma) = \begin{cases} c_k^1, & \sigma \in \sigma_k'\sigma_k''\,, \\ 0, & \sigma \in \gamma''\,, \end{cases} \quad C_2(\sigma) = \begin{cases} c_k^2, & \sigma \in \sigma_k'\sigma_k''\,, \\ 0, & \sigma \in \gamma''\,, \end{cases} \end{aligned}\right\} \tag{55}$$

σ_k', σ_k'' $(k = 1, \dots, n)$ are the points on γ, corresponding to the points of division of the contour L, $U\sigma_k'\sigma_k'' = \gamma'$, $U\sigma_k''\sigma_{k+1}' = \gamma''$, $\sigma_{n+1}' = \sigma_1'$, $f_1(\sigma), f_2(\sigma)$ are the given functions; the real constants c_k^1, c_k^2 $(k = 1, \dots, n)$ have to be determined. We fix one pair of constants $(c_n^1 = c_n^2 = 0)$.

Introducing the vectors $\boldsymbol{\varrho}(\zeta) = (\varrho_1(\zeta), \varrho_2(\zeta))$, $\boldsymbol{f}(\sigma) = (f_1(\sigma), f_2(\sigma))$, $\boldsymbol{C}(\sigma) = (C_1(\sigma), C_2(\sigma))$ and the matrix $\boldsymbol{A}(\sigma) = (A_{\alpha\beta}(\sigma))$, $\alpha, \beta = 1, 2$, the boundary condition (55) can be rewritten as

$$\operatorname{Re}[\boldsymbol{A}(\sigma)\, \boldsymbol{\varrho}(\sigma)] = \boldsymbol{f}(\sigma) + \boldsymbol{C}(\sigma)\,, \qquad \sigma \in \gamma\,. \tag{56}$$

The Riemann-Hilbert boundary value problem (56) is connected with the problem of linear conjugation for the piecewise holomorphic vectors (N. P. VEKUA [1], MUSHELIŠVILI [2])

$$\left.\begin{aligned} &\boldsymbol{\varrho}^+(\sigma) = a(\sigma)\, \boldsymbol{\varrho}^-(\sigma) + \boldsymbol{b}(\sigma)\,, && \sigma \in \gamma\,,\\ &a(\sigma) = -\boldsymbol{A}^{-1}(\sigma)\, \overline{\boldsymbol{A}(\sigma)}\,, && \boldsymbol{b}(\sigma) = \boldsymbol{A}^{-1}(\sigma)\,[\boldsymbol{f}(\sigma) + \boldsymbol{C}(\sigma)]\,. \end{aligned}\right\} \tag{57}$$

The solvability conditions of problem (57), and consequently of problem (56), are of the form

$$\int_\gamma \boldsymbol{Q}(\sigma)\,[\boldsymbol{X}^+(\sigma)]^{-1}\, \boldsymbol{A}^{-1}(\sigma)\,[\boldsymbol{f}(\sigma) + \boldsymbol{C}(\sigma)]\, d\sigma = 0\,, \tag{58}$$

where $\boldsymbol{Q} = (Q_1, Q_2)$, Q_1, Q_2 are arbitrary polynomials of degree $n - 2$ (for $n = 1$, $Q_1 = Q_2 = 0$) and $\boldsymbol{X}(\zeta)$ is a so-called *canonical matrix* of problem (57); in our case $\boldsymbol{X}(\zeta)$ is constructed explicitly.

The conditions (58) are equivalent to the conditions of the form

$$\int_0^{2\pi} \boldsymbol{R}_k(\sigma)\,[\boldsymbol{f}(\sigma) + \boldsymbol{C}(\sigma)]\, d\theta = 0\,, \qquad \sigma = e^{i\theta}\,, \qquad k = 1, \dots, 2n - 2\,, \tag{59}$$

where $\boldsymbol{R}_k(\sigma)$ are the real vectors which are constructed explicitly by means of the matrices $\boldsymbol{X}$ and $\boldsymbol{A}$.

The conditions (59) represent the real system of $2n - 2$ linear equations with respect to $2n - 2$ real constants c_k^1, c_k^2 $(k = 1, \dots, n - 1)$. On the basis of the uniqueness theorem this system is uniquely solvable.

The method of integral equations (Fredholm as well as singular ones) is the general method of studying the boundary value problems also in the case of an anisotropic body. The integral equations for the solution of the plane problems of the theory of elasticity for an anisotropic body were constructed in the works of I. N. VEKUA [1], MIHLIN [1], ŠERMAN [2] and others.

20.6. Concluding remarks

For the first time the application of the theory of analytic functions to the plane problem of the theory of elasticity was given in the work of KOLOSOV [2], which was published in 1909.

This direction has been found very useful; by means of the theory of analytic functions we succeeded in solving the fundamental problems of the plane theory of elasticity.

The direction achieved far going development in MUSHELIŠVILI's investigations which were summarized in his monograph [3].

At present we have vast literature concerning the application of the theory of analytic functions to the two-dimensional problems of mathematical theory of elasticity; except the above mentioned book of MUSHELIŠVILI we confine ourselves to the references of the monographs of KOLOSOV [2, 4], LEHNICKIĬ [1, 2], GALIN [1], KALANDIJA [1], SAVIN [1, 2], SNEDDON and BERRY [1]. In the above mentioned books one can find a detailed bibliography on the given subject.

21. Some Applications of Generalized Analytic Functions in the Shell Theory

E. I. Obolašvili

21.1. Introduction

The methods of the theory of analytic functions of a complex variable have long been used both in theoretical and applied disciplines, in particular, in the theory of partial differential equations, the plane theory of elasticity, hydromechanics and so on.

The analytic functions are also efficiently used in the shell theory. Thus, the calculations of the membrane equilibrium of a 2nd order shell of positive curvature are reduced to boundary value problems for analytic functions (see Vlasov [1], Gol'denveĭzer [1], Flügge [1], Sokolovskiĭ [1, 2]).

Applications of generalized analytic functions have a more recent history. The theory of generalized analytic functions was first substantiated about 50 years ago by Carleman [2]. Later I. N. Vekua [5] and Bers [1] constructed the general theory of generalized analytic functions. The theory of generalized analytic functions has now quite a wide range of applications. It is closely related with quasi-conformal mappings, the surface theory, the theory of elasticity, the shell theory, gas dynamics and so on.

Special mention should be made of the fact that by the methods of the theory of generalized analytic functions I. N. Vekua considerably broadened and deepened the study of the problems arising when we deal with the state of stressed membrane equilibrium of convex shells and their relation with the problem of infinitesimal surface bendings. These topics were investigated with sufficient completeness by I. N. Vekua in his monographs [4, 5], where with the aid of the theory of generalized analytic functions he succeded in solving a number of essentially important problems of the shell theory.

In the monograph [1] of Aleksandrov and Solov'ev the theory of generalized analytic functions is used for the solution of axially symmetric problems of elasticity.

In this work we do not consider many applications of generalized analytic functions but concentrate our attention on a few applications in the membrane shell theory. In the first place, a system of equilibrium equations of a membrane shell is derived from the general system of equilibrium equations of an elastic body and then the general complex Vekua representations are obtained for the components of the stress tensor and the displacement vector of the membrane shell of positive curvature.

The boundary value problems for membrane shells of positive curvature are reduced to the boundary value problems for generalized analytic functions. In particular, for these functions we consider the boundary value problem in which the linear conjugation conditions are given on a part of the boundary and the Riemann-Hilbert type conditions on the remainder. For the analytic functions this problem was investigated by us and for the general case by Tutschke.

In addition, we consider the boundary value problems for shells of positive curvature with cuts. The boundary value problems are obtained in a particular case when the shell is of 2nd order. The effective solutions are given for some domains cut along the finite number of segments of a straight line or arcs of a circle.

21.2. Surface stressed state of an elastic body and the membrane equilibrium of a shell

Let (x_1, x_2, x_3) be the coordinates of the point x in a continuous medium with respect to some curvilinear coordinate system, $\boldsymbol{R}$ the radius vector of the same point. $\boldsymbol{R}_i = \partial \boldsymbol{R}/\partial x_i$ $(i = 1, 2, 3)$ will be the basic vectors of the coordinate system under consideration, $\boldsymbol{R}_i \boldsymbol{R}_k = g_{ik}$ the covariant components of the metric tensor.

In what follows it will always be assumed that the Latin indices have the range 1, 2, 3, while the Greek indices have the range 1, 2. Besides, we will use the generally accepted summation convention of Einstein, consisting in adding over the repeated (dummy) indices. The contravariant components g^{ik} of the metric tensor and the conjugate basic vectors $\boldsymbol{R}^i$ are respectively defined by the equalities

$$g_{ij} g^{ik} = \boldsymbol{R}_j \boldsymbol{R}^k = \delta_j^k , \tag{1}$$

$$\boldsymbol{R}^i = g^{ik} \boldsymbol{R}_k , \tag{2}$$

where δ_j^k are the Kronecker symbols.

By $\boldsymbol{F}_{(n)}$ we denote the force applied to an area with the unit normal $\boldsymbol{n}$ and calculated per unit area. $\boldsymbol{F}_{(i)}$ and $\boldsymbol{F}^{(i)}$ will be the forces when $\boldsymbol{n}$ has the direction $\boldsymbol{R}_i$ and $\boldsymbol{R}^i$, respectively. It is known (Sneddon and Berry [1]) that in transforming the coordinate system the values

$$\boldsymbol{F}^i = \boldsymbol{F}^{(i)} \sqrt{g^{ii}} , \qquad \boldsymbol{F}_i = \boldsymbol{F}_{(i)} \sqrt{g_{ii}} \tag{3}$$

are transformed as contravariant and covariant components, respectively, of the first-rank tensor. In other words, if $\bar{x}_i$ are the coordinates of the point in another coordinate system and $\bar{\boldsymbol{F}}^i$, $\bar{\boldsymbol{F}}_i$ are the corresponding values in this system, then

$$\bar{\boldsymbol{F}}^i = \boldsymbol{F}^k \frac{\partial \bar{x}_i}{\partial x_k} , \qquad \bar{\boldsymbol{F}}_i = \boldsymbol{F}_k \frac{\partial x_k}{\partial \bar{x}_i} . \tag{4}$$

$\boldsymbol{F}^i$ and $\boldsymbol{F}_i$ are called the *contravariant* and *covariant components* of the stress force, respectively. They are not vectors in the usual sense, since they depend on the choice of the coordinate system. $\boldsymbol{F}_{(n)}$ is represented by the Cauchy formula

$$\boldsymbol{F}_{(n)} = \boldsymbol{F}^i n_i = \boldsymbol{F}_i n^i , \tag{5}$$

where n_i and n^i are the covariant and contravariant components of the vector $\boldsymbol{n}$, i.e.

$$\boldsymbol{n} = n_i \boldsymbol{R}^i = n^i \boldsymbol{R}_i . \tag{6}$$

The equalities (1), (5), (6) give

$$\boldsymbol{F}^i = g^{ik} \boldsymbol{F}_k , \qquad \boldsymbol{F}_i = g_{ik} \boldsymbol{F}^k . \tag{7}$$

Expanding $\boldsymbol{F}^i$, $\boldsymbol{F}_i$ in $\boldsymbol{R}^i$ or $\boldsymbol{R}_i$, we have

$$\boldsymbol{F}^i = F^{ik} \boldsymbol{R}_k = F_k^i \boldsymbol{R}^k ,$$

$$\boldsymbol{F}_i = F_{ik} \boldsymbol{R}^k = F_i^k \boldsymbol{R}_k , \tag{8}$$

F^{ik}, F_{ik} and F_i^k are called the *contravariant*, *covariant* and *mixed components* of the stress tensor, respectively. As it is assumed that the resultant moment of external forces is zero, the stress tensor is symmetric. The assumption that the resultant vector of external forces is zero enables us to obtain the system of equilibrium equations of a continuous medium in an arbitrary coordinate system (SNEDDON and BERRY [1])

$$\nabla_i F^{ik} + X^k = 0\,, \tag{9}$$

or

$$\nabla_i F_k^i + X_k = 0\,, \tag{10}$$

where X^k and X_k are the contravariant and covariant components of the volume force, ∇_i is the covariant differentiation operation, i.e.

$$\nabla_i F^{ik} = \frac{\partial F^{ik}}{\partial x_i} + G^i_{ij} F^{jk} + G^k_{ij} F^{ij}\,, \tag{11}$$

G^k_{ij} are the Christoffel symbols of the second kind:

$$G^k_{ij} = g^{kl} G_{ij,l} = \frac{1}{2} g^{kl} \left(\frac{\partial g_{il}}{\partial x_j} + \frac{\partial g_{jl}}{\partial x_i} - \frac{\partial g_{ij}}{\partial x_l} \right). \tag{12}$$

Let S be a smooth surface, $\boldsymbol{n}$ the unit normal vector; (x_1, x_2) the curvilinear coordinates of the point of S. It is assumed that each point x of the given elastic body can be unambiguously characterized by three values (x_1, x_2, x_3), where $|x_3|$ is the distance from the point x to the surface S, (x_1, x_2) are the coordinates of the base of the normal drawn from the point x to S. The radius vector of the base of the normal will be denoted by $\boldsymbol{r}$. Then, obviously,

$$\boldsymbol{R} = \boldsymbol{r} + x_3 \boldsymbol{n}\,, \qquad g_{\alpha 3} = \boldsymbol{R}_\alpha \boldsymbol{n} = 0\,, \qquad g_{33} = 1\,,$$
$$g_{\alpha\beta} = a_{\alpha\beta} - 2 b_{\alpha\beta} x_3 + b^\nu_\alpha b_{\nu\beta} (x_3)^2\,, \tag{13}$$

where $a_{\alpha\beta}$, $b_{\alpha\beta}$ are the coefficients of the first and second quadratic forms of S.

Definition 21.1. The stressed state of an elastic body will be called the *surface stressed state, parallel to* S, if at every point of the body the conditions

$$F^{3k} = F^k_3 = 0\,, \qquad k = 1, 2, 3\,, \tag{14}$$

which are equivalent to the conditions

$$F_{k3} = F^3_k = 0\,, \tag{15}$$

are satisfied.

From the law of transformation of the 2nd rank tensor components one may easily see that assumptions (14) are invariant with respect to the transformation of the coordinates of the type

$$\bar{x}_\alpha = \bar{x}_\alpha(x_1, x_2)\,, \qquad \alpha = 1, 2\,, \tag{16}$$
$$\bar{x}_3 = x_3\,.$$

By virtue of (11), (12), (14) we have

$$\nabla_3 F^{3\alpha} \equiv 0\,, \qquad \nabla_i F^{i3} \equiv G^3_{\alpha\beta} F^{\alpha\beta}\,. \tag{17}$$

From (9) we now obtain the system of equilibrium equations of an elastic body when it is in the surface stressed state

$$\nabla_\alpha F^{\alpha\beta} + X^\beta = 0\,, \qquad \beta = 1, 2\,,$$
$$G^3_{\alpha\beta} F^{\alpha\beta} + X^3 = 0\,. \tag{18}$$

In the case when S is a plane, we have the classical plane stressed state of an elastic body and the corresponding system of equations is obtained from (18). For the plane $G^3_{\alpha\beta} = 0$ and therefore $X^3 = 0$ is the condition of the plane stressed state of the elastic body (Muskhelišvili [3]).

Thus for the surface stressed state of the elastic body we have the system (18) with the unknown $F^{\alpha\beta}$. This system includes x_3 as a parameter, which means that for fixed x_3 we obtain a two-dimensional problem (Obolašvili [7]).

Let the elastic body have the form of a thin shell the thickness of which is $2h$ and the middle surface is S, i.e. $-h \leqq x_3 \leqq h$. By virtue of (12), (13) we get

$$G^3_{\alpha\beta} = G_{\alpha\beta,3} = b_{\alpha\beta} - b^\nu_\alpha b_{\nu\beta} x_3 \,. \tag{19}$$

Assume h to be so small that we may take the following approximate equalities

$$G^3_{\alpha\beta} = b_{\alpha\beta}\,, \qquad g_{\alpha\beta} \cong a_{\alpha\beta}\,, \qquad G_{\alpha\beta,\nu} = \Gamma_{\alpha\beta,\nu}\,, \qquad G^\nu_{\alpha\beta} \cong \Gamma^\nu_{\alpha\beta}\,, \tag{20}$$

where $\Gamma^\nu_{\alpha\beta}$ are Christoffel symbols of the 2nd kind on the surface S, calculated by the metric tensor $a_{\alpha\beta}$. Taking into account equations (20), from (18) we obtain the system of equations of the classical theory of membrane shells

$$\begin{aligned} &\nabla_\alpha F^{\alpha\beta} + X^\beta = 0\,, \\ &b_{\alpha\beta} F^{\alpha\beta} + X^3 = 0\,, \end{aligned} \tag{21}$$

where

$$\nabla_\alpha F^{\alpha\beta} = \frac{\partial F^{\alpha\beta}}{\partial x_\alpha} + \Gamma^\alpha_{\alpha\nu} F^{\nu\beta} + \Gamma^\beta_{\alpha\nu} F^{\alpha\nu}\,, \tag{22}$$

and

$$\Gamma^\alpha_{\alpha\beta} = \frac{1}{\sqrt{a}} \frac{\partial \sqrt{a}}{\partial x_\beta}\,. \tag{23}$$

$a = |a_{\alpha\beta}|$ is the discriminant of the principal metric tensor of the surface.

If by analogy with the case of a plane stressed state we assume that the third equation from system (21) is the condition imposed on X^3 in order to obtain a membrane state of the shell, then we will have the variant differing from the classical membrane state of the shell. It is clear that this variant may be investigated by the methods used in the plane theory of elasticity.

In the classical membrane shell theory system (21) is considered as a system of three equations with three unknown $F^{\alpha\beta}$, i.e. the stresses are completely defined from the condition of the shell equilibrium without taking into account the interdependence of stresses and strains (Flügge [1], Leipholz [1]). This theory is frequently used for the approximate calculation of shells.

Below we will consider the classical case for which the generalized analytic functions are used.

The law expressing the state of an isotropic homogeneous elastic body, i.e. Hooke's law, in an arbitrary curvilinear coordinate system takes the form

$$F^k_i = \lambda\theta g^k_i + 2\mu e^k_i \tag{24}$$

where λ, μ are the Lamé constants, e^k_i the mixed components of the strain tensor, u_i the components of the displacement,

$$2e_{ik} = \nabla_i u_k + \nabla_k u_i\,, \qquad \theta = \nabla_i u^i = e^i_i\,, \tag{25}$$

By virtue of (14), (24) we have

$$e^3_3 = \frac{\lambda}{\lambda + 2\mu} e^\alpha_\alpha\,, \qquad F^\beta_\alpha = \frac{2\mu\lambda}{\lambda + 2\mu} g^\beta_\alpha e^\nu_\nu + 2\mu e^\beta_\alpha \tag{26}$$

21.3. General complex Vekua representations for stresses and displacements of a membrane shell of positive curvature

Let the curvature of the shell middle surface S be positive. Then on S there exists the conjugate isometric coordinate system (x_1, x_2). In this coordinate system for the coefficients of the second basic quadratic form of the surface we have the equalities

$$b_{12} = 0\,, \qquad b_{11} = b_{22} \equiv D > 0\,. \tag{27}$$

Then from the third equation of (20) follows

$$F^{22} = -F^{11} - \frac{1}{D} X^3\,. \tag{28}$$

By virtue of equalities (23), (24), (28) the system of the first two equations of (23) takes the form

$$\begin{aligned}
&\frac{\partial \sqrt{a}\, F^{11}}{\partial x} + \frac{\partial \sqrt{a}\, F^{12}}{\partial y} + (\Gamma^1_{11} - \Gamma^1_{22}) \sqrt{a}\, F^{11} + 2\Gamma^1_{12} \sqrt{a}\, F^{12} = X\,,\\
&\frac{\partial \sqrt{a}\, F^{12}}{\partial x} - \frac{\partial \sqrt{a}\, F^{11}}{\partial y} + (\Gamma^2_{11} - \Gamma^2_{22}) \sqrt{a}\, F^{11} + 2\Gamma^2_{12} \sqrt{a}\, F^{12} = Y\,,
\end{aligned} \tag{29}$$

where $x \equiv x_1$, $y \equiv x_2$,

$$X = \sqrt{a}\, X^1 - \frac{1}{D} \Gamma^1_{22} X^3\,,$$

$$Y = \sqrt{a}\, X^2 - \frac{1}{D} \Gamma^2_{22} X^3 - \frac{\partial \dfrac{\sqrt{a}}{D} X^3}{\partial y}\,. \tag{30}$$

Due to (27) the Codazzi equations (I. N. Vekua [4]) have the form

$$\frac{\partial \ln D}{\partial x} = \Gamma^2_{12} - \Gamma^1_{22}\,, \qquad \frac{\partial \ln D}{\partial y} = \Gamma^1_{12} - \Gamma^2_{11}\,. \tag{31}$$

Summing the corresponding equations (24) and (31), we get

$$\frac{\partial \ln \sqrt{a}\, D}{\partial x} = \Gamma^1_{11} - \Gamma^1_{22} + 2\Gamma^2_{12}\,, \qquad \frac{\partial \ln \sqrt{a}\, D}{\partial y} = \Gamma^2_{22} - \Gamma^2_{11} + 2\Gamma^1_{12}\,. \tag{32}$$

With the last equations taken into account system (29) may be rewritten in the form

$$\begin{aligned}
&\frac{\partial p F^{11}}{\partial x} + \frac{\partial p F^{12}}{\partial y} + p a_1 F^{11} + p a_2 F^{12} = pX\,,\\
&\frac{\partial p F^{12}}{\partial x} - \frac{\partial p F^{11}}{\partial y} + p a_2 F^{11} - p a_1 F^{12} = pY\,,
\end{aligned} \tag{33}$$

where

$$a_1 = \frac{1}{2} \frac{\partial \ln \sqrt{a}\, D}{\partial x} - 2\Gamma^2_{12}\,, \qquad a_2 = -\frac{1}{2} \frac{\partial \ln \sqrt{a}\, D}{\partial y} + 2\Gamma^1_{12}\,; \qquad p = \sqrt{a^{3/2} D}\,. \tag{34}$$

Multiplying the second equation from (33) by $-\mathrm{i}$ (i is the imaginary unity) and summing with the first equation, we obtain the complex representation of the system (33)

$$\frac{\partial W}{\partial \bar{z}} + A\overline{W} = Z\,, \tag{35}$$

where

$$W = p(F^{11} - \mathrm{i}F^{12})\,, \tag{36}$$

$$A = \tfrac{1}{2}(a_1 - \mathrm{i}a_2)\,, \qquad Z = \tfrac{1}{2}p(X - \mathrm{i}Y)\,, \tag{37}$$

$$2\frac{\partial}{\partial \bar{z}} = \frac{\partial}{\partial x} + \mathrm{i}\frac{\partial}{\partial y}\,,$$

Thus by virtue of the equations (28), (36) the stress tensor components are represented in the form

$$\begin{aligned} F^{11} &= \frac{1}{p}\,\mathrm{Re}\,[W] + F_0^1\,, \\ F^{22} &= -\frac{1}{p}\,\mathrm{Re}\,[W] - F_0^1 - \frac{1}{D}X^3\,, \\ F^{12} &= -\frac{1}{p}\,\mathrm{Im}\,[W] + F_0^2\,, \end{aligned} \tag{38}$$

where W is the generalized analytic function and the solution of the equation

$$\frac{\partial W}{\partial \bar{z}} + A\overline{W} = 0\,, \tag{39}$$

while F_0^α are defined by means of partial solutions of the equation (35). W is called the *complex stress function.* X^i are the continuous functions in the domain under consideration.

We know the theorem: $A \equiv 0$ is the necessary and sufficient condition for the middle surface of the shell to be the 2nd order surface of positive curvature. The sufficiency of this condition was shown in I. N. VEKUA [5], while the necessity in VLASOV [1], FLÜGGE [1]. It is interesting to define a class of surfaces of positive curvature for which

$$A = \mathrm{const} \neq 0\,.$$

After defining the stress tensor components the strain tensor components are defined by equations (26). After that by virtue of (20), (25), (27) we will have the equations for the displacement components

$$\begin{aligned} &\frac{\partial u_1}{\partial x} - \Gamma_{11}^\alpha u_\alpha - Du_3 = e_{11}\,, \\ &\frac{\partial u_2}{\partial y} - \Gamma_{22}^\alpha u_\alpha - Du_3 = e_{22}\,, \\ &\frac{\partial u_1}{\partial y} + \frac{\partial u_2}{\partial x} - 2\Gamma_{12}^\alpha u_\alpha = 2e_{12}\,. \end{aligned} \tag{40}$$

In view of (32) this system may be written as

$$\frac{\partial q u_1}{\partial x} - \frac{\partial q u_2}{\partial y} - a_1 q u_1 - a_2 q u_2 = q(e_{11} - e_{22}),$$

$$\frac{\partial q u_1}{\partial y} + \frac{\partial q u_2}{\partial x} - a_2 q u_1 + a_1 q u_2 = 2 q e_{12}.$$

Hence for the complex function

$$V = q(u_1 + iu_2) \tag{41}$$

we get the equation

$$\frac{\partial V}{\partial \bar{z}} - \bar{A}\bar{V} = \frac{1}{2} q(e_{11} - e_{22} + 2ie_{12}), \tag{42}$$

where

$$q = (\sqrt{a}\, D)^{-1/2}.$$

The displacement components are thus represented in the form

$$u_1 = \frac{1}{q} \operatorname{Re}[V] + u_{10}, \qquad u_2 = \frac{1}{q} \operatorname{Im}[V] + u_{20}, \tag{43}$$

where $q(u_{10} + iu_{20})$ is the partial solution of equation (42) and V is the solution of the homogeneous equation

$$\frac{\partial V}{\partial \bar{z}} - \bar{A}\bar{V} = 0. \tag{44}$$

V is called the *complex displacement function.* Since the equations (39) and (44) are adjoint, their arbitrary solutions, continuous in the closed domain G, satisfy the Green identity (I. N. VEKUA [4])

$$\operatorname{Re}\,[i \int_{\partial G} WV\, dz] = 0, \tag{45}$$

where ∂G is the boundary of the domain G, i.e. the boundary of the shell consisting of the finite number of smooth closed curves.

Let us assume now that the volume force is zero. Then, taking into account (28), (36), (41), the Green identity (45) may be written in the form

$$\int_{\partial G} F^{\alpha\beta} u_\beta n_\alpha\, ds = \int_{\partial G} F_{(n)} U\, ds = 0, \tag{46}$$

i.e. (45) expresses the following mechanical fact: in the absence of the volume force and deformation, when the shell is in the stressed membrane equilibrium state, the work of the contour force applied to the shell boundary is zero at the displacement points of the boundary.

21.4. Boundary value problems in the theory of shells

In the classical membrane theory of shells of positive curvature two boundary conditions are given on either edge of a shell; one condition is static, while the other geometric. The static condition is used for defining the stress tensor $F^{\alpha\beta}$ and the geometric condition for defining the displacement vector (I. N. VEKUA [4, 5, 7], VLASOV [1], GOL'DENVEIZER [1]). Due to the equations (35), (36), (41), (42) the problem of defining

$F^{\alpha\beta}$ and the problem of defining the displacement vector are in terms of mathematics one and the same problem. Therefore below we will consider only the problem of defining $F^{\alpha\beta}$, i.e. only the static condition will be taken into account.

Let $\boldsymbol{l} = l_\alpha \boldsymbol{r}^\alpha = l^\alpha \boldsymbol{r}_\alpha$ be the unit vector of some direction passing through the point (x_1, x_2) of the middle surface S and lying on its tangential plane; $\boldsymbol{t} = t_\alpha \boldsymbol{r}^\alpha = t^\alpha \boldsymbol{r}_\alpha$ is the unit vector of the direction which is tangential to the surface S and perpendicular to $\boldsymbol{l}$. By virtue of the equations (5), (8), (15) we have

$$\boldsymbol{F}_{(l)} = F^{\alpha\beta} \boldsymbol{r}_\alpha l_\beta . \tag{47}$$

Writing $\boldsymbol{F}_{(l)}$ in the form

$$\boldsymbol{F}_{(l)} = N_l \, \boldsymbol{l} + S_l \, \boldsymbol{t}, \tag{48}$$

we obtain by virtue of (47)

$$N_l = \boldsymbol{F}_{(l)} \, \boldsymbol{l} = F^{\alpha\beta} l_\alpha l_\beta , \qquad S_l = \boldsymbol{F}_{(l)} \, \boldsymbol{t} = F^{\alpha\beta} t_\alpha l_\beta . \tag{49}$$

The substitution of (38) into the last equations results in the following representations

$$\begin{aligned} N_l &= \frac{1}{p} \operatorname{Re} \left[(l_1 + il_2)^2 \, W(z) \right] + N_0 , \\ S_l &= - \frac{1}{p} \operatorname{Im} \left[(l_1 + il_2) \, (l^1 - il^2) \, W \right] + S_0 . \end{aligned} \tag{50}$$

The contravariant components of vector (47) are represented in the form

$$\begin{aligned} T^1 &\equiv F^{1\beta} l_\beta = \frac{1}{p} \operatorname{Re} \left[(l_1 + il_2) \, W \right] + T_0^1 , \\ T^2 &\equiv F^{2\beta} l_\beta = - \frac{1}{p} \operatorname{Im} \left[(l_1 + il_2) \, W \right] + T_0^2 - \frac{l_2}{D} X^3 . \end{aligned} \tag{51}$$

Hence

$$T^1 - iT^2 = \frac{1}{p} (l_1 + il_2) \, W + T_0^1 - iT_0^2 + \frac{il_2}{D} X^3$$

follows where N_0, S_0, T_0^α are defined by means of the partial solution of the non-homogeneous equation (35) and therefore we consider them known.

Problem 1. It is required to define the internal forces $F^{\alpha\beta}$ with N_l or S_l given on the shell boundary.

Let G be a domain in the plane of the complex variable $z = x + iy$, corresponding to the middle surface of the shell, ∂G the boundary of G. By virtue of (50), when, for example, N_l is given, this problem is reduced to the problem of defining the generalized function W analytic in G, provided we have the Riemann-Hilbert condition:

$$\operatorname{Re} \left[(l_1 + il_2)^2 \, W(t) \right] = f(t) , \qquad t \in \partial G . \tag{52}$$

The existence of the solution of the last problem was discussed with sufficient completeness in the monograph [5] of I. N. Vekua. In the particular case, when the shell is of the 2nd order positive curvature, W is, as indicated in the previous section, the analytic function of one complex variable. Then, as known from Mushelišvili [2], problem (52) is solved effectively for sufficiently many domains. Unlike the analytic functions, for the generalized analytic functions the boundary value problem (52) is solved effectively in very particular cases. If A is a real constant, the solution of equation (39) in the domain $y > 0$, which for $y = 0$ satisfies the condition Re $[W] = f(x)$, is represented in the form (see OBOLAŠVILI [4])

$$W = \int_{-\infty}^{\infty} f(t)\, M(x - t, a)\, dt\,, \tag{53}$$

where $M(x, y) = AH_0^{(1)}(2iA\varrho) + \dfrac{\partial H_0^{(1)}(2iA\varrho)}{\partial z}$, $2\dfrac{\partial}{\partial z} = \dfrac{\partial}{\partial x} - i\dfrac{\partial}{\partial y}$, $\varrho^2 = x^2 + y^2$, $H_0^{(1)}$ is the Hankel function of the zeroth order.

Formula (53) is called the *Schwarz formula* for the generalized analytic function. Taking into account the asymptotic estimate in the neighbourhood of $z = 0$

$$iH_0^{(1)}(iz) = -\frac{2}{\pi}\ln z + \dots,$$

where the points stand for the summands which remain finite for $z = 0$, we obtain from (53) by passing to the limit as $A \to 0$ the well-known Schwartz formula for holomorphic functions.

Let us now consider a shell cut along the finite number of arcs of the curve. Let the domain G cut along the finite number of segments of the straight line or arcs of the circle correspond to this shell in the plane $z = x + iy$. Also, let on both sides of the cut one of the following values be given:

(1) $N_l^{\pm}$,

(2) $S_l^{\pm}$,

(3) N_l^{+}, S_l^{-},

(4) $N_l^{\pm}$ on one part of the cut, $S_l^{\pm}$ on the remaining part;

here the signs $(+)$ and $(-)$ denote the boundary values on the left-hand and right-hand side of the cut, respectively, while on the closed boundary of the shell N_l or S_l is given. For an arbitrary shell of positive curvature the solution of this problem cannot be solved effectively, since the corresponding problem for the generalized analytic function is not solved in quadratures. The theoretical investigation of the problem is certainly interesting but here we will limit our discussion to the shells of the 2nd order, for which the problem in question is reduced to the problem for holomorphic functions $\Phi = u + iv$.

Let G be one of the following domains: (1) a circular domain ($|z| < 1$) cut along the finite number of segments with diameter $y = 0$;

(2) a semi-circular domain ($|z| < 1, y > 0$) cut along the finite number of segments of the y-axis;

(3) a half-plane ($x > 0$) cut along the finite number of segments of the straight line $y = 0$ or arcs of the semicircle ($|z| = 1$, $x > 0$).

The boundaries of these domains, excluding the segments and arcs, will be denoted by ∂G and the set of segments or arcs by L. Without loss of generality a homogeneous condition, say, Re $[\Phi(t)] = 0$, may be considered on the boundary ∂G. Then by virtue of the Riemann-Schwartz symmetry principle, the boundary value problems for the domain G are reduced to the boundary value problems for the infinite plane Ω cut along the finite number of segment L of the straight line $y = 0$ (or arcs of the circle).

Below we give the solution of the following boundary value problems.

Problem 2. Define the function $\Phi = u + iv$, holomorphic in Ω and vanishing at infinity, if any of the following conditions is satisfied on L

(2_1) $u^\pm = f^\pm$,
(2_2) $u^+ = f^+$, $v^- = f^-$,
(2_3) $u^\pm = f^\pm$ on L_1, and $v^\pm = f_1^\pm$ on L_2, where $L_1 + L_2 = L$.

Here $f^\pm$, $f_1^\pm$ are the Hölder-continuous functions given on L. Φ is assumed to be continuously extensible everywhere on L except, perhaps, its ends c_k in the neighbourhood of which

$$|\Phi(z)| < \frac{\text{const}}{|z - c_k|^\alpha}, \qquad \alpha < 1 .$$

The problem of defining the harmonic in Ω function by the condition (2_1) was first solved in quadratures by Keldyš and Sedov [1]. Later, Mushelišvili [2] solved this problem in the generalized sense, in particular, in the sense of separating the set of solutions into classes and finding the necessary and sufficient conditions of solvability.

We present the solutions of problem 2 with the conditions (2_2), (2_3) when L is a set of segments of the straight line. The case when L is a set of arcs of the circle is solved quite similarly.

Consider the function holomorphic in Ω

$$\Psi(z) = \overline{\Phi(\bar{z})} . \tag{54}$$

Then the boundary conditions (2_2) are reduced to the boundary conditions for two analytic functions

$$\begin{aligned} \Phi^+(t) &= -\Psi^-(t) + 2f^+ , \\ \Psi^+(t) &= \Phi^-(t) - 2if^- , \qquad t \in L , \end{aligned} \tag{55}$$

while from the boundary conditions (2_3) we get

$$\begin{aligned} \Phi^+(t) &= -\Psi^-(t) + 2f^+ , \\ \Psi^+(t) &= -\Phi^-(t) + 2f^- , \end{aligned} \quad t \in L_1 , \qquad \begin{aligned} \Phi^+ &= \Psi^- + 2if_1^+ , \\ \Psi^+ &= \Phi^- - 2if_1^- , \end{aligned} \quad t \in L_2 . \tag{56}$$

It is easy to show that if $(\Phi(z), \Psi(z))$ is an arbitrary solution of the problem (55) or (56), then $\Phi_1(z) = \overline{\Psi(\bar{z})}$, $\Psi_1(z) = \overline{\Phi(\bar{z})}$ will also be its solution. Therefore, the solution for which the condition (54) is satisfied may always be constructed; in particular, the solution will have the form

$$\tfrac{1}{2}\,[\Phi(z) + \overline{\Psi(\bar{z})}] , \qquad \tfrac{1}{2}\,[\Psi(z) + \overline{\Phi(\bar{z})}] .$$

In the case of the conditions (55) consider the holomorphic in Ω functions

$$\Phi_1(z) = \Phi(z) + i\Psi(z) , \qquad \Phi_2(z) = \Phi(z) - i\Psi(z) . \tag{57}$$

The conditions (55) then give

$$\begin{aligned} \Phi_1^+(t) &= i\Phi_1^-(t) + 2(f^+ + f^-) , \\ \Phi_2^+(t) &= -i\Phi_2^-(t) + 2(f^+ - f^-) , \qquad t \in L . \end{aligned} \tag{58}$$

We have obtained two independent linear conjugation problems for one function holomorphic in Ω. For their solutions in quadratures see Mushelišvili [2]. Then Φ is defined by the equations (57).

In the case of the conditions (56) the functions

$$\Phi_1(z) = \Phi(z) + \Psi(z) , \qquad \Phi_2(z) = \Phi(z) - \Psi(z) \tag{59}$$

are considered. Then from the conditions (56) we get

$$\begin{aligned} \Phi_1^+(t) &= -\Phi_1^-(t) + 2(f^+ + f^-)\,, \\ \Phi_2^+ &= \Phi_2^-(t) + 2(f^+ - f^-)\,, \end{aligned} \quad t \in L_1\,,$$

$$\begin{aligned} \Phi_1^+ &= \Phi_1^- + 2\mathrm{i}(f_1^+ - f_1^-)\,, \\ \Phi_2^+ &= -\Phi_2^- + 2\mathrm{i}(f_1^+ + f_1^-)\,, \end{aligned} \quad t \in L_2\,. \tag{60}$$

In this case Φ_1 and Φ_9 are also represented in quadratures (MUSHELIŠVILI [2]), while Φ is defined by the equations (59).

It should be noted that if in (39) $A =$ const and the infinite plane is cut along the positive axis, then, using the conditions (2_1), the solution of problem 2 for the generalized function, analytic in Ω, is effectively constructed by the Wiener-Hopf method. It is interesting to investigate the construction of the solution of this problem by the conditions (2_2) or (2_3).

Consider now the case when X^k have discontinuities of the 1st kind along some lines $L_1, L_2, \dots, L_n$ lying on the middle surface. Let the lines $\Gamma_1, \Gamma_2, \dots, \Gamma_r$ correspond to $L_1, L_2, \dots, L_n$ on the plane of the complex variable $z = x + iy$ and let Ω be the domain, lying in the same plane and corresponding to the middle surface of the shell. The boundary ∂G of the domain Ω corresponds to the shell edge L. Inside ∂G are the lines $\Gamma_1, \Gamma_2, \dots, \Gamma_n$.

Problem 3. It is required to define the stressed membrane equilibrium state of the shell when, for example, N_l is given on the shell boundary and the condition of the force continuity

$$(T^1 - \mathrm{i}T^2)^+ = (T^1 - \mathrm{i}T^2)^- \tag{61}$$

is given along the lines $L_1, L_2, \dots, L_n$. Due to (50), (51) this problem is reduced to the following one:

Define in the domain Ω the piecewise generalized analytic function with the jump lines $\Gamma_0 = \Gamma_1 + \dots + \Gamma_n$ by the boundary conditions

$$\mathrm{Re}\,[(l_1 + il_2)^2\, W(t)] = f(t)\,, \qquad t \in \partial G\,, \tag{62}$$

$$W^+(t) = W^-(t) + g(t)\,, \qquad t \in \Gamma_0\,. \tag{63}$$

Boundary value problems of the membrane shells consisting of pieces of different shells of positive curvature are also reduced to the above problem. Instead of the condition (63) we consider the more general condition

$$W^+(t) = G(t)\, W^-(t) + g(t)\,, \qquad t \in \Gamma_0\,, \tag{64}$$

where $G \neq 0$ is a Hölder-continuous function given on Γ_0. Thus we have the conditions of the linear conjugation problem on the internal contours and the condition of the Riemann-Hilbert problem on the external contour. The solution of this problem is effectively constructed only in the case when W is a function, piecewise holomorphic in Ω, i.e. when the shell of positive curvature is of the 2nd order ($A \equiv 0$).

Indeed, any function, piecewise holomorphic in Ω, with the jump lines Γ_0, which satisfies condition (64), is represented in the form

$$W(z) = \frac{\chi(z)}{2\pi \mathrm{i}} \int\limits_{\Gamma_0} \frac{g(t)\, \mathrm{d}t}{X^+(t)\,(t - z)} + \chi(z)\, \Phi(z) \tag{65}$$

where $\Phi(z)$ is an arbitrary holomorphic function inside ∂G, $\chi(z)$ the canonical solution corresponding to the problem (64). To define $\Phi(z)$ we substitute (65) in condition

(62) and obtain the Riemann-Hilbert problem for the domain Ω. If Ω is a circular domain or is conformally mapped on a circular domain, then Φ is represented in quadratures (see Mushelišvili [2]).

If the shell is closed we will have the problem with the boundary conditions (63), which is sufficiently completely investigated (see I. N. Vekua [5], Mushelišvili [2]).

The existence of the solution of the problem with the boundary conditions (62), (64) was investigated in Tutschke [15] for piecewise generalized analytic functions.

As an example the shell will be considered, the middle surface of which is a spherical belt of unit radius bounded by the parallels $\vartheta = \vartheta_0$ and $\vartheta = \vartheta_1$ $(0 < \vartheta_0 \leqq \vartheta \leqq \vartheta_1 < \pi)$. On the surface of the sphere we consider a net of the lines (x, y),

$$x = \tan\frac{\vartheta}{2}\cos\psi\,, \qquad y = \tan\frac{\vartheta}{2}\sin\psi\,, \tag{66}$$

where ϑ, ψ are the geographical coordinates of the point on the surface. For the system (x, y) the coefficients of the 1st and 2nd basic quadratic forms of the surface will have the form

$$a_{11} = a_{22} = b_{11} = b_{22} = \frac{4}{(1 + x^2 + y^2)^2}\,, \qquad a_{12} = b_{12} = 0\,.$$

For the parallel the direction of the tangential normal is defined by the equations

$$l_1 + il_2 = \frac{2z}{(1 + z\bar{z})\sqrt{z\bar{z}}}\,; \qquad l^1 + il^2 = \frac{z(1 + z\bar{z})}{2\sqrt{\bar{z}z}}\,. \tag{67}$$

In the plane $z = x + iy$ we will have the one-to-one correspondence of the circular ring Ω $(\tan\vartheta_0/2 \leqq |z| \leqq \tan\vartheta_1/2)$ to the spherical belt, on the boundary of which the following boundary conditions are given

$$N_l = f_1(\psi) \quad \text{for} \quad \vartheta = \vartheta_1\,, \qquad S_l = f_2(\psi) \quad \text{for} \quad \vartheta = \vartheta_0\,. \tag{68}$$

Besides, along the lines $L_1, \dots, L_n$, to which in the domain Ω correspond the lines $\Gamma_1, \dots, \Gamma_n$, we have the stress continuity conditions (61). By virtue of equations (50), (51), (67) the boundary conditions (68), (61) are reduced to the boundary conditions for the function Φ which is piecewise holomorphic in Ω

$$\begin{aligned} &\operatorname{Re}[t^2\Phi(t)] = f_1(t)\,, \qquad |t| = \tan\frac{\vartheta_1}{2}\,(\Gamma^1)\,,\\ &\operatorname{Im}[t^2\Phi(t)] = f_2(t)\,, \qquad |t| = \tan\frac{\vartheta_0}{2}\,(\partial G)\,, \end{aligned} \tag{69}$$

$$\Phi^+ = \Phi^- + g(t)\,, \qquad t \in \Gamma_0 \equiv \Gamma_1 + \dots + \Gamma_n\,. \tag{70}$$

Φ, satisfying condition (70), is represented in the form

$$\Phi(z) = \frac{1}{2\pi i}\int_{\Gamma_0}\frac{g(t)\,dt}{t - z} + \Psi(z)\,,$$

where $\Psi(z)$ is an arbitrary function holomorphic in Ω. Then, by virtue of (69), we obtain the boundary conditions for defining Ψ:

$$\begin{aligned} &\operatorname{Re}[t^2\Psi(t)] = f(t)\,, \qquad t \in \Gamma^1\,,\\ &\operatorname{Im}[t^2\Psi(t)] = \omega(t)\,, \qquad t \in \partial G\,. \end{aligned} \tag{71}$$

This problem for a circular ring may be solved by different ways. For example, by the method of integral equations it is established that the problem always has the unique solution. Besides, it was found that in a doubly connected domain with the mixed conditions (68) the membrane equilibrium stressed state can always be defined without any additional conditions.

It should be noted that a detailed investigation of the problem of obtaining a membrane equilibrium stressed state in the multiply connected domain with N_t or S_t given everywhere on the boundary is contained in I. N. Vekua [5].

The generalized analytic functions in the shell theory have applications beyond the membrane theory. For example, in I. N. Vekua [7], Kaldani [1] one can find their applications for defining the stressed state of convex composite shells on the basis of equilibrium equations of the new variant of the shell theory.

Finally, we mention that in higher-dimensional Euclidean spaces certain analogies of the generalized Cauchy-Riemann system may be constructed, whose solutions have properties analogous to those of the generalized analytic functions. Especially, two analogous systems have been constructed; first the system

$$\operatorname{div} \boldsymbol{U} + (\boldsymbol{A}\boldsymbol{U}) = 0\,, \qquad \operatorname{rot} \boldsymbol{U} + [\boldsymbol{U} \times \boldsymbol{B}] = 0\,,$$

where $\boldsymbol{A}$, $\boldsymbol{B}$ are given vector functions, $\boldsymbol{U} = (U_1, U_2, U_3)$ is an unknown vector function, called *generalized potential vector*; and second the generalized Moisil-Theodorescu system

$$\operatorname{div} \boldsymbol{U} + (\boldsymbol{A}\boldsymbol{U}) = 0\,, \qquad \operatorname{grad} \varphi + \operatorname{rot} \boldsymbol{U} + [\boldsymbol{U} \times \boldsymbol{B}] + \varphi \boldsymbol{A} = 0\,,$$

where φ, $\boldsymbol{U}$ unknown. A special case of this system is the system of the equations of a stationary irrotational motion of a fluid

$$\operatorname{div} \varrho \boldsymbol{U} = 0\,, \qquad \operatorname{rot} \boldsymbol{U} = 0\,.$$

Thus, without doubt the investigation of the mentioned systems remains important, see Obolašvili [3, 5, 6].

Further references

I. N. Vekua [6].

22. Complex Analysis in Probability Theory[1])

H.-J. ROSSBERG and B. JESIAK

22.1. Introduction

Probability theory investigates random mass-scale phenomena with mathematical methods; famous examples are the law of large numbers and the central limit theorem (CLT). In the present paper we first explain some basic concepts of this branch of mathematics; after that we describe several parts of probability theory in which complex analysis, in particular the theory of complex functions, plays an important role. An exhaustive treatment of the subject is impossible. For instance, we cannot enter into a discussion of stochastic processes; for this we refer the reader to DYM and McKEAN [1] and the survey articles ROSSBERG [10] and TAKÁCS [1]. Particularly important monographs on analytic methods of probability theory are those of KAWATA [1], LINNIK and OSTROVSKIĬ [1], LUKACS [3, 6], RAMACHANDRAN [1]. Analytic methods of probability theory are also in the foreground of the discussions in LAHA and ROHATGI [1]. See also the survey articles LUKACS [1 and 4]. A more detailed exposition of the very recent results contained in 22.3.—22.6. and 22.8. is to be found in ROSSBERG, JESIAK, and SIEGEL [3].

22.1.1. Random variables and distribution functions (d.f.). Let us be given a probability space $(\Omega, \mathfrak{A}, \mathsf{P})$, i.e. a fundamental set Ω of "elementary events" ω, a σ-Algebra $\mathfrak{A}$ of subsets of Ω, and a probability measure defined on all sets A of $\mathfrak{A}$. Such a space is the abstract model of the phenomena to be studied and underlies every probabilistic investigation. In science, technics, and every day life we often meet random variables such as measured values, meteorologic or biologic dates; by means of a probability space they can be described in the shape of a well-known mathematical concept: A *random variable* is an $\mathfrak{A}$-measurable function $X: \Omega \to \mathbb{R}^1$.

According to this definition we can introduce the function

$$F(x) := \mathsf{P}\big(\omega : X(\omega) < x\big)\,, \qquad x \in \mathbb{R}^1; \tag{1}$$

it is called the *distribution function* (d.f.) of the random variable X. For this we will briefly write $X \sim F$ and say that X is *subject* to F. We have the following characterization theorem for the d.f.:

---

[1]) This paper was thoroughly discussed in the Leipzig seminar of analytic probability theory. The authors want to express their sincere thanks in particular to D. BISCHOFF, M. DEWESS, G. LAUE, M. RIEDEL, and G. SIEGEL for reading drafts of the paper and making valuable remarks. M. DEWESS, moreover, contributed 22.11. and we thank her heartily for this.

Theorem 22.1.1. *A function $F: \mathbb{R}^1 \to [0, 1]$ is a d.f. if and only if it possesses the following properties: It is non-decreasing, left-continuous, and has the limits $F(-\infty) = 0$, $F(\infty) = 1$.*

There exists a mechanical interpretation of an arbitrary d.f. F. Namely, it distributes the mass 1 over $\mathbb{R}^1$ such that the portion $F(b) - F(a)$ is contributed to the interval $[a, b)$. To a certain extent such a mass distribution can be characterized by its centre of gravity a and its moment of inertia. Analogously in probability theory we use

a) the mathematical expectation of a random variable X and its d.f. F,

$$\mathsf{E}X := \int_{-\infty}^{\infty} x \, \mathrm{d}F(x) = a ,$$

b) the variance of X and F,

$$\operatorname{Var} X := \mathsf{E}(X - a)^2 = \int_{-\infty}^{\infty} (x - a)^2 \, \mathrm{d}F(x) = \sigma^2 .$$

Moreover, the moments $m_k = \mathsf{E}X^k$ $(k = 1, 2, \ldots)$ are of great use.

Examples 1. The best known d.f. is the normal d.f.

$$\mathsf{P}(X < x) = \Phi_{a,\sigma^2}(x) := \frac{1}{\sqrt{2\pi}\,\sigma} \int_{-\infty}^{x} \exp\left(\frac{(u - a)^2}{2\sigma^2}\right) \mathrm{d}u \tag{2}$$

with the parameters a and σ^2 introduced above. Its importance has two reasons. On one hand it is the limit d.f. in the CLT. On the other hand Φ_{a,σ^2} has an abundance of simple characteristic properties; examples will be given in 22.4 and 22.10. For the sake of brevity we write Φ for $\Phi_{0,1}$.

2. The Poisson distribution P_λ is intimately connected with the Poisson limit theorem for probabilities of *rare events* and the Poisson process. It attributes a positive mass to the points 0, 1, 2, ... such that

$$\mathsf{P}(X = k) := \mathrm{e}^{-\lambda} \frac{\lambda^k}{k!} \qquad (k = 0, 1, \ldots)$$

where $\lambda = \mathsf{E}X = \operatorname{Var} X > 0$.

3. The (negative) exponential d.f.

$$\mathsf{P}(X < x) := 1 - \mathrm{e}^{-ax} \qquad (x \geqq 0)$$

plays an important role in Markov chains, reliability and queuing theory; here $a^{-1} = \mathsf{E}X$.

4. A constant a is a degenerated random variable; it is subject to the degenerated d.f.

$$\varepsilon_a(x) := \begin{cases} 0 & \text{if} \quad x \leqq a , \\ 1 & \text{if} \quad x > a . \end{cases}$$

The left extremity $\inf \big(x : F(x) > 0\big)$ of a d.f. F is briefly denoted by lext (F). For instance we have lext $(\Phi) = -\infty$, lext $(P_\lambda) = 0$.

Probability theory is intimately connected with measure theory, but it is no part of it. It is the notion of independence of several random variables $X_1, \ldots, X_n$ that is characteristic for probability theory since it is not of interest in measure theory. In analogy with (1) we can consider the common d.f.

$$F(x_1, \ldots, x_n) := \mathsf{P}(X_1 < x_1, \ldots, X_n < x_n)$$

of the random variables $X_1, \dots, X_n$. They are said to be *independent*, if for all $(x_1, \dots, x_n) \in \mathbb{R}^n$ we have $F(x_1, \dots, x_n) = \mathsf{P}(X_1 < x_1) \dots \mathsf{P}(X_n < x_n)$.

We often meet the sum of two random variables $X_i \sim F_i$ $(i = 1, 2)$. This motivates the great importance of the composition $F_1 * F_2$ of two d.f. Namely, we have

Theorem 22.1.2. *If the independent random variables X_1 and X_2 are subject to the d.f. F_1 and F_2, respectively, then*

$$\mathsf{P}(X_1 + X_2 < x) = \int_{-\infty}^{\infty} F_1(x-u)\, F_2(\mathrm{d}u) = \int_{-\infty}^{\infty} F_2(x-u)\, F_1(\mathrm{d}u)\,.$$

22.1.2. The use of characteristic functions (c.f.). With an arbitrary d.f. F we can define the *Fourier-Stieltjes transform*

$$f(t) = \int_{-\infty}^{\infty} \mathrm{e}^{\mathrm{i}tx}\, \mathrm{d}F(x)\,, \qquad t \in \mathbb{R}^1\,,$$

which is called the *characteristic function* (c.f.) of F. It is uniformly continuous on $\mathbb{R}^1$ and satisfies $f(0) = 1$, $|f(t)| \leqq 1$, $f(t) = \overline{f(-t)}$ for all $t \in \mathbb{R}^1$.

To the four d.f. mentioned above there correspond the c.f.

$$f_1(t) = \exp\left(\mathrm{i}at - \frac{\sigma^2}{2} t^2\right), \qquad f_2(t) = \exp\left(\lambda(\mathrm{e}^{\mathrm{i}t} - 1)\right),$$

$$f_3(t) = (1 - \mathrm{i}t/a)^{-1}\,, \qquad f_4(t) = \exp\,(\mathrm{i}at)\,.$$

The following four theorems are basic for the applications of c.f.

Theorem 22.1.3 (uniqueness theorem). *Distinct d.f. have distinct c.f.*

There exist several proofs. In 22.6. we quote the more general theorem 22.6.7 which can be derived from the uniqueness theorem for Laplace-Stieltjes transforms by means of the Phragmén-Lindelöf theory.

By virtue of theorem 22.1.3 investigations of unknown d.f. can be replaced by the study of the corresponding c.f. This amounts to tackling the problems under consideration by means of quite different tools.

It is often not possible to calculate f if F is given or to find F explicitely if f is known. Hence there arises the question: Given a class of d.f. F with a certain property $\mathfrak{P}_F$; which property $\mathfrak{P}_f$ do the corresponding c.f. share — and vice versa? Important contributions to this field are the theorems 22.1.4, 22.1.5, and corollary 22.2.1.

Theorem 22.1.4. *If the d.f. F possesses the moment m_k then the corresponding c.f. f has the derivative $f^{(k)}$ of order k. The converse is true only for $k = 2l$ even. In either case we have*

$$\mathrm{i}^k m_k = \mathrm{i}^k \mathsf{E} X^k = f^{(k)}(0)\,.$$

This result is impleasant for k odd. In 22.8. we show that for non-negative random variables moments of an odd order exist under a surprisingly weak condition on f.

Theorem 22.1.5 (composition theorem). *A d.f. F is the composition $F = F_1 * F_2$ of the d.f. F_1 and F_2 if and only if for the corresponding c.f. we have $f = f_1 f_2$.*

In view of this result c.f. are used particularly often, when sums of independent random variables are involved.

Let now $\{F_n\}$ stand for a sequence of d.f. It is said to *converge weakly* to the d.f. F,

$$F_n \xrightarrow{w} F\,, \tag{4}$$

if $\lim_{n\to\infty} F_n(x) = F(x)$ in all continuity points of F. If (4) is granted then we can replace F_n $(n \gg 1)$ by F which can sometimes be studied more easily, see the CLT and 22.9. In the following theorem $\{f_n\}$ denotes the sequence of c.f. corresponding to $\{F_n\}$, while f corresponds to the d.f. F.

Theorem 22.1.6 (continuity theorem). *We have* (4) *if, and only if* $f_n \to f$ *pointwise.*

In view of theorems 22.1.5 and 22.1.6 the theory of c.f. has been particularly successfully applied in the following important branch of probability theory.

22.1.3. The theory of sums of independent random variables. We consider the sequence $\{X_n\}$ of independent random variables and form the sums

$$S_n = X_1 + \ldots + X_n \tag{5}$$

asking for the asymptotic behaviour of their d.f. as $n \to \infty$. In this theory, particular attention has been paid to the law of large numbers and the CLT, and interest in these subjects does not cease in our time. The first tells us under which conditions the well-known stability of the arithmetic means S_n/n for $n \gg 1$ occurs such that we obtain results which are nearly independent of chance. A modern representation was given by Révész [1], but complex analysis is not used in this field the principal importance of which can hardly be exaggerated. On the other hand, the different versions of the CLT contain conditions under which

$$F_n(x) := \mathsf{P}\left(\frac{S_n - A_n}{B_n} < x\right) \xrightarrow{w} \Phi(x)\,, \tag{6}$$

where $\{B_n\}$ $(B_n > 0)$ and $\{A_n\}$ are sequences of numbers appropriately chosen. The CLT explains the well-known fact that random variables important in practice are often approximately subject to Φ_{a,σ^2}. As an example we mention the errors in measurements. They are the result of a great many of independent influences each of which gives only a very small contribution. This property of the summands can be described by

$$\lim_{n\to\infty} \max_{1\leq k\leq n} \mathsf{P}(|X_k/B_n| \geqq \varepsilon) = 0 \qquad \forall \varepsilon > 0\,. \tag{7}$$

The summands are then called *infinitely small.*

A particularly well-known version of the CLT concerns the case in which $X_k \sim G_k$, the variances $\operatorname{Var} X_t = \sigma_k^2$ exist, and $\mathsf{E}X_k = 0$ $(k = 1, 2, \ldots)$. Putting $B_n^2 := \sigma_1^2 + \ldots + \sigma_n^2$ we get

$$\operatorname{Var}\frac{S_n - A_n}{B_n} = 1 \qquad (n = 1, 2, \ldots); \tag{8}$$

this means that no proper part of the mass distributed over $\mathbb{R}^1$ according to F_n can escape to infinity as $n \to \infty$.

Theorem 22.1.7 (Lindeberg-Feller). *Assume* (8). *The summands* X_k/B_n $(k = 1, \ldots, n)$ *are infinitely small and we have* (6) *if, and only if,*

$$\lim_{n\to\infty} \sum_{k=1}^{n} \int_{|x|\geqq\varepsilon} x^2 \,\mathrm{d}G_k(x) = 0 \qquad \forall \varepsilon > 0\,. \tag{9}$$

The condition (9) is the famous Lindeberg condition. It implies a uniform smallness of X_k/B_n which is stronger than (7); namely from (9) it follows that

$$\lim_{n\to\infty} \mathsf{P}\left(\max_{1\leq k\leq n}\left|\frac{X_k}{B_n}\right| \geqq \varepsilon\right) = 0\,.$$

Thus in the framework of our model we have an intuitive explanation for the fact that we often meet d.f. in praxis which resemble Φ_{a,σ^2}: Uniform smallness of the summands as described by (9) is sufficient for (6). Limit d.f. $G \neq \Phi_{a,\sigma^2}$ can appear on the right-hand side of (6), but only under more restricting conditions.

This way the theoretical and practical importance of the CLT is obvious. But, moreover, in the long history of this theorem research on more and more general versions demanded new ideas and methods. Therefore work connected with the CLT has greatly influenced the development and sharpening of the tools now generally used in probability theory. In 22.5. we show that the CLT goes on playing this stimulating role and that complex analysis has an increasing importance in this field.

22.1.4. Infinite divisibility. With the aim of generalizing the classical theory of sums the so-called *triangular array* was introduced. By this we mean a double sequence of random variables

$$X_{n1}, X_{n2}, \dots, X_{nk_n} \qquad (n = 1, 2, \dots; k_n \to \infty)$$

which are independent in every row, i.e. for n fixed. Then we introduce the sums

$$\Sigma_n := X_{n1} + \dots + X_{nk_n} - r_n \qquad (n = 1, 2, \dots) \tag{10}$$

where $\{r_n\}$ is a sequence appropriately chosen. We obtain $\Sigma_n = (S_n - A_n)/B_n$ as a particular case for $X_{nk} = X_k/B_n$, $r_n = A_n/B_n$, $k_n = n$.

We now consider another particular case of the triangular array assuming that the random variable ξ can be decomposed in n independent and identically distributed summands as

$$\xi = \xi_{n1} + \dots + \xi_{nn}\,.$$

If $\xi \sim \Psi$, $\xi_{nk} \sim \Psi_n$, then Ψ is the n-fold composition

$$\Psi = \Psi_n * \dots * \Psi_n\,, \tag{11}$$

and for the corresponding c.f. we have by theorem 22.1.5 $\psi = \psi_n^n$. If Ψ permits the decomposition (11) for all $n \geqq 2$, then Ψ and ψ are called *infinitely divisible* (i.d.). The four examples of d.f. and c.f. given above share this property.

By (11) it is obvious that every i.d.d.f. Ψ is the limit d.f. for a special triangular array. The fundamental importance of the class $\mathfrak{G}$ of i.d.d.f. is a consequence of the fact that under a natural assumption no other limit d.f. can appear. This is expressed by

Theorem 22.1.8 (Hinčin). *If $\{X_{nk}\}$ are infinitely small, i.e. if*

$$\lim_{n\to\infty} \max_{1\leqq k\leqq k_n} \mathsf{P}(|X_{nk}| \geqq \varepsilon) = 0$$

for all $\varepsilon > 0$ then the class of all possible limit d.f. for sums of the form (10) *coincides with $\mathfrak{G}$.*

I.d.c.f. can be characterized by the canonical representation formula.

22.1.5. Canonical representation formula (Lévy-Hinčin). *The c.f. f is i.d. if, and only if, it permits the representation*

$$\log f(t) = \mathrm{i}at - \frac{\sigma^2}{2}t^2 + \int\limits_{u\neq 0} \left(\mathrm{e}^{\mathrm{i}tu} - 1 - \frac{\mathrm{i}tu}{1+u^2}\right)\frac{1+u^2}{u^2}\,\mathrm{d}R(u) \tag{12}$$

where $a \in \mathbb{R}^1$, $\sigma^2 \geqq 0$, *and* R *is (except for a non-negative factor) a d.f. The representation is unique.*

If we consider $(S_n - A_n)/B_n$ instead of Σ_n and if all summands are identically distributed, then the class of possible limit d.f. is considerably narrower. These d.f. are called *stable* and we discuss them in 22.3.

In our time the i.d.d.f. play an important role also in other parts of probability theory. The classical theory of infinite divisibility focussed attention to c.f., but in our time interest shifted to the corresponding, d.f.; see the survey article STEUTEL [1] and 22.4.

22.1.6. The theory of complex functions. Many important d.f. and c.f. such as the examples given above are analytically continuable into certain domains. We will explain a number of basic concepts and results connected with this fact. A personal point of view cannot be denied since we emphasize the great methodological importance of the theory of PHARGMEN and LINDELÖF the use of which is unfortunately sporadic in the international literature. The very last section is devoted to a theory stimulated by an apparently pure analytic remark of E. BOREL which could be given a precise meaning only in the framework of probability theory.

We will need the following domains of the $z = t + iy$-plane,

$$H_+ := \{z : y > 0\}\,, \qquad D := \{z : y = 0\}\,, \qquad S_{ab} := \{z : -a < y < b\}\,,$$

where $0 \leqq a, b \leqq \infty$. Moreover, we use the following classes of functions

$$M_\pm := \{\varphi : \text{analytic in } H_\pm, \text{ continuous in } H_\pm + D, |\varphi(t)| \leqq 1\}\,,$$
$$N_\pm := \{\varphi \in M_\pm : |\varphi(z)| \leqq 1 \text{ in } H_\pm\}\,.$$

22.2. Analytic characteristic functions (a.c.f.) and ridge functions

Let again f stand for the c.f. of the d.f. F. Introducing the functions

$$f_-(z) := \int_{-\infty}^{-0} e^{izu}\, dF(u) \in N_-\,, \qquad f_+(z) = \int_{-0}^{\infty} e^{izu}\, dF(u) \in N_+$$

we get the decomposition

$$f(t) = f_-(t) + f_+(t)\,, \qquad t \in \mathbb{R}^1\,. \tag{13}$$

If $F(0) = 0$, then $f = f_+$; this case is considered in 22.8. Since F is monotone the following lemma from the theory of Laplace-Stieltjes transforms plays a considerable role, see WIDDER [1].

Lemma 22.2.1. *The line of convergence of the integrals defining* $f_\pm$ *has a singularity on the imaginary axis.*

This fact and theorem 22.2.1 motivate that a c.f. f is said to be *analytic* (a.c.f.), if its restriction to some interval $|t| < r$ is continuable into the circle about 0 with radius r. In view of lemma 22.2.1 and (13) this case occurs if and only if 0 is no singularity for $f_\pm$. Hence we immediately obtain

Theorem 22.2.1. *To every a.c.f.* f *there corresponds a strip* S_{ab}, $0 < a, b \leqq \infty$, *in which it is analytic; the points* ib *and* $-ia$ *— if they are finite — are singularities of* f. *In* S_{ab} *we have the representation*

$$f(z) = \int_{-\infty}^{\infty} e^{izu}\, dF(u)\,. \tag{14}$$

Since d.f. are the proper object of interest the following corollary is important.

Corollary 22.2.1. *The c.f. f is analytic if and only if for some $r > 0$ as $x \to \infty$*

$$T_F(x) := 1 - F(x) + F(-x) = O(e^{-rx}) . \tag{15}$$

The proof uses the fact that the integral (14) exists for $z = iy$, $y < r$, if, and only if, (15) is satisfied.

From (14) we obtain the following important facts.

Corollary 22.2.2. *Every a.c.f. f has a domain of regularity which is symmetric with respect to the imaginary axis; in this domain we have $f(z) = \overline{f(-\bar{z})}$.*

Corollary 22.2.3. *An a.c.f. f possesses the ridge property*

$$f(iy) = \max_{-\infty<t<\infty} |f(t + iy)| , \qquad t \in \mathbb{R}^1 . \tag{16}$$

From the ridge property we obtain by the maximum modulus principle the

Corollary 22.2.4. *For an a.c.f. f we have for $0 < r < \min(a, b)$*

$$\max_{|z|=r} |f(z)| = \max \big(f(ir), f(-ir)\big) .$$

The a.c.f. f is integral if $a = b = \infty$. This is the case if and only if (15) holds for all $r > 0$. Then the growth of f has an intimate connection with the asymptotic behaviour of T_F which must then be described more precisely than in (15); see e.g. Ramachandran [1], Jakovleva [1], Dewess and Riedel [1], Dewess [2]. In view of corollary 22.2.4 in these investigations the behaviour of the real function $f(iy)$ is considered.

A value-theoretical result due to Marcinkiewicz [1] has become famous. Its proof if based on Hadamard's factorization theorem and Borel's estimate of a canonical product.

Theorem 22.2.2. *An integral function of finite order $\varrho > 2$ whose exponent of convergence is less than ϱ cannot be a c.f.*

Let now P_m stand for a polynomial of degree m so that the c.f. f_1 of Φ_{a,σ^2} (see 22.1.) is $f_1 = \exp(P_2)$. Then we immediately obtain a useful consequence of theorem 22.2.2.

Theorem 22.2.3 (Marcinkiewicz). *The function* $\exp(P_m)$ $(m > 2)$ *cannot be a c.f.*

The importance of this result is due to two reasons.

a) With the aim of characterizing Φ_{a,σ^2}, in a number of proofs it is shown that the c.f. f of the d.f. F under consideration has the form $f = \exp(P_m)$. Then it follows that $m = 2$, since $m = 1$ corresponds to the degenerated case $F = \varepsilon_a$, see 22.10.

b) The theorem was stimulating, and there exists a great many of generalizations and modifications, see Lukacs [3, 6] and Ramachandran [1]. More recent results will be quoted presently.

Special classes of a.c.f. f deserve particula rattention. An interesting theory concerns the class $\mathfrak{D}$ (Dantzig), see Lukacs [6]. An a.c.f. f belongs to $\mathfrak{D}$, if $\varphi(t) := f^{-1}(it)$, $|t| < \min(a, b)$, defines another a.c.f. Note that in view of theorem 22.2.1 an a.c.f. f given on a finite interval is uniquely defined on $\mathbb{R}^1$.

It happens that f_- is analytic on a circle K about 0, but f_+ has not this property. Then (14) clearly exists in a strip S_{0b} $(0 < b \leqq \infty)$. In this case f is called a *boundary c.f.* By virtue of (14) many results valid for a.c.f. carry over to boundary c.f. C.f. of non-negative random variables are either a.c.f. or boundary c.f., for details see Lukacs [3, 6].

The ridge property of a.c.f. suggested to consider functions f analytic in S_{ab} (or S_{0b}) satisfying (16); they are called *ridge functions*. A.c.f. and boundary c.f. form proper partial sets of this class. These investigations show, for instance, which properties of a.c.f. are consequences of (16). E.g. for integral ridge functions satisfying $f(0) = 1$, the function $f(iy)$, $y \in \mathbb{R}^1$, proves real. Hence the above corollary 22.2.2 remains true, but for its proof we need Schwarz's reflection principle.

By a well-known theorem of Phragmén-Lindelöf integral ridge functions have at least order 1. Indeed all orders $\varrho \geqq 1$ occur. But the situation changes radically, if we make assumptions on the location of zeros. Already MARCINKIEWICZ (loc. cit.) proved his theorem 22.2.2 for ridge functions without formulating it in this more general way. Further results are contained in LINNIK and OSTROVSKIĬ [1]. We quote two more recent propositions.

Theorem 22.2.4. *An entire ridge function f of finite order having only real zeros permits the representation*

$$f(z) = c \exp(-dz^2 + ibz) \prod_k (1 - z^2/a_k^2),$$

where $a_k > 0$, $\sum_k a_k^{-2} < \infty$, $\operatorname{Im} b = 0$, $d \geqq 0$, $c \neq 0$ *real* (GOL'DBERG' and OSTROWSKIĬ [2]).

This means that f has at most the order 2.

Theorem 22.2.5 *Let f stand for a ridge function of finite order ϱ defined in H_+; assume that the exponent of convergence for the zeros contained in $\{z : y \geqq a > 0\}$ is $\varrho_1 < \varrho$. Then $\varrho \leqq 3$* (KAMYNIN [1]).

This theorem is sharp, note the example $f(z) = \exp(iz^3)$.

OSTROVSKIĬ [2] considered also the following classes $\mathfrak{T}_1$ and $\mathfrak{T}_2$ of integral functions in which the entire ridge functions form a proper subset. With the notation $M(r) := \max_{|z|=r} |g(z)|$ all functions $g \in \mathfrak{T}_1$ satisfy

$$g(t + iy) = M(|y|)$$

while for all $g \in \mathfrak{T}_2$

$$\operatorname{Re} g(t + iy) \leqq M(|y|).$$

We have the relation $\mathfrak{T}_1 \subset \mathfrak{T}_2$.

KAMYNIN and OSTROVSKIĬ [1] started the study of integral functions which are non-negative definite on $\mathbb{R}^1$, the case of integral functions of n variables is also considered.

22.3. Analytic distribution functions (a.d.f.)

Now we turn to the question under which conditions a d.f. F proves continuable. If not otherwise stated the following results are contained in JESIAK [1, 2].

Definition 22.3.1. A d.f. F is said to be *analytic* if there exists a function R of the omplex variable $z = x + iy$ with the following properties:

(i) R is analytic in an open and simply connected domain $A \subseteqq H_+$ the boundary of which contains the real axis $\mathbb{R}^1$,

(ii) R is continuous on $\mathbb{R}^1$ where $R(x) = F(x)$.

(In the following we denote the analytic continuation R of F also by F.) As a first example of an a.d.f. we mention the normal distribution Φ.

Also the other stable distributions (see 22.1) are interesting in our context. They constitute a proper subset of the i.d.d.f. The canonical representation (12) has then the simple form

$$\log f(t) = iat - c\,|t|^\alpha \left[1 + i\beta \frac{t}{|t|}\,\omega(t, \alpha)\right] \tag{17}$$

where a, c, α, β are real numbers satisfying $c \geqq 0$, $0 < \alpha \leqq 2$, $|\beta| \leqq 1$, and

$$\omega(t, \alpha) := \begin{cases} \tan \dfrac{\pi\alpha}{2} & \text{if } \alpha \neq 1\,, \\ \dfrac{2}{\pi} \log |t| & \text{if } \alpha = 1\,. \end{cases}$$

Note, that the c.f. of Φ_{a,σ^2} is a special case of (17):

$$\omega(t, 2) \equiv 0\,, \qquad c = \sigma^2/2\,, \qquad \alpha = 2\,.$$

Obviously, a and c are scale parameters. Therefore we denote a stable d.f. and its c.f. with essential parameters α and β by $\Psi_{\alpha,\beta}$ and $\psi_{\alpha,\beta}$, respectively.

From (17) we obtain

$$|f(t)| = |\psi_{\alpha,\beta}(t)| \leqq e^{-c|t|^\alpha}\,, \tag{18}$$

so that $f \equiv \psi_{\alpha,\beta}$ is absolutely integrable on $\mathbb{R}^1$. Therefore the well-known Fourier inversion formula

$$p_{\alpha,\beta}(x) = \frac{1}{2\pi} \int_{-\infty}^{\infty} e^{-itx}\,\psi_{\alpha,\beta}(t)\,dt \tag{19}$$

is available where $p_{\alpha,\beta}$ is the density of $\Psi_{\alpha,\beta}$. Now integrations along appropriate contours permit us to derive integral representations as well as series expansions for $p_{\alpha,\beta}$ (see LUKACS [3, 6]. They yield the following result.

Theorem 22.3.1. *The density* (19) *has the following analytic properties*:

(i) *If* $1 < \alpha \leqq 2$ *then* $p_{\alpha,\beta}$ *is an integral function of order* $\dfrac{\alpha}{\alpha - 1}$.

(ii) *If* $\alpha = 1$, $\beta \neq 0$ *then* $p_{\alpha,\beta}$ *is an integral function of infinite order.*

(iii) *The density* $p_{1,0}$ *is a rational function with poles at* $z_{1,2} = \pm ic$.

(iv) *For* $0 < \alpha < 1$ *we have the representation*

$$p_{\alpha,\beta}(x) = \begin{cases} \dfrac{1}{\pi x}\,\Phi_1(x^{-\alpha}) & \text{if } x > 0\,, \\ \dfrac{1}{\pi\,|x|}\,\Phi_2(|x|^{-\alpha}) & \text{if } x < 0\,, \end{cases}$$

where $\Phi_{1,2}$ *are integral functions of order* $1/1 - \alpha$.

Accordingly, theorem 22.3.1 tells us that stable d.f. are analytic d.f. in the sense of definition 22.3.1 if and only if $1 \leqq \alpha \leqq 2$. The if-part of this statement follows also from (18) and theorem 22.3.2.

Since an a.d.f. continuously assumes real values on $\mathbb{R}^1$ it follows immediately from definition 22.3.1 that an a.d.f. F is analytic in the domain $B = A \cup A' \cup \mathbb{R}^1$ (where

$A' = \{z : \bar{z} \in A\}$) and that $F(z) = \overline{F(\bar{z})}$ $(z \in B)$ (by Schwarz's reflexion principle). Hence the domain of regularity, the zeros, and the singularities of F are located symmetrically with respect to $\mathbb{R}^1$. Note that B need not contain a horizontal strip (contrary to a.c.f.); this can easily be verified by examples using Mittag-Leffler's theorem.

The tail behaviour of the c.f. f has a strong influence on the analytic properties of the corresponding a.d.f.

Theorem 22.3.2. *Assume that for some* $\mu \geqq 0$

$$k_\mu := \varliminf_{t \to \infty} \frac{-\log |f(t)|}{t^{1+\mu}} > 0 . \tag{20}$$

Then F is an a.d.f. whose domain of regularity contains the strip $S_{k_0} := \{z : |\mathrm{Im}\, z| < k_0\}$. *For $z \in S_{k_0}$ the d.f. F and its density p have the integral representations*

$$F(z) = F(0) + \frac{1}{2\pi \mathrm{i}} \int_{-\infty}^{\infty} \frac{1 - \mathrm{e}^{-\mathrm{i}tz}}{t} f(t)\, \mathrm{d}t \quad \text{and} \quad p(z) = \frac{1}{2\pi} \int_{-\infty}^{\infty} \mathrm{e}^{-\mathrm{i}tz} f(t)\, \mathrm{d}t , \tag{21}$$

respectively. If (20) *holds for some $\mu > 0$ then F is integral with the order* $1 \leqq \varrho \leqq 1 + 1/\mu$; *if* $\varrho = 1 + 1/\mu$, *then the type τ of F is not greater than* $\mu[(1 + \mu)^{1+\mu} k_\mu]^{-1/\mu}$.

Boas and Smithies [1] showed that the conditions $k_0 > 0$ $(k_0 = \infty)$ are not necessary for a d.f. F to be analytic (integral); on the other hand these conditions cannot be weakened without additional assumptions. So we have $k_0 = \pi/2$ and $k_0 = 0$ for the integral d.f. $F_1(x) = \exp(-\mathrm{e}^{-x})$ and $F_2(x) = \exp(1 - \mathrm{e}^{\mathrm{e}^{-x}})$, respectively.

These results can be improved if we consider the class of *symmetric* d.f. which are defined by

$$F(x) = 1 - F(-x + 0) \qquad (x \in \mathbb{R}^1) .$$

This implies that the c.f. f is real. Let $\mathfrak{P}$ be the class of all *symmetric* d.f. satisfying $f(t) \geqq 0$ $(t \in \mathbb{R}^1)$. Then we have a certain analogy to theorem 22.2.1.

Theorem 22.3.3. *Assume that $F \in \mathfrak{P}$ is analytic in the disk $\{|z| < s\}$. Then F is analytic in the strip $S_s = \{z : |\mathrm{Im}\, z| < s\}$. In S_s the integral representations* (21) *hold and the density p has the ridge property. Further, in this case we have necessarily $k_0 > 0$ and the points $z_{1,2} = \pm \mathrm{i}k_0$ (if finite) are singularities of F.*

The proof of theorem 22.3.3 uses the fact that the assumptions imply that F possesses a density p which coincides with a certain c.f. (but for a positive factor). This way we can use the theory of a.c.f. sketched in 22.2.

The class $\mathfrak{P}$ plays a certain role for the study of symmetrizized d.f. F^s. Let X, X' be two independent random variables with the common d.f. F. Then the random variable $Y = X - X'$ has the symmetric d.f. $F^s(x) := F(x) * [1 - F(-x)]$ and the c.f. $f^s(t) = |f(t)|^2 \geqq 0$, i.e. we have $F^s \in \mathfrak{P}$. The trick of studying F^s instead of F is widely used. Further, all symmetric i.d.d.f. trivially belong to $\mathfrak{P}$ since their c.f. have no zeroes.

Next we turn to the analytic character of convolutions the study of which led to surprising results. So the known heuristic rule "a convolution is 'not worse' than 'the best' of its components" fails in this case. For example, the symmetrizations F_i^s $(i = 1, 2)$ of the integral d.f. F_i considered above are no entire functions (F_1^s is analytic in $|\mathrm{Im}\, z| < \pi$, F_2^s has a singularity at $z = 0$, i.e. F_2^s is even not analytic in the sense of definition 22.3.1). In this context the following notion proves very useful:

Definition 22.3.2. The strip S_s is said to be a *strip of boundedness* of the a.d.f. F if for any $s_1 < s$ there exists a constant $C(s_1)$ such that

$$|F(z)| \leqq C(s_1) \quad \text{for} \quad |\mathrm{Im}\, z| \leqq s_1 .$$

Now simple arguments of complex analysis yield the

Theorem 22.3.4. *Assume that S_s is a strip of boundedness of the a.d.f. F_1. Then for any d.f. F_2 the convolution $F = F_1 * F_2$ is analytic in S_s and S_s is a strip of boundedness of F. Further, we have the integral representation*

$$F(z) = \int_{-\infty}^{\infty} F_1(z-u)\, \mathrm{d}F_2(u) , \qquad z \in S_s .$$

This theorem is basic for the following one in the proof of which we make use of the method of symmetrization and the results about the class $\mathfrak{P}$ as well.

Theorem 22.3.5. (i) *The d.f. F is analytic with strip of boundedness if and only if $k_0 > 0$. In this case S_{k_0} is a strip of boundedness of F.*

(ii) *If $F^s(x) = F(x) * [1 - F(-x)]$ is analytic in S_s, then F is analytic with strip of boundedness $S_{s/2}$.*

Part (i) of theorem 22.3.5 gives a better insight into the meaning of the quantity k_0 (of theorem 22.3.2); part (ii) tells us that the existence of a strip of boundedness assumed in theorem 22.3.4 is even necessary if we consider symmetrized d.f. (this result is due to Raĭkov [1]).

An example for an a.d.f. without strip of boundedness is

$$F_2(x) = \exp\,(1 - \mathrm{e}^{\mathrm{e}^{-x}}) .$$

From the general results sketched up till now we can easily derive consequences concerning the analytic behaviour of an i.d.d.f. F. Of course, it depends on the parameter σ^2 and Lévy's spectral function R introduced in (12). If $\sigma^2 > 0$, then F has a normal component and in view of theorem 22.3.4 it is an integral d.f. For the case $\sigma^2 = 0$ Jesiak [1] gave conditions on R ensuring that F is analytic.

22.4. Continuation of distribution functions

Let $\mathfrak{B}$ be a given class of d.f., $G \in \mathfrak{B}$, and $S \subset R^1$. Assume that for a certain d.f. $F \in \mathfrak{B}$ we have $F(x) = G(x)$ $(x \in S)$. We want to find conditions

a) either on G (if $S \subset R^1$ is fixed), or

b) on S (if $G \in \mathfrak{B}$ is fixed),

such that $F \equiv G$ follows. As an example for the case a) we quote the following theorem (Rossberg [6]) where $S = (-\infty, 0)$ and $\mathfrak{B}$ the class of i.d. d.f

Theorem 22.4.1. *Let F be i.d. and assume that $F(x) = \Phi(x)$ $(x \leqq 0)$. Then it follows that $F \equiv G$.*[1])

In the proof of theorem 22.4.1 we make use of the canonical representation (12) for f which is — in view of $F(x) = \Phi(x)$ $(x \leqq 0)$ — even valid in H_+. From this very

[1]) Conjectured by A. N. Kolmogorov about 25 years ago; proved as an essential tool for the proof of theorem 22.5.1. — A multivariate extension is due to Siegel [2] who used the *method of projections* to reduce the task to theorem 22.4.1., cf. 22.11.

assumption it can also be shown that

$$\log f(iy) = \frac{y^2}{2}(1 + o(1)) \qquad (y \to \infty).$$

It turns out that such a growth is only possible if $R \equiv 0$, $a = 0$, $\sigma^2 = 1$. This is the assertion. This theorem was the starting point of continuation theory, but the assumption "i.d." proved to be too strong. It suffices that f has no zeros, see ROSSBERG [13].

The next result is due to IBRAGIMOV [1]; it represents an essential generalization of theorem 22.4.1.

Theorem 22.4.2. *Let F, G be i.d.d.f. satisfying*

$$0 < G(x) = o(e^{-r|x|}) \quad \text{for all} \quad r > 0, \quad x \to -\infty \tag{22}$$

and

$$F(x) = G(x) \quad \text{for} \quad x \leqq 0. \tag{23}$$

Then it follows that $F \equiv G$.

Simple examples show that assumption (22) is essential, see loc. cit.

The main tool of the proof is Nevanlinna's theory of meromorphic functions: In view of (22) the c.f. g is analytic in H_+ (see 22.2) and by (23) f has the same property. Since f and g are i.d.c.f. they have no zeros in H_+. So we obtain the deficiencies $\delta(f/g, 0) = \delta(f/g, \infty) = 1$. Now by Nevanlinna's second main theorem the sum of all deficiencies is not greater than 2, hence $\delta(f/g, 1) = 0$, i.e. $\delta(f - g)/g, 0) = \delta(f/g - 1, 0) = \delta(f/g, 1) = 0$. This means that the function $f - g$ has "many" zeroes in H_+. On the other hand, according to (23) we have $|f - g| \leqq 2$. Both these consequences are compatible only if $f - g \equiv 0$.

In the proof just sketched the infinite divisibility of F, G is used only for the conclusion that f, g have no zeroes in H_+. Therefore it is not surprising that we can replace the assumption of infinite divisibility by some restrictions on the zeroes of f, g in H_+. OSTROVSKIĬ [5] proved such a generalization of theorem 22.4.2; a modification of theorem 22.4.2 is due to JESIAK [6].

Next we turn to a continuation theorem concerning the Poisson distribution P_λ, see ROSSBERG, JESIAK, and SIEGEL [1].

Theorem 22.4.3. *Let F be i.d. and assume that $F(x) = P_\lambda(x)$ for $x \leqq 1 + \tau$ ($\tau > 0$, arbitrary). Then we have $F \equiv P_\lambda$.*

An essential tool for the proof of theorem 22.4.3 is a result from the theory of Phragmén-Lindelöf (cf. theorem 22.6.2). Further continuation theorems about the class of i.d.d.f. with $S = (-\infty, 0)$ are based on the following general result for Fourier-Stieltjes transforms (see JESIAK [5], RIEDEL [1]).

Lemma 22.4.1. *Let F, G be d.f. with $G(x) > 0$ for all x. Assume that g is analytic in H_+ and that $\overline{\lim}_{x\to-\infty} F(x)/G(x) < \infty$. Then f is also analytic in H_+ and we have*

$$\lim_{x\to-\infty} \frac{F(x)}{G(x)} \leqq \underline{\lim}_{y\to\infty} \frac{f(iy)}{g(iy)}, \qquad \overline{\lim}_{y\to\infty} \frac{f(iy)}{g(iy)} \leqq \overline{\lim}_{x\to-\infty} \frac{F(x)}{G(x)}.$$

As a consequence of lemma 22.4.1 we quote the following result which is a generalization of theorem 22.4.1. Theorem 22.4.4 was the first result that led beyond continuation theory (RIEDEL [1]).

Theorem 22.4.4. *Let F be i.d. and assume that*

$$\lim_{x \to -\infty} \frac{F(x)}{\Phi(x)} = 1 . \tag{24}$$

Then we have $F \equiv \Phi$.

Generalizations are contained in Rossberg and Siegel [1] and Jesiak, Rossberg, and Siegel [1].

If $\mathfrak{V}$ is the class of i.d.d.f. and S a bounded set then examples show that we cannot expect general continuation theorems without further assumptions. For instance, if S is a finite interval and if the increment of G on S is "not too large" then we can always construct an i.d.d.f. $F \not\equiv G$ coinciding with G on S. But if we restrict the class $\mathfrak{V}$ we get continuation theorems even with a countable set S. We give an example where $\mathfrak{V}$ is the class of all stable d.f.; it is due to Jesiak, Rossberg, and Zolotarev, for details see Jesiak [3].

Theorem 22.4.5. *Let F, G be stable d.f. If $F(x) = G(x) \neq 0, 1$ on an infinite (bounded or unbounded) point set $S = \{x_i\}$, then $F \equiv G$.*

The proof is based on the analytic properties of stable d.f. surveyed in 22.3. and on known asymptotic formulas as well.

Furthermore, there are continuation theorems for the class $\mathfrak{P}$ introduced in 22.3.; we quote the following result which is due to Siegel [1].

Theorem 22.4.6 *Let be $G \in \mathfrak{P}$ and I a point set with limit point $x = 0$. Assume that G satisfies a Lipschitz condition, i.e.*

$$G(x) - G(-x) < Bx \qquad (x > 0, B > 0) .$$

If F is a symmetric analytic d.f. with $F(x) = G(x)$ $(x \in S)$ then $F \equiv G$.

The method applied in the proof of theorem 22.4.6 is the same as that used for the proof of the following interesting lemma (Ibragimov and Linnik [1]).

Lemma 22.4.2. *Let φ be the c.f. of Φ, f an arbitrary c.f. and S a point set with limit point $t = 0$. Then from $f(t) = \varphi(t)$ $(t \in S)$ it follows that $F \equiv \Phi$.*

The present state of continuation theory is surveyed in Rossberg, Jesiak, and Siegel [1]. Very recent contribution are Blank [1 and 2] and Blank and Ostrovskiĭ [1], where deep generalizations to theorem 22.4.4 are given.

22.5. Limit theorems for sums of independent random variables

The continuation theory for d.f. arose from the needs of a new branch of the theory of sums of random variables which is characterized by the assumption of *restricted convergence*; with the notation introduced in 22.1. this means

$$F_n(x) \xrightarrow{w} \hat{F}(x) , \qquad x \in S , \tag{25}$$

where $\{F_n\}$ is a sequence of d.f., $S \subset R^1$ is an appropriate set, and $\hat{F}: S \to [0, 1]$ is monotone; here isolated points of S are continuity points of F by convention. This concept is useful since it is often possible to conclude from (25) that weak convergence of $\{F_n\}$ to a certain d.f. F takes place on the line. Under reasonable assumptions each problem of this type is inevitably connected with the question if there is a unique i.d.d.f. F that continues $\hat{F}$ to R^1. This is the reason why continuation theory has been developing together with this new theory of sums the present state of which is surveyed in Rossberg, Jesiak, and Siegel [2]. Of course, particular, attention has been paid to i.d. continuations. This way complex analysis plays a more important role here than in the classical theory of sums.

The most elegant results obtained so far concern the CLT. It was also a new version of this theorem (theorem 22.5.1) that initiated the new development so that the CLT keeps its stimulating influence on probability theory also in our time. The decisive idea is due to V. M. ZOLOTAREV who proposed to prove theorem 22.5.1 at a visit to the Karl Marx University Leipzig; see ROSSBERG and SIEGEL [2].

Theorem 22.5.1. *Assume* $S := (-\infty, 0)$ *and* $\hat{F}(x) := \Phi(x)$, $x \in S$. *Let* $X_1, X_2, \ldots$ *be independent and identically distributed random variables. Then* (6) *follows from* (25).

The general plan for the proof is typical. From Helly's selection theorem we see that there exists a subsequence $\{n'\}$ and a monotone function F such that the weak convergence

$$F_{n'} \xrightarrow{w} F \tag{26}$$

takes place. Now we are faced with two problems.

1. Compactness problem for $\{F_n\}$: $F(\infty) = 1$? (If this is true then F is automatically i.d.). This is a probabilistic problem.

2. Continuation problem: Does there exist only one i.d.d.f. which continues $\Phi(x)$, $x < 0$? This is an analytic problem with an affirmative answer, see theorem 22.4.1.

Theorem 22.5.2. has consequences for the theory of moderate deviations where now only one-sided assumptions on the convergence can be made, see AMOSOVA [1 and 2].

Now we turn to a generalization of theorem 22.5.1, see ROSSBERG [11].

Theorem 22.5.2. *Let the summands* $X_1, X_2, \ldots$ *be independent and identically distributed and assume* (25) *for* $S = (-\infty, 0)$, $\hat{F} > 0$, $\hat{F}(-\infty) = 0$. *Then there exists a stable d.f.* $\Psi_{\alpha\beta}$ *such that*

$$F_n \xrightarrow{w} \Psi_{\alpha\beta}\,.$$

Let us assume that the compactness problem of theorem 22.5.2 is solved. We enter into the discussion of the continuation problem. For this we have to distinguish the following cases.

a) If F satisfies (22) then the desired result follows from theorem 22.4.2.

b) Otherwise it can be shown that we have the representation

$$F = \Psi_{\alpha\beta} * F_1\,, \qquad f = \psi_{\alpha\beta} f_1\,,$$

where the d.f. $\Psi_{\alpha\beta}$ and its c.f. $\psi_{\alpha\beta}$ are stable and α is uniquely determined by $\hat{F}$ while the d.f. F_1 has a c.f. f_1 which is continuable to H_+.

b_1) $1 \leqq \alpha \leqq 2$. In this case $\Psi_{\alpha\beta}$ is an a.d.f. by (18) and theorem 22.3.2. Since $|f| = |\psi_{\alpha\beta}|\,|f_1| \leqq |\psi_{\alpha\beta}|$ we see from theorem 22.3.2 that F, too, is an a.d.f., i.e. the continuation is unique.

b_2) $0 < \alpha < 1$. In this case the canonical representation (17) shows that the c.f. $\psi_{\alpha\beta}$ is not continuable from $\mathbb{R}^1$ into H_+; but it can be continued into H_+ from the positive half line. Hence the c.f. $f = \psi_{\alpha\beta} f_1$ of F has also this property. Assume now that there were two different continuations F and F' of $\hat{F}$. Then

$$\Delta(t) := f(t) - f'(t) = \int_0^\infty e^{itx}\, d\big(F(x) - F'(x)\big)$$

is continuable from $\mathbb{R}^1$ to H_+ and $\Delta(iy)$ is real. Accordingly, continuing $f - f'$ from the positive half line we obtain a representation of $\Delta(iy)$ which contains the parameters of $\psi_{\alpha\beta}$ whence it can be shown that $f \equiv f'$. This terminates the proof.

In more complicated situations we cannot simply assume (25) and conclude that complete convergence takes place. Let us, for instance, consider a triangular array as introduced in 22.1. such that $\Sigma_n \sim H_n$, $X_{nk} \sim F_{nk}$. Under certain classical assumptions (A), say, we have

$$H_n \xrightarrow{w} G \tag{27}$$

where G is a certain a.d.f.

Let now $S \subset \mathbb{R}^1$ stand for a set with a finite limit point and assume

$$\lim_{n\to\infty} H_n(x) = G(x)\,, \qquad x \in S\,, \tag{28}$$

i.e. restricted convergence on S. Suppose that the compactness problem is solved. We are concerned with the continuation problem. By means of appropriate assumptions on $\{F_{nk}\}$ it can be ensured that all limit d.f. for $\{H_n\}$ are analytic. Then clearly by (28) there is only one limit d.f. for converging subsequences $\{H_n\}$, whence (27) follows. This idea was carried out in Jesiak [4] in such a way that a certain part of the classical assumption (A) was replaced by (28).

In the CLT it was possible to obtain the convergence (6) from assumptions which are even still weaker than (25), see Riedel [5]. We illustrate this by a new version of the Lindeberg-Feller theorem.

For a brief formulation of the result we introduce the functions

$$\underline{F}(x) := \varliminf_{n\to\infty} F_n(x)\,, \qquad \overline{F}(x) := \varlimsup_{n\to\infty} F_n(x)\,, \qquad x \in \mathbb{R}^1\,.$$

It is the salient feature of the following theorem that convergence of $\{F_n\}$ is not assumed in any point x. A more general version is contained in Jesiak, Rossberg, and Siegel [1] and Rossberg and Siegel [1].

Theorem 22.5.3. *Assume* (8). *If X_k/B_n are infinitely small and*

$$\lim_{x\to-\infty} \frac{\underline{F}(x)}{\Phi(x)} = \lim_{x\to-\infty} \frac{\overline{F}(x)}{\Phi(x)} = 1$$

then the Lindeberg condition (9) *follows.*

The proof follows the plan described after theorem 22.5.1. In this case the compactness of $\{F_n\}$ follows from $\operatorname{Var} S_n = 1$. The uniqueness of the limit d.f. is ensured by theorem 22.4.4.

The reader will already feel that complex analysis plays a more important role in the new theory of sums than in the classic theory. We illustrate this with another example where Vitali's convergence theorem can be applied, see Jesiak and Rossberg [1]. For this purpose we again consider sums of the form (10) where $\Sigma_n \sim H_n$.

Theorem 22.5.4. *Let $S \subset \mathbb{R}^1$ be a set with a finite limit point. Assume, moreover,* $\operatorname{Var} \Sigma_n \leqq 1$ $(n \geqq 1)$ *and*

$$|f_{nk}(t)| \leqq e^{-c_{nk}|t|}\,, \qquad t \geqq 1\,,$$

where f_{nk} stands for the c.f. of F_{nk}, $c_{nk} \geqq 0$,

$$c := \varliminf_{n\to\infty} \sum_{k=1}^{k_n} c_{nk} > 0\,.$$

Then from (28) *it follows that there exists an a.d.f. G such that* (27) *holds.*

Proof. By assumption the c.f. of Σ_n satisfies, for sufficiently large n,

$$|\mathsf{E}\, e^{it\Sigma_n}| = \prod_{k=1}^{k_n} |f_{nk}(t)| \leqq e^{-c|t|/2}, \qquad |t| \geqq 1.$$

From this it can be shown that Σ_n is subject to an a.d.f. with strip of boundedness. Accordingly, Vitali's theorem is at hand. It implies the assertion except for the properties $G(-\infty) = 0$, $G(\infty) = 1$ of the limit function. But this follows from $\operatorname{Var} \Sigma_n \leqq 1$.

Finally we refer the reader to the so-called *theory of large deviations* which constitutes a classical part of the theory of sums. Here it is the saddlepoint method as explained in EVGRAFOV [1] and RÜHS [1] that is effectively used for the approximate calculation of definite integrals. This way for instance W. RICHTER obtained his significant local theorems which are now to be found in IBRAGIMOV and LINNIK [1].

22.6. The theory of Phragmén and Lindelöf

22.6.1. The purely analytic theory. It was the fundamental discovery of PHRAGMÉN and LINDELÖF that for all functions $R \in M_+$ there exists the following alternative. Either they belong to N_+ or they grow at least like the exponential function; consider for instance $R(z) = \exp(icz)$ for $c \geqq 0$ and $c < 0$, respectively. This alternative is made precisely in the following classical elementary result. Note that for functions $R \in M_+$ of the exponential type

$$k_R := \overline{\lim_{y\to\infty}} \frac{\log |R(iy)|}{y} \tag{29}$$

always exists finitely.

Theorem 22.6.1. *A function $R \in M_+$ of the exponential type with $k_R \leqq 0$ belongs to N_+.*

In our time there exist a great many of conditions restricting the growth of functions $R \in M_+$ such that $R \in N_+$ follows, see for instance the books by CARTWRIGHT [1] and EVGRAFOV [1] and the papers LEHMANN and ROSSBERG [1], ROSSBERG and LEHMANN [1]; for an introduction to this theory see ROSSBERG, JESIAK, and SIEGEL [3].

For probability theory the following recent results are of particular interest. In this field equation (29) — when applied to c.f. — permits a simple interpretation: Let F be a d.f. such that $\operatorname{lext}(F) > -\infty$; then its c.f. f belongs to the exponential type in H_+ and $k_f = -\operatorname{lext}(F)$.

Theorem 22.6.2. *Let us be given the functions $\varphi \in N_+$ ($\varphi \not\equiv 0$) and $R \in M_+$. If*

$$\varphi R \in N_+$$

then R belongs to the exponential type. If, moreover, $k_\varphi = 0$ then $R \in N_+$ (ROSSBERG [2, 12] and RIEDEL [2]).

This proposition is particularly useful for probability theory since φ may be a c.f. General considerations on its consequences were made in ROSSBERG [5, 8] and TUTSCHKE [3].

Theorem 22.6.3. *Assume that $R \in M_+$ has the properties $k_R = 0$ and*

$$|R(t + iy)| \leqq \exp(cy^2), \qquad y \geqq 0,$$

for some $c > 0$; then $R \in N_+$ (LINNIK and OSTROVSKIĬ [1]).

Theorem 22.6.4. *Assume that $g_\pm \in N_\pm$, $g_\pm(0) > 0$, $k_{g\pm} = 0$. Consider an integral function g without zeros satisfying*

$$\frac{gg_+}{D} \in N_+ , \qquad \frac{gg_-}{D} \in N_-$$

for some $D > 0$. Then $g \equiv$ const[1]) (SEIFFERT and RIEDEL [1]).

This theorem is trivial (by virtue of theorem 22.6.2) only if $|g(t)| \leqq 1$.

To our view the importance of the Phragmén-Lindelöf theory has been underestimated so far. In order to facilitate applications it would be worth-while to collect all results together and to classify their proofs. An idea for this is explained in ROSSBERG [8]. An analysis of different restrictions on the growth of functions $R \in M_+$ is also contained in ROSSBERG [5].

Next we want to present another theorem. If $R \in M_+$ and $k_R = 0$ in (29) then we will say that *R belongs to the minimal type of order* 1 *on the ray* $\pi/2$ or briefly that *R has the growth* (1, 0) *on* $\pi/2$. According to theorem 22.6.1 a function $R \in M_+$ of the exponential type with growth (1, 0) on $\pi/2$ belongs to N_+. Mapping theorem 22.6.1 to a quarter plane we see that a function possessing at most order two and minimal type is bounded. Applying this fact to both quarter planes of H_+ we obtain

Lemma 22.6.1. *If $R \in M_+$ belongs at most to the minimal type of order* 2 *and has growth* (1, 0) *on* $\pi/2$, *then $R \in N_+$* (LEHMANN and ROSSBERG [1]).

A mapping of lemma 22.6.1 to the quarter planes of H_+ shows that $R \in M_+$ may belong to the minimal type of order 4; if it has growth

$$(1, 0) \quad \text{on} \quad \pi/2 , \qquad (2, 0) \quad \text{on} \quad \pi/2 \pm \pi/4 \tag{30}$$

then $R \in N_+$.

Obviously this *method of successive mapping* can be continued; loc. cit. it was not brought to the very end. Namely we finally obtain

Theorem 22.6.5. *Let the function $R \in M_+$ have growth* (30) *and, moreover,*

$$(4, 0) \quad on \quad \frac{\pi}{2} \pm (2k+1)\frac{\pi}{8} \qquad (k = 0, 1) ,$$

$$(8, 0) \quad on \quad \frac{\pi}{2} \pm (2k+1)\frac{\pi}{16} \qquad (k = 0, 1, 2, 3) ,$$

$$\vdots$$

Then either $R \in M_+$ or R has infinite order.

Proof. Assume that $R \in M_+$ has at most order $\varrho < \infty$. Choose a natural number r such that $2^r > \varrho$. Then the first r conditions of the theorem imply $R \in N_+$.

It is worth noticing that this theorem cannot be improved by mapping on the quarter planes of H_+ as described above. It would be interesting to find out which of the known theorems of the Phragmén-Lindelöf type for H_+ are consequences of theorem 22.6.5.

22.6.2. Applications. Unfortunately the use of the Phragmén-Lindelöf theory in the international literature on probability theory is sporadic. More systematicly it was applied by Leipzig authors and it proved to be a powerful tool in the following fields:

[1]) loc. cit. it is also assumed that g_+ has no zeros, but this assumption is superfluous as is seen from the proof.

a) Queuing theory: Compare the survey article ROSSBERG [10] and 22.9.

b) Characterization theory for d.f.: Compare the survey articles ROSSBERG [7, 9] and 22.10.

c) C.f. of non-negative random variables: See e.g. LAUE [1].

d) Continuation theory for d.f.: Compare the survey article ROSSBERG, JESIAK, and SIEGEL [1] and 22.4.

In this chapter we wish to give only the following details. From theorem 22.6.1 a slight generalization of a result proved in ROSSBERG [7] can easily be established; namely we have the

Theorem 22.6.6. *Let* $r\colon \mathbb{R}^1 \to \mathbb{R}^1$ *be a function of bounded variation,* $r(-\infty) = 0$, *and assume that*

$$R(t) := \int_{-\infty}^{\infty} e^{itu}\, dr(u)$$

is continuable to H_+ *and belongs to the exponential type. If*

$$R(iy) = O(e^{-\tau y}) \quad as \quad y \to \infty \tag{31}$$

then $r(x) = 0$, $x < \tau$.

The proof amounts to showing that $\int_{-\infty}^{\tau} e^{itu}\, dr(u)$ is integral and bounded everywhere.

As an immediate consequence we obtain the following generalization of the uniqueness theorem 22.1.3 for c.f.

Theorem 22.6.7. *Let* F *and* G *be two d.f. with c.f.* f *and* g, *respectively. We have*

$$F(x) = G(x)\,, \qquad x < \tau\,,$$

if and only if the difference $R := f - g$ *possesses an analytic continuation to* H_+ *which belongs to the exponential type and satisfies* (31).

Now we turn to an application of theorem 22.6.2, see ROSSBERG and DUC [1].

Lemma 22.6.2. *Consider the functions* P_+^*, R_+^*, $Q_+^* \in M_+$ *and* $R_- \in M_-$ *and assume that* Q_+^*/R_+^* *has no poles in* H_+ *and approaches its boundary values continuously. If*

$$P_+^*(t)\, \big(R_+^*(t) + R_-^*(t)\big) = Q_+^*(t), \qquad -\infty < t < \infty \tag{32}$$

then R_-^* *is continuable to an integer function of order* 1, *and for its type* τ *we have* $0 \leqq \tau \leqq |k_{p_+^*}|$.

The analytic continuation can be carried out by means of Schwarz' reflection principle; then it is easily seen that in H_+ the assumptions of theorem 22.6.2 are satisfied.

Lemma 22.6.2 permits to prove the

Theorem 22.6.8. *Let* P *stand for a d.f. such that* lext $(P) = 0$ *while* r *is like in theorem* 22.6.6. *If*

$$Q(x) := \int_{-\infty}^{\infty} P(x - u)\, dr(u) = 0\,, \qquad x < \tau\,,$$

then $r(x) = 0$, $x < \tau$.

For an application of this result see 22.10.

22.7. Factorization of characteristic functions

By the composition theorem 22.1.5 we see that certain c.f. f can be represented as

$$f = f_1 f_2 \tag{33}$$

where f_1 and f_2 are non-degenerated c.f. (compare example 4 in 22.1.). Such c.f. f are called *decomposable*, f_1 and f_2 are the factors. The i.d.c.f. form a particular important subclass of the decomposable c.f. It is quite natural to ask which c.f. are decomposable and which properties have the factors.

As examples for important results typical for this field — the so-called arithmetic of c.f. — we quote two analogous theorems which proved very stimulating.

Lemma 22.7.1. *Let the c.f. f be analytic in S_{ab} $(a, b > 0)$. If it permits the factorization* (33) *then the factors, too, are analytic in S_{ab}.*

The proof makes use of corollary 22.2.1 and the connection which exists between the asymptotic behaviour of the d.f. F, F_1, F_2 corresponding to f, f_1, f_2, respectively.

Theorem 22.7.1 (Cramèr). *The normal d.f. Φ_{a,σ^2} possesses only normal components.*

For the proof we start from the assumption (compare example 1)

$$\exp\left(iat - \frac{\sigma^2}{2} t^2\right) = f_1(t)\, f_2(t)\,.$$

By lemma 22.7.1 f_1 and f_2 are entire c.f. without zeros; further, it can be shown that their orders cannot exceed 2. Now Hadamard's factorization theorem tells us that

$$\log f_j(t) = a_{0j} + a_{1j}t + a_{2j}t^2$$

and elementary considerations show that $a_{0j} = 0$, a_{1j} purely imaginary, and $a_{2j} < 0$ $(j = 1, 2)$. From the view-point of continuation theory theorem 22.7.1 was generalized in Rossberg [13]; for related results see Titov [1].

Theorem 22.7.2 (Raĭkov). *The Poisson d.f. P_λ possesses only Poisson components.*

By example 2 we have to start with the assumption $\exp\left(\lambda(e^{it} - 1)\right) = f_1(t)\, f_2(t)$.

By lemma 22.7.1 the factors are integral, but in this case they may have infinite order. On the other hand it is clear that F_1 and F_2 are lattice d.f. and this facilitates the proof.

Deeper tools are needed if we want to factorize the c.f.

$$\exp\left(\lambda(e^{it} - 1) + iat - \frac{\sigma^2}{2} t^2\right)$$

since in this case the factors are neither necessarily of finite order nor do they necessarily belong to lattice d.f. As a tool the theorem is needed which states that a non-constant periodic integral function is a trigonometric polynomial, see Lukacz [1].

But there exist indecomposable c.f., and therefore c.f. can have indecomposabl factors. For these cases we have the important

Theorem 22.7.3 (Hinčin). *A c.f. having no indecomposable factors is i.d.*

Surprisingly enough in spite of this result there exist i.d.c.f. having indecomposable factors. This is due to the fact that there does not exist a uniqueness theorem for factorizations of c.f. Therefore the subclass I_0 was introduced which contains all i.d.c.f. having no indecomposable factors. This class proved very interesting and complex analysis is widely used for its investigation; see the monographs cited in

22.1. and the survey article OSTROVSKIĬ [4]. With respect to its generality we mention a very recent result due to FRYNTOV and ČISTJAKOV [1]; the class L of i.d.d.f. occuring here is well-known to experts and we do not define it.

Theorem 22.7.4. *Assume that $f \in L$ and that in* (12) *we have $\sigma^2 = 0$, while R satisfies as $x \to \infty$*

$$R(\infty) - R(x) + R(-x) = O(e^{-x \ln x});\ ^{1)}$$

then $f \in I_0$.

In the proof the authors make use of new results belonging to the theory of Phragmén-Lindelöf.

Another application of this very theory (namely of the theorems 22.6.2 and 22.6.3) is connected with the question under which conditions the factorization of a c.f. is unique. Let us consider a c.f. f given only on an interval $I_T := (t : |t| \leqq T)$. If f is an a.c.f. or a boundary c.f. then the continuation of f is trivially unique. But there exist examples of c.f. $f_1 \not\equiv f_2$ which coincide on I_T. Then clearly

$$ff_1 = ff_2$$

for all c.f. f vanishing outside I_T (Hinčin's phenomenon, see LUKACS [3, 6]); obviously for the corresponding d.f. we have

$$(F * F_1)(x) = (F * F_2)(x)\,. \tag{34}$$

But under which conditions does (34) imply $F_1 \equiv F_2$? It is the salient feature of the following theorem that (34) is used only on a half axis, see ROSSBERG and DUC [1].

Theorem 22.7.5. *Assume* (34) *for $x \leqq \tau$ and, moreover, $F(x) > 0$ for $x \in \mathbb{R}^1$. If*

$$F(x) = O(e^{xr})\,, \qquad x \to -\infty\,,$$

for all $r > 0$ and

$$|F_1(x) - F_2(x)| = O(e^{-x^2})\,, \qquad x \to -\infty\,,$$

then $F_1 \equiv F_2$.

More results on Hinčin's phenomenon are surveyed in LUKACS [3, 6]. The recent development of factorization theory is also surveyed in CUPPENS [3].

22.8. Characteristic functions of non-negative random variables

If lext $(F) = x_0 > -\infty$, then the Laplace-Stieltjes transform

$$\hat{f}(s) = \int_{x_0}^{\infty} e^{-sx}\, dF(x)\,, \qquad \operatorname{Re} s \geqq 0\,, \tag{35}$$

exists and we obviously have $f(t) = \hat{f}(-it)$. It is remarkable that it suffices for many purposes to introduce the Laplace-Stieltjes transform (35) as a real function for $s \geqq 0$. Analogously

$$\omega(s) := \int_0^{\infty} e^{-sx}\, U(dx)\,, \qquad s > 0\,,$$

can be introduced for measures U attributing a finite mass to every finite interval of $\mathbb{R}^1$. This is carried out in FELLER [1] and a number of theorems are proved this

[1]) In an older theorem of OSTROVSKIĬ $O(e^{-x^2})$ is assumed.

way including a uniqueness theorem, a continuity theorem, and real Tauberian theorems. We quote a further result.

Theorem 22.8.1. *A function $g\colon [0, \infty) \to (0, 1]$ is the Laplace-Stieltjes transform of a d.f. satisfying $F(0) = 0$ if and only if it is completely monotone and $g(0) = 1$.*

In several probabilistic papers the Poisson formula for functions harmonic in H_+ is used. It can be applied also to c.f. f; we need only put $s = -iz$ in (35) so that

$$f(z) = \int_{x_0}^{\infty} e^{izx}\, dF(x)\,, \qquad z \in H_+ + R^1\,. \tag{36}$$

Then, if $x_0 = 0$, by the Poisson formula we have (see LAUE [1])

$$f(z) = \frac{2z}{\pi i} \int_0^{\infty} \frac{\operatorname{Re} f(u)}{u^2 - z^2}\, du\,, \qquad z \in H_+.$$

and (if $F(+0) = 0$)

$$f(z) = \frac{2}{\pi} \int_0^{\infty} \frac{u \operatorname{Im} f(u)}{u^2 - z^2}\, du\,, \qquad z \in H_+\,.$$

These formulas permit to obtain interesting conclusions such as new conditions for the existence of moments (LAUE [1, 4]); compare theorem 22.1.4 which is unsatisfactory for odd orders.

Theorem 22.8.2. *The non-negative random variable X possesses the moment of order $2k+1$ if the derivative $f^{(2k)}$ of order $2k$ exists and*

$$\left|\frac{\operatorname{Im} f^{(2k)}(t)}{t}\right| < C\,, \qquad t \neq 0\,.$$

This result is particularly elegant for $k = 0$ since in this case the criterion concerns the c.f. $f = f^{(0)}$ of X itself. Note that theorems 22.1.4 and 22.8.2 imply the interesting

Corollary 22.8.1. *Let X be a non-negative random variable. Then $\mathsf{E}X^k$ exists if and only if $f^{(k)}(0)$ exists.*

Poisson formulas yield also inversion formulas for c.f. f in which either only $\operatorname{Re} f$ or only $\operatorname{Im} f$ occur (see LAUE [5]).

Theorem 22.8.3. *Let f be the c.f. f of a non-negative random variable X. Suppose that either $\operatorname{Re} f$ or $\operatorname{Im} f$ is absolutely integrable over R^1.*

Then the corresponding d.f. F is absolutely continuous and the formulas

$$p(x) = \frac{2}{\pi} \int_0^{\infty} \operatorname{Re} f(u) \cos ux\, du\,, \qquad x \geqq 0\,,$$

$$p(x) = \frac{2}{\pi} \int_0^{\infty} \operatorname{Im} f(u) \sin ux\, du\,, \qquad x \geqq 0\,,$$

respectively, express its density p in terms of f.

Finally we direct the reader's attention to an unsolved problem. If $x_0 = 0$ then the c.f. (36) permits the representation $f = \psi\chi$ where $\chi \equiv 1$, if $f \neq 0$ in H_+; otherwise we have $\psi \neq 0$ in H_+ and

$$\chi(z) = \prod_k \frac{z - a_k}{z - \bar{a}_k} \frac{z + \bar{a}_k}{z + a_k}$$

is meromorphic, where a_k are the zeroes of f in H_+ (ROSSBERG [2]). The function $\chi \not\equiv 1$ is never a c.f., but in all known examples (see e.g. LAUE [3]) the factor ψ is a c.f.; hence we suppose that this is generally true.

If such a theorem holds then the influence of the zeros of c.f. f becomes clearer since there is a simpler c.f. ψ which corresponds to f uniquely. Then the natural question arises which properties does F share with Ψ? For instance, since χ has all derivatives at 0 corollary 22.8.1 implies

Corollary 22.8.2. *F possesses the moment of order k if and only if so does Ψ.*

In view of $|\chi(t) \equiv 1$ also the asymptotic behaviour of $|f|$ and $|\psi|$ is the same.

For further results on c.f. of the form (36) see LAUE [6, 7].

22.9. Wiener-Hopf methods

We return to the sums (5) assuming that the summands are subject to the common d.f. F. Then S_n describes the *random walk* of a particle on $\mathbb{R}^1$ whose n-th jump is X_n and leads to S_n. For such a stochastic process there exists a simple alternative. Either with probability 1 we have a drift $S_n \to \pm\infty$ or with probability 1 every level $c \in \mathbb{R}^1$ will be crossed infinitely often. In the latter case we have a random oscillation (FELLER [1]; see also BINGHAM [1]).

If $S_n \to -\infty$ then the random variable $m := \max(0, S_1, S_2, \ldots) \sim W$ obviously exists. Clearly $W(x) = 0$, $x < 0$; but $W(x)$, $x > 0$, is the unique d.f. satisfying the Wiener-Hopf integral equation

$$W(x) = (F * W)(x), \qquad x > 0. \tag{37}$$

The random walk permits a number of interesting applications, for instance in queuing theory where the following model GI/G/1 plays an important role. There is one server and a sequence of *customers* enter the serving system. The *interarrival times* α_i between two successive customers are independent and subject to the common d.f. A. Moreover, the *service times* β_i of the customers are independent and subject to another common d.f. B. A customer finding the server idle is served immediately, otherwise he enters a queue and has to wait up to the moment when all customers are served who arrived before him.

An interesting problem arises when we consider the d.f. $\{W_n\}$ of the consecutive *waiting times* of the customers. Under natural conditions it turns out that they weakly tend to the d.f. W mentioned above where

$$F := B * A_-$$

and $A_-(x) := 1 - A(-x + 0)$. For the c.f. corresponding to F, A, B we have clearly

$$f(t) = b(t)\, a(-t). \tag{38}$$

In the general theory of the equation (37) an auxiliary function H is introduced such that equation (37) can be continued to $\mathbb{R}^1$; unfortunately H lacks any particular meaning in general. But in this very case we can rewrite (37) in the shape

$$W(+0)\left(U_-(x) - \varepsilon_0(x)\right) + W(x) = W * F(x), \qquad x \in \mathbb{R}^1; \tag{39}$$

here $W(+0)$ stands for the probability of no waiting while $U_-(x) = P(-Y < x)$ where Y denotes the length of an idle time of the server. Obviously $U_-(x) = 1$ for $x > 0$ so that the equations (37) and (39) coincide for $x > 0$.

For the c.f. corresponding to W, U_-, and F we have from (39)

$$W(+0)(u_- - 1) + w(1 - f) = 0 . \tag{40}$$

In nearly all applications either a or b is a rational c.f. Accordingly by (38) f is continuable either to H_+ or to H_-. All these cases can easily be treated with the Schwarz' reflection principle. To fix ideas we assume that b is rational. Then, locating the poles of w by means of Rouché's theorem, we obtain *conservation theorems* even if b is unknown. We cite an example, see ROSSBERG [1 and 10].

Theorem 22.9.1 *If B is a mixture of exponential d.f. then so is W.*

The essential means for tackling the general case is the famous Wiener-Hopf factorization. In probability theory the formula

$$1 - f = (1 - u_-)(1 - u_+) \tag{41}$$

is valid for every d.f. It is of great interest since the c.f. $u_\pm$ can be interpreted in the random walk considered above; u_- occurs also in (40) (FELLER [1]). Clearly there is an intimate connection between the d.f. F, U_+, U_-, see ROGOZIN [1], LAUE [2], MÜLLER [1], DEWESS [3].

Now we directly obtain the solution of (40) in the shape

$$w = W(+0)(1 - u_+)^{-1} .$$

So it is by a stroke of good fortune that equation (40) has such a simple shape that we need not carry out the second step necessary in the general theory, namely construct an integral function which proves to be bounded everywhere. But this ingenious trick has been adapted to various probabilistic problems, for instance to the derivation of further factorization identities, see BOROVKOV [1, 2] and 22.10.

Survey articles on Wiener-Hopf methods in probability theory are GREENWOOD [1], GREENWOOD and SHAKED [1], MALIŠEV [1]. An application in characterization theory is given in ROSSBERG [4].

The two functions $u_\pm$ occurring in (41) play an essential role in many problems connected with the random walk $\{S_n\}$. Necessary and sufficient conditions for them to be rational were given by BOROVKOV [1] and ROSSBERG [3] using quite different methods.

22.10. Characterization theory

We consider the independent random variables $X_1, \ldots, X_n \sim F$. The distribution theory of mathematical statistics studies random variables $S(X_1, \ldots, X_n)$ and is particularly interested in the d.f. G they are subject to if F is given explicitly. Sometimes different functions $S_i(X_1, \ldots, X_n)$ $(i = 1, 2)$ are needed which prove either identically distributed or independent for special choice of F. Results of such types are important for statistical tests.

Conversely characterization theory makes assumptions like $S \sim G$ where S and G are given explicitely. Then the problem is to determine F (except for a few parameters). In certain cases it is also the independence of S_1 and S_2 that permits to determine F.

The following simple examples are due to BERNSTEIN and PÓLYA, respectively; see RÉNYI [1].

Theorem 22.10.1. *Assume* Var $X_1 < \infty$. $S_1 := X_1 + X_2$ *and* $S_2 := X_1 - X_2$ *are independent if and only if* $F \equiv \Phi_{a,\sigma^2}$.

Theorem 22.10.2. *Assume* Var $X_1 < \infty$. $S_1 := X_1$ *and* $S_2 := 2^{-1/2}(X_1 + X_2)$ *are identically distributed if and only if* $F \not\equiv \Phi_{a,\sigma^2}$.

These propositions proved stimulating since it is natural to ask if they are still true when we replace S_1 and S_2 by more general linear combinations of $X_1, \dots, X_n$. Theorem 22.10.1 was generalized by DARMOIS [1] and SKITOVIČ [1]. The proof of their result given in LUKACS and LAHA [1] uses the method of finite differences which is essentially real, but finally Marcinkiewicz' theorem 22.2.2 is applied.

On the other hand theorem 22.10.2 was modified by MARCINKIEWICZ [1] in the following way where even $n = \infty$ is admitted.

Theorem 22.10.3. *Assume that all moments of* X_1 *exist If then*

$$L_1 := \sum_{j=1}^{\infty} a_j X_j \quad \text{and} \quad L_2 := \sum_{j=1}^{\infty} b_j X_j$$

exist and are identically distributed, then either $L_1 = L_2$ *or* $X_1 = \Phi_{a,\sigma^2}$.

The proof makes use of the theorems 22.7.1 and 22.2.3, but otherwise uses real analysis.

The most important monographs on characterization theory are KAGAN, LINNIK, and RAO [1] and LUKACS and LAHA [1]. More recent results are contained in GYIRES [1] and GALAMBOS and KOTZ [1].

In this brief survey we want to direct the reader's attention to the role that can be played by the theory of Phragmén-Lindelöf. As indicated by the theorems cited above characterization theory makes statistical assumptions from which it is not obvious that they imply strong conditions on F. It is only finally that uniqueness of F turns out (except for the parameters).

In the Phragmén-Lindelöf theory we have an analogous situation. The assumption $g \in M_+$ and some restrictions on the increase of g do not trivially imply that $g \in N_+$, i.e. that the class of functions g is narrow. Clearly the best results are obtained if g is entire and also $g \in N_-$.

This analogy seems to be far-fetched but it can really be used. As a first example we quote the simple but deep result of SEIFFERT and RIEDEL [1].

Theorem 22.10.4. *Consider the non-negative random variables* Y_1 *and* Y_2 *subject to the d.f.* F_1 *and* F_2, *respectively, whose c.f. have no zeros in* $H_+ + \mathbb{R}^1$. *If*

$$Y_1 - Y_2 \sim H$$

where H *is a given d.f., then* F_1 *and* F_2 *are uniquely determined.*

The assumption can directly be reformulated as the assumptions of theorem 22.6.4.

Further we refer to ROSSBERG [7] where two statistical problems are reduced to theorem 22.6.6 while a third one is solved directly by means of theorem 22.6.2.

But there are also other cases where the statistical assumptions cannot immediately be reformulated in terms of the Phragmén-Lindelöf theory, and it is only in the course of the proof that there arises the occasion of applying results of that theory. We are going to list a few examples.

For $n < \infty$ it was LINNIK [1 and 2] who generalized theorem 22.10.3. Namely, he made no assumptions on moments, but supposed only

$$\max |a_i| \neq \max |b_i| .$$

His method was essentially simplified by RIEDEL [3]. Using theorem 22.6.2 this author changed the plan of the proof. Namely, he proved that all moments of F exist so that theorem 22.10.3 yields the desired result.

Theorems 22.10.2 and 22.10.3 concern linear forms of independent random variables. These results can be extended to stochastic integrals which depend on a family of random variables $X(t)$ $(t \geqq 0)$ called a *stochastic process*. The random variable $X(t_2) - X(t_1)$ is called its *increment* over the interval $[t_1, t_2]$. A process $X(t)$ is called *homogeneous with independent increments* if the distribution of the increment $X(t + s) - X(t)$ depends only on s and if the increments over non-overlapping integrals are independent. A process $X(t)$ is said to be *continuous* if for every t $(\geqq 0)$ $X(t + s)$ tends to $X(t)$ in probability as $s \to 0$. A continuous homogeneous process with independent increments is called a *Wiener process* if the distribution of the increments is normal and $X(0) = 0$.

For a homogeneous continuous process with independent increments $X(t)$ $(t \geqq 0)$ and for a continuous function $a(t)$ defined in $[A, B]$ the stochastic integral $\int_A^B a(t)\, \mathrm{d}x(t)$ is defined as limit of Riemann sums in the sense of convergence in probability. RIEDEL [4] gave a characterization of Wiener processes by identically distributed stochastic integrals. In order to formulate the result put $S(z) := \int_A^B (|a_1(t)|^z - |a_2(t)|^z)\, \mathrm{d}t$ where $a_1(t)$, $a_2(t)$ are continuous in $[A, B]$.

Theorem 22.10.5. *Suppose that $S(z)$ is analytic in a neighbourhood of the origin. The conditions*

(i) *$X(t)$ is a Wiener process and*

(ii) *for a continuous, homogeneous process with independent increments $X(t)$ the integral $\int_A^B \alpha_i(t)\, \mathrm{d}X(t)$ $(i = 1, 2)$ are identically distributed*

are equivalent if and only if $S(x)$ $(0 < x \leqq 2)$ has only a zero at $x = 2$.

For more results relevant in this connection see RIEDEL [6, 7].

22.11. Multivariate problems

In our time more and more problems which were first posed and solved for random variables are carried over to random vectors $\boldsymbol{X} = (X_1, \dots, X_n)$ i.e. to vectors whose components are random variables.

Many concepts and results well-known in the one-dimensional case can more or less easily be extended to n dimensions. For instance a d.f. F can be introduced by

$$F(x_1, \dots, x_n) := \mathsf{P}(X_1 < x_1, \dots, X_n < x_n) . \tag{42}$$

It defines an n-variate finite measure p on the class B_n of all Borel sets of $\mathbb{R}^n$ which satisfies $p(\mathbb{R}^n) = 1$; it is called an *n-variate probability law.*

Let p_1 and p_2 be two such laws. Then the law

$$p(B) = \int_{\mathbb{R}^n} p_1(B - \boldsymbol{x})\, p_2(\mathrm{d}\boldsymbol{x}) , \qquad B \in B_n . \tag{43}$$

is called the *convolution* of p_1 and p_2; here $B - \boldsymbol{x}$ stands for the set

$$(\boldsymbol{y} \in \mathbb{R}^n : \exists \boldsymbol{u} \in B : \boldsymbol{y} = \boldsymbol{u} - \boldsymbol{x})$$

where $\boldsymbol{x} = (x_1, \dots, x_n) \in \mathbb{R}^n$. We write $p = p_1 * p_2$. (43) is said to be a *decomposition* of p, p_1 and p_2 are the factors. As in the univariate case one can raise the problem of finding the factors of a given law p.

With the complex vectors $\boldsymbol{x} = (x_1, \dots, x_n)$, $\boldsymbol{y} = (y_1, \dots, y_n)$ we define the scalar product

$$(\boldsymbol{x}, \boldsymbol{y}) := \sum_{j=1}^{n} x_j \bar{y}_j$$

and the vector

$$\operatorname{Im} \boldsymbol{y} := (\operatorname{Im} y_1, \dots, \operatorname{Im} y_n) .$$

The Fourier-Stieltjes transform of an n-variate probability law p,

$$f(t) := \int_{\mathbb{R}^n} e^{i(t, \boldsymbol{x})} \, p(d\boldsymbol{x}) , \qquad t \in \mathbb{R}^n , \tag{44}$$

is called the *c.f.* of p. For f the uniqueness theorem, the convolution theorem, and the continuity theorem can be derived.

The analytical properties of n-variate c.f. are determined to a great extent by the analytical properties of projections. Let be $\boldsymbol{e} \in \mathbb{R}^n$, $|\boldsymbol{e}| = 1$, and $B \subset \mathbb{R}^1$ an arbitrary Borel set. Then the univariate law

$$p_{\boldsymbol{e}}(B) := \mathsf{P}\big((\mathrm{N}, \boldsymbol{e}) \in B\big) \tag{45}$$

is called the *projection* of the n-variate law p on the vector $\boldsymbol{e}$. Obviously, denoting by f and $f_{\boldsymbol{e}}$ the c.f. of p and $p_{\boldsymbol{e}}$, respectively, we have

$$f_{\boldsymbol{e}}(t) = f(t\boldsymbol{e}) , \qquad t \in \mathbb{R}^1 .$$

The univariate c.f. $f_{\boldsymbol{e}}$ is called the *projection* of f on the vector $\boldsymbol{e}$. In many cases we can restrict ourselves to the study of projections. It is to be noted that an n-variate d.f. F $(n > 1)$ need not be determined by the projections on n fixed linear independent vectors, but the projections on all n-dimensional vectors determine F uniquely.

For the study of a.c.f. we need the space $\mathbb{C}^n$ of all complex vectors $\boldsymbol{z} = (z_1, \dots, z_n)$. Let $G \subset \mathbb{R}^n$ be a convex open set. Then we say that

$$(\boldsymbol{z} \in \mathbb{C}^n : \operatorname{Im} \boldsymbol{z} \in G) \tag{46}$$

is the *convex tube* with basis G. Now we can formulate the n-variate version of theorem 22.2.1, see CUPPENS [1].

Theorom 22.11.1 *Let f be an a.c.f. Then it is analytic on a convex tube* (46) *and admits the representation*

$$f(\boldsymbol{z}) = \int_{\mathbb{R}^n} e^{i(\boldsymbol{z}, \boldsymbol{x})} \, p(d\boldsymbol{x}) \tag{47}$$

on this tube. G is the interior of the set

$$(\boldsymbol{y} \in \mathbb{R}^n : \int_{\mathbb{R}^n} e^{-i(\boldsymbol{y}, \boldsymbol{x})} \, p(d\boldsymbol{x}) < \infty) .$$

If γ is a point of the boundary of G, then $i\gamma$ is a singular point of f.

From this some corollaries analogous to those in 22.2. follow, for instance we have the ridge property.

Corollary 22.11.1. *For the a.c.f.* (47) *we have for any* $\boldsymbol{x} \in \mathbb{R}^n$, $\boldsymbol{y} \in G$

$$|f(\boldsymbol{x} + \mathrm{i}\boldsymbol{y})| \leqq f(\mathrm{i}\boldsymbol{y}) .$$

From LINNIK and OSTROVSKIĬ [1] we quote

Theorem 22.11.2. *Let* $\boldsymbol{e}_1, \dots, \boldsymbol{e}_n \in \mathbb{R}^n$ *be n linear independent vectors and p an n-variate probability law with c.f. f. Assume that the projections* $f_{\boldsymbol{e}_1}, \dots, f_{\boldsymbol{e}_n}$ *of f are analytic in the strips* $a_j < \operatorname{Im} t < b_j$ $(a_j < 0 < b_j;\ j = 1, \dots, n)$, *respectively. Then f is an a.c.f. whose basis is the interior of the convex hull of the vectors* $a_i\boldsymbol{e}_i$ *and* $b_i\boldsymbol{e}_i$ $(i = 1, \dots, n)$.

For one-dimensional entire c.f. f the asymptotic behaviour is related to the behaviour of T_F, see 22.2. These results can be extended to the n-dimensional case, see DEWESS [1], SCHOPF [2].

Next we mention decomposition problems as discussed in 22.7., n-variate i.d. laws and their c.f. can be defined in the same manner as in 22.1. The canonical representation formula (12) now reads

$$\log f() = \mathrm{i}(\boldsymbol{a},\boldsymbol{t}) - \frac{1}{2}(\boldsymbol{T},\boldsymbol{t}) + \int\limits_{\mathbb{R}^n\setminus\{0\}} \left(\mathrm{e}^{\mathrm{i}(\boldsymbol{t},\boldsymbol{x})} - 1 - \frac{\mathrm{i}(\boldsymbol{t},\boldsymbol{x})}{1 + \|\boldsymbol{x}\|^2}\right) R(\mathrm{d}\boldsymbol{x})$$

where $\boldsymbol{a}$ is a fixed vector and T is a linear non-negative symmetric operator with finite trace.

Lemma 22.7.1 as well as theorems 22.7.1–22.7.3 can be carried over to n dimensions. There are also papers devoted to the class I_0^n of i.d.c.f. having no indecomposable factors.

Finally we emphasize that it is not in all cases a more or less technical affair to carry one-dimensional results over to n dimensions, and it is often the step from $n = 1$ to $n = 2$ which is decisive. A further serious problem arises when theorems valid for n dimensions are to be extended to a Hilbert space.

Recently SIEGEL [2] succeeded in carrying theorem 22.4.1 over to an arbitrary $n \leqq \infty$. He assumed that the i.d.d.f. F coincides with the normal d.f. Φ on a half-space and proved that $F \equiv \Phi$ using the method of projections. It is an important open problem if the assertion remains true if F and Φ coincide on a subspace narrower than a half-space.

For details on multi-variate problems we refer the reader to CUPPENS [1]. Parts of the books LINNIK and OSTROVSKIĬ [1] and LAHA and ROHATGI [1] are also devoted to this field. Recent results on decomposition problems are contained in CUPPENS [2].

22.12. Random power series

It was BOREL [1] who considered the question under which conditions the circle of convergence is a natural boundary for a power series. From the condition which he found, he concluded, "que, si les coefficients sont quelconques, le cercle de convergence est une coupure. Dire que les coefficients sont quelconques, c'est, en effet, dire que (sauf la condition qui résulte de ce que le rayon de convergence est donné) les valeurs des n premiers coefficients n'ont aucune influence sur les valeurs des suivants."

To make this statement, in particular the concepts "quelconques" and "n'ont aucune influence", precise we have to consider power series with random coefficients[1]).

[1]) They may, of course, be complex, i.e. of the shape $X + \mathrm{i}Y$, where X and Y are real random variables.

This way it turns out in which sense it happens rather often that the circle of convergence of a power series is the natural border of an analytic function. Borel's remark is, in fact, the origin of a considerably vast field of modern probability theory which treats objects belonging to the theory of complex functions with probabilistic technics. A detailed survey is contained in LUKACS [5].

22.12.1. Infinite series of random variables. Infinite series $\sum a_i$ of measurable functions $\{a_i\}$ are well-known objects. If $\{a_i\}$ are random variables defined on a probability space $(\Omega, \mathfrak{A}, \mathsf{P})$, then $\sum a_i$ converges at random and the following fact is basic. There do exist random events $E \in \mathfrak{A}$ with a structure such that $\mathsf{P}(E)$ can only be 0 or 1. This is distinctly expressed by Kolmogorov's *zero-one law*, see RÉNYI [1], FELLER [1].

Theorem 22.12.1. *Let be given a sequence $\{A_i\}$ of independent random events from $\mathfrak{A}$. With $B^{(n)} \subseteqq \mathfrak{A}$ we denote the smallest σ-Algebra containing $A_n, A_{n+1}, \ldots$ If $C \in \mathfrak{A}$ is a random event such that $C \in B^{(n)}$ $(n = 1, 2, \ldots)$ then either $\mathsf{P}(C) = 0$ or $= 1$.*

Namely it turns out that such an event C is independent of itself, i.e. $\mathsf{P}(C) = \mathsf{P}(CC) = \mathsf{P}^2(C)$.

The assumptions of theorem 22.12.1 imply that an arbitrary finite subset of $\{A_i\}$ has no influence on the occurence of C. An example for such an event is contained in the following lemma which gives a criterion to decide between the two possibilities occurring in theorem 22.12.1.

Lemma 22.12.1 (BOREL-CANTELLI). *Let the random events $\{A_i\}_1^\infty$ be independent and put*

$$C := \left(\textit{infinitely many } A_n \ (n \geqq 1) \textit{ occur}\right).$$

Then $\mathsf{P}(C) = 0$ *if and only if* $\sum_{i=1}^{\infty} \mathsf{P}(A_i) < \infty$.

The intuitive meaning of this criterion is obvious. If $\mathsf{P}(A_i) \to 0$ sufficiently fast then with probability 1 only finitely many A_i will occur.

The following result is a consequence of theorem 22.12.1 due to the fact that the convergence of an infinite series is not influenced by the values of finitely many partial sums.

Corollary 22.12.1. *The series*

$$S := \sum_{k=1}^{\infty} b_k$$

where b_k are independent random variables can converge only with probability 0 *or* 1.

This rises the question for criteria to decide which case of the alternative will occur. A particularly important answer is contained in Kolmogorov's so-called *three-series theorem*; see RÉNYI [1].

22.12.2. Random power series. With a given sequence $\{a_i\}$ of random variables we consider the formal series

$$f(z, \omega) := \sum_{k=0}^{\infty} a_k(\omega) z^k \tag{48}$$

where z is an arbitrary complex number. It is called a *random power series with the coefficients* $\{a_k\}$. Hence for z fixed f is an infinite series of random variables, while for ω fixed f is a power series with radius of convergence r,

$$r^{-1}(\omega) := \overline{\lim_{n\to\infty}} \sqrt[n]{|a_n(\omega)|} . \tag{49}$$

The finite sums

$$U_n(z, \omega) := \sum_{k=0}^{n} a_k(\omega)\, z^k \qquad (n = 0, 1, \ldots)$$

are random variables; they are called *partial sums of the random power series.* The random power series f is said to *converge with probability* 1, if the sequence $\{U_n\}$ converges with probability 1.

The radius of convergence (49) is obviously a random variable. If a_n are independent then r is not influenced by the values of finitely many a_n. Due to this fact it can be shown from theorem 22.12.1 that the probability

$$\mathsf{P}(r < x) \tag{50}$$

assumes only the values 0 and 1. If it vanishes identically we have $\mathsf{P}(r = \infty) = 1$; otherwise (50) represents a degenerated d.f., i.e. r is a constant.

The following theorem is a consequence of the three-series theorem mentioned above. It shows that we have $r = 1$ under natural assumptions.

Theorem 22.12.2. *Assume that in the random power series* (48) *the absolute values* $|a_n|$ *are independent and identically distributed with common d.f. F. Then*

$$\mathsf{P}\big(r(\omega) = 1\big) = 1 \quad \textit{if} \quad \int_1^\infty \log x \, \mathrm{d}F(x) < \infty ,$$

$$\mathsf{P}\big(r(\omega) = 0\big) = 1 \quad \textit{if} \quad \int_1^\infty \log x \, \mathrm{d}F(x) = \infty .$$

On the other hand we meet quite a different situation if the coefficients a_n depend on each other. For an example consider a random variable $X \sim F$ where $F(+0) = 0$. Putting $a_n := X^{-n}$ we see from (49) that $r = X$, i.e. $r \sim F$. Hence every d.f. F such that $F(+0) = 0$ describes the behaviour of a radius of convergence.

22.12.3. On Borel's problem. Let be given a random power series f with independent coefficients $\{a_i\}$ and the radius of convergence r. Obviously the circle of convergence is a natural boundary or not, independent of every finite set $(a_0, \ldots, a_n)$. As a consequence the possibility to continue f beyond its circle of convergence is also a random event with probability 0 or 1. The problem is now under which conditions it is the first case that occurs.

A very simple solution was given by Steinhaus [1]. He started from a fixed power series

$$g(z) := \sum_{n=0}^{\infty} c_n z^n \tag{51}$$

where $|c_n| = 1$ so that the radius of convergence is also 1. Then he introduced the random power series

$$f(z, \omega) := \sum_{n=0}^{\infty} c_n \, \mathrm{e}^{2\pi i \varphi_n} \, z^n$$

where φ_n are independent random variables uniformly distributed on (0, 1).

Theorem 22.12.3. *Under the above assumption the singularities of f are dense on the unity circle with probability* 1.

To provide a more general solution we consider the random power series (48).

Let $\{a_n'\}$ be a sequence of random variables such that the a_n and a_n' are independent and identically distributed for each n. Putting $\hat{a}_n := a_n - a_n'$ we can introduce a symmetrized random power series

$$f^s := \sum_{n=0}^{\infty} \hat{a}_n z^n$$

corresponding to f.

Lemma 22.12.1. *Given the random power series* (48), *there exists a fixed series* (51) *such that $f - g$ and f^s have the same radius of convergence.*

Now we are in a position to formulate a final result.

Theorem 22.12.4. *Let f be a random and g be a fixed power series such that $f - g$ and f^s have the same finite radius. Then the random power series $f - g$ is not continuable beyond its circle of convergence.*

22.12.4. Random integral functions. A random power series f is said to be a *random integral function* if $\mathsf{P}(\overline{\lim}\, \sqrt[n]{|a_n|} = 0) = 1$. In this case the radius r in (49) is almost certainly infinite. Order and type of fixed integral functions $g(z)$ can be expressed by the asymptotical behaviour of the coefficients. In the very same way we can define — for a given random power series f — the random variables

$$\varrho(\omega) := \overline{\lim_{n\to\infty}} \frac{n \log n}{-\log |a_n|}$$

and — provided that $\varrho(\omega) < \infty$ —

$$\tau(\omega) := \frac{1}{\mathrm{e}\, \varrho(\omega)} \overline{\lim_{n\to\infty}} \left[n\, |a_n(\omega)|^{\varrho/n}\right].$$

Modern investigations on ϱ and τ concern the case in which a_n may depend on each other, see e.g. WALK [1].

Further references

GNEDENKO [1], GNEDENKO and KOLMOGOROV [1], LUKACS [2], RAMACHANDRAN [2], ROSSBERG and JESIAK [1], ZOLOTAREV [1].

Bibliography[1])

ABDURAHMANOV, R. (Абдурахманов, Р.)

[1] Оператор Коши и краевые задачи Римана и Гильберта в некоторых пространствах обобщенных функций. I, II. Izv. Akad. Nauk Tadž. SSR, otd. fiz.-mat. geol.-him. nauk 1975, **3** (57) (1976), 16–24; **2** (60) (1976), 3–18.

ADAM, J. C., A. GOURDIN-SERVENIER, J. P. NEDELEC, and P. A. RAVIART

[1] Study of an implicit scheme for integrating Maxwell's equations. Comp. Meth. in Appl. Mech. and Eng. **22** (1980), 327–346.

AGMON, S.

[1] Multiple layer potentials and the Dirichlet problem for higher order elliptic equations in the plane. Comm. Pure Appl. Math. **10** (1957), 179–239.

[2] Lectures on elliptic boundary value problems. Van Nostrand, New York, Toronto, London, Melbourne 1965.

AGMON, S., A. DOUGLIS, and L. NIRENBERG

[1] Estimates near the boundary for solutions of elliptic partial differential equations satisfying general boundary conditions I, II. Comm. Pure Appl. Math. **12** (1959), 623–727; ibid. **17** (1964), 35–92.

AHLFORS, L. V.

[1] Über eine Methode in der Theorie der meromorphen Funktionen. Soc. Sci. Fenn. Comment. Phys.-Math. **8:10** (1935), 14 pp.

[2] Zur Theorie der Überlagerungsflächen. Acta Math. **65** (1935), 157–194.

[3] The theory of meromorphic curves. Acta Soc. Fenn. A (2) **3**, **4** (1941), 31 pp.

[4] Complex analysis. McGraw-Hill, New York, Toronto, London 1953 (2nd ed. 1966).

[5] Lectures on quasiconformal mappings. Van Nostrand, Princeton (N.J.) 1966.

[6] Conditions for quasiconformal deformations in several variables. *In*: Contributions to analysis (a collection of papers dedicated to Lipman Bers), Academic Press, New York 1974, 19–25.

[7] On a class of quasiconformal mappings. Sitz. Österr. Akad. Wiss., Math.-nat. Klasse, Abt. II, **185** (1976), 1–3, 5–10.

[8] Das mathematische Schaffen Rolf Nevanlinnas. Ann. Acad. Sci. Fenn. A I **2** (1976), 1–15.

AĬSENBERG, L. A., and A. P. JUŽAKOV (Айзенберг, Л. А., и А. П. Южаков)

[1] Интегральные представления и вычеты в многомерном комплексном анализе. Nauka, Novosibirsk 1979.

AKBAROV, R. (Акбаров, Р.)

[1] О краевой задаче Римана с заданными главными частями. Dokl. Akad. Nauk Tadžik. SSR **21** (1978) 3, 3–6.

[2] Задача сопряжения с заданными частями для векторов. Dokl. Akad. Nauk Tadžik. SSR **21** (1978) 11, 3–6.

[3] Задача Гильберта для кусочно-аналитической функции с заданными главными частями. Dokl. Akad. Nauk BSSR **22** (1978), 588–591.

AKSENT'EV, L. A., JU. E. HOHLOV, and E. A. ŠIROKOVA (Аксентьев, Л. А., Ю. Е. Хохлов и Е. А. Широкова)

[1] О единственности решения внешней обратной краевой задачи. Mat. Zametki **24** (1978), 319–330.

[1]) The transcription of the names is made as in the Mathematical Reviews (also in such cases, in which other transcriptions are used).

AKSENT'EV, L. A., N. B. IL'INSKIĬ, M. T. NUŽIN, R. B. SALIMOV, and G. G. TUMAŠEV (Аксентьев, Л. А., Н. Б. Ильинский, М. Т. Нужин, Р. Б. Салимов и Г. Г. Тумашев)
[1] Теория обратных краевых задач для аналитических функций и ее приложения. Itogi Nauki Tehn., Ser. Mat. Anal. **18** (1980), 67–124.
ALEHNO, A. G. (Алехно, А. Г.)
[1] О краевой задаче Карлемана с бесконечным индексом произвольного степенного порядка. Dokl. Akad. Nauk BSSR **22** (1978), 873–876.
[2] Об однородных краевых задачах со сдвигом в случае бесконечного индекса. Izv. Akad. Nauk BSSR, Ser. fiz.-mat. Nauk **133** (1978) 5.
[3] О краевой задаче Карлемана с бесконечным индексом для полуплоскости в классе функций вполне регулярного роста. Dokl. Akad. Nauk BSSR **23** (1979), 402–405.
[4] О краевой задаче Римана с конечным числом точек завихрения. Dokl. Akad. Nauk BSSR **23** (1979), 1069–1072.
[5] Об однородных краевых задачах со сдвигом в случае конечного числа точек завихрения. Dokl. Akad. Nauk BSSR **24** (1980), 206–209.
[6] Неоднородная краевая задача Римана с бесконечным индексом на контуре Ляпунова. Izv. Akad. Nauk BSSR, Ser. Fiz.-Mat. Nauk **1** (1980), 51–58.
ALEKNA, P. JU. (Алекна, П. Ю.)
[1] Краевая задача Гильберта с бесконечным индексом логарифмического порядка для полуплоскости. Litov. Mat. Sb. **17** (1977) 1, 5–12 (Engl. transl.: Lith. Math. J. **17** (1977), 1–6).
[2] Необходимые и достаточные условия разрешимости неоднородной краевой задачи Римана с минус-бесконечным индексом логарифмического порядка $\min(\alpha, \beta) \geqq 1$ для полуплоскости. Litov. Mat. Sb. **18** (1978) **3**, 5–14.
ALEKSANDROV, A. JA., and JU. I. SOLOV'EV (Александров, А. Я., и Ю. И. Соловьев)
[1] Постранственные задачи теории упругости. Nauka, Moskva 1978.
ALEKSANDRIJA, G. H. (Александрия, Г. Х.)
[1] Об одной задаче линейного сопряжения с заданными смещениями в классе обобщенных аналитических функций. Soobšč. Akad. Nauk Gruzin. SSR **21** (1958), 257–262.
AMOSOVA, N. N. (Амосова, Н. Н.)
[1] О вероятностях умеренных уклонений сумм независимых случайных величин. Teor. verojatn. prim. **29** (1979), 858–865.
[2] Узкие интегральные зоны нормального притяжния. Lietuvos. mat. rinkinys **XXI** (1981), 1–15.
ANDERSON, J. M.
[1] Asymptotic values of meromorphic functions of smooth growth. Glasgow Math. J. **20** (1979), 155–162.
ANDERSON, J. M., and J. CLUNIE
[1] Entire functions of finite order and lines of Julia. Math. Z. **112** (1969), 59–73.
ANDREOTTI, A., and J. ŁAWRYNOWICZ
[1] On the generalized complex Monge-Ampère equations on complex manifolds and related questions Bull. Acad. Polon. Sci., Sér. Sci. Math. Astronom. Phys. **25** (1977), 943–948.
[2] The generalized complex Monge-Ampère equation and a variational capacity problem. Bull. Acad. Polon Sci., Sér. Sci. Math. Astronom. Phys. **25** (1977), 339–347.
ANGER, G. (ed.)
[1] Inverse and improperly posed problems in Differential Equations. Akademie-Verlag, Berlin 1979.
ANSELO NE P. M.
[1] Collectively Compact Operator Approximation Theory. Prentice Hall, London 1971.
ARNOLD, D. N., and W. L. WENDLAND
[1] On the asymptotic convergence of collocation methods. Preprint Nr. 665, TH Darmstadt, Dept. Math., 1982. Math. Comp. (*to appear*).
ARZANOV, G. V. (Арзанов, Г. В.)
[1] О разрешимости однородной нелинейной краевой задачи степенного типа. Izv. vysš. učebn. Zaved., Mat. 1978, **195** (1978) 8, 8–18.

Atkinson, K. A.
[1] A Survey of Numerical Methods for the Solution of Fredholm Integral Equations of the Second Kind. SIAM, Philadelphia 1976.
Avanissian, V., and A. Traore
[1] Sur les fonctions polyanalytiques de plusieurs variables. C.R. Acad. Sci. Paris, AB **286** (1978) 17, 743–746.
Babuška, I., and A. K. Aziz
[1] Survey lectures on the mathematical foundations of the finite element method. *In*: A. K. Aziz(ed.): The Mathematical Foundation of the Finite Element Method with Applications to Partial Differential Equations. Academic Press, New York 1972, 3–359.
Baernstein, A.
[1] Proof of Edrei's spread conjecture. Proc. London Math. Soc. **26** (1973) 3, 418–434.
Baker, I. N.
[1] Linear Picard sets for entire functions. Math. Nachr. **64** (1974), 263–276.
Balk, M. B.(Балк, М. Б.)
[1] Теорема Пикара для целых бианалитических функций. Dokl. Akad. Nauk SSSR **152** (1963), 1282–1285.
[2] Вырожденные бианалитические отображения. Izv. Akad. Nauk Arm. SSR, fiz.-mat. Nauki **17** (1964) 2, 9–15.
[3] Бианалитические функции с неизолированными точками. Izv. Akad. Nauk Arm. SSR, fiz.-mat. Nauki **17** (1964) 3, 7–19.
[4] Большая теорема Пикара для целых полианалитических функций. Usp. Mat. Nauk **20** (1965) 2, 159–165.
[5] Теоремы единственности для полианалитических функций. Izv. AN Arm. SSR, fiz.-mat. Nauki **18** (1965) 3, 3–14.
[6] Полианалитические функции постоянного модуля. Litov. Mat. Sb. **6** (1966) 1, 31–36.
[7] Целые полианалитические функции с ограниченным множеством нулей. Izv. Akad. Nauk Arm. SSR, Matematika **1** (1966) 5, 341–357.
[8] О значениях, принимаемых целой полианалитической функцией. Dokl. Akad. Nauk SSSR **167** (1966), 12–15.
[9] Основная теорема алгебры для полианалитических полиномов. Litov. Mat. Sb. **8** (1968) 4, 401–404.
[10] О множестве нулей неприводимой полианалитической функции. Sb. Mat. Anal. Teor. Funkc., Moskva 1974, 3rd ed., 146–153.
[11] О факторизации полианалитической функции в окрестности ее изолированной особой точки. Izv. Akad. Nauk Arm. SSR, Mat. **10** (1975) 5, 389–397.
[12] О квазинормальных семействах полианалитических функций. Sib. Mat. Ž. **18** (1977), 1211–1219.
[13] Об одной теореме пикаровского типа для мероморфных полианалитических функций. Sb. Mat. Anal. Teor. Funkc., Moskva **9** (1978), 116–122.
Balk, M. B., and A. A. Gol'dberg (Балк, М. Б., и А. А. Гольдберг)
[1] Усиленный вариант большой теоремы Пикара для полианалитических функций. Ukrain. Mat. Ž. **28** (1976), 435–442.
[2] О функциях, полианалитических в круге. Izv. vysš. učebn. Zaved., Mat. **180** (1977) 5, 3–14.
[3] Об одном усилении большой теоремы Пикара для бианалитических функций. Sb. Mat. Anal. Teor. Funkcii, Moskva **7** (1978), 82–88.
Balk, M. B., and N. F. Manuilov (Балк, М. Б., и Н. Ф. Мануилов)
[1] О факторизации целой полианалитической функции нескольких переменных, имеющей ограниченное множество нулей. Sb. Teor. Funkc., Funkc. Anal. Prilož., Har'kov **31** (1979), 10–13.
Balk, M. B., and A. A. Poluhin (Балк, М. Б., и А. А. Полухин)
[1] Предельное множество бианалитических функций в ее изолированной особой точке. Smol. Mat. Sb. **3** (1970), 3–12.
Balk, M. B., and M. F. Zuev (Балк, М. Б., и М. Ф. Зуев)
[1] О полианалитических функциях Usp. Mat. Nauk **25** (1970) 5, 203–226.
[2] О числе значений, принимаемых целой полианалитической функцией неизолированно. Izv. Akad. Nauk Arm. SSR, Matematika **7** (1972) 5, 313–324.

BANK, S.
[1] On the solutions of algebraic differential equations whose coefficients are entire functions of finite order. Ann. Mat. Pura Appl. (IV) **83** (1969), 175–183.
[2] On meromorphic solutions of algebraic differential equations. Comment. Math. Helv. **44** (1969), 401–409; and Addendum, ibid. **45** (1970), 506.
[3] On the growth of certain meromorphic solutions of arbitrary second order algebraic differential equations. Proc. Amer. Math. Soc. **25** (1970) 4, 791–797.
[4] A general theorem concerning the growth of solutions of first-order algebraic differential equations. Compositio Math. **25** (1972), 61–70.
[5] On determining the growth of meromorphic solutions of algebraic differential equations having arbitrary entire coefficients. Nagoya Math. J. **49** (1973), 53–65.
[6] Some results on analytic and meromorphic solutions of algebraic differential equations. Adv. Math. **15** (1975), 41–62.
[7] On the existence of meromorphic solutions of differential equations having arbitrarily rapid growth. J. reine angew. Math. **288** (1976), 176–182.
[8] On the growth of meromorphic solutions of linear differential equations having arbitrary entire coefficients. Ann. Mat. Pura Appl. **107** (1976), 279–289.

BANK, S., and R. KAUFMAN
[1] On Meromorphic Solutions of First-Order Differential Equations. Comment. Math. Helv. **39** (1956) 51, 289–299.
[2] On Briot-Bouquet differential equations and a question of Einar Hille. Math. Z. **177** (1981), 549–559.
[3] On the Order of Growth of Meromorphic Solutions of First-Order Differential Equations. Math. Ann. **241** (1979), 57–67.
[4] On the growth of meromorphic solutions of the differential equation $(y')^m = R(z, y)$. Acta Math. **144** (1980), 223–248.

BANK, S., and I. LAINE
[1] On the growth of meromorphic solutions of linear and algebraic differential equations. Math. Scand. **40** (1977), 119–126.

BATEMAN, H., and A. ERDÉLYI (eds.)
[1] Higher transcendental functions. II. McGraw Hill, New York, Toronto, London 1953.

BATES, R. H. T., and D. J. N. WALL
[1] Null field approach to scalar diffraction. Phil. Trans. Royal Soc. London **287** (1977), 45–114.

BAUER, K. W.
[1] Über eine der Differentialgleichung $(1 \pm z\bar{z})^2 w_{z\bar{z}} \pm n(n+1) w = 0$ zugeordnete Funktionentheorie. Bonner math. Schriften (1965) 23.
[2] Zur Verallgemeinerung des Poissonschen Satzes. Ann. Acad. Sci. Fenn. AI **437** (1968), 3–27.
[3] Über Differentialgleichungen der Form $F(z, \bar{z}) w_{z\bar{z}} - n(n+1) w = 0$. Monatsh. Math. **75** (1971), 1–13.
[4] Differentialoperatoren bei partiellen Differentialgleichungen. Ber. Ges. Math. Datenverarb. Bonn **77** (1973), 7–17.
[5] Eine verallgemeinerte Darboux-Gleichung I. Monatsh. Math. **80** (1975), 1–11.
[6] Bestimmung und Anwendung von Vekua-Resolventen. Monatsh. Math. **85** (1977), 89–97.

BAUER, K. W., and H. FLORIAN
[1] Bergman-Operatoren mit Polynomerzeugenden. *In*: GILBERT, R. P., and R. WEINACHT (eds.): Function Theoretic Methods in Differential Equations. Research Notes in Mathematics, vol. 8. Pitman, London, San Francisco, Melbourne 1976.

BAUER, K. W., and S. RUSCHEWEYH
[1] Differential Operators for Partial Differential Equations and Function Theoretic Applications. Lecture Notes in Mathematics, vol. 791. Springer, Berlin, Heidelberg, New York 1980.

BEDFORD, E., and B. A. TAYLOR
[1] Some potential theoretic properties of plurisubharmonic functions (*to appear*).

BEGEHR, H.
[1] Beitrag zum Satz von Picard-Borel-Nevanlinna auf Riemannschen Flächen. Dissertation, FU Berlin (West) 1968.

[2] Zur Nullpunktabhängigkeit der Nevanlinnaschen Defekte. Math. Z. **106** (1968), 374–378.
[3] Über Defektbegriffe in der Theorie der meromorphen Funktionen. Math. Z. **116** (1970), 349–354.
[4] Zur Wertverteilung approximativ analytischer Funktionen. Habilitationsschrift, FU Berlin (West) 1970; Arch. Math. Basel **23** (1972), 41–49.
[5] Die logarithmische Methode in der Wertverteilungstheorie pseudoanalytischer Funktionen. Ann. Acad. Fenn. A I **549** (1973), 17 pp.
[6] Boundary value problems of entire index for generalized hyper-analytic functions. Topics in Complex Analysis, Lect. Univ. Delaware 1977, 13–23.
[7] Boundary value problems for systems with Cauchy-Riemannian main part. Romanian-Finnish Seminar on Complex Analysis. Bucharest 1980. Lecture Notes in Mathematics, Springer, Berlin, Heidelberg, New York (*to appear*).

BEGEHR, H., and R. P. GILBERT
[1] Über das Randwertnormproblem für ein nichtlineares elliptisches System. *In*: MEISTER, V. E., N. WECK, and W. L. WENDLAND (eds.) [1, 112–121].
[2] Das Randwert-Normproblem für ein fastlineares elliptisches System und eine Anwendung. Ann. Acad. Sci. Fenn. A I **3** (1977), 179–184.
[3] Randwertaufgaben ganzzahliger Charakteristik für verallgemeinerte hyperanalytische Funktionen. Appl. Anal. **6** (1977), 189–205.
[4] Piecewise continous solutions of pseudoparabolic equations in two space dimensions. Proc. Royal Soc. Edinburgh **81A** (1978), 153–173.
[5] On Riemann boundary value problems for certain linear elliptic systems in the plane. J. Diff. Eq. **32** (1979), 1–14.
[6] Solvability conditions for Riemann-Hilbert problems associated with generalized hyperanalytic systems.

BEGEHR, H., and G. N. HILE
[1] Nonlinear Riemann boundary value problems for a nonlinear elliptic system in the plane. Math. Z. **179** (1982), 241–261.
[2] Riemann boundary value problems for nonlinear elliptic systems. Complex Variables — Theory and Appl. **1** (1983), 239–261.

BEGEHR, H., and G. C. HSIAO
[1] Nonlinear boundary value problems for a class of elliptic systems. *In*: TUTSCHKE, W. (ed.): Komplexe Analysis und ihre Anwendung auf partielle Differentialgleichungen. Martin-Luther-Universität Halle-Wittenberg 1980, 90–102.
[2] On nonlinear boundary value problems for an elliptic system in the plane. *In*: EVERETT, N., and B. SLEEMAN (eds.): Proc. conf. ord. part. diff. eq. Dundee 1980, Lecture Notes in Math., vol. **846**. Springer, Berlin 1981, 55–63.
[3] Nonlinear boundary value problems of Riemann-Hilbert type. Contemporary Math. **11** (1982), 139–153.
[4] A priori estimates for elliptic systems (*to appear*).
[5] The Hilbert Boundary Value Problem for Nonlinear Elliptic Systems. Proc. Royal Soc. Edinburgh (*to appear*).

BEHNKE, H., and F. SOMMER
[1] Theorie der analytischen Funktionen einer komplexen Veränderlichen. 2nd ed. Springer, Berlin, Göttingen, Heidelberg 1962.

BERGER, M. S., and M. SCHECHTER
[1] On the solvability of semilinear operator equations and elliptic boundary value problems. Bull. Amer. Math. Soc. **78** (1972) 5, 741–745.

BERGMAN, S.
[1] Integral operators in the theory of linear partial differential equations. Ergebniss der Mathematik, N.F., vol. 23. Springer, Berlin, Göttingen, Heidelberg 1961 (Russ transl.: Mir, Moskva 1964).
[2] On stream functions of compressible fluid flows. Ber. Ges. Math. Datenverarb. Bonn **77** (1973), 19–27.

BERKER, R.
[1] Intègration des équations du mouvement d'un fluide visqueux incompressible. *In*: FLÜGGE, S. (ed.): Handbuch der Physik, vol. VIII/2. Springer, Berlin, Göttingen, Heidelberg 1963, 1–384.

Bers, L.
[1] Theory of pseudo-analytic functions. New York 1953 (hect.).
[2] Mathematical aspects of subsonic and transsonic gas dynamics. John Wiley, London 1958.
[3] Univalent solutions of linear equations. Comm. Pure Appl. Math. **6** (1953), 513–526.

Bers, L., F. John, and M. Schechter
[1] Partial differential equations. Interscience publishers. New York, Chichester, Toronto, Sidney 1964 (Russ. transl.: Mir, Moskva 1966).

Bers, L., and L. Nirenberg
[1] On a representation theorem for linear elliptic systems with discontinuous coefficients and its applications. Atti del Convegno internazionale sulle Equazioni alle derivate parziali, Trieste 1954, 111–140.
[2] On linear and nonlinear elliptic boundary value problems in the plane. Atti del Convegno internazionale sulle Equazioni alle derivate parziali, Trieste 1954, 141–167.

Beurling, A., and J. Deny
[1] Dirichlet spaces. Proc. Nat. Acad. Sci. U.S.A. **45** (1959), 208–215.

Bhatnagar, S. C., and R. P. Gilbert
[1] Bergman type operators for pseudoparabolic equations in several space variables. Fac. Sci. Natur. Univ. Kiril et Metodij, Skopje, Annuaire, Math. Phys. Chim. **27** (1976/77) 4, 5–27.
[2] Integral operators generating solutions of $\Delta_3^2 u_t + A\Delta_3 u_t + B\Delta_3 u + Cu = 0$. Glasnik Mat. Ser. III **12** (32) (1977), 31–48.
[3] Bergman type operators for pseudoparabolic equations in several space variables. Math. Nachr. **76** (1977), 61–68.
[4] Constructive methods for solving $\Delta_4^2 u + A\Delta_4 u_t + Bu_t + C\Delta_4 u + Du = 0$. Math. Nachr. **80** (1977), 105–113.

Bicadze, A. V. (Бицадзе, А. В.)
[1] Об единственности решения задач Дирихле для эллиптических уравнений с частными производными. Usp. Mat. Nauk **3** (1948), 211–212.
[2] О так называемых площадно-моногенных функциях. Dokl. Akad. Nauk SSSR **59** (1948) 8, 1385–1388.
[3] Краевые задачи для эллиптических уравнений второго порядка. Nauka, Moskva 1966.

Bieberbach, L.
[1] Theorie der gewöhnlichen Differentialgleichungen. Auf funktionentheoretischer Grundlage dargestellt. Grundlehren der mathematischen Wissenschaften, vol. 66, 2nd ed. Springer, Berlin, Göttingen, Heidelberg, New York 1965.

Binderman, Z.
[1] On some Hilbert's problem with fixed singularity. Politechn. Łódz, Zeszyty nauk. **245** (1976) 8, 35–42.
[2] On some problem with fixed singularities of Vekua. Demonstratio Math. **10** (1977), 129–138.
[3] On some problem of Hilbert-Vekua type for infinite sequences of functions. Politechn. Łódz, Zeszyty nauk. **254** (1977), 5–21.

Bingham, N. H.
[1] Wiener-Hopf and related methods in probability. *In*: Brannan, D. A., and J. G. Clunie (eds.): Aspects of contemporary complex analysis, Academic Press, London 1980.

Birkhoff, G. D.
[1] A theorem on matrices of analytic functions. Math. Ann. **74** (1913), 122–133.

Bischoff, G.
[1] Zur numerischen Lösung von Randwertproblemen partieller Differentialgleichungen elliptischen Typs. Beitr. Anal. **12** (1978), 29–45.

Blank, N. M. (Бланк, Н. М.)
[1] Об однозначной определенности функций ограниченной вариации и функции распределения асимптотическим поведением при $x \to -\infty$. *In*: Problemy ustoĭčivosti stohastičeskih modeleĭ, Moskva 1981, 10–15.
[2] Об условиях однозначной определенности функций ограниченной вариации и функции распределения значениями на полупрямой. Teor. verojatn. prim. **XXVII** (1982), 784–787.

Blank, N. M., and V. Ostrovskiĭ (Бланк, Н. М., и В. Островский)
[1] О функциях ограниченной вариации, близких к нормальной функции распределения. *In*: Teorija operatorov funkcionalnyh prostranstvah i ee priloženija, Kiev 1981.
Boas, R. P., and F. Smithies
[1] On the characterization of a distribution function by its Fourier transform. Amer. J. Math. **60** (1938), 523–531.
Böhmer, K.
[1] Über lineare Differentialgleichungen mit Lösungen von endlicher Wachstumsordnung. Arch. Math. Basel **22** (1971), 394–400.
[2] Charakterisierung von linearen Differentialgleichungen durch Wachstumsbedingungen I; II. Arch. Math. Basel **26** (1975), 64–74; 170–177.
Boĭkov, I. V. (Бойков, И. В.)
[1] О приближенном решении сингулярных интегральных уравнений.Dokl. Akad. Nauk SSSR **203** (1972), 511–514.
Bojarski, B. V. (Боярский, Б. В.)
[1] Обобщенные решения системы дифференциальных уравнений первого порядка эллиптического типа с разрывными коэффициентами. Mat. Sb. **43** (85) (1957) 4, 451–503.
[2] Об устойчивости задачи Гильберта для голоморфного вектора. Soobšč. Akad. Nauk Gruzin. SSR **21** (1958), 391–398.
[3] Об особом случае задачи Римана-Гильберта. Dokl. Akad. Nauk SSSR **119** (1958), 411–414.
[4] Задача Римана-Гильберта для голоморфного вектора. Dokl. Akad. Nauk SSSR **126** (1959), 695–698.
[5] Абстрактная задача линейного сопряжения и фредгольмовы пары подпространств. *In*: Sb. Differencialnye i integralnye uravnenija. Kraevye zadači. Trudy inst. priklad. mat. im. I. N. Vekua, Tblissi 1979.
[6] Теория обобщенного аналитического вектора. Ann. Polon. Math. **17** (1966), 281–320.
[7] Исследования по уравнениям эллиптического типа на плоскости и граничным задачам теории функций. Thesis (habilitation), Moskva 1960.
[8] Анализ разрешимости граничных задач теории функций. vol. Issledovanija po sovremennym problemam teorii funkciĭ kompleksnogo peremennogo, Moskva **1961**, pp. 57.
[9] Subsonic flow of compressible fluid. *In*: The Mathematical Problems in Fluid Mechanics, Polish Acad. Sci., Warszawa 1967, 9–32.
Bojarski, B. V., and T. Iwaniec
[1] Quasiconformal mappings and nonlinear elliptic equations in two variables. I, II. Bull. Acad. Polon. Sci., Sér. Math. Astronom. Phys. **12** (1974), 473–478; 479–484.
[2] Topics in quasiconformal theory in several variables. *In*: Ławrynowicz, J., and O. Lehto (eds.): Proc. of the First Finnish-Polish Summer School in Complex Analysis at Podlesice II. Uniwersytet Łódzki, Łódz 1978, 21–44.
Bombieri, E., E. de Giorgi, and E. Giusti
[1] Minimal cones and the Bernstein problem. Invent. Math. **7** (1969), 243–268.
Borel, E.
[1] Sur les series de Taylor. C.R. Acad. Sci. Paris **123** (1896), 1051–1052.
[2] Sur les zéros des fonctions entieres. Acta Math. **20** (1897), 357–396.
Borgnis, F. E., and C. H. Papas
[1] Electromagnetic Waveguides and Resonators. *In*: Flügge, S. (ed.): Handbuch der Physik, vol. XVI. Springer, Berlin, Göttingen, Heidelberg 1958, 285–422.
Borovkov, A. A. (Боровков, А. А.)
[1] Вероятностные процессы в теории массового обслуживания. Nauka, Moskva 1972.
[2] Wahrscheinlichkeitstheorie. Akademie-Verlag, Berlin 1976.
Bosch, W., and P. Krajkiewicz
[1] The big Picard theorem for polyanalytic functions. Proc. Amer. Math. Soc. **26** (1970) 1, 145–150.
Bott, R., and S. S. Chern
[1] Hermitian vector bundles and the equidistribution of the zeros of their holomorphic sections. Acta Math. **114** (1965), 71–112.

BOUTROUX, P.
[1] Leçons sur les fonctions définies par les équations différentielles du premier ordre. Gauthier-Villars, Paris 1908.
[2] Recherches sur les transcendentes de M. Painlevé et l'étude asymptotique des équations différentielles du second ordre. Ann. Sci. Ecole Norm. Sup. (3) **30** (1913), 255–375; ibid. **31** (1914), 99–159.

BRACKX, F.
[1] Non-(k)-monogenic points of functions of a quaternion variable. *In*: MEISTER, V. E., N. WECK, and W. L. WENDLAND (eds.) [1, 138–149].
[2] The behaviour at infinity of (k)-monogenic functions of a quaternion variable. Simon Stevin **52** (1978), 49–60.
[3] The exponential function of a quaternion variable. Appl. Anal. **8** (1979), 265–278.

BRACKX, F., R. DELANGHE, and F. SOMMEN
[1] Clifford Analysis. Pitman, Boston, London, Melbourne 1982.

BRAMBLE, J., and A. SCHATZ
[1] Rayleigh-Ritz-Galerkin methods for Dirichlet's problem using subspaces without boundary conditions. Comm. Pure Appl. Math. **23** (1970), 653–675.

BREBBIA, C. A. (ed.)
[1] Boundary Élément Methods. Proc. Third Intern. Sem. Irvine (Cal.) 1981, Springer, Berlin, Heidelberg, New York 1981.

BRELOT, M.
[1] Éléments de la théorie classique du potentiel. Centre de Documentation, Paris 1959.

BRIOT, CH., and J.-C. BOUQUET
[1] Intégration des équations différentielles au moyen des fonctions elliptiques. J. Ecole Polytechnique **21:36** (1856).
[2] Recherches sur les propriétés des fonctions définies par des équations différentielles. J. Ecole Imp. Polytechnique **21** (1856), 133–197.
[3] Théorie des fonctions elliptiques. 2nd ed. Paris 1875.

BROWDER, F. E.
[1] Approximation by solutions of partial differential equations. Amer. J. Math. **84** (1962), 134–160.

BROWN, P. M.
[1] Integral operators for fourth order linear parabolic equations. SIAM J. Math. Anal. **9** (1978), 121–141.

BROWN, P. M., and R. P. GILBERT
[1] Constructive methods for higher order analytic Sobolev-Galpern equations. Bull. Math. Soc. Sci. Math. RSR, NS **19**(67) (1975) 3/4, 213–225.

BROWN, P. M., R. P. GILBERT, and G. C. HSIAO
[1] Constructive function theoretic methods for fourth order pseudoparabolic equations in two space variables. Rend. Math., VI. Ser. **8** (1975), 935–951.

BUCHANAN, J.
[1] Generalized hyperanalytic function theory. Ph. D. Dissertation, University of Delaware 1979.

BUCHANAN, J., and R. P. GILBERT
[1] The Hilbert problem for hyperanalytic functions. Appl. Analysis **11** (1980/81), 303 to 323.
[2] First order elliptic systems. Academic Press, New York, London (*to appear*).

BURGATTI, P.
[1] Sulla funzioni analitiche d'ordini n. Boll. Unione Mat. Ital. **1** (1922).

CALDERÓN, A. P.
[1] Uniqueness in the Cauchy problem for partial differential equations. Amer. J. Math. **80** (1958), 16–36.

CALDERÓN, A. P., C. P. CALDERÓN, E. FABES, M. JODEIT, and N. M. RIVIERE
[1] Applications of the Cauchy integral on Lipschitz curves. Bull. Amer. Math. Soc. **84** (1978), 287–290.

CARATHÉODORY, C.
[1] Über das Schwarzsche Lemma bei analytischen Funktionen von zwei komplexen Veränderlichen. Math. Ann. **97** (1926), 76–98.

[2] Über eine spezielle Metrik, die in der Theorie der analytischen Funktionen auftritt. Atti Pontificia Acad. Nuovi Lincei **80** (1927), 135–141.
[3] Funktionentheorie. I, II, Birkhäuser, Basel 1950.

CARLEMAN, T.
[1] Sur la résolution de certaines équations intégrales. Ark. Math. Astr. Fys. **16** (1922) 26.
[2] Sur les systémes linéaires aux dérivées partielles du premier ordre à deux variables. C.R. Acad. Sci. Paris **197** (1933).

CARTAN, E.
[1] Sur la géometrie pseudo-conforme des hypersurface des deux variables complexes I. Ann. Mat. Pura Appl. (4) **11** (1932), 17–90 (*reprinted in*: Oeuvres II, 2, 1231–1304); II. Ann. Scuola Norm. Sup. Pisa (2) **1** (1932), 333–354 (*reprinted in*: Oeuvres III, 2, 1217–1238).
[2] Les systémes différentiels extérieurs et leurs applications géométriques. Hermann, Paris 1945.
[3] Séminaire ENS Paris 1951/52. Reprint, Benjamin 1967.
[4] Séminaire ENS Paris 1953/54. Reprint, Benjamin 1967.

CARTWRIGHT, M. L.
[1] Integral functions. Cambridge University Press, Cambridge 1962.

ČEREDNIČENKO, V. G. (Чередниченко, В. Г.)
[1] Обратная задача для интеграла типа Коши. Diff. Uravn. **13** (1977), 174–176 (Engl. transl.: Diff. Eq. **13** (1977), 121–123).

CERNECKIĬ, V. A., and V. D. DIDENKO (Цернецкий, В. А., и В. Д. Диденко)
[1] Краевая задача Римана с комплексной ортогональной матрицей. Mat. Zametki **23** (1978), 405–416.

CHAKRABARTI, A., and S. CHAKRABARTI
[1] A note on certain singular integral equations with kernels $K(z, \xi)/(z - \xi)$. Indian J. pure appl. Math. **8** (1977), 327–334.

CHAZY, J.
[1] Sue les équations différentielles du troisiéme ordre et d'ordre superior dont l'intégrale générale a des points critiques fixes. Acta Math. **34** (1911), 317–385.

CHERN, S. S.
[1] Complex analytic mappings of Riemann surfaces. I. Amer. J. Math. **82** (1960), 323–337.
[2] Complex manifolds without potential theory. Van Nostrand, Princeton (N.J.) 1967.

CHERN, S. S., H. I. LEVINE, and L. NIRENBERG
[1] Intrinsic norms on a complex manifold. *In*: SPENCER, D. C., and S. IYNAGA (eds.): Global analysis (papers in honour of KODAIRA), Univ. of Tokyo Press and Princeton Univ. Press, Tokyo 1969, 119–139; *reprinted in*: CHERN, S. S.: Selected papers, Springer Berlin, Heidelberg, New York 1978, 371–391.

CHERN, S. S., and J. K. MOSER
[1] Real hypersurfaces in complex manifolds. Acta Math. **133** (1974), 219–271.

CHRISTIANSEN, S.
[1] A review of some integral equations for solving the Saint-Venant torsion problem. The Danish Center of Applied Mathematics and Mechanics. The Technical University of Denmark, Report (1976) 109.

CHRISTIANSEN, S., and H. RASMUSSEN
[1] Numerical solution for two-dimensional annular electrochemical machining problems. The Technical University of Denmark, Report (1973) 53.

CIARLET, P. G.
[1] The Finite Element Method for Elliptic Problems. North Holland, Amsterdam 1978.

CIARLET, P. G., and P. A. RAVIART
[1] The combined effect of curved boundaries and numerical integration in isoparametric finite element methods. *In*: AZIZ, K. (ed.): The Mathematical Foundation of the Finite Element Method with Applications to Partial Differential Equations. Academic Press, New York 1972, 409–474.

CIECHANOWICZ-HALKA, B.
[1] Ein nichtlineares Randproblem der flachen Elastizitätstheorie. Demonstratio Math. **11** (1978), 583–590.

Čibrikova, L. I. (Чибрикова, Л. И.)
[1] Граничные задачи теории аналитических функций на римановых поверхностях. Itogi Nauki Tehn., Ser. Mat. Anal. **18** (1980), 3–66 (Engl. transl.: J. Sov. Math. **18** (1980), 3–66).
Čibrikova, L. I., and V. V. Sil'vestrov (Чибрикова, Л. И., и В. В. Сильвестров)
[1] К вопросу об эффективности решения краевой задачи Римана для автоморфных функций. Izv. vysš. učebn. Zaved., Mat. **12(199)** (1978), 117–121.
Clements, D. L., and C. Rogers
[1] On the Bergman Operator method and anti-plane contact problems involving an inhomogeneous half-space. SIAM J. Appl. Math. **34** (1978), 764–773.
Clunie, J.
[1] On integral and meromorphic functions. J. London Math. Soc. **37** (1962), 17–27.
[2] On a result of Hayman. J. London Math. Soc. **47** (1967), 389–392.
Collingwood, E. F., and A. J. Lohwater
[1] The theory of cluster sets. Cambridge University Press, Cambridge 1966.
Colton, D. L.
[1] On the inverse scattering problem for axially symmetric solutions of the Helmholtz equation. Quart. J. Math. **22** (1971), 125–130.
[2] Bergman operators for elliptic equations in four independent variables. SIAM J. Math. Anal. **3** (1972), 401–412.
[3] On the analytic theory of pseudoparabolic equations. Quart. J. Math. **23** (1972), 179–192.
[4] Integral operators for elliptic equations in three independent variables I, II. Appl. Anal. **4** (1974), 77–95; 283–295.
[5] Solution of boundary value problems by the method of integral operators. Pitman, London, San Francisco, Melbourne 1975.
[6] Complete families of solutions for parabolic equations with analytic coefficients. SIAM J. Math. Anal. **6** (1975), 937–947.
[7] Partial differential equations in the complex domain. Pitman, London, San Francisco, Melbourne 1976.
[8] Integral operators and inverse problems in scattering theory. *In*: Meister, V. E., N. Weck, and W. L. Wendland (eds.) [1, 17–28].
[9] The approximation of solutions to initial boundary value problems for parabolic equations in one space variable. Quart. Appl. Math. **33** (1976), 377–386.
[10] The solution of initial boundary value problems for parabolic equations by the method of integral operators. J. Diff. Eq. **26** (1977), 181–190.
[11] A reflection principle for solutions to the Helmholtz equation and an application to the inverse scattering problem. Glasgow Math. J. **18** (1977), 125–130.
[12] The solution of problems in radiowave propagation by the method of parabolic equations and transformation operators. Appl. Anal. **8** (1978), 11–22.
[13] Analytic Theory of Partial Differential Equations. Pitman, London 1980.
[14] Schwarz reflection principles for solutions of parabolic equations. Proc. Amer. Math. Soc. **82** (1981), 87–94.
Colton, D. L., and A. Kirsch
[1] Stable methods for solving the inverse scattering problem for a cylinder. Proc. Royal Soc. Edinburgh **89A** (1981), 181–188.
Colton, D. L., and R. E. Kleinman
[1] The direct and inverse scattering problems for an arbitrary cylinder. Dirichlet boundary conditions. Proc. Royal Soc. Edinburgh **86A** (1980), 29–42.
Colton, D. L., and W. Wendland
[1] Constructive methods for solving the exterior Neumann problem for the reduced wave equation in a spherically symmetric medium. Proc. Royal Soc. Edinburgh **75A** (1975/76), 99–106.
Courant, R., and D. Hilbert
[1] Methods of Mathematical Physics, vol. II. Interscience Publishers, New York 1961.
Cousin, P.
[1] Sur les fonctions de n variables complexes. Acta Math. **19** (1895), 1–62.
Čoriev, H. (Чориев, Х.)
[1] Регулярная устойчивость граничной задачи Риманна и вычисление частичных индекксв. Taškent. Gos. Univ. Naučn. Tr. **548** (1977), 120–123.

COROI-NEDELCU, M.
[1] Equations aux dérivées spatiales. Revue Roumaine Math. pur. appl. **13** (1968), 1285-1291.
CUPPENS, R.
[1] Decomposition of multivariate probabilities. New York 1975.
[2] Recent results on the decomposition of multivariate probabilities. *In*: PARUCKI, R. (ed.): Multivariate analysis V, Proc. 5th Int. Symp., Pittsburgh 1978, 273–286.
[3] The arithmetic of distribution functions. *In*: LAHA, R. G., and V. K. ROHATGI (eds.): Contributions to probability. Academic Press, New York 1981, 101–114.
CZERNER, P.
[1] Komplexe Methoden in der globalen Auflösungstheorie für algebraische Abbildungen der Form $G(z, \bar{z}, w) = 0$. Beitr. Anal. **10** (1977), 79–89.
[2] Eine zweistufige Reduktionstheorie für algebraische Gleichungen der Form $G(z, \bar{z}, w, \bar{w}) = 0$. Math. Nachr. **87** (1979), 109–119.
DANG, D. Q., and D. H. NORRIE
[1] A finite element method for the solution of a singular integral equations. Comp. Math. Appl. **4** (1978), 219–224.
DANILJUK, I. I. (Данилюк, И. И.)
[1] О задаче Гильберта с измерыми коэффициентами. Sib. Mat. Ž. **1** (1960), 171 to 197.
DARMOIS, G.
[1] Analyse générale des liaisons stochastique. Revue Intern. Statist. Inst. **21** (1953), 2–8.
DEAUX, R.
[1] Introduction to the geometry of complex numbers. F. Ungar, New York 1955 (transl. from French).
DECEWICZ, G.
[1] Non-continuous and nonlinear Hilbert-Haseman problem for a system of arbitrary power. Demonstratio Math. **11** (1978), 613–633.
DELANGHE, R.
[1] On regular-analytic functions with values in a Clifford algebra. Math. Ann. **185** (1970), 91–111.
[2] On the singularities of functions with values in a Clifford algebra. Math. Ann. **196** (1972), 293–319.
[3] On regular functions with values in topological modules over a Clifford algebra. Bull. Soc. Math. Belgique **25** (1973), 131–138.
DELANGHE, R., and F. BRACKX
[1] Hypercomplex function theory and Hilbert modules with reproducing kernels. Proc. London Math. Soc., Ser. 3, **37** (1978), 545–576.
DENTONI, P., and G. P. RIZZA
[1] Una nuova classe di funzioni in un'algebra reale. Rend. Ist. Mat. Univ. Trieste **4** (1972), 171–181
DENTONI, P., and M. SCE
[1] Funzioni regolari nell'algebra di Cayley. Rend. Sem. Mat. Padova **50** (1973), 251–267.
DEWESS, M.
[1] A note on the asymptotic behaviour of an entire characteristic function of a multivariate probability law. J. Multivar. Analysis **7** (1977), 608–613.
[2] The tail behaviour of a distribution function. Math. Nachr. **81** (1978), 217–231.
[3] On applications of dominated variation to queuing theory. Math. Operationsforsch. Statist., Ser. Optim. **9** (1978), 601–609.
DEWESS, M., and M. RIEDEL
[1] The connection between the proximate order of an entire characteristic function and the corresponding distribution function. Czechoslovak. Math. J. **27** (**102**) (1977), 173–185.
DIDENKO, V. D., and N. JA. TIHONENKO (Диденко, В. Д., и Н. Я. Тихоненко)
[1] О приближенном решении матричной задачи Римана. Izv. vyss. učebn. Zaved., Mat. **212** (1980) 1, 16–19.
DIKANSKIĬ, A. S. (Диканский, А. С.)
[1] Сопряженные краевые задачи к эллиптическим дифференциальным и псевдодифференциальным задачам в ограниченной области. Mat. Sb. **91** (1973), 62–77.

DINGHAS, A.
[1] Vorlesungen über Funktionentheorie, Springer, Berlin, Göttingen, Heidelberg 1961.
[2] Über den Picard-Borelschen Satz und die Nevanlinnaschen Ungleichungen. J. Math. Pur. Appl. **42** (1963) 9, 223–251.
[3] Über das gegenseitige Verhältnis der Methoden von F. Nevanlinna und Ahlfors in der Wertverteilungstheorie. Revue Roumaine Math. pur. appl. **18** (1973), 1555–1569.
[4] Wertverteilung meromorpher Funktionen in ein- und mehrfach zusammenhängenden Gebieten. Lecture Notes in Mathematics, vol. 783. Springer, Berlin, Heidelberg, New York 1980.
DÖRRIE, H.
[1] Einführung in die Funktionentheorie. Oldenbourg, München 1951.
DOUGLIS, A.
[1] A function-theoretic approach to elliptic systems of equations in two variables. Comm. Pure Appl. Math. **6** (1953), 259–289.
DUFRESNOY, J.
[1] Théorie nouvelle de familles complexes normales; applications à l'étude des fonctions algébroides. Ann. Sci. École Norm. Sup. **61** (1944) 3, 1–44.
DUGUÉ, D.
[1] Le défaut au sens de M. Nevanlinna dépend de l'origine choisie. C.R. Acad. Sci. Paris **225** (1947), 555–556.
DRASIN, D.
[1] The inverse problem in the Nevanlinna theory. Acta Math. **138** (1977), 83–151.
DYBIN, V. B., and V. N. GAPONENKO (Дыбин, В. Б., и В. Н. Гапоненко)
[1] Краевая задача Римана на прямой с бесконечным числом особенностей. Sib. Mat. Ž. **17** (1976), 705–706.
DYM, H., and H. P. MCKEAN
[1] Gaussian processes, function theory and the inverse spectral problem. Academic Press, New York, San Francisco, London 1976.
DŽRBAŠJAN, M. M. (Джрбашян, М. М.)
[1] Интегральные преобразования и представления функций в комплексной области. Nauka, Moskva 1966.
[2] Теория факторизации функций, мероморфных в круге. Mat. Sb., n. Ser. **79(121)** (1969) 4, 517–615 (Engl. transl.: Math. USSR Sb. **8** (1969), 493–522).
DŽRBAŠJAN, M. M., and V. S. ZAHARJAN (Джрбашян, М. М., и В. С. Захарян)
[1] Граничные свойства подклассов мероморфных функций ограниченного вида. Izv. Akad. Nauk SSSR, Ser. Mat. **34** (1970), 1262–1339; **35** (1971), 727.
DŽURAEV, A. (Джураев, А.)
[1] Системы уравнений составного типа. Nauka, Moskva 1972.
EDENHOFER, J.
[1] Analytische Funktionen auf Algebren. Dissertation, Technische Universität München 1973.
EFENDIEV, M. A. (Эфендиев, М. А.)
[1] Нелинейная задача Гильберта для мероморфных функций в круге. Izv. Akad. Nauk Azerbaidž. SSR, Ser. fiz.-tehn. mat. Nauk **5** (1976), 53–58.
[2] Нелинейная задача Гильберта в кольце. Dokl. Akad. Nauk Azerbaidž. SSR **32** (1976) 11, 3–5 (1977).
EICHHORN, W.
[1] Funktionentheorie in Algebren über dem reellen Zahlkörper und ihre Anwendung auf partielle Differentialgleichungen. Dissertation, Universität Würzburg 1961.
[2] Funktionalgleichungen in reellen Vektorräumen und verallgemeinerte Cauchy-Riemannsche Differentialgleichungen, speziell die Weylsche Gleichung des Neutrinos. Aequationes Math. **5** (1970), 255–267.
EISENSTAT, S. C.
[1] On the rate of convergence of the Bergman-Vekua metohd for the numerical solution of elliptic boundary value problems. SIAM J. Math. Anal. **11** (1974), 654–680.
EVGRAFOV, M. A. (Евграфов, М. А.)
[1] Асимптотические оценки и целые функции. Moskva 1979.
FABER, K.
[1] Konforme Abbildungen der hyperbolisch-euklidischen Ebene durch hyperkomplexe Funktionen. Math. Z. **84** (1964), 254–262.

FEDOROV, V. S. (Федоров, В. С.)
[1] О коэффициентах Фурье. Mat. Sb. **8(50)** (1940), 41–56.
[2] Моногенность. Mat. Sb. **18(60)** (1946), 353–378.

FELLER, W.
[1] Introduction to probability theory. 2nd ed. Wiley-Chapman, New York 1971.

FICHERA, G.
[1] Linear elliptic equations of higher order in two independent variables and singular equations. *In*: LANGER, R. (ed.): Part. Diff. Eq. and Mech. Conf. Univ. Wisconsin Press, 1961, 55–80.

FICHERA, G., and P. RICCI
[1] The single layer potential approach in the theory of boundary value problems for elliptic equations. *In*: MEISTER, V. E., N. WECK, and W. L. WENDLAND (eds.) [1, 39–50].

FIX, G. J., and G. STRANG
[1] An Analysis of the Finite Element Method. Prentice Hall, Englewood Cliffs (N.J.) 1973.

FIX, G. J., M. D. GUNZBURGER, and R. A. NICOLAIDES
[1] On infite element methods of the least squares type. Comp. Math. Appl. **5** (1979), 87–98.

FLORIAN, H.
[1] Normale Integraloperatoren. Monatsh. Math. **69** (1965), 18–29.

FLORIAN, H., and R. HEERSINK
[1] Ein verallgemeinertes Riemann-Hilbertsches Randwertproblem. Appl. Anal. **8** (1979), 277–284.

FLÜGGE, W.
[1] Statik und Dynamik der Schalen. Springer, Berlin, Göttingen, Heidelberg 1957.

FORSTER, O.
[1] Einführung in die komplexe Analysis mehrerer komplexer Veränderlicher. Vorlesungsskripte, hrsg. vom Fachbereich Mathematik der Universität Regensburg 1973.
[2] Riemannsche Flächen. Springer, Berlin, Heidelberg, New York 1977.

FRANK, G.
[1] Über ganze Funktionen, die einer linearen Differentialgleichung genügen. Manuscripta Math. **4** (1971), 215–253.
[2] Zur lokalen Werteverteilung der Lösungen linearen Differentialgleichungen. Manuscripta Math. **6** (1972), 381–404.
[4] Eine Vermutung von Hayman über Nullstellen meromonpher Funktionen. Math. Z. **149** (1976), 29–36.

FRANK, G., and W. HENNEKEMPER
[1] Einige Ergebnisse über die Werteverteilung meromorpher Funktionen. Resultate Math. **4** (1981), 39–54.

FRANK, G., W. HENNEKEMPER, and G. POLLOCZEK
[1] Nullstellen meromorpher Funktionen und deren Ableitungen. Math. Ann. **225** (1977), 145–154.

FRANK, G., and E. MUES
[1] Über den Index der Lösungen linearer Differentialgleichungen. Manuscripta Math. **5** (1971), 155–163.
[2] Differentialpolynome. Vortragsreihe, Oberwolfach 1979.

FRANK, G., and H. WITTICH
[1] Zur Theorie linearer Differentialgleichungen im Komplexen. Math. Z. **130** (1973), 363–370.

FRYNTOV, A. E., and G. P. ČISTJAKOV (Фрынтов, А. Е., и Г. П. Чистяков)
[1] О безгранично делимых распределениях класса I_0. Sb. Teor. Funkc., Funkc. Anal. Prilož., Kiev **32** (1979) 5, 91–97.

FUCHS, L.
[1] Über Differentialgleichungen, deren Integrale feste Verzweigungspunkte besitzen. Sitzungsber. Akad. Berlin (1884), 699–720.
[2] Gesammelte Werke. Bd. I, II, III, Berlin 1904.

FUCHS, W. H. J.
[1] Developments in the classical Nevanlinna theory of meromorphic funktions. Bull. Amer. Math. Soc. **73** (1967), 275—291.
[2] The development of the theory of deficient values. Ann. Acad. Sci. Fenn. AI **7** (1982).

FUETER, R.
[1] Funktionentheorie im Hyperkomplexen. Vorlesungsausarbeitung, Math. Inst. Univ. Zürich 1949.
[2] Analytische Funktionen einer Quaternionenvariablen. Comment. Math. Helv. **4** (1932), 9—20.
[3] Reguläre Funktionen einer Quaternionenvariablen. Vorlesungsausarbeitung, Math. Inst. Univ. Zürich 1940.

FUKS, B. A. (Фукс, Б. А.)
[1] Введение в теорию аналитических функций многих комплексных переменных. Fizmatgiz, Moskva 1962.
[2] Специальные главы теории аналитических функций многих комплексных переменных. Fizmatgiz, Moskva 1962.

FUKUSHIMA, M.
[1] Dirichlet forms and Markov processes. North-Holland, Amsterdam, Oxford, New York/Kodansha Ltd., Tokyo 1980.

GABDULHAEV, B. G. (Габдулхаев, Б. Г.)
[1] Об одном прямом методе решения интегралных уравнений. Izv. vysš. učebn. Zaved., Mat. **3** (1965), 51—60.
[2] Прямые методы решения некоторых операторных уравнений, I. Izv. vysš. učebn. Zaved., Mat. **11** (1971), 33—44.

GABDULHAEV, B. G., and N. N. DUŠKOV (Габдулхаев, Б. Г., и Н. Н. Душков)
[1] Метод механических квадратур для сингулярных интегральных уравнений. Izv. vysš. učebn. Zaved., Mat. **12** (1974), 3—14.

GABRINOVIČ, V. A. (Габринович, В. А.)
[1] Краевая задача типа Карлемана для полианалитических функций. Izv. Akad. Nauk BSSR, Ser. fiz.-mat. Nauk **3** (1977), 48—57.
[2] О краевой задаче типа Карлемана для метааналитических функций. Dokl. Akad. Nauk BSSR **21** (1977), 112—115.

GACKSTATTER, F.
[1] Gesamtdefekt und Krümmungsmittelwerte bei meromorphen Funktionen. Sitzungsber. Bayer. Akad. Wiss., Math. Nat. Kl. (1970), 79—102.
[2] Die Gaußsche und mittlere Krümmung der Realteilflächen in der Theorie der meromorphen Funktionen. Math. Nachr. **54** (1972), 211—227.

GACKSTATTER, F., and I. LAINE
[1] Zur Theorie der gewöhnlichen Differentialgleichungen im Komplexen. Ann. Polon. Math. **38** (1980), 259—287.

GAHOV, F. D. (Гахов, Ф. Д.)
[1] О краевой задаче Римана. Mat. Sb. **2(44)** (1937), 673—683.
[2] Краевая задача Римана для системы n пар функций. Usp. Mat. Nauk **7** (1952), 3—54.
[3] Краевые задачи. Nauka, Moskva 1963 (Engl. transl.: Pergamon Press, Oxford 1966).

GAIER, D.
[1] Integralgleichungen erster Art und konforme Abbildung. Math. Z. **147** (1976), 113 to 129.

GALAMBOS, J., and S. KOTZ
[1] Characterizations of probability distributions. Lecture Notes in Mathematics, vol. 675. Springer, Berlin, Heidelberg, New York 1978.

GALIN, L. A. (Галин, Л. А.)
[1] Контактные задачи теории упругости и вязкоупругости. Nauka, Moskva 1980.

GAMBIER, R.
[1] Sur les équations différentielles du second ordre et du premier degré dont l'intégrale générale est à points critiques fixes. Acta Math. **33** (1910), 1—55.

GARABEDIAN, P. R.
[1] Partial differential equations with more than two independent variables in the complex domain. J. Math. Mech. **9** (1969), 241–271.
GAVDZINSKIĬ, V. V., and A. P. NEČAEV (Гавдзинский, В. В., и А. П. Нечаев)
[1] Краевая задача для пары функций аналитических в несовпадающих областях. Ukrain. Mat. Ž. **31** (1979), 658–665.
GAVEAU, B., and J. ŁAWRYNOWICZ
[1] Espaces de Dirichlet invariant biholomorphes et capacités associées. Bull. Acad. Polon. Sci., Sér. Sci. Math. **30** (1982), 63–69.
[2] Intégrale de Dirichlet sur une variété complexe. Séminaire Pierre Lelong — Henri Skoda (Analyse), Années 1980/81 Lecture Notes in Mathematics, vol. 919. Springer, Berlin, Heidelberg, New York 1982, 131–166.
GAVRILOV, S. K. (Гаврилов, С.К.)
[1] Задача типа Калемана с разрывным сдвигом на римановой поверхности с краем. Mat. Anal. Prilož. ,Rostov. Univ. **6** (1974), 182–192.
[2] Смешанная краевая задача с разрывным сдвигом. Izv. vysš. učebn. Zaved., Mat. **161** (1975) 10, 86–88.
[3] Об одной смешанной краевой задаче на римановой поверхности с краем. Izv. vysš. učebn. Zaved., Mat. **176** (1977) 1, 31–42.
GEHRING, F. W.
[1] The asymptotic values for analytic functions with bounded characteristic. Quart. J. Math. **9** (1958) 2, 282–289.
GILBERT, R. P.
[1] Function theoretic methods in partial differential equations. Academic Press, New York, London 1969.
[2] Constructive Methods for Elliptic Equations. Lecture Notes in Mathematics, vol. 365. Springer, Berlin, Heidelberg, New York 1974.
[3] Nonlinear boundary value problems for elliptic systems in the plane. Proc. Intern. Conf. Nonlinear Systems Applicat. Techn. Rep. **63** (1976), Univ. of Delaware.
[4] Verallgemeinerte hyperanalytische Funktionentheorie. *In*: TUTSCHKE, W. (ed.): Komplexe Analysis und ihre Anwendungen auf partielle Differentialgleichungen, vol. 2. Martin-Luther-Univ. Halle-Wittenberg 1980, 124–145.
GILBERT, R. P., and C. N. HILE
[1] Generalized Hypercomplex Function Theory. Trans. Amer. Math. Soc. **195** (1974), 1–29.
[2] Hypercomplex function theory in the sense of L. Bers. Math. Nachr. **72** (1976), 187–200.
[3] Hilbert function modules with reproducing kernels. Nonlinear Anal. **1** (1977), 135 to 150.
GILBERT, R. P., and G. C. HSIAO
[1] Constructive function theoretic methods for higher order pseudoparabolic equations. *In*: MEISTER, V. E., N. WECK, and W. L. WENDLAND (eds.) [1, 51–67].
GILBERT, R. P., and D. K. KUKRAL
[1] A function theoretic method for $\Delta_4^2 u + Q(x)\, u = 0$. Ann. Math. Pura Appl. CIV (1975), 31–42.
GILBERT, R. P., and P. LINZ
[1] The numerical solution of some elliptic boundary value problems by integral operator methods. *In*: COLTON, D. L., and R. P. GILBERT (eds.): Constructive and Computational Methods for Differential and Integral Equations. Lecture Notes in Mathematics, vol. 430. Springer, Berlin, Heidelberg, New York 1974, 237–252.
GILBERT, R. P., and G. F. ROACH
[1] Constructive methods for meta- and pseudoparabolic systems. Bull. Math. Soc. Sci. Math. RSR, NS **20** (68) (1976), 97–109.
GILBERT, R. P., and M. SCHNEIDER
[1] Generalized meta- and pseudoparabolic equations in the plane. *In*: Sb. Kompleksnyĭ analiz i ego priloženija. Nauka, Moskva 1978, 160–172.
GILBERT, R. P., and W. WENDLAND
[1] Analytic, generalized hyperanalytic function theory and an application to elasticity. Proc. Royal Soc. Edinburgh **73A** (1975), 317–331.

GIRAULT, V.
[1] A mixed finite element method for the stationary Stokes equations. SIAM J. Num. Anal. **15** (1978), 534—555.
GIRAULT, V., and P. A. RAVIART
[1] An analysis of a mixed finite element method for the Navier-Stokes equations. Num. Math. **33** (1979), 235—271.
GLOWACKI, Z. ST.
[1] A generalized compound Hilbert-Riemann problem for a system of functions. Demonstratio Math. **10** (1977), 523—538.
GNEDENKO, B. V.
[1] Lehrbuch der Wahrscheinlichkeitsrechnung. 7th ed. Akademie-Verlag, Berlin 1979 (transl. from Russian).
GNEDENKO, B. V., and A. N. KOLMOGOROV
[1] Grenzwertverteilungen von Summen unabhängiger Zufallsgrößen. 2nd ed. Akademie-Verlag, Berlin 1960 (transl. from Russian).
GOEBEL, M., and L. v. WOLFERSDORF
[1] Optimale Steuerprobleme bei Noetherschen Operatorgleichungen II, III. Math. Nachr. **75** (1976), 91—99; **82** (1978), 77—85.
GOGONEA, S.
[1] Sur quelques problemes aux limites de type Hilbert pour le plan muni de coupures rectilignes alignees. Bull. Math. Soc. Sci. Math. RSR, NS **20** (68) (1976), 111—124.
GOHBERG, I. C. (Гохберг, И. Ц.)
[1] О числе решений однородного сингулярного интегрального уравнения с непрерывными коэффициентами. Dokl. Akad. Nauk SSSR **122** (1958), 327—330.
GOHBERG, I. C. (GOCHBERG, I. Z.), and I. A. FELDMAN
[1] Faltungsgleichungen und Projektionsverfahren zu ihrer Lösung. Akademie-Verlag, Berlin 1974 (transl. from Russian).
GOHBERG, I. C., and M. G. KREĬN (Гохберг, И. Ц., и М. Г. Крейн)
[1] Системы интегральных уравнений на полупрямой с ядрами, зависящими от разности аргументов. Usp. Mat. Nauk **13** (1958) 2, 3—72.
GOHBERG (GOCHBERG), I., and N. KRUPNIK
[1] Einführung in die Theorie der eindimensionalen singulären Integraloperatoren. Birkhäuser, Basel, Boston, Stuttgart 1979.
GOL'DBERG, A. A. (Гольдберг, А. А.)
[1] О дефектах мероморфных функций. Dokl. Akad. Nauk SSSR **98** (1954), 893 to 895.
[2] О однозначных решениях дифференциальных уравнений первого порядка. Ukrain. Mat. Ž. **8**:**3** (1956), 254—261.
[3] Мероморфные функции. Itogi Nauki, Ser. Mat. Anal. **10** (1973), 5—97.
[4] О росте мероморфных решений дифференциальных уравнений второго порядка. Diff. Uravn. **14** (1978), 824—829 (Engl. transl.: Diff. Eq. **14** (1978), 584—588).
[5] К вопросу о связи между дефектом и уклонением мероморфней функции. Sb. Teor. Funkc., Funkc. Anal. Prilož. **29** (1978), 31—35.
GOL'DBERG, A. A., and I. V. OSTROVSKIĬ (Гольдберг, А. А., и И. В. Островский)
[1] Распределение значений мероморфных функций. Nauka, Moskva 1970.
[2] О росте целых хребтовых функций с действительными нулями. Mat. fiz. i funkc. analiz., vyp. U, Har'kov.
GOL'DENVEĬZER, A. L. (Гольденвейзер, А. Л.)
[1] Теория упругих тонких оболочек. Nauka, Moskva 1979.
GOLDSCHMIDT, B.
[1] Lösung des 1. Cousinschen Problems für Vekuasche Differentialgleichung. Math. Nachr. **67** (1975), 237—246.
[2] Funktionentheoretische Eigenschaften der Lösungen der Vekuaschen Differentialgleichung mit Koeffizienten in $L^{loc}(G)$. Math. Nachr. **84** (1978), 229—246.
[3] Funktionentheoretische Eigenschaften verallgemeinerter analytischer Vektoren. Math. Nachr. **90** (1979), 59—90.
[4] Verallgemeinerte analytische Vektoren im R^n. Habilitationsschrift, Martin-Luther-Univ. Halle-Wittenberg 1980.
[5] A Cauchy integral formula for a class of elliptic systems of partial differential equations of first order in the space. Math. Nachr. **108** (1982), 167—178.

[6] Existence and representation of solutions of a class of elliptic systems of partial differential equations of first order in the space. Math. Nachr. **108** (1982), 159–166.

Golubev, V. V. (Golubew, W. W.)

[1] Vorlesungen über Differentialgleichungen im Komplexen. VEB Deutscher Verlag der Wissenschaften, Berlin 1958 (transl. from Russian).

Gorodža, L. V., Ju. P. Emec, M. F. Žukova, and È. I. Zverovič (Городжа, Л. В, Ю. П. Емец, М. Ф. Жукова, и Э. И. Зверович)

[1] О применений обобщенной краевой задачи Римана к расчету электрических полей. Dokl. Akad. Nauk BSSR **23** (1979), 118–120.

Grauert, H., and G. Müllich

[1] Vektorbündel vom Rang 2 über dem n-dimensionalen komplex projektiven Raum. Manuscripta Math. **16** (1975), 75–100.

Grauert, H., and H. Reckziegel

[1] Hermitesche Metriken und normale Familien holomorpher Abbildungen. Math. Z. **89** (1965), 108–125.

Grauert, H., and R. Remmert

[1] Theory of Stein Spaces. Springer, Berlin, Heidelberg, New York 1979.

Greenwood, P.

[1] Wiener-Hopf methods. Adv. Appl. Prob. **7** (1975), 767–785.

Greenwood, P., and M. Shaked

[1] Fluctuation of random walk in R^d and storage systems. Adv. Appl. Prob. **9** (1977), 566–587.

Griffiths, Ph., and J. King

[1] Nevanlinna theory and holomorphic mappings between algebraic varieties. Acta Math. **130** (1973), 145–220.

Grišin, A. F. (Гришин, А. Ф.)

[1] О сравнении дефектов $\delta_p(a)$. Teor. Funkc., funkc. Anal. Prilož. **25** (1976), 56–66.

Grötzsch, H.

[1] Über die Verzerrung bei schlichten nichtkonformen Abbildungen und eine damit zusammenhängende Erweiterung des Picardschen Satzes. Ber. Verh. Sächs. Akad. Wiss. Leipzig **80** (1928), 503–507.

Gromak, V. I.

[1] One-parameter systems of solutions of Painlevé's equation. Diff. Uravn. **14** (1978), 2131–2135.

Grothendieck, A.

[1] Sur la classification des fibrés holomorphes sur la sphere de Riemann. Amer. J. Math. **79** (1957), 121–138.

Grudskiĭ, S. M., and V. B. Dybin (Грудский, С. М., и В. Б. Дыбин)

[1] Краевая задача Римана с разрывами почти-периодического типа у ее коэффициента. Dokl. Akad. Nauk SSSR **237** (1977), 21–24.

Gürsey, F., and H. C. Tze

[1] Complex and quaternionic analyticity in chiral and gauge theories. I. Ann. Phys. **128** (1980), 29–130.

Gunning, R. C., and H. Rossi

[1] Analytic functions of several complex variables. Prentice-Hall, Englewood Cliffs (N.J.) 1965.

Guz, E.

[1] On generalized Hilbert's problem containing constant strong singularity. Politechn. Łódz, Zeszyty nauk. **245** (1977), 23–29.

[2] On certain boundary problem of Riemann type containing the constant singularity. Politechn. Łódz, Zeszyty nauk. **301** (1977), 23–32.

Gyires, B. (ed.)

[1] Analytic function methods in probability theory. Colloquium mathematica societatis János Bolyai, vol. 21. North Holland, Amsterdam, New York 1980.

Haack, W.

[1] Allgemeine Randwertprobleme für Differentialgleichungen vom elliptischen Typus. Math. Nachr. **7** (1952), 1–30.

Haack, W., and W. Wendland

[1] Vorlesungen über partielle und Pfaffsche Differentialgleichungen. Birkhäuser, Basel 1969 (Engl. transl.: Pergamon Press, Oxford 1971, London 1972).

HABETHA, K.

[1] Über die Werteverteilung in Winkelräumen. Math. Z. **77** (1961), 453–467.

[2] Über die Werteverteilung pseudoanalytischer Funktionen. Ann. Acad. Sci. Fenn. A **406** (1967), 20 pp.

[3] Über lineare elliptische Differentialgleichungssysteme mit analytischen Koeffizienten. Ber. Ges. Math. Datenverarb. Bonn **77** (1973), 68–89.

[4] On Zeros of elliptic systems of first order in the plane. *In*: GILBERT, R. P., and R. J. WEINACHT (eds.): Function Theoretic Methods in Differential Equations. Research Notes 8. Pitman, London 1976, 45–62.

[5] Eine Bemerkung zur Funktionentheorie in Algebren. *In*: MEISTER, V. E., N. WECK, and W. L. WENDLAND (eds.) [1, 502–509].

HAĆ, A., and R. SŁYSZ

[1] A compound problem of Hilbert-Riemann type for an infinite system of analytic functions. Demonstratio Math. **11** (1978), 105–118.

[2] A non-linear Hilbert-Riemann problem with boundary condition of the class L_p^α Demonstratio Math. **12** (1979), 359–372.

HADAMARD, J.

[1] Essai sur l'étude des fonctions données par leur développement de Taylor. Thèse de Doctorat. J. Math. Pur. Appl. **8** (1892) 4, 101–186.

[2] Etude sur les propriétés des fonctions entières et en particulier d'une fonction considerée par Riemann. J. Math. Pur. Appl. **9** (1893) 4, 171–215.

HAEFELI, H. G.

[1] Hyperkomplexe Differentiale. Comment. Math. Helv. **20** (1947), 382–420.

[2] Konformitätsbedingungen bei vierdimensionalen Abbildungen. Ann. Acad. Sci. Fenn. A I **250** (1958) 11, 8 pp.

AF HÄLLSTRÖM, G.

[1] Über meromorphe Funktionen mit mehrfach zusammenhängenden Existenzgebieten. Dissertation, Univ. Helsinki 1939; Acta Acad. Aboensis, Math. Phys. **12** (1939) 8, 100 pp.

[2] Wertverteilungssätze pseudomeromorpher Funktionen. Acta Acad. Aboensis, Math. Phys. **21** (1958) 9, 23 pp.

HANSEN, E. B.

[1] Numerical solution of integro-differential and singular integral equations for plate bending problems. J. Elasticity **6** (1976), 39–56.

HASEMAN, C.

[1] Anwendung der Theorie der Integralgleichung auf einige Randwertaufgaben der Funktionentheorie. Dissertation, Göttingen 1907; Math. Ann. **66** (1909), 258–272.

HAUSDORFF, F.

[1] Zur Theorie der Systeme komplexer Zahlen. Ber. Verh. Königl. Sächs. Gesellsch. Wiss. Leipzig, math.-phys. Kl. **52** (1900), 43–61.

HAYMAN, W. K.

[1] An integral function with a defective value that is neither asymptotic nor invariant under change of origin. J. Math. Soc. London **28** (1953), 369–376.

[2] Picard values of meromorphic functions and derivatives. Ann. Math. **70** (1959) 2, 9–42.

[3] Meromorphic functions. Oxford University Press, Oxford 1964.

[4] On the Valiron deficiencies of integral functions of infinite order. Ark. Mat. **10** (1972), 163–172.

[5] Some achievements of Nevanlinna theory. Ann. Acad. Sci. Fenn. AI **7** (1982) 1, 65–71.

HEERSINK, R.

[1] Partial Differential Equations with Complex Variables. Matem. Balkanica **4.42** (1974), 245–250.

[2] Über Lösungsdarstellungen und funktionentheoretische Methoden bei elliptischen Differentialgleichungen. Math.-stat. Sekt. Forschungszentrum Graz, Ber. **67** (1976).

[3] Characterization of Certain Differential Operators in the Solution of Linear Partial Differential Equations. Glasgow Math. J. **17** (1976), 83–88.

[4] Über einige Verallgemeinerungen eines elliptischen Randwertproblems von Hadamard. Acta Phys. Austriaca **52** (1980), 277–291.

HEIN, J. C.
[1] Eine Integralgleichungsmethode zur Lösung eines gemischten Randwertproblems der beliebig belasteten Kreiszylinderschale mit Ausschnitten glatter Berandung. Dissertation, Techn. Hochsch. Darmstadt 1979.

HEINZ, W.
[1] Ein Verfahren zur Bestimmung der isolierten Nullstellen von Polynomen in z und $\bar{z}$. Math. Nachr. **65** (1975), 219–222.

HELLWIG, G.
[1] Das Randwertproblem eines linearen elliptischen Systems. Math. Z. **56**, (1952), 388–408.
[2] Partial Differential Equations. Blaisdell Publ., New York 1964.

HEROLD, H.
[1] Differentialgleichungen im Komplexen. Studia Math. 2. Vandenhoeck und Ruprecht, Göttingen 1975.

HESTENES, D.
[1] Space-time algebra. Gordon and Breach, New York, London, Paris 1966.
[2] Vectors, spinors and complex numbers in classical and quantum physics. Amer. J. Phys. **39** (1971), 1013–1027.

HILBERT, D.
[1] Über eine Anwendung der Integralgleichungen auf ein Problem der Funktionentheorie. Verh. III. Int. Math. Kongr., Heidelberg 1904.
[2] Grundzüge einer allgemeinen Theorie der linearen Integralgleichungen. Teubner, Leipzig 1912, 2nd ed. 1924.

HILE, G.
[1] Hypercomplex function theory applied to partial differential equations. Ph. D. dissertation, Indiana Univ., Bloomington 1972.
[2] Representations of solutions of a special class of first-order systems. J. Diff. Eq. **25** (1977), 410–424.
[3] Elliptic systems in the plane with first-order terms and constant coefficients. Comm. Part. Diff. Eq. **3** (1978), 949–977.

HILE, G., and M. PROTTER
[1] Unique continuation property and Cauchy problem for first-order systems of partial differential equations. Comm. Part. Diff. Eq. **1** (1976), 136–151.
[2] Maximum principles for a class of first-order elliptical systems. J. Diff. Eq. **24** (1977), 136–151.

HILL, C. D.
[1] A method for the construction of reflection laws for a parabolic equation. Trans. Amer. Math. Soc. **133** (1968), 357–372.

HILLE, E.
[1] Lectures on Ordinary Differential Equations. Addison-Wesley, Reading (Mass.) 1969.
[2] Finiteness of the order of meromorphic solutions of some non-linear ordinary differential equations. Proc. Roy. Soc. Edinburgh sect. A, **72**:**29** (1973/74), 331–336.
[3] Zero point problems for linear differential equations of the second order. Mat. Tidsskrift, B (1927), 25–44. *See* HILLE, E. [10].
[4] On some generalizations of the Malmquist theorem. Math. Scand. **39** (1976), 59–79.
[5] Ordinary Differential Equations in the Complex Domain. Wiley-Interscience, New York 1976.
[6] Some remarks on Briot-Bouquet differential equations I. Comment. Math. Univ. Carolinae XXI (1978), 119–132.
[7] Some remarks on Briot-Bouquet differential equations II. J. Math. Anal. Appl. **65** (1978), 572–585.
[8] Second order Briot-Bouquet differential equations. Acta Sci. Math. **40** (1978), 63–72.
[9] Higher order Briot-Bouquet differential equations. Ark. Mat. **16** (1978), 271–286.
[10] Classical analysis and functional analysis. Selected papers (KALLMAN, R. R., ed.). Mathematicians of Our Time, vol. 11. The MIT Press, Cambridge (Mass.), London 1975.

HODGE, V. W.
[1] Theorie und Anwendungen harmonischer Integrale. Teubner, Leipzig 1958.

Hoefer, E. T.
[1] Approximation to the solutions of linear partial differential equations given by Bergman operators. Ph. D. thesis, State Univ. of New York, Buffalo 1973.

Hörmander, L.
[1] On a theorem of Grace. Math. Scand. **2** (1954), 55—64.
[2] An introduction to complex analysis in several variables. Van Nostrand, Princeton (N.J.) 1966 (Russ. transl.: Mir, Moskva 1968).
[3] Linear Partial Differential Operators. Springer, Berlin, Heidelberg, New York 1969.

Hsiao, G. C., P. Kopp, and W. L. Wendland
[1] A Galerkin collocation method for some integral equations of the first kind. Computing **25** (1980), 89—130.

Hsiao, G. C., and R. C. MacCamy
[1] Solution of boundary value problems by integral equations of the first kind. SIAM Review **15** (1973), 449—481.

Hsiao, G. C., and W. L. Wendland
[1] A finite element method for some integral equations of the first kind. J. Math. Anal. Appl. **58** (1977), 449—481.
[2] The Aubin-Nitsche lemma for integral equations. J. Integr. Equa. **3** (1981), 299—315.

Huckemann, F.
[1] Über den Einfluß von Randstellen Riemannscher Flächen auf die Wertverteilung. Math. Z. **65** (1956), 240—282.

Hvedelidze, B. V. (Хведелидзе, Б. В.)
[1] О краевой задаче Пуанкаре теории логарифмического потенциала для многосвязной области. Soobšč. Akad. Nauk Gruzin. SSR **2** (1941), 571—578; 865—872.
[2] Линейные разрывные граничные задачи теории функций, сингулярные интегральные уравнения и некоторые их приложения. Trudy Tbil. Mat. Inst. **23** (1956), 3—158.
[3] Метод интегралов типа Коши в разрывных граничных задачах теории голоморфных функций одной комплексной переменной. Itogi Nauki Tehn. **7** (1975), 5—162 (Engl. transl.: J. Sov. Math. **7** (1977), 309—415).

Hvedelidze, B. V., and G. F. Mandžavidze (Хведелидзе, Б. В., и Г. Ф. Манджавидзе)
[1] О задаче Римана-Привалова с непрерывными коэффициентами. Dokl. Akad. Nauk. SSSR **123** (1958) 5.

IAEA (International Atomic Energy Agency)
[1] Complex Analysis and its Applications. Lectures at an international seminar course I, II, III. Vienna 1976.

Ibragimov, I. A. (Ибрагимов, И. А.)
[1] Об определении безгранично делимой функции распределения по ее значениям на полупрямой. Teor. verojatn. prim. **22** (1977), 393—399.

Ibragimov, I. A., and Ju. V. Linnik (Ибрагимов, И. А., и Ю. В. Линник)
[1] Независимые и стационарно связанные величины. Moskva 1965.

Iftimie, V.
[1] Fonctions hypercomplexes. Bull. math. Soc. Sci. Math. RSR, NS **9** (**57**) (1965), 279—332.

Ikebe, Y., T. Y. Li, and F. Stenger
[1] The numerical solution of the Hilbert problem. Theory Approx., Proc. Conf. Calgary 1975 (1976), 338—358.

Il'inskiĭ, N. V., V. M. Fomin, and E. G. Šešukov (Ильинский, Н. В., В. М. Фомин и Е. Г. Шешуков)
[1] Применение метода Бергмана-Назарова к решению уравнений нелинейной теории фильтрации. Vyčislit. priklad. Mat., meždvedomst. naučn. Sb. **17** (1972), 3—18.

Ince, E. L.
[1] Ordinary differential equations. Longmans, Green and Co., Dover, New York 1944.

Isahanov, R. S. (Исаханов, Р. С.)
[1] О некоторых граничных задачах теории аналитических функций. Trudy Tbil. Mat. Inst. Akad. Nauk. Gruzin. SSR 1976, 62—80.
[2] Об одной граничной задаче для голоморфных функций. Soobšč. Akad. Nauk Gruzin. SSR **84** (1976), 573—576.

[3] Граничная задача для нескольких функций, голоморфных в различных областях. Soobšč. Akad. Nauk Gruzin. SSR **85** (1977), 313–316.

[4] Линейная граничная задача для нескольких голоморфных функций. Trudy Tbil. Mat. Inst. **58** (1978), 106–122.

ISHIBARA, S.

[1] Distributions with complex structure. Kodai Math. J. **1** (1978), 264–276.

IVANOV, V. V. (Иванов, В. В.)

[1] Некоторые свойства особых интегралов типа Коши и их приложения. Dokl. Akad. Nauk SSSR **121** (1958), 793–794.

[2] The theory of approximate methods and their application to the numerical solution of singular integral equations. Noordhoff, Leyden 1976.

JABLONSKIĬ, A. I. (Яблонский, А. И.)

[1] Дифференциальные уравнения с решениями без особых точек в конечной плоскости. Diff. Uravn. **1** (1965), 67–78.

JACENKO, S. A. (Яценко, С. А.)

[1] Обобщенная краевя задача Газемана на замкнутой римановой поверхности. Ukrain. Mat. Ž. **28** (1976), 708–710 (Engl. transl.: Ukrain. Math. J. **28** (1976), 550–552).

JACKO, A. I., and S. I. JACKO (Яцко, А. И., и С. И. Яцко)

[1] Обобщенная краевая задача Римана с кусочно-непрерывными коэффициентами. Ukrain. Mat. Ž. **30** (1978), 645–653.

[2] К теории обобщенной краевой задачи Римана с кусочно-непрерывными коэффициентами. *In*: Theoretical and applied problems for differential equations and algebra. Collect. sci. Works. Kiev, 1978, 250–253.

JAKOVLEVA, N. I. (Яковлева, Н. И.)

[1] О росте целых характеристических функции вероятностных законов. Teor. funkc. anal. prilož. **15** (1972), 43–49.

JANK, G.

[1] On Integro-Differential-Operators for partial differential equations. *In*: GILBERT, R. P., and R. J. WEINACHT (eds.): Function theoretic methods in differential equations. Research Notes 8. Pitman, London, San Francisco, Melbourne 1976.

JAWSON, M. A., and G. T. SYMM

[1] Integral Equation Methods in Potential Theory and Elastostatics. Academic Press, London 1977.

JESIAK, B.

[1] On analytic distribution functions and analytic properties of infinitely divisible distribution functions. Teor. verojatn. prim. **24** (1979), 825–831.

[2] *See* [1] (extension), Preprint, Karl-Marx-Universität Leipzig 1979.

[3] A uniqueness theorem for stable laws. Math. Nachr. **92** (1979), 243–246.

[4] Limit theorems under the assumption of restricted convergence on a set with a finite limit print. Math. Nachr. **95** (1980), 187–194.

[5] On the unique determination of Levy's spectral functions and parameters of infinitely divisible distributions. Teor. verojatn. prim. **25** (1981), 160–165.

[6] A characterization theorem for a class of distribution functions. Studia Sci. Math. Hung. (submitted).

JESIAK, B., and H.-J. ROSSBERG

[1] New versions of the Lindeberg-Feller theorem. Teor. verojatn. prim. **26** (1981), 858–862.

JESIAK, B., H.-J. ROSSBERG, and G. SIEGEL

[1] On the central limit theorem for the triangular array and the Lindeberg-Feller theorem. Preprint AdW der DDR. Wiss. Sitz. zur Stochastik 07/79 (1979).

JOHN, F.

[1] Plane waves and spherical means applied to partial differential equations. Interscience Publ., New York, London 1955.

[2] Continuation and reflection of solutions of partial differential equations. Bull. Amer. Math. Soc. **63** (1957), 327–344.

JUHANONOV, N. N. (Юханонов, Н. Н.)

[1] Об одной обобщенной краевой задаче сопряжения и ее приближенном решении. Dokl. Akad. Nauk Tadž. SSR **21** (1978) 7, 18–22.

JULIA, G.
[1] Lecons sur le fonctions uniformes a point singulier essentiel isole. Gauthier-Villars, Paris 1923.
JUNG, H. W. E.
[1] Einführung in die Theorie der algebraischen Funktionen einer Veränderlichen. Walter de Gruyter, Berlin, Leipzig 1923.
JUNGHANNS, P.
[1] Kollokationsverfahren zur näherungsweisen Lösung singulärer Integralgleichungen mit unstetigen Koeffizienten. Math. Nachr. **102** (1981), 17–24.
JUNGHANNS, P., and B. SILBERMANN
[1] Zur Theorie der Näherungsverfahren für singuläre Integralgleichungen auf Intervallen. Math. Nachr. **103** (1981), 199–244.
[2] Lokale Theorie des Kollokationsverfahrens zur näherungsweisen Lösung singulärer Integralgleichungen. Preprint P-Math-10/82, AdW der DDR, Inst. f. Math., Berlin 1982.
KAC, B. A. (Кац, Б. А.)
[1] О некоторых задачах граничного сопряшения аналитических функций. Sib. Mat. Ž. **17** (1976), 1271–1288.
[2] Краевая задача с невзаимооднозначными сдвигами в случае многосвязной области. Izv. vysš. učebn. Zaved., Mat. **176** (1977) 1, 134–136.
[3] О краевой задаче с невзаимооднозначными сдвигами на счетном множестве окружностей. Izv. vysš. učebn. Zaved., Mat. **201** (1979) 2, 66–69. (Engl. transl.: Sov. Math. Dokl. **23** (1979), 48–50).
[4] О краевой задаче Римана с коэффициентом, допускающим разрывы колебательного типа. Dokl. Akad. Nauk SSSR **244** (1979), 521–525. (Engl. transl.: Sov. Math. Dokl. **20** (1979), 77–81).
[5] О числе линейно независимых решений краевой задачи с невзаимооднозначными сдвигами. Sib. Mat. Ž. **21** (1980), 207–211.
KADEN, F.
[1] Lösung partieller komplexer Differentialgleichungen in lokalkonvexen Räumen. Math. Nachr. **97** (1980), 253–269.
KÄHLER, E.
[1] Theorie der Systeme von Differentialgleichungen. Leipzig 1934.
KAGAN, A. M., JU. V. LINNIK, and S. R. RAO (Каган, А. М., Ю. В. Линник и С. Р. Рао)
[1] Характеризационные задачи математической статистики. Nauka, Moskva 1972.
KAKIČEV, V. A. (Какичев, В. А.)
[1] Методы решения краевых задач линейного сопряжения для функций голоморфных в бицилиндрических областях, II. Teor. func., funkc. anal. prilož. **24** (1975), 30–41.
[2] Двухмерная краевая задача типа задачи Римана, содержащая интегро-дифференциальные операторы. Izv. Akad. Nauk Arm. SSR, Mat. **12** (1977), 189–203.
KAKIČEV, V. A., and V. M. ŠELKOVIČ (Какичев, В. А., и В. М. Шелкович)
[1] Решение краевых задач теории аналитических функций многих переменных в алгебрах Владимирова. Mat. Zametki **22** (1977), 51–60.
KALANDIJA, A. I. (Каландия, А. И.)
[1] Математические методы двумерной упругости. Moskva 1973 (Engl. transl.: Mathematical Methods of Two-dimensional Elasticity. Mir, Moscow 1975).
KALDANI, N. V.
[1] Determination of the stressed state of convex composite shells. *In*: Theory of Shells, North-Holland 1978.
KAMYNIN, I. P. (Камынин, И. П.)
[1] Обобщение теоремы Марцинкиевича о целых характеристических функциях вероятностных распределений. Zap. naučn. sem. LOMI **85** (1979), 94–103.
KAMYNIN, I. P., and I. V. OSTROVSKIĬ (Камынин, И. П., и И. В. Островский)
[1] О нулевых множествых целых эрмитовопозитивных функций. Dokl. Akad. Nauk USSR **7** (1980), 20–23.
KAPLAN, W.
[1] Analytic ordinary differential equations in the large. *In*: Proc. U.S.-Jap. Seminar on Differential and Functional equations. Minneapolis 1967.

Kasimov, F. A. (Касимов, Ф. А.)
[1] Общее комплексное представление решения одного класса дифференциальных уравнений в частных производных второго порядка. Azerbaidž. Gos. Univ. Učen. Zap., Ser. Fiz.-Mat. Nauk **3** (1972), 41–45.

Kato, T.
[1] Perturbation theory for linear operators. Springer, Berlin, Heidelberg, New York 1966.

Kawata, T.
[1] Fourier analysis in probability theory. Academic Press, New York, London 1972.

Kazidzhanova, D. (Казидзханова, Д.)
[1] Общая задача сопряжения для обобщенной системи с вырожденной матрицей краевых целовий. Dokl. Akad. Nauk Tadž. SSR **22** (1979), 643–647.

Kaziev, A. (Казиев, А.)
[1] Обобщенная краевая задача со сдвигом для эллиптического уравнения с неаналитическим коэффициентом. Dokl. Akad. Nauk Tadž. SSR **20** (1979) 2, 10–13.

Keldyš, M. V., and L. I. Sedov (Келдыш, М. В., и Л. И. Седов)
[1] Эффективные решения некоторых краевых задач для гармонических функций. Dokl. Akad. Nauk USSR XXI (1937) 1.

Klink, H.-K.
[1] Komplexe Methoden in der Kontrolltheorie, Untersuchungen optimaler Lösungen des Dirichletproblems elliptischer Systeme erster Ordnung in der Ebene mit Steuerfunktion. Diss. A, TH Magdeburg 1978.
[2] Über komplexe Methoden in der Kontrolltheorie: Ein Dirichletproblem. Math. Nachr. **96** (1980), 53–81.
[3] Komplexe Methoden in der Kontrolltheorie II (*to appear*).
[4] Komplexe Methoden in der Kontrolltheorie: Die Riemann-Hilbertsche Randwertaufgabe (*to appear*).

Kluge, R. (ed.)
[1] Theory of Nonlinear Operators. Proc. of the fifth international summer school. Abh. Akad. Wiss. DDR, Abt. Math., Nat., Techn. **6N** (1978). Akademie-Verlag, Berlin 1978.

Kneser, H.
[1] Funktionentheorie. Vandenhoeck und Ruprecht, Göttingen 1958.

Knopp, K.
[1] Funktionentheorie Teil I, II. 6th ed. Walter de Gruyter, Berlin 1944.

Kobayashi, S.
[1] Invariant distances on complex manifolds and holomorphic mappings. J. Math. Soc. Japan **19** (1967), 460–480.
[2] Hyperbolic manifolds and holomorphic mappings. Marcel Dekker, New York 1970.
[3] Transformation groups in differential geometry. Springer, Berlin, Heidelberg, New York 1972.
[4] Intrinsic distances, measures and geometric function theory. Bull. Amer. Math. Soc. **82** (1976), 357–416.

Kobayashi, S., and K. Nomizu
[1] Foundations of differential geometry I, II. Interscience Publ., New York, London, Sydney 1963, 1969.

Kočura, A. I. (Кочура, А. И.)
[1] Задача Римана с обобщенных коэффициентом. Ukrain. Mat. Ž. **29** (1977), 817 to 823.

Kolosov, G. V. (Колосов, Г. В.)
[1] Sur les problèmes d'élasticité à deux dimensions. C.R. Acad. sci. **146** (1908) 10, 522–525; **148** (1909) 19, 1242–1244; **148'** (1909) 25, 1706.
[2] Об одном приложении теории функций комплексного переменного к плоской задаче математической теории упругости. Jur'ev 1909.
[3] О сопряженных дифференциальных уравнениях с частными производными. Trudy èlektrotehn. Inst., Petrograd 1914, II, 179–199.
[4] Применение комплексных диаграмм и теории функций комплексной переменной к теории упругости. Leningrad, Moskva 1935.

KOMIN, R.
[1] Komplexe Funktionen einer Variablen und ihre isolierten Singularitäten. *In*: NAAS, I. (ed.), [1, 3–30].

KOOHARA, A.
[1] Similarity principle of the generalized Cauchy-Riemann equations for several complex variables. J. Math. Soc. Japan **23** (1971), 213–249.

KOPPELMAN, W.
[1] The Riemann-Hilbert problem for finite Riemann surfaces. Comm. Pure Appl. Math. **12** (1959), 13–35.
[2] Singular integral equations, boundary value problems and the Riemann-Roch Theorem. J. Math. Mech. **10** (1961), 247–277.

KOSTJUKOVIČ, M. E. (Костюкович, М. Е.)
[1] К вопросу о совместности одного уравнения класса Фукса с некоторыми уравнениями Пенлеве. Diff. Uravn. **15** (1979), 1135–1137.

KOVERNYĬ, N. S., and S. A. JACENKO (Коверный, Н. С, и С. А. Яценко)
[1] Обобщенная краевая задача Римана для сложного контура. Ukrain. Mat. Ž. **32** (1980), 384–388.

KRACHT, M. W.
[1] Integraloperatoren für die Helmholtzgleichung. ZAMM **50** (1970), 389–396.
[2] Über Bergman-Operatoren für lineare partielle Differentialgleichungen zweiter Ordnung. Habilitationsschrift, Universität Düsseldorf 1974.
[3] Spezielle Bergman-Operatoren nebst Anwendungen. *In*: MEISTER, V. E., N. WECK, and W. L. WENDLAND (eds.) [1, 327–341].

KRACHT, M. W., and E. KREYSSIG
[1] Constructive methods for certain Bergman operators. Appl. Anal. **8** (1979), 232–252.

KRAJKIEWICZ, P.
[1] Exceptional values of analytic functions. Proc. Amer. Math. Soc. **40** (1973) 1, 159 to 165.
[2] Polyanalytic functions with exceptional values. Trans. Amer. Math. Soc. **197** (1974), 181–210.
[3] A new proof of a theorem of Saxer. Ann. Polon. Math. **3** (1977), 293–309.

KRAJKIEWICZ, P., and W. BOSCH
[1] Polyanalytic functions with equal modulus. Proc. Amer. Math. Soc. **23** (1969) 1, 127–132.

KREYSZIG, E.
[1] On Bergman operators for partial differential equations in two variables. Pacific J. Math. **36** (1971), 201–208.

KRISZTEN, A.
[1] Hyperkomplexe und pseudoanalytische Funktionen. Comment. Math. Helv. **26** (1952), 6–35.

KRUGLOV, V. È. (Круглов, В. Э.)
[1] Абелевы дифференциалы и уравнение поверхности даны через группу циклических перестановок. Soobšč. Akad. Nauk Gruzin. SSR **92** (1978), 537–540.

KRUŠKAL', S. L. (Крушкаль, С. Л.)
[1] Квазиконформные отображения и римановы поверхности. Nauka, Sib. otdel., Novosibirsk 1975.

KRUŠKAL', S. L., B. N. APANASOV, and N. A. GUSEVSKIĬ (Крушкаль, С. Л., Б. Н. Апанасов, и Н. А. Гусевский)
[1] Униформизация и клейновы группы. Novosib. Gos. Univ., Novosibirsk 1979.

v. KRZYWOBLOCKI, M. Z.
[1] Bergman's integral operator method in the theory of wave propagation through a stratified troposphere. Nonlinear Vibrations Problems, Inst. of Basic Techn. Prob., PAN, Dep. Vibr., Warsaw, **6** (1966), 83–129.
[2] Bergman's and Gilbert's operators in elasticity, electromagnetism, fluid dynamics, wave mechanics. *In*: GILBERT, R. P., and R. G. NEWTON (eds.): Analytic methods in Mathematical Physics. New York 1970, 207–247.

KÜHN, E.
[1] Über die Funktionentheorie und das Ähnlichkeitsprinzip einer Klasse elliptischer Differentialgleichungssysteme in der Ebene. Dissertation, Dortmund 1974.

Kürcz, A.
[1] Zur Lösung parabolischer Gleichungen und Systeme mittels Integraloperatoren. Wiss. Z. Techn. Hochsch. Karl-Marx-Stadt **20** (1978), 499–509.
Kukral, D. K.
[1] On a Bergman-Whittaker type operator in five or more variables. Proc. Amer. Math. Soc. **99** (1973) 1, 122–124.
Kujala, R. O., and A. L. Vitter III (eds.)
[1] Value distribution theory. Proc. Tulane Univ. Programm. (Value distribution theory in complex analysis and related topics in differential geometry.) Marcel Dekker, New York, Part A 1974; Part B 1973.
Kulagina, M. F. (Кулагина, М. Ф.)
[1] О разрешимости неоднородной задачи Римана в случае счетного множества контуров. Izv. vysš. učebn. zaved., Mat. **177** (1977), 131–134.
[2] Задача Римана в случае счетного множества контуров в классе обобщенных аналитических функций. Izv. vysš. učebn. zaved., Mat. **218** (1980), 82–85.
Kuprin, V. A. (Куприн, В. А.)
[1] Ряды Фабера и задачи линейного сопряжения. Izv.vysš. učebn. zaved., Mat. **212** (1980), 20–26
Kuroda, T.
[1] On analytic functions on some Riemann surfaces. Nagoya Math. J. **10** (1956), 27–50.
Kutateladze, G. A. (Кутателадзе, Г. А.)
[1] Задача Римана-Гильберта для кругово кольца. Soobšč. Akad. Nauk Gruzin. SSR **94** (1979), 293–296.
Kuz'mina, A. G. (Кузьмина, А. Г.)
[1] К вопросу о построении кусочно-голоморфной функции со счетным множеством особых линий. Izv. vysš. učebn. zaved., Mat. **2** (**165**) (1976), 31–32.
Kveselava, D. A. (Квеселава, Д. А.)
[1] Riemann-Hilbert problem for a multiply-connected domain. Soobšč. Akad. Nauk Gruzin. SSR **6** (1945), 581–590 (in Georgian).
Kvitko, A. N. (Квитко, А. Н.)
[1] К теории краевой задачи Римана для обобщенных аналитических функций в случае сложного контура на плоскости. Sib. Mat. Ž. **20** (1979), 659–663.
Ladyženskaja, O. A. (Ladyshenskaya, O. A.)
[1] The Mathematical Theory of Viscous Incompressible Flow. Gordon and Breach, New York 1969.
Laha, R. G., and V. K. Rohatgi
[1] Probability theory. John Wiley & Sons, New York 1979.
Laine, I.
[1] On the behaviour of the solutions of some first order differential equations. Ann. Acad. Sci. Fenn. AI **497** (1971), 2–26.
[2] Admissible solutions of Riccati differential equations. Publ. Univ. Joensuu, Ser. B (1972) 1, 8 pp.
[3] Admissible solutions of some generalized algebraic differential equations. Publ. Univ. Joensuu, Ser. B (1974) 10, 6 pp.
Laine, I., O. Lehto, and T. Sorvali (eds.)
[1] Complex Analysis, Proc. Joensuu 1978, Lecture Notes in Mathematics, vol. 747. Springer, Berlin, Heidelberg, New York 1979.
Lammel, E.
[1] Über eine zur Laplaceschen Differentialgleichung in drei Veränderlichen gehörige Funktionentheorie. I. Math. Z. **51** (1949), 658–689.
Lanckau, E.
[1] Zur Lösung gewisser partieller Differentialgleichungen mittels parameterabhängiger Bergman-Operatoren. Nova Acta Leopoldina, Neue Folge 201, **36** (1971).
[2] Allgemeine exakte Integration einer partiellen Differentialgleichung 6. Ordnung aus der Turbulenztheorie. Wiss. Z. Techn. Hochsch. Karl-Marx-Stadt **19** (1977), 542 to 431.
[3] Explizite Lösungen für Rand- und Anfangswertaufgaben bei Gleichungen gemischten Typs mit drei unabhängigen Veränderlichen, Rostocker Math. Koll., Tagung Elliptische Differentialgleichungen, Rostock 1977, 143–150.

[4] Bergmansche Integraloperatoren für dreidimensionale Gleichungen und Gleichungssysteme. Beitr. Anal. **12** (1978), 99–112; **13** (1979), 83–87.
[5] Integraloperatoren für Differentialgleichungen gemischten Typs mit drei unabhängigen Veränderlichen. Wiss. Z. Techn. Hochsch. Karl-Marx-Stadt **20** (1978), 723–726.
[6] Exakte Behandlung von Filtrationsproblemen mit nichtlinearem Filtrationsgesetz. *In*: VIII. Int. Kongr. Anw. Math. in Ing. wiss., Weimar 1978, vol. 2, 319–323.
[7] General Vekua Operators. *In*: Ławrynowicz, J. (ed.) [5, 301–311].
[8] Behandlung pseudoparabolischer Gleichungen mit funktionentheoretischen Methoden. Beitr. Anal. **16** (1981), 87–96.
[9] On the representation of Bergman-Vekua operators in the three-dimensional case. Z. Anal. Anw. **2**(1) (1983), 1–10.

Laue, G.
[1] Zur Theorie der charakteristischen Funktionen nichtnegativer Zufallsgrößen. *In*: Transactions of the sixth Prague conference, Prague 1973, 1–18.
[2] Results on the theory of characteristic functions of non-negative random variables and their application to the factorization $\varepsilon_0 - F = (1 - \zeta)(\varepsilon_0 - H^+) * (\varepsilon_0 - H^-)$. Math. Nachr. **80** (1977), 279–288.
[3] The importance of the zeros of characteristic functions of positive random variables. Preprint, Karl-Marx-Univ. Leipzig 1979.
[4] Remarks on the relation between fractional moments and derivations of characteristic functions. J. Appl. Prob. **17** (1980), 456–466.
[5] Remarks on the theory of characteristic functions of non-negative random variables, Sankhya **45** (1983), A, Pt. 1.
[6] Characteristic functions of non-negative random variables: The connection between Re f and Im f. Transactions of the ninth Prague Conference 1982, 1983.
[7] Existence and representation of density functions. Math. Nachr. (*to appear*).

Lavrent'ev, M. A. (Лаврентьев, М. А.)
[1] О построении потока, обтекающего дугу заданной формы. Trudy ZAGI **118** (1932), 3–56.
[2] Sur une classe de représentations continues. Rec. Math. **48** (1935), 407–423.

Lavrent'ev, M. A., and B. V. Šabat (Лаврентьев, М. А., и Б. В. Шабат)
[1] Проблемы гидродинамики и их математические модели. Nauka, Moskva 1973.

Ławrynowicz, J.
[1] Condenser capacities and an extension of Schwarz's lemma for hermitian manifolds. Bull. Acad. Polon. Sci., Sér. Sci. Math. Astronom. Phys. **23** (1975), 839–844.
[2] On a class of capacities on complex manifolds endowed with an hermitian structure and their relation to elliptic and hyperbolic quasiconformal mappings, (a). Ann. Polon. Math. **33** (1976), 178 (Abstract); (b) Dissertationes Math. **166** (1980), 48 pp. (in extenso).
[3] On holomorphic continuability of quasiregular functions. *In*: Laine, I., O. Lehto, and T. Sorvali (eds.) [1].
[4] On biholomorphic continuability of regular quasiconformal mappings. *In*: Ławrynowicz, J. (ed.) [5, 326–346].
[5] Analytic functions. Proc. Conf. Kozubnik (Poland) 1979. Lecture Notes in Mathematis, vol. 798. Springer, Berlin, Heidelberg, New York 1980.

Lax, P. D.
[1] A stability theory of abstract differential equations and its application to the study of local behavior of solutions of elliptic equations. Comm. Pure Appl. Math. **9** (1956), 747–766.

Lax, P. D., and R. S. Phillips
[1] Scattering theory. Academic Press, New York 1967.

Le-din' Zon
[1] Об одном случае решения в замкнутном виде задачи Гильберта для вектор-функции. Mat. Anal. Prilož., Rostov. Univ. **5** (1974), 189–193.

Lehnickiĭ, S. G. (Лехницкий, С. Г.)
[1] Анизотропные пластинки. GITTL, Moskva, 2nd ed. 1957.
[2] Теория упругости анизотропного тела. Nauka, Moskva 1977 (Engl. transl.: Mir, Moscow 1981).

LEHMANN, I., and H.-J. ROSSBERG
[1] Elementary generalization of a theorem of Phragmén-Lindelöf with applications. Math. Nachr. **70** (1976), 87–93.
LEHTO, O.
[1] Sur le théorie des fonctions méromorphes à charactéristique bornée. C.R. Acad. Sci. Paris **236** (1953), 1943–1945.
[2] Value distribution of meromorphic functions. *In*: Proc. Rolf Nevanlinna Sympos. complex analysis. Silivri 1976, 37–45.
[3] On the birth of the Nevanlinna theory. Ann. Acad. Sci. Fenn. AI **7** (1982).
LEHTO, O., and K. I. VIRTANEN
[1] Quasikonforme Abbildungen. Springer, Berlin, Heidelberg, New York 1965 (2nd ed.: Quasiconformal mappings in the plane, ibid. 1973).
LEIPHOLZ, H.
[1] Theory of elasticity. Noordhoff, Leyden 1974.
LEVIN, B. JA.
[1] Distribution of Zeros of Entire Functions. Amer. Math. Soc., Providence (R.I.) 1964.
LEWY, H.
[1] On the reflection laws of second order differential equations in two independent variables. Bull. Amer. Math. Soc. **65** (1959), 37–58.
LINNIK, JU. V. (Линник, Ю. В.)
[1] Линеарные формы и статистические критерии, I, II. Ukrain. Mat. Ž. **5** (1953), 207–243; 247–290 (Engl. transl.: Selected Translat. Math. Stat. and Prob., Amer. Math. Soc., Providence (R.I.) **3** (1963), 1–90).
[2] Independently and equally distributed statistics and some statistical and analytical problems connected with them. Mimeographed notes, Calcutta 1954.
LINNIK, JU. V., and I. V. OSTROVSKIĬ (Линник, Ю. В., и И. В. Островский)
[1] Разложения случайних величин и векторов. Nauka, Moskva 1972.
LI, MING-ZHONG
[1] On the generalized Riemann-Hilbert boundary value problem of a system of second order linear elliptic equations. Faudn J. (1979), 63–72 (Chinese).
LIONS, J. L.
[1] Quelques Méthodes de Résolution des Problèmes aux Limites Non Linèaires. Dunod, Paris 1969.
LITVINČUK, G. S. (Литвинчук, Г. С.)
[1] Краевые задачи и сингулярные интегральные уравнения со сдвигом. Nauka, Moskva 1977.
LOUNESTO, P.
[1] Spinor valued regular functions in hypercomplex analysis. Report, Helsingin tekn. Korkeakolou Mat. Laitos A 154 (1979), 79 pp.
LOVE, A. E. H.
[1] A Treatise on the Mathematical Theory of Elasticity. Dover, New York 1944.
LU, CHIEN-KE
[1] The periodic Riemann boundary value problem and its applications to the theory of elasticity. Acta Math. Sinica **13** (1963), 343–383 (Chinese) (Engl. transl.: Chinese Math. **4** (1964), 372–422).
LUBOWICZ, H., and B. WIEPRZKOWICZ
[1] Approximate solution of the generalized linear Hilbert problem for a system of functions. Demonstratio Math. **10** (1977), 101–111.
LUKACS, E.
[1] Analytical methods in probability theory. *In*: Symposium on Probability Methods in Analysis. Lecture Notes in Mathematics, vol. 31. Springer, Berlin, Heidelberg, New York 1967, 208–238.
[2] Stable distributions and their characteristic functions. Jahresber. DMV **71** (1969) 84–114.
[3] Characteristic functions. 2nd ed. London 1970.
[4] A survey of the theory of characteristic functions. Adv. Appl. Prob. **4** (1972), 1–38.
[5] Stochastic convergence. Academic Press, New York, London 1975.
[6] Development in characteristic function theory. Griffith, London 1983.
LUKACS, E., and R. G. LAHA
[1] Applications of characteristic functions. Griffith, London 1964.

Lukaševič, N. A. (Лукашевич, Н. А.)
[1] Элементарные решения некоторых уравнений Пенлеве. Diff. Uravn. **1** (1965), 731–735.
[2] К теории четвертого уравнения Пенлеве. Diff. Uravn. **3** (1967), 771–780.
[3] О решениях пятого уравнения Пенлеве. Diff. Uravn. **4** (1968), 1413–1420.
[4] К теории уравнения Пенлеве. Diff. Uravn. **6** (1970), 425–430.

Lukaševič, N. A., and A. I. Jablonskiĭ (Лукашевич, Н. А., и А. И. Яблонский)
[1] Об одном классе решений шестого уравнения Пенлеве. Diff. Uravn. **3** (1967), 520–523.

Mächler, H.
[1] Sätze vom Kaplanschen Typ für partielle Differentialgleichungssysteme erster Ordnung in der Ebene. Dissertation, Univ. Halle 1978.

Magomedov, G. A., and V. P. Palamodov (Магомедов, Г. А., и В. П. Паламодов)
[1] Обобщенные аналитические функции многих переменных. Mat. Sb. **106** (148) (1978) 4, 515–543.

Magnaradze, L. G. (Магнарадзе, Л. Г.)
[1] Основные задачи плоской теорий упругости для контуров с угловыми точками. Trudy Tbil. Mat. Inst. IV (1938).

Malgrange, B.
[1] Existence et approximation des solutions des équations aux derivées partielles et des équations de convolutions. Ann. Inst. Fourier **6** (1956), 271–355.

Malmquist, J.
[1] Sur les fonctions à un nombre fini de branches définies par les équations différentielles du premier ordre. Acta Math. **36** (1913), 297–343.

Malonek, H. (Малонек, X.)
[1] К вопросу о справедливости теорем типа Фрагмена-Линделефа для решений некоторых комплексных дифференциальных неравенств. Izv. Akad. Nauk Arm. SSR, Mat., XIV, **3** (1979), 192–200.
[2] Über die Gültigkeit von Sätzen von Phragmen-Lindelöfschen Typ bei reellen partiellen Differentialgleichungen und ihre Herleitung durch komplexe Methoden. Dissertation A, Univ. Halle 1980.
[3] Beweis von Eigenschaften der Lösungen reeller partieller Differentialgleichungen höherer Ordnung durch Faktorisierung des Hauptteils. Math. Nachr. **96** (1980), 17–26.

Malyšev, V. A. (Малышев, В. А.)
[1] Уравнения Винера-Хопфа и их применения в теории вероятностей. Itogi Nauki Tehn. **13** (1976), 5–35.

Mamourian, A.
[1] General transmission and boundary value problems for first-order elliptic equations in multiply-connected plane domains. Demonstratio Math. **12** (1979), 785–802.

Mandžavidze, G. F. (Манджавидзе, Г. Ф.)
[1] Приближенное решение граничных задач теории аналитических функций. *In*: Markuševič, A. I. (ed.): Исследования по современным проблемам теории функций комплексного переменного. Moskva 1960, 365–370.
[2] О применении теории обобщенных аналитических функций к граничной задаче сопряжения со смещением. Dokl. Akad. Nauk SSSR **237** (1977), 1285–1288 (Engl. transl.: Sov. Math. Dokl. **18** (1977), 1554–1558).
[3] О приведении граничной задачи линейного сопряжения со смещением к задаче линейного сопряжения. Dokl. Akad. Nauk SSSR **234** (1977), 758–760 (Engl. transl.: Sov. Math. Dokl. **18** (1977), 763–766).
[4] О семействе краевых задач линейного сопряжении. Soobšč. Akad. Nauk Gruzin. SSR **95** (1979), 289–292.

Marcinkiewicz, J.
[1] Sur une propriété de la loi de Gauss. Math. Z. **44** (1938), 612–618.

Marin, S. P.
[1] A finite element method for problems involving the Helmholtz equation in two dimensional exterior regions. Ph. D. Thesis, Department of Mathematics, Carnegie-Mellon University, Pittsburgh (Pennsylvania) 1978.

Markušević, A. I., and L. A. Markušević (Маркушевич, А. И., и Л. А. Маркушевич)
[1] Введение в теорию аналитических функции. Prosvečenie, Moskva 1977.
Martirosjan, G. A. (Мартиросян, Г. А.)
[1] Задача Пуанкаре для эллиптической системи на плоскости. Izv. Akad. Nauk Arm. SSR, Mat. **12** (1977), 310–324.
Matos, M. C. (ed.)
[1] Infinite dimensional holomorphy and applications. Proc. of an International Symposium. North-Holland, Amsterdam, New York, Oxford 1977.
Meister, V. E., N. Weck, and W. L. Wendland (eds.)
[1] Function Theoretic Methods for Partial Differential Equations. Lecture Notes in Mathematics, vol. 561. Springer, Berlin, Heidelberg, New York 1976.
Meunargija, O. V. (Меунаргия, О. В.)
[1] О построении эффективного решения задачи линейного сопряжения для нескольких неизвестных функций. Teor. mat. fiz. **37** (1978), 423–426 (Engl. transl.: Theor. math. phys. **37** (1979), 1116–1118).
Mihaĭlov, L. G. (Михайлов, Л. Г.)
[1] A new class of singular integral equations and its applications to differential equations with singular coefficients. Wolters-Noordhoff, Groningen 1970.
[2] Об условиях совместности и многообразии решений обобщенной системы Коши-Римана со многими переменными. Dokl. Akad. Nauk SSSR **249** (1979) 6, 1313–1317.
Mihlin, S. G. (Михлин, С. Г.)
[1] Плоская деформация в анизотропной среде. Trudy Seismolog. inst. AN SSSR **76** (1936).
[2] Сингулярные интегральные уравнения с непрерывыми коэффициентами. Dokl. Akad. Nauk SSSR **59** (1948), 435–438.
Mihlin (Michlin), S. G., and S. Prössdorf
[1] Singuläre Integraloperatoren. Akademie-Verlag, Berlin 1980.
Millar, R. F.
[1] Singularities of two-dimensional exterior solutions of the Helmholtz equation. Proc. Cambridge Phil. Soc. **69** (1971), 175–188.
[2] The singularities of solutions to analytic elliptic boundary value problems. *In*: Meister, V. E., N. Weck, and W. L. Wendland (eds.) [1, 73–87].
v. Mises, R., and K. O. Friedrichs
[1] Fluid Dynamics. Springer, Berlin, Heidelberg, New York 1971.
Mitchell, J.
[1] Approximation to the solution of linear partial equations given by Bergman integral operators. Ber. Ges. Math. Datenverarb. Bonn **77** (1973), 97–107.
[2] Properties of solutions of linear partial differential equations given by integral operators. *In*: Meister, V. E., N. Weck, and W. L. Wendland (eds.) [1, 348–361].
Mkojan, P. H. (Мкоян, П. Х.)
[1] Один специальный случай задачи Римана для счетного множества контуров с двумя предельными точками. Izv. Vysš. Učebn. Zaved., Mat. **213** (1980), 66–69.
Mohon'ko, A. Z., and V. D. Mohon'ko (Мохонько, А. З., и В. Д. Мохонько)
[1] Оценки неванлинновских характеристик некоторых классов мероморфных функций и их приложения к дифференциальным уравнениям. Sib. Mat. Ž. **15** (1974), 1305–1322 (Engl. transl.: Sib. Math. J. **15** (1974), 921–934.)
Moisil, G. C.
[1] Les fonctions monogénes dans les espace à plusieurs dimensions. *In*: Comptes rendus du Congrés des Sociétés savantes de Paris et des Départements, tenu à Paris en 1931, Section des Sciences, 1931, 129–138.
Moisil, G. C., and N. Théodorescu
[1] Fonctions holomorphes dans l'éspace. Mathematica (Cluj) **5** (1931), 142–159.
Monahov, V. N. (Монахов, В. Н.)
[1] Краевые задачи со свободными границами для эллиптических систем уравнений. Nauka, sib. otdel., Novosibirsk 1977.

MONTEL', P. (Монтель, П.)
[1] Нормальные семейства аналитических функций. GITTL, Moskva, Leningrad 1936.
MOREV, I. A. (Морев, И. А.)
[1] О функциях, моногенных в смысле В. С. Федорова. Učen. Zap. Moskov. Obl. pedag. Inst. Trudy Kafedr. Mat. **188** (1967), 173–179.
MORROW, J., and K. KODAIRA
[1] Complex manifolds. Holt, Rinehart and Winston, New York, London 1971.
MÜLLER, C.
[1] Radiation patterns and radiation fields. J. Rat. Mech. Anal. **4** (1955), 235–246.
MÜLLER, J.
[1] Conservation theorems for the factorization identity $\varepsilon_0 - K = (\varepsilon_0 - H) * (\varepsilon_0 - U)$. Math. Nachr. **78** (1977), 263–266.
MUES, E.
[1] Über faktorisierbare Lösungen von Riccatischen Differentialgleichungen. Math. Z. **121** (1971), 145–156.
[2] Über lineare Differentialgleichungen, deren Lösungen sich als zusammengesetzte Funktionen $f \circ g$ mit gemeinsamem g darstellen lassen. Arch. Math. (Basel) **27** (1977), 187–192.
[3] Über die Nullstellen homogener Differentialpolynome. Manuscripta Math. **23** (1978), 325–341.
[4] Zur Wertverteilung von Differentialpolynomen. Arch. Math. (Basel) **32** (1979), 55–67.
[5] Über ein Problem von Hayman. Math. Z. **164** (1979), 239–259.
MUES, E., and R. REDHEFFER
[1] On the growth of logarithmic derivatives. J. London Math. Soc. 8 (1974), 412-425.
MUES, E., and N. STEINMETZ
[1] The theorem of Tumura-Clunie for meromorphic functions. J. London Math. Soc. **23** (1981), 1. 113–122.
MURTAZAEV, D. (Муртазаев, Д.)
[1] Об одной краевой задаче со сдвигом для аналитических функций. Dokl. Akad. Nauk Tadž. SSR **21** (1978) 1, 7–10.
[2] Задача Римана-Гильберта для системи уравнений первого порядка составного типа с двумя двойными характеристиками. Izv. Akad. Nauk Tadž. SSR, Otd. fiz.-mat. geol.-him. nauk 1978, **4** (1978) 18–24.
MUSHELIŠVILI (MUSCHELISCHWILI), N. I. (Мусхелишвили, Н. И.)
[1] Приложение интеграла типа Коши к одному классу сингулярных интегральных уравнений. Trudy Tbil. Mat. Inst. **10** (1941), 1–43.
[2] Сингулярные интегральные уравнения. Граничные задачи теории функций и некоторые их приложения к математической физике. Nauka, Moskva 1946. 2nded. Moskva 1962 (Engl. transl.: Noordhoff, Groningen 1953; German transl.: Akademie-Verlag Berlin 1956).
[3] Некоторые основные задачи математической теории упругости. Nauka, Moskva 1966.
MUSHELIŠVILI, N. I., and N. P. VEKUA (Мусхелишвили, Н. И., и Н. П. Векуа)
[1] Краевая задача Римана для нескольких неизвестных функций. Trudy Tbil. Mat. Inst. **12** (1943), 1–46.
NAAS, J.
[1] Beiträge zur komplexen Analysis und deren Anwendungen in der Differentialgeometrie. Akademie-Verlag, Berlin 1974.
[2] Eine Darstellung der Flächen in R^3 durch die Lösungen der inhomogenen Cauchy-Riemannschen Differentialgleichung. Math. Nachr. **47** (1970), 3–12.
NAAS, J., and TUTSCHKE, W.
[1] Komplexe Analysis und ihr Allgemeinheitsgrad. Math. Nachr. **75** (1976), 127–131.
[2] Some probabilistic aspects in partial complex differential equations. *In*: Sb. Kompleksnyĭ analiz i ego priloženija. Nauka, Moskva 1978, 409–412.
[3] On the error in the approximate solution of boundary value problems of nonlinear first order differential equations in the plane. Appl. Anal. **7** (1978), 239–246.
[4] Complex methods in nonlinear analysis. Banach Center Publications, vol. 11 (Complex Analysis).

[5] Geometrie und komplexe Differentiationen. I. Verallgemeinerungen holomorpher Strukturen, insbesondere auf den Sphären S^{2n}. Math. Nachr. **39** (1969), 183–193.

NAAS, J., and H. L. SCHMIDT
[1] Mathematisches Wörterbuch. Unveränderter Nachdruck der 3. Aufl. Akademie-Verlag, Berlin/Teubner, Stuttgart 1979.

NASHED, M. Z. (ed.)
[1] Ill-Posed Problems: Theory and Practice. Proc. Symp. Univ. of Delaware, October 2–6, 1979.

NAZAROV, G. I. (Назаров, Г. И.)
[1] О сходимости ряда Бергмана для функции тока при сверхзвуковом движении Газа. Učen. zap. Univ. Tomsk **73** (1973), 50–55.

NAZAROV, G. I., A. A. PUČKOV, and N. G. PUČKOVA (Назаров, Г. И., А. А. Пучков и Н. Г. Пучкова)
[1] Дифференциальные и интегральные формы решения дифференциальных уравнений четвертого порядка. Čisl. metody meh. splošn. sredy 7, 6, Gazovaja din. 1976, 96–110.

NECAEV, A. P., and A. V. PROČENKO (Нецаев, А. П., и А. В. Проченко)
[1] Индекс краевой задачи, краевое условие которой содержит производные. Dokl. Akad. Nauk Ukrain. SSR, Ser. A (1978), 596–599.

NEDELEC, J. C.
[1] Approximation des équations intégrales en mécanique et en physique. Lecture Notes, Centre de Mathématiques Appliques. Ecole Polytechnique, Palaiseau, France, 1977.

NEDELEC, J. C., and J. PLANCHARD
[1] Une méthode variationelle d'élements finis pour la résolution numerique d'un probléme extérieur das R^3. Revue Franc. Automatique, Inf. Rech. Oper. R **3** (1973), 105–129.

NEITTAANMÄKI, P., and J. SARANEN
[1] On finite element approximation of the gradient for solution of Poisson's equation. Num. Math. **37** (1981) 3, 333–337.

v. NEUMANN, J.
[1] Sur la théorie des jeux. C.R. Acad. Sci. Paris **186** (1928), 1689–1691; a detailed version: Zur Theorie der Gesellschaftsspiele. Math. Ann. **100** (1928), 295–320 (reprinted in: Collected works VI, 1–26).

NEVANLINNA, F.
[1] Über die Anwendung einer Klasse von uniformisierenden Transzendenten zur Untersuchung der Wertverteilung analytischer Funktionen. Acta Math. **50** (1927), 159 to 188.

NEVANLINNA, F., and R. NEVANLINNA
[1] Über die Eigenschaften einer analytischen Funktion in der Umgebung einer singulären Stelle oder Linie. Acta Soc. Sci..Fenn. **50** (1922) 5, 46 pp.

NEVANLINNA, R.
[1] Zur Theorie der meromorphen Funktionen. Acta Math. **46** (1925), 1–99.
[2] Über die Eigenschaften meromorpher Funktionen in einem Winkelraum. Acta Soc. Sci. Fenn. **50** (1925) 12, 45 pp.
[3] Le théorème de Picard-Borel et la thèorie des fonctions méromorphes. Gauthier-Villars, Paris 1929.
[4] Eindeutige analytische Funktionen. Springer, Berlin, Göttingen, Heidelberg 1936.
[5] Uniformisierung. Springer, Berlin, Göttingen, Heidelberg 1953.
[6] Analytic functions. Springer, Berlin, Heidelberg, New York 1970.
[7] Über rationale Approximation meromorpher Funktionen mit dem totalen Defekt 2. Bonner math. Schriften **121** (1980), 33–40.

NGOAN, W., and I. V. OSTROVSKIĬ (Нгоан, В., и И. В. Островский).
[1] О логарифмической производной мероморфной функции. Dokl. Akad. Nauk Arm. SSR **41** (1965), 272–277.

NGUYEN THUA HOP (Нгуен Тхыа Хоп)
[1] О нормальной разрешимости задачи Дирихле для систем эллиптического типа А. В. Бицадзе. Dokl. Akad. Nauk SSSR **167** (1966), 982–984.
[2] Systems of Green functions of the Riemann-Hilbert problem for generalized analytic functions. Acta math. Vietnam **2** (1977) 1, 96–115.

NIKOLAUS, J.
[1] Über ganze Lösungen linearer Differentialgleichungen. Arch. Math. (Basel) **18** (1967), 618–626.
NIRENBERG, L.
[1] An application of generalized degree to a class of nonlinear problems. *In*: Coll. Analyse Fonctionnelle, Liège 1972, 57–74.
NITSCHE, J.
[1] Das erste Randwertproblem eines linearen elliptischen Differentialgleichungssystems. Math. Nachr. **7** (1952), 31–33.
[2] Untersuchungen über die linearen Randwertprobleme linearer und quasilinearer elliptischer Differentialgleichungssysteme. Math. Nachr. **14** (1955), 75–127; 157 to 182.
[3] Zur Konvergenz des Ritzschen Verfahrens und der Fehlerquadrat-Methode. *In*: COLLATZ, L. (ed.): Numerische Mathematik, Differentialgleichungen, Approximationstheorie, Internat. Ser. Num. Math., vol 9. Birkhäuser, Basel 1968, 97–103.
[4] Zur Konvergenz von Näherungsverfahren bezüglich verschiedener Normen. Num. Math. **15** (1970), 224–228.
NOBLE, B.
[1] Error analysis of collocation methods for solving Fredholm integral equations. *In*: MILLER, J. (ed.): Topics in Numerical Analysis. Royal Irish Acad. Conf. Num. Anal. Academic Press, London 1973, 211–232.
NOETHER, F.
[1] Über eine Klasse singulärer Integralgleichungen. Math. Ann. **82** (1921), 42–63.
NOSHIRO, K.
[1] Cluster sets. Springer, Berlin, Göttingen, Heidelberg 1960.
OBNOSOV, JU. V. (Обносов, Ю. В.)
[1] К решению линейной задачи Гильберта в одном особом случае. Izv. vysš. učebn. zaved., Mat. **208** (1979) 9, 29–40.
OBOLAŠVILI, E. I. (Оболашвили, Е. И.)
[1] Безмоментные оболочки положительной кривизни, на которых действуют разрывные внешние силы. Trudy Konf. po teorii plastin obolocek, Kazan 1960.
[2] Решение некоторых граничных задач для обобщенных аналитических функций. Diff. Uravn. **10** (1974), 112–116.
[3] Пространственный аналог обобщенных аналитических функций. Soobšč, Akad. Nauk. Gruzin. SSR **73** (1974) 1, 21–24.
[4] Преобразование Фурье и его применения в теории упругости. Tbilissi 1979.
[5] Пространственные обобщенные векторы. Diff. Uravn. **11** (1975) 1, 108–115.
[6] Обобщенная система Коши-Римана в многомерном пространстве. Trudy. Tbil. Mat. Inst. **LVIII** (1978), 168–174.
[7] Some generalizations of the plane theory of elasticity and their relation to the shell theory. *In*: Theory of shells, North-Holland, Amsterdam 1978.
ODEN, J. T., and J. K. LEE
[1] Theory of mixed and hybrid finite element approximations in linear elasticity. *In*: GERMAIN, P., and B. NAYROLES (eds.): Applications of Methods of Functional Analysis to Problems in Mechanics. Lecture Notes in Mathematics, vol. 503. Springer, Berlin, Heidelberg, New York 1976, 90–109.
OHTSUKA, M.
[1] Nevanlinna's main theorems on Riemann surfaces. Proc. Coll. 348 on value distribution of meromorphic functions and holomorphic curves. Research Inst. Math. Sci. 1979, 18–109.
OKADA, M.
[1] Potentiels kählériens et espaces de Dirichlet. C.R. Acad. Sci. Paris, Sér. I, **292** (1981), 159–161.
OKONEK, CH., M. SCHNEIDER, and H. SPINDLER
[1] Vectorbundels on Complex Projective Spaces. Birkhäuser, Boston, Basel, Stuttgart 1980.
ORTON, M.
[1] Hilbert transforms, Plemelj relations and Fourier transforms of distributions. SIAM J. Math. Anal. **4** (1973), 656–670.

[2] Hilbert boundary value problems — a distributional approach. Proc. Royal Soc. Edinburgh **77A** (1977), 193—208.

Osgood, F.

[1] Lehrbuch der Funktionentheorie. I, II_1 (2nd ed.), II_2. Teubner, Leipzig, Berlin 1907, 1929, 1932.

Ostrovskiĭ, I. V. (Островский, И. В.)

[1] Связь роста мероморфной функции с распределением ее значений по аргументам. Izv. Akad. Nauk SSSR, ser. Mat. **25** (1961) 2, 277—328.

[2] О целых функциях, удовлетворяющих некоторым специальным неравенствам, связанных с теорией характеристических функции. Zap. meh.-mat. fak. HGU i HMO **29** (1963), 145—168 (Engl. transl.: Selected Translations Math. Statistics and Proc. Amer. Math. Soc., Providence (R.I.) **7** (1967), 203—234).

[3] Об одной теореме М. Б. Балка. *In*: Sb. Mat. fiz. i funkc. anal. (Sb. naučn. trudov FTINT Akad. Nauk USSR) vyp. I, Har'kov 1969, 191—202.

[4] The arithmetic of probability distributions. J. Multivar. Anal. **7** (1977), 475—490.

[5] Об одном классе функций ограниченной вариации на прямой, определяемых своими значениями на полупрямой. Zap. naučn. sem. LOMI **92** (1979) (Issled. po lin. oper. i teorii funkc. **IX**).

Ovsjannikov, L. V. (Овсянников, Л. В.)

[1] Задача Коши в шкале Банаховых пространств аналитических функций. *In*: Trudy simpoziuma po mehanike splošnoĭ sredy i rodstvennym problemam analiza. Tbilissi, 23.—29. 9. 1971, vol. II. Mecniereba, Tbilissi 1974, 219—229 (vol. I, Tbilissi 1973).

Painlevé, P.

[1] Lecons sur la théorie analytique des équations différentielles. Professées a Stockholm 1895. Hermann, Paris 1897.

[2] Sur les équations différentielles du second ordre ou d'ordre superieur dont l'intégrale générale est uniforme. Acta Math. **25** (1902), 1—85.

Pančenko, I. F. (Панченко, И. Ф.)

[1] Краевая задача Римана для n пар функций на римановой поверхности в пространстве $L_p^n(\Gamma)$ в случае сложного контура. *In*: Investigations in functional analysis and differential equations. Math. Sci., interuniv. Work Collect., Kišinev 1978, 68—75.

Pascali, D.

[1] Polinoamele areolare zi devzoltarea Almunsi in plan. Comun. Acad. Rep. Popul. Roumaine **10** (1960)4, 257—262.

[2] Однородная задача Гильберта для голоморфных кватернионов. Acad. Rep. Popul. Roumaine, Revue Math. pur. appl. **8** (1963), 653—659.

[3] Vecteures analytiques généralises. Revue Roumaine Math. pur. appl. **10** (1965), 779—808.

Pčel'nik, V. K. (Пчельник, В. К.)

[1] Краевая задача Карлемана с разрывным квазиконформым сдвигом. Dokl. Akad Nauk BSSR **23** (1979), 976—978.

Petrenko, V. P. (Петренко, В. П.)

[1] Изучение структуры множества положительных отклонений мероморфных функций. Izv. Akad. Nauk SSSR **33** (1969), 1330—1348; **34** (1970), 31—56. (Engl. transl.: Math. USSR Izv. **3** (1969), 1251—1270; **4** (1970), 31—57).

[2] О величинах отклонений и дефектах QПМ-функций. Sib. Mat. Ž. **13** (1972), 823—840 (Engl. transl.: Sib. Math. J. **13** (1972), 571—584).

[3] Рост целых кривых конечного нижнего порядка. Mat. Sb. (N.S.) **97** (139) (1975), 469—492 (Engl. transl.: Math. USSR Sb. **26** (1975), 427—448).

[4] Рост мероморфных функций. Višča škola, Har'kov 1978.

Petrov, V. A. (Петров, В. А.)

[1] Граничные теоремы единственности для полианалитеческих функций. Litov. Mat. Sb. **6**, (1966) **4**, 585—589.

Pfluger, A.

[1] Zur Defektrelation ganzer Funktionen endlicher Ordnung. Comment. Math. Helv. **19** (1946), 91—104.

Picard, É.

[1] Mémoire sur les fonctions entières. Ann. École Norm. **9** (1880), 145—166.

PICHORIDES, S. K.
[1] On the best values of the constante in the theorems of M. Riesz, Zygmund and Kolmogorov. Studia Math. XLIV (1972), 165–179.
PIERZCHALSKI, A.
[1] On quasi- and pseudo-conformal deformations (*to appear*).
PLEMELJ, I.
[1] Riemannsche Funktionenscharen mit gegebener Monodromiegruppe. Monatsh. Math. Phys. **19** (1908), 205–210.
POGORZELSKI, W.
[1] Integral equations and their applications, vol. 1, PWN, Warszawa 1966.
POINCARÉ, H.
[1] Sur les fonctions entières. Bull. Soc. Math. France **11** (1883), 136–144.
POKAZEEV, V. I. (Показеев, В. И.)
[1] Об одном классе функций Аппелля для двоякопериодических групп. Izv. vysš. učebn. zaved., Mat. **206** (1979) 7, 90–93.
POLOŽIĬ, G. N. (Положий, Г. Н.)
[1] Обобщение теории аналитических функций комплексного переменого. Izd. Kievskogo universiteta, Kiev 1965.
POMMERENKE, C.
[1] Univalent Functions. Vandenhoeck und Ruprecht, Göttingen 1975.
PORTEOUS, I. R.
[1] Topological geometry. Van Nostrand Reinhold, London 1969.
PORTMANN, W. O.
[1] A derivative for Hausdorff analytic functions. Proc. Amer. Math. Soc. **10** (1959), 101–105.
PRABHU, N. U.
[1] Wiener-Hopf techniques in queuing theory. *In*: CLARKE, A. B. (ed.): Mathematical Methods in Queuing Theory. Lecture Notes in Economics and Mathematical Systems, vol. 98. Springer, Berlin, Heidelberg, New York 1974, 81–90.
PRIMAČUK, L. P. (Примачук, Л. П.)
[1] Задача Гильберта для внешности пересечения вдоль прямой линии с подстановочными матрицами. Akad. Nauk. Arm. SSR, Dokl. **65** (1977), 115–118.
[2] Об одном интегрируемом случае задачи Римана с подстановочной матрицей. Dokl. Akad. Nauk BSSR **22** (1978), 310–313.
[3] Нелинейная задача сопряжения для голоморфного вектора на разомкнутых дугах. Dokl. Akad. Nauk BSSR **24** (1980), 101–104.
PRIVALOV (PRIWALOW), I. I. (Привалов, И. И.)
[1] Об одной граничной задаче в теории аналитических функций. Mat. Sb. **41** (1934), 519–526.
[2] Граничные свойства аналитических функций. 2nd ed. Gostehizdat, Moskva, Leningrad 1950 (German transl.: Randeigenschaften analytischer Funktionen. 2nd ed. VEB Deutscher Verlag der Wissenschaften, Berlin 1956).
[3] Введение в теорию функций комплексного переменного. 11th ed. Nauka, Moskva 1967 (German transl.: Einführung in die Funktionentheorie, I (3rd ed.), II (2nd ed.), III (2nd ed.). Teubner, Leipzig 1967, 1965, 1966).
PRÖSSDORF, S.
[1] Einige Klassen singulärer Gleichungen. Akademie-Verlag, Berlin/Birkhäuser, Basel 1974.
[2] On approximation methods for the solution of one dimensional singular integral equations. Appl. Anal. **7** (1978), 259–270.
[3] Zur Splinekollokation für lineare Operatoren in Sobolewräumen. *In*: Recent Trends in Mathematics, Reinhardsbrunn 1982. Teubner-Texte zur Mathematik, Bd. 50. Teubner, Leipzig 1983, 251–262.
PRÖSSDORF, S., and G. SCHMIDT
[1] Notwendige und hinreichende Bedingungen für die Konvergenz des Kollokationsverfahrens bei singulären Integralgleichungen. Math. Nachr. **89** (1979), 203–215.
[2] A finite element collocation method for singular integral equations. Math. Nachr. **100** (1981), 33–60.

[3] Finite element collocation method for systems of singular integral equations. Preprint P-Math.- 26/81, AdW der DDR, Inst. für Math., Berlin 1981.

PRÖSSDORF, S., and B. SILBERMANN

[1] О сходимости методов редукции и коллокации для систем сингулярных интегральных уравнений. Dokl. Akad. Nauk SSSR **226** (1976) 3, 516–519.

[2] Общие теоремы сходимости проекционных методов для операторных уравнении в банаховых пространствах. Dokl. Akad. Nauk SSSR **230** (1976) 3, 527–529.

[3] Projektionsverfahren und die näherungsweise Lösung singulärer Gleichungen. Teubner, Leipzig 1977.

[4] Über Näherungsverfahren zur Lösung singulärer Gleichungen. 7. Tagung über Probleme und Methoden der Mathematischen Physik. Techn. Hochsch. Karl-Marx-Stadt. Tagungsberichte 1979, 95–112 (II. Teil).

PÜNGEL, J.

[1] Zur Darstellung von Lösungsmengen partieller Differentialgleichungen. Math.-stat. Sektion, Forschungszentrum Graz, Ber. 72 (1977).

[2] Lineare Abbildungen zwischen Lösungsmengen partieller Differentialgleichungen im Komplexen. Math.-stat. Sektion, Forschungszentrum Graz, Ber. 91 (1978).

PJATECKIĬ-ŠAPIRO (PYATETZKI-SHAPIRÓ), I. I. (Пятецкий-Шапиро, И. И.)

[1] Геометрия классических областей и теория автоморфных функций. Fizmatgis, Moskva 1961 French. transl.: Géometrie des domaines classiques et théorie des fonctions automorphes, Dunod, Paris 1966; Engl. transl.: Gordon and Breach, New York, London, Paris 1969).

RAGGER, H.

[1] Zur variationellen Formulierung elliptischer Randwertprobleme. Diplomarbeit, Techn. Univ. Graz 1979.

RAĬKOV, D. A. (Райков, Д. А.)

[1] О компонировании аналитических функций распределения. Dokl. Akad. Nauk SSSR **23** (1939), 511–514.

RAMACHANDRAN, B.

[1] Advanced theory of characteristic functions. Statistical Publ. Soc., Calcutta 1967.

[2] On some fundamental lemmas of Linnik. *In*: GYIRES, B. (ed.) [1, 293–306].

RATHSFELD, A.

[1] Eine Splinekollokationsmethode für singuläre Integralgleichungen mit unstetigen Koeffizienten. Math. Nachr. (*to appear*).

REIMANN, H. M.

[1] Ordinary differential equations and quasiconformal mappings. Inventiones Math. **33** (1976), 247–270.

REINARTZ, E.

[1] Eine konstruktive Methode zur Lösung elliptischer Randwertprobleme. Ber. Ges. Math. Datenverarb. Bonn **94** (1974).

RELLICH, F.

[1] Über die ganzen Lösungen einer gewöhnlichen Differentialgleichung erster Ordnung. Math. Ann. **117** (1940), 587–598.

[2] Elliptische Funktionen und die ganzen Lösungen von $y'' = f(y)$. Math. Z. **47** (1940), 153–160.

RÉNYI, A.

[1] Wahrscheinlichkeitsrechnung. VEB Deutscher Verlag der Wissenschaften, Berlin 1962.

RÉVÉSZ, P.

[1] The laws of large numbers. Academic Press, New York, London 1968.

DE RHAM, G.

[1] Variétés differentiable. Herman, Paris 1955.

RICCI, P.

[1] Sui potenziale di semplice strato per le equazioni ellittiche di ordine superiore in due variabili. Rend. Mat. **7** (1974), 1–39.

RICKMAN, S.

[1] Value distribution of quasiregular mappings. Ann. Acad. Sci. Fenn. A.I. **7** (1982).

RICKMAN, S., and P. MATTILA

[1] Averages of the counting of a quasiregular map. Acta Math. **143** (1979), 273–305.

RIEDEL, M.
[1] On the one-sided tails of infinitely divisible distributions. Math. Nachr. **70** (1976), 155–163.
[2] Some theorems of the Phragmén-Lindelöf theory for subharmonic functions. Math. Nachr. **78** (1977), 13–20.
[3] On a characterization of the normal distribution by identically distributed linear statistics. Preprint, Karl-Marx-Univ. Leipzig 1979; J. multivar. Anal. 1983 (*to appear*).
[4] Characterization of stable processes by identically distributed stochastic integrals. Adv. Appl. Prob. **12** (1980), 689–709.
[5] A new version of the central limit theorem. Teor. verojatn. prim. **22** (1977), 187.
[6] Representation of the characteristic function of a stochastic integral. J. appl. Prob. **17** (1980), 448–455.
[7] Characterization of stable processes by identically distributed stochastic integrals. Adv. appl. Prob. **12** (1980), 689–709.

RIEMANN, B.
[1] Grundlagen für eine allgemeine Theorie der Funktionen einer veränderlichen komplexen Größe. Inauguraldissertation, Göttingen 1851. *In*: Gesammelte Werke. Teubner, Leipzig 1892, 3–45.
[2] Zwei allgemeine Sätze über lineare Differentialgleichungen mit algebraischen Koeffizienten. *In*: Gesammelte Werke. Teubner, Leipzig 1892, 379–390.

RIESZ, F., and B. SZ.-NAGY
[1] Leçons d'analyse fonctionelle. Akadémiai Kiadó, Budapest 1952 (German transl.: Vorlesungen über Funktionalanalysis. VEB Deutscher Verlag der Wissenschaften, Berlin 1956).

RIZZA, G. B.
[1] Contributi recenti alla teoria delle funzioni nelle algebre. Rend. Sem. mat. fis. Milano **43** (1973), 45–54.
[2] Monogenic functions on the real algebras and conformal mappings. Boll. Unione Mat. Ital., Ser. 4, **12** Suppl. Fac. 3 (1975), 437–450.

RODIN, JU. L. (Родин, Ю. Л.)
[1] Об условиях разрешимости краевых задач Римана и Гильберта на римановых поверхностях. Dokl. Akad. Nauk SSSR **129** (1959), 1234–1237.
[2] Структура общего решения краевой задачи Римана для голоморфного вектора на компактной римановой поверхности. Dokl. Akad. Nauk SSSR **232** (1977), 1019–1022 (Engl. transl.: Sov. Math. Dokl. **18** (1977), 201–205).

RODIN, JU. L., and A. JURAKULOV (Родин, Ю. Л., и А. Юракулов)
[1] Краевая задача Римана для обобщенных аналитических функций с сингулярным коэффициентом на компактной римановой поверхности. Soobšč. Akad. Nauk Gruzin. SSR **96** (1979), 21–24.

RÖHRL, H.
[1] Holomorphic fibre bundles over Riemann surfaces. Bull. Amer. Math. Soc. **68** (1962) 3, 125–160.

ROGOZIN, B. A. (Рогозин, Б. А.)
[1] О распределении величины первого перескока. Teor. verojatn. prim. **9** (1964), 498–515.

ROGOZIN, S. V. (Рогозин, С. В.)
[1] Об однородной краевой задаче Римана с бесконечным индексом и неограничном модулем коэффициента для полуплоскости в исключительном случае. Izv. Akad. Nauk BSSR, fiz.-mat. Nauk **117** (1978) 2.

ROGOZIN, S. V., and M. È. TOLOČKO (Рогозин, С. В., и М. Э. Толочко)
[1] Однородная краевая задача Римана с бесконечным индексом для полуплоскости в исключительном случае. Izv. Akad. Nauk BSSR, fiz.-mat. Nauk (1978) 3, 5–10.

ROGOŽINA, I. S. (Рогожина, И. С.)
[1] Задача Гильберта для кусочно-аналитической функции. Kabardino-Balk. Gos. Univ., Ucen. Zap., Ser. fiz.-mat. **19** (1963), 259–263.

ROHDE, H. W.
[1] Zur Verallgemeinerung des Nevanlinnaschen Defektbegriffes durch Dinghas. Math. Ann. **200** (1973), 141–150.

ROSCULET, M. N.
[1] Functii monogene pe algebre comutative. Editura Acad. Republ. Social. Romania, Bucuresti 1975.

ROSSBERG, H.-J.
[1] Eine neue Methode zur Behandlung der Integralgleichung von Lindley und ihre Verallgemeinerung durch Finch. EID **3** (1967), 215–238.
[2] Eigenschaften der charakteristischen Funktionen von einseitig beschränkten Zufallsgrößen und ihre Anwendung auf ein Charakterisierungsproblem der mathematischen Statistik. Math. Nachr. **37** (1968), 37–57.
[3] Ausbeutung einer bekannten Wiener-Hopf-Faktorisierung beim Wartemodell $G/H/1$ und einer mit ihm zusammenhängenden Irrfahrt. Math. Operationsforsch. Statist. **2** (1971), 129–146.
[4] Characterization of the exponential and Pareto distribution by means of some properties which the differences and quotients of order statistics are subject to. Math. Operationsforsch. Statist. **3** (1972), 207–216.
[5] Funktionentheoretische Möglichkeiten und Resultate, angeregt durch Probleme der Wahrscheonlichkeitsrechnung. Mitteilungen Math. Ges. DDR (1974) 4, 62–72.
[6] On a problem of Kolmogorov concerning the normal distribution function. Teor. verojatn. prim **19** (1974), 824–828.
[7] An extension. of the Phragmén-Lindelöf theory which is relevant for characterization theory *In*: PATIL, G. P., S. KOTZ, and J. K. ORD (eds.): Statistical distributions in scientific work, vol. 3. Dordrecht, Boston 1975, 87–96.
[8] New viewpoints, results and problems in the theory of Phragmén-Lindelöf. Ann. Polon. Math. **33** (1976), 91–100.
[9] Phragmén-Lindelöf theorems in probability theory. Adv. Appl. Prob. **10** (1977), 270–272.
[10] Kombinatorische und funktionentheoretische Methoden in der Bedienungstheorie *In*: GNEDENKO, B. V., and D. KÖNIG (eds.). Handbuch der Bedienungstheorie, vol. I. Akademie-Verlag, Berlin **1983**, **307**–331.
[11] Limit theorems for identically distributed summands assuming the convergence of the distribution functions on a half axis. Teor. verojatn. prim. **24** (1979), 692–709.
[12] Non-elemenary generalization of a theorem of Polya-Szegö from the theory of Pragmén-Lidelöf. Math. Nachr. **82** (1978) 269–276.
[13] Generalization of Cramér's and Linnik's factorization theorems in the framework of the continuation theory of distribution functions. Preprint, Karl-Marx-Univ. Leipzig; Z. Anal. Anw. (*submitted*).

ROSSBERG, H.-J., and CHU DUC
[1] Uniqueness theorems for components of a convolution given on a half-line. Math. Operationsforsch. Statist. **11** (1980), 361–371.

ROSSBERG, H.-J., and B. JESIAK
[1] On the unique determination of stable distribution functions. Math. Nachr. **82** (1978), 297–300.

ROSSBERG, H.-J., B. JESIAK, and G. SIEGEL
[1] Continuation of a distribution functions. *In*: GANI, J., and V. K. ROHATGI (eds.): Contribution to probability. New York 1981, 29–48.
[2] A new branch in the theory of sums of independent random variables. Serdica **7** (1981), 93–106.
[3] Analytic methods of Probability Theory. Akademie-Verlag, Berlin (*to appear*).

ROSSBERG, H.-J., and I. LEHMANN
[1] Über Wachstumsbeschränkungen für analytische Funktionen, die die Beschränktheit dieser Funktionen zur Folge haben. Math. Nachr. **61** (1974), 259–270.

ROSSBERG, H.-J., and G. SIEGEL
[1] One sided characterization of the normal distribution in the set of infinite divisible distributions. Teor. verojatn. prim. **26** (1981), 400–407.
[2] Continuation of convergence in the central limit theorem. Teor. verojatn. prim. **20** (1975), 885–887.

LE ROUX, M. N.
[1] Equations intégrales pour le probléme du potential électrique dans le plan. C.R. Acad. Sci. A (1974), 278.

[2] Méthode d'élement finis pour la résolution numérique de problémes extérieurs en dimension deux. Revue Franc. Automatique, Inf. Rech. Oper. R **11** (1977), 27–60.

RÜHS, F.

[1] Funktionentheorie. 3rd ed. VEB Deutscher Verlag der Wissenschaften, Berlin 1976.

RÜPRICH, W.

[1] Das Rand-Sprungwertproblem für implizite partielle komplexe Differentialgleichungssysteme in der Ebene. Math. Nachr. **101** (1981), 309–319.

RUNDELL, W., and M. STECHER

[1] A method of ascent for parabolic and pseudoparabolic partial differential equations. SIAM J. Math. Anal. **7** (1976), 898–912.

[2] A Runge approximation and unique continuation theorem for pseudoparabolic equations. SIAM J. Math. Anal. **9** (1978), 1120–1125.

RUSAK, V. N., and M. A. ŠEŠKO (Русак, В. Н., и М. А. Шешко)

[1] Сходимость приближенного решения краевой задачи Римана. Izv. Akad. Nauk BSSR, Ser. fiz.-mat. Nauk **1** (1977), 25–33.

RUSCHEWEYH, S.

[1] Gewisse Klassen verallgemeinerter analytischer Funktionen. Bonner math. Schriften **39** (1969).

[2] Operatoren bei partiellen Differentialgleichungen; Eigenschaften der Lösungen. Vortragsauszug, Steierm. Math. Symposium 1970.

[3] Über Funktionen mit einem Maximumprinzip und eine Klasse von Differentialoperatoren in der Funktionentheorie. Habilitationsschrift, Univ. Bonn 1971.

RYSJUK, N. A. (Рысюк, Н. А.)

[1] О нелинейной краевой задаче типа задачи конечного автомата. Dokl. Akad. Nauk BSSR **21** (1977), 299–301.

ŠABAT, B. V. (Шабат, Б. В.)

[1] Введение в комплексный анализ. Nauka, Moskva 1976.

SALIMOV, JA. Š. (Салимов, Я. Ш.)

[1] Об одной нагруженной задаче Гильберта. Izv. Akad. Nauk Azerbaidž. SSR, Ser. fiz.-tehn. mat. Nauk **5** (1978), 58–62.

SAMELSON, H.

[1] On de Rham's theorem. Topology **6** (1967) 4, 427–432.

SARANEN, J.

[1] Über die Approximation der Lösungen der Maxwellschen Randwertaufgabe mit der Methode der finiten Elemente. Appl. Anal. **10** (1980), 15–30.

[2] A least squares approximation method for first order elliptic systems in the plane. Appl. Anal. (*to appear*).

SARIO, L., and K. NOSHIRO

[1] Value distribution theory. Van Nostrand, Toronto, New York, London 1966.

SATO, M., et al.

[1] Studies in Holomorphic Quantum Fields. Proc. Japan Acad. Sci. 1977–1980.

SAVIN, C. N. (Савин, Г. Н.)

[1] Концентрация напряжений около отверстий. Gostehizdat, Moskva, Leningrad 1951.

[2] Распределение напряжений около отверстий. Naukova dumka, Kiev 1968.

SAXER, W.

[1] Über eine Verallgemeinerung des Satzes von Schottky. Compositio Math. **1** (1934), 207–216.

SCHEIDEGGER, A. E.

[1] Hydrodynamics in Porous Media. *In*: FLÜGGE, S. (ed.): Handbuch der Physik, vol. VIII/2. Springer, Berlin, Göttingen, Heidelberg 1963, 625–662.

SCHEFFERS, G.

[1] Verallgemeinerung der Grundlage der gewöhnlichen complexen Funktionen, I. II. Ber. Verh. Sächs. Ges. Wiss., Leipzig, math.-phys. Kl. **45** (1893), 828–848; **46** (1894), 120–134.

SCHIFFER, M., and D. C. SPENCER

[1] Functionals of finite Riemann surfaces. Princeton University Press, Princeton (N.J.) 1954.

SCHMIDT, G.
[1] On spline collocation for singular integral equations. Preprint P-Math-13/82, AdW der DDR, Inst. f. Math., Berlin 1982; Math. Nachr. (*to appear*).
SCHOPF, G.
[1] Das Nullstellengebilde von Potenzreihen in z und $\bar{z}$. Math. Nachr. **78** (1977), 319–326.
[2] О сопряженных порядках и сопряженных типах убывания многомерного вероятностного закона. Teor. verojatn. prim. **24** (1979), 204–207.
SCHRYER, N. L.
[1] Constructive approximation of solutions to linear elliptic boundary value problems. SIAM J. Numer. Anal. **9** (1972), 546–572.
SCHWARTZ, L.
[1] Théorie des distributions. Nouvelle édition, Hermann, Paris 1966.
SCHWARZ, H. A.
[1] Zur Integration der partiellen Differentialgleichung Δu. J. reine angew. Math. **74** (1872), 218–253.
SEDOV, L. I. (Седов, Л. И.)
[1] Плоские задачи гидродинамики и аэродинамики. Nauka, Moskva 1980.
SEIFFERT, I., and M. RIEDEL
[1] Uniqueness of two factorization problems. Teor. verojatn. prim. **24** (1979), 207–211.
SEIFERT, H., and W. THRELFALL
[1] Lehrbuch der Topologie. Teubner, Leipzig 1934.
SEIF-MSHIMBA, A.
[1] Construction of the Solution to the Dirichlet Boundary Value Problem Belonging to $W_{1/p}(G)$ for systems of General (Linear or Nonlinear) Partial Differential Equations in the Plane. Math. Nachr. **99** (1980), 145–163.
SEĬFULLAEV, R. K. (Сейфуллаев, Р. К.)
[1] Краевая задача Римана на негладкой разомкнутой кривой. Mat. Sb. (N.S.) **112** (154) (1980), 147–161.
SELBERG, H. L.
[1] Über die Werteverteilung der algebroiden Funktionen. Math. Z. **31** (1930), 709 – 728.
[2] Algebroide Funktionen und Umkehrfunktionen Abelscher Integrale. Arch. Norske Vid. Akad. Oslo, Mat.-nat. Kl. **8** (1934), 1–72.
ŠERMAN, D. I. (Шерман, Д. И.)
[1] Смешанная задача статической теории упругости для плоских многосвязных областей Dokl. Akad. Nauk SSSR, t. XXVIII (1940) 1, 29–32.
[2] К решению плоской задачи теории упругости для анизотропной зреды. Priklad. mat. i meh., t. VI (1942) **6**.
ŠEŠKO, M. A. (Шешко, М. А.)
[1] Приближенное решение краевой задачи Римана с разрывными коэффициентами. Izv. vysš. učebn. Zaved., Mat. **191** (1978), 91–103.
SHEA, D. F.
[1] On the Valiron deficiencies of meromorphic functions of finite order. Trans. Amer. Math. Soc. **124** (1966), 201–227.
SIEGEL, G.
[1] Uniqueness of symmetric distribution functions defined on bounded sets and its applications. Teor. verojatn. prim. **24** (1979), 831–834.
[2] Characterization of a Gaussian distribution in the set of infinitely divisible distributions on a Hilbert space. Preprint, Leipzig 1980; Math. Nachr. (*to appear*).
SIMIONESCU, C., and CL. SIMIONESCU
[1] Complex analysis methods in control theory. Bul. Univ. Braşov, Ser. C, Mat. Fiz. Chim., **19** (1977), 71–88.
SIMIONESCU, CL.
[1] On a control problem in the theory of whirls. Bul. Univ. Braşov, Ser. C, Mat. Fiz. Chim. **20** (1978), 125–132.
SIMONENKO, I. B. (Симоненко, И. Б.)
[1] Краевая задача Римана с измеримым коэффициентом. Dokl. Akad. Nauk SSSR **135** (1960), 538–541 (Engl. transl.: Sov. Math. Dokl. **1** (1960), 1295–1298).

[2] Краевая задача Римана для n пар функций с измеримыми коэффициентами и ее применение к исследованию сингулярных интегралов в пространствах $\mathcal{L}_p$ с весами. Dokl. Akad. Nauk SSSR **141** (1961), 36—39 (Engl. transl.: Sov. Math. Dokl. **2** (1961), 1391—1394).

[3] Краевая задача Римана для n пар функций с измеримыми коэффициентами и ее применение к исследованию сингулярных интегралов в пространствах $\mathcal{L}_p$ с весами. Izv. Akad. Nauk SSSR, Ser. Mat. **28** (1964), 277—306.

[4] Некоторые общие вопросы теории краевой задачи Римана. Izv. Akad. Nauk SSSR, Ser. Mat. **32** (1968), 1138—1146.

Simons, J.

[1] Minimal varieties in Riemannian manifolds. Ann. Math. **88** (1968), 62—105.

Skitovič, V. P. (Скитович, В. П.)

[1] Об одном свойстве нормального распределения. Dokl. Akad. Nauk SSSR **89** (1953), 217—219.

Sleeman, B. D.

[1] The three dimensional inverse scattering problem for the Helmholtz equation. Proc. Cambridge Phil. Soc. **73** (1973), 477—488.

Slizkiĭ, V. I. (Слизкий, В. И.)

[1] К теории Нётера обобщенной задачи Карлемана для двух функций для многосвязной области. Sib. Mat. Ž. **17** (1976), 1422—1426.

[2] О теории Нётера обобщенной краевой задачи карлемана для двух функций для многосвязной области. Ukrain. Mat. Ž. **29** (1977), 694—697.

[3] Об одной краевой задаче для двух функций на римановой поверхности. Ukrain. Mat. Ž. **31** (1979), 588—592.

Sneddon, J. N., and D. S. Berry

[1] The classical theory of elasticity. Springer, Berlin, Göttingen, Heidelberg 1958.

Snyder, H.

[1] Monogonic functions and residue theory in a harmonic algebra. Bull. Amer. Math. Soc. **59** (1953), 547—548.

[2] A Hypercomplex Function-Theory Associated with Laplace's Equation. Mathematische Forschungsberichte. VEB Deutscher Verlag der Wissenschaften, Berlin 1968.

Sobolev, S. L. (Соболев, С. Л.)

[1] Об одной предельной задаче теории логарифмического потенциала и ее применение к отражению плоских упругих волн. Trudy Seism. inst. **11** (1930), 1—18.

Sobrero, L.

[1] Theorie der ebenen Elastizität unter Benützung eines Systems hyperkomplexer Zahlen. Hamburger Math. Einzelschriften. Teubner, Leipzig 1934.

Sofronov, I. D. (Софронов, И. Д.)

[1] К приближенному решению сингулярных интегральных уравнений. Dokl. Akad. Nauk SSSR **III** (1956) 1, 37—39.

Sohockiĭ, Ju. V. (Сохоцкий, Ю. В.)

[1] Об определенных интегралов в функциях, употребляемых при разложениях в ряды. St. Petersburg 1873.

Sokolovskiĭ, V. V. (Соколовский, В. В.)

[1] О безмоментных оболочках вращения. Prikl. mat. meh., t. **I** (1938), vyp. 3.

[2] Уравнения равновесия безмоментных оболочек. Prikl. mat. meh., t. **7** (1943), vyp. 1.

Soldatov, A. P. (Солдатов, А. П.)

[1] Краевая задача линейного сопряжения теории функций. Izv. Akad. Nauk SSSR, Ser. Mat. **43** (1979), 184—202.

Sommen, F.

[1] Randwaarden van monogene functies en Fourier transformaties. Proefschrift, Fac. Wetensch., Rijksuniv. Gent 1980.

Lê hûng Són

[1] Fortsetzungssätze vom Hartogsschen Typ für verallgemeinerte analytische Funktionen mehrerer komplexer Variabler. Math. Nachr. **93** (1979), 177—186.

[2] Verfahren zur Fortsetzung verallgemeinerter analytischer Funktionen mehrerer komplexer Variabler. Beitr. Anal. **13** (1979), 89—109.

SPRÖSSIG, W.
[1] Methods of function theory for the solution of non-linear partial differential equations. *In*: KLUGE, R. (ed.) [1, 399–402].
STEIN, E. M., and G. WEISS
[1] Generalization of the Cauchy-Riemann equations and representation of the rotation group. Amer. J. Math. **90** (1968), 163–196.
STEINHAUS, H.
[1] Über die Wahrscheinlichkeit dafür, daß der Konvergenzkreis einer Potenzreihe ihre natürliche Grenze ist. Math. Z. **31** (1930), 408–414.
STEINMETZ, N.
[1] Bemerkung zu einem Satz von Yosida. *In*: LAINE, I., O. LEHTO, and T. SORVALI (eds.) [1, 369–377].
[2] Zur Theorie der binomischen Differentialgleichungen. Math. Ann. **244** (1979), 263 to 274.
[3] Über die faktorisierbaren Lösungen gewöhnlicher Differentialgleichungen. Math. Z. **170** (1980), 169–180.
[4] Ein Malmquistscher Satz für algebraische Differentialgleichungen erster Ordnung. J. reine angew. Math. **316** (1980), 44–53.
[5] Über das Anwachsen der Lösungen homogener algebraischer Differentialgleichungen zweiter Ordnung. Manuscripta Math. **32** (1980), 303–308.
[6] Über die Nullstellen von Differentialpolynomen. Math. Z. **176** (1981), 255–264.
STEPHAN, E., and W. L. WENDLAND
[1] Remarks to Galerkin and least squares methods with finite elements for general elliptic problems. *In*: EVERITT, W. M., and B. D. SLEEMAN (eds.): Ordinary and Partial Differential Equations. Lecture Notes in Mathematics, vol. 564. Springer, Berlin, Heidelberg, New York 1976, 461–471. Manuscripta Geodaetica **1** (1976), 93–123.
STEUTEL, F. W.
[1] Some recent results in infinite divisibility. Stochastic processes and their applications **1** (1973), 125–143.
STOILOV, S. (Стоилов, С.)
[1] Лекции о топологических принципах теории аналитических функций. Nauka, Moskva 1964.
STOLL, W.
[1] Value Distribution of Holomorphic Maps into Compact Complex Manifolds. Lecture Notes in Mathematics, vol. 135. Springer, Berlin, Heidelberg, New York 1970.
[2] Value distribution theory. Part B. Deficient and Bezout estimates. *In*: KUJALA, R. O., and A. L. VITTER III (eds.) [1].
[3] Value Distribution on Parabolic Spaces. Lecture Notes in Mathematics, vol. 600. Springer, Berlin, Heidelberg, New York 1977.
STOUT, L. E.
[1] The boundary values of holomorphic functions of several complex variables. Duke Math. J. **44** (1977) 1, 105–108.
STRANG, G.
[1] The finite element method and approximation theory. *In*: HUBBARD, B. (ed.): Numerical Solution of Partiel Differential Equations II, Synspade 1970. Academic Press, New York 1971, 547–583.
STRELIC, Š. I. (Стрелиц, Ш. И.)
[1] Асимитотические свойства аналитических решений дифференциальных уравнений. Mintas, Vilnius 1972.
[2] A remark on meromorphic solutions of differential equations. Proc. Amer. Math. Soc. **65** (1977), 255–261.
SUNDARARAMAN, D.
[1] Moduli, deformations and classifications of compact complex manifolds. Research Notes in Mathematics, vol. 45. Pitman, Boston, London, Melbourne 1980.
SYMM, G. T.
[1] An integral equation method in conformal mapping. Numer. Math. **9** (1966), 250–258.
SZNITKO, W.
[1] Remarks on a discontinuous Hilbert boundary problem. Demonstratio Math. **12** (1979), 289–294.

TANAKA, N.
[1] On the pseudo-conformal geometry of hypersurfaces of the space of complex variables. J. Math. Soc. Japan **14** (1962), 397–429.

TAKÁCS, L.
[1] Combinatorial and analytic methods in the theory of queues. Adv. Appl. Prob. **7** (1975), 607–635.

TARSKI, J.
[1] Analytic functions in quantum field theory. *In*: IAEA [1, vol. III, 181–192].

TEICHERT-GAIDA, A.
[1] On some compound non-linear boundary value problem. Demonstratio math. **9** (1976), 409–421.
[2] A nonlinear problem of Sjöstrand-v. Wolfersdorf's type. Demonstratio math. **11** (1978), 395–413.

TEMAM, R.
[1] Navier Stokes Equations. North Holland, Amsterdam, New York, Oxford 1977.

THÉODORESCU, N.
[1] La derivée aréolaire et ses applications à la physique mathematique. Thèse, Paris 1931.

TIHONENKO, N. JA. (Тихоненко, Н. Я.)
[1] О приближенном решении сингулярных интегральных уравнений и краевых задач со сдвигом. Dokl. Akad. Nauk SSSR **230** (1976), 291–294 (Engl. transl.: Sov. Math. Dokl. **17** (1976), 1310–1314).
[2] Приближенное решение двухэлементных краевых задач со сдвигом. Ukrain. Mat. Ž. **29** (1977), 77–88 (Engl. transl.: Ukrain. Math. J. **29** (1977), 58–67).

TIMOFEEV, E. K. (Тимофеев, Е. К.)
[1] К теории краевых задач линейного сопряжения со смещениями. Mat. Anal. Prilož., Rostov. Univ., **7** (1975), 69–77.
[2] Граничная задача линейного сопряжения со смещением. Sib. Mat. Ž. **17** (1976), 392–401 (Engl. transl.: Sib. Math. J. **17** (1976), 306–313).

TITOV, A. N. (Титов, А. Н.)
[1] Об определении свертки одинаковых функций распределения по ее значениям на полупрямой. Teor. verojatn. prim. **XXVI** (1981), 610–611.

TJURIKOV, E. V. (Тюриков, Е. В.)
[1] Нелинейная краевая задача Римана-Гильберта для квазилинейных эллиптических систем. Dokl. Akad. Nauk SSSR **247** (1979) 5, 1068–1072 (Engl. transl.: Sov. Math. Dokl. **20** (1979), 863–866).

TODA, H.
[1] Sur les combinaisons exceptionelles de fonctions holomorphes. Applications aux fonctions algébroides. Thohoku Math. J. **22** (1970), 290–319.
[2] On a modified deficiency of meromorphic functions. Thohoku Math. J. **22** (1970), 635–658.

TOKIBETOV, Ž. A. (Токибетов, Ж. А.)
[1] О формулах Коши и Помпейю для обобщенных аналитических функций. Mat. Zametki **12** (1972) 3, 263–268.

TREVES, F.
[1] Basic linear differential equations. Academic Press, New York, San Francisco, London 1975.

VAN TRIAN, T.
[1] On a certain generalized boundary value problem of function theory. Demonstratio Math. **10** (1977), 261–274.

TRIEBEL, H.
[1] Fourier Analysis and Function Spaces. Teubner, Leipzig 1977.

TUNGATAROV, A. T. (Тунгатаров, А. Т.)
[1] Об одном классе иррегулярных обобщенных систем Коши-Римана. Izv. Akad. Nauk Kazan. SSR, Ser. fiz.-mat. **5** (1977), 56–61.

TUTSCHKE, W.
[1] Grundlagen der Funktionentheorie. VEB Deutscher Verlag der Wissenschaften, Berlin 1967.

[2] Das Reziprozitätstheorem für eine Klasse pseudoholomorpher Funktionen mehrerer komplexer Variabler. Sitzungsber. Sächs. Akad. Wiss. Leipzig (1969), 1–19.
[3] Über eine spezielle Form des Maximumprinzips für holomorphe Funktionen im Zusammenhang mit einem Satz von Rossberg. Ann. Polon. Math. **22** (1970), 341–143.
[4] Die Cauchysche Integralformel für morphe Funktionen mehrerer komplexer Variabler. Math. Nachr. **54** (1972), 385–391.
[5] Über Fixpunktmethoden bei partiellen komplexen Differentialgleichungen. *In*: NAAS, J. (ed.): [1], 31–41.
[6] Über die Untersuchung von Problemen der optimalen Steuerung in der Ebene mittels komplexer Methoden (Pontrjaginsches Maximumprinzip für partielle Differentialgleichungen). Math. Nachr. **60** (1974), 201–214.
[7] Ein Differenzierbarkeitssatz für approximativ analytische Funktionen. Math. Z. **142** (1975), 27–31.
[8] Die neuen Methoden der komplexen Analysis und ihre Anwendung auf nichtlineare Differentialgleichungssysteme. Sitzungsber. Akad. Wiss. DDR **17 N** (1976). Akademie-Verlag, Berlin 1976.
[9] Lösung nichtlinearer partieller Differentialgleichungssysteme erster Ordnung in der Ebene durch Verwendung einer komplexen Normalform. Math. Nachr. **75** (1976), 283–298.
[10] Partielle komplexe Differentialgleichungen in einer und in mehreren komplexen Variablen. VEB Deutscher Verlag der Wissenschaften, Berlin 1977.
[11] Vorlesungen über partielle Differentialgleichungen. Teubner-Texte, Teubner, Leipzig 1978.
[12] Задача Римана-Гильберта для нелинейных систем дифференциальных уравнений в плоскости. *In*: Sb. Kompleksnyĭ analiz i ego priloženija. Nauka, Moskva 1978, 537–542.
[13] Ein Optimalitätskriterium für Lösungen des Dirichletproblems bei elliptischen Systemen erster Ordnung mit Parameterfunktion. Beitr. Anal. **12** (1978), 19–27.
[14] Solutions with prescribed periods on the boundary components for non-linear elliptic systems of first order in multiply connected domains in the plane. Martin-Luther-Univ. Halle, Preprint Nr. 27 (1979), 3–9.
[15] Reduction of the problem of linear conjugation for first order nonlinear elliptic systems in the plane to an analogous problem for holomorphic functions. *In*: ŁAWRYNOWICZ, J. (ed.) [5, 446–455].
[16] О задаче Римана-Гильберта для нелинейных систем уравнений первого порядка эллиптического типа. *In*: Sb. Differencialnye i integralnye uravnenija. Kraevye zadači. Trudy inst. priklad. mat. im. I. N. Vekua, Tbilissi 1979, 257–261.
[17] Elliptic Evolution Processes. Math. Nachr. **100** (1981), 21–31.
[18] Задача с начальными значениями для обобщенных аналитических функций, зависящих от времени (обобщения теорем Коши-Ковалевской и Хольмгрена). Dokl. Akad. Nauk **262** (1982) 5, 1081–1085 (Engl. transl.: Soviet Math. Dokl. **25** (1982) 1, 201–205).

USMANOV, Z. D. (Усманов, З. Д.)
[1] О бесконечно малых изгибаниях поверхностей положительной кривизный с изолированной точкой уплощения. Mat. Sb. **83** (**125**) (1970), 596–615 (Engl. transl.: Math. USSR Sb. **12** (1970), 595–614).
[2] К вопросу о деформации поверхности с точкой уплощения. Mat. Sb. **89** (**131**) (1972), 61–82 (Engl. transl.: Math. USSR Sb. **18** (1972), 61–81).
[3] Исследование обобщенных аналитических функций с неподвижной особой точкой. Izv. Tadž. Akad. Nauk. **43** (1972) 1, 3–6.
[4] К теории уравнения $\partial_{\bar z}\Phi - \frac{\lambda}{2\bar z}\bar\Phi = 0$. Dokl. Tadž. Akad. Nauk **15** (1972) 5, 13–16.
[5] Об одном классе обобщенных систем Коши-Римана с сингулярной точкой. Sib. Mat. Ž. **14** (1973), 1076–1087. (Engl. transl.: Sib. Math. J. **14** (1973) 5, 1076 to 1087).

USMANOV, N. (Усманов, Н.)
[1] Особые случаи общей граничной задачи линейного сопряжения в параболическом случае. Dokl. Akad. Nauk Tadž. SSR **20** (1977) 7, 11–14.

Valiron, G.
[1] Fonctions entières et fonctions méromorphes d'une variable. Mémorial des sciences math. 2. Gauthier-Villars, Paris 1925.
[2] Fonctions entières d'ordre fini et fonctions meromorphes. Inst. Math. Univ., Genève 1960.
[3] Sur les fonctions entières vérifiant une classe d'équations différentielles. Bull. Soc. Math. France **51** (1923), 33–45.
[4] Fonctions analytiques et équations différentielles. J. Math. Pur. Appl. **31** (1952) 9, 293–303.

Vekua, I. N. (Векуа, И. М.)
[1] Приложение метода акад. Н. И. Мусхелишвили к решению граничных задач плоской теории упругости анизотропной среды. Soobšč. Akad. Nauk Gruzin. SSR **1** (1940) 10.
[2] Об одной линейной граничной задаче Римана. Trudy Tbil. Mat. Inst. **11** (1942), 109–139.
[3] Новые методы решения эллиптических уравнений. GITTL, Moskva, Leningrad 1948 (Engl. transl.: New methods for solving elliptic differential equations. Amsterdam 1967).
[4] Системы дифференциальных уравнений первого порядка эллиптического типа и граничные задачи. Mat. Sb. **31** (73) (1952) 2, 217–314.
[5] Обобщенные апалитические функции. Fizmatgiz, Moskva 1959 (Engl. transl.: Generalized analytic functions. Reading 1962; German transl.: Verallgemeinerte analytische Funktionen. Akademie-Verlag, Berlin 1963).
[6] Основы тензорного анализа и теории ковариантов. Nauka, Moskva 1978.
[7] О некоторых новых применениях обобщенных уравнений Коши-Римана в теории оболочек. *In*: Trudy Vsesojuzn. konfer. po uravn. s častn. proizvod., Moskva 1978.
[8] Об одном свойстве обобщенной системы Коши-Римана. Soobšč. Akad. Nauk Gruzin. SSR **14** (1953), 449–453.

Vekua, N. P. (Векуа, Н. П.)
[1] Системы сингулярных интегральных уравний. Nauka, Moskva 1970.
[2] О некоторых обобщенных граничных задачах Гильберта для нескольких неизвестных функции с малым параметром и связанных с ними интегральных и интегро-дифференциальных уравнениях. Trudy Tbil. Mat. Inst. **49** (1975), 19–38.
[3] Об одной граничной задаче Римана-Гильберта для систем аналитических функций. Soobšč. Akad. Nauk Gruzin. SSR **88** (1977), 273–276.
[4] О некоторых мероморфных решениях однородной граничной задачи Римана-Гильберта для нескольких неизвестных функций. Soobšč. Akad. Nauk Gruzin. SSR **89** (1978), 17–20.
[5] Неоднородная граничная задача Римана-Гильберта для систем аналитических функций. Soobšč. Akad. Nauk Gruzin. SSR **89** (1978), 273–276.

Vinogradov, V. S. (Виноградов, В. С.)
[1] Об одной краевой задаче для линейных эллиптических систем дифференциальных уравнений первого порядка на плоскости. Dokl. Akad. Nauk SSSR **118** (1958), 1059–1062.
[2] Об ограниченности решений краевых задач для линейных эллиптических систем уравнений на плоскости. Dokl. Akad. Nauk SSSR **121** (1958), 399–402.
[3] Об одной задаче для квазилинейных систем уравнений на плоскости. Dokl. Akad. Nauk SSSR **121** (1958), 579–581.
[4] Об одной граничной задаче для эллиптической системы специального вида. Diff. Uravn. **7** (1971), 1226–1234; 1341.
[5] Об общей граничной задаче для эллиптической системы первого порядка на плоскости. Diff. Uravn. **7** (1971) 8, 1440–1448.
[6] О разрешимости одного сингулярного интегрального уравнения. Dokl. Akad. Nauk SSSR **241** (1978) 2, 272–274.

Vladimirov, V. S. (Владимиров, В. С.)
[1] Методы теории функций многих комплексных переменных. Nauka, Moskva 1964.
[2] Линейные пассивные системы. Teor. mat. fiz. **1** (1969), 67–94.

[3] Уравнения матемтаической физики. 2nd ed. Nauka, Moskva 1971.
[4] Многомерные линейные пассивные системы. *In*: Sb. Mehanika splošnoĭ sredy i rodstvennye problemy analiza. Nauka, Moskva 1972.
[5] Holomorphic functions of several complex variables with non-negative imaginary part and some applications. *In*: IAEA [1, vol. III, 259–287].

Vlasov, V. Z. (Власов, В. З.)
[1] Общая теория оболочек и ее приложение в технике. Gostehizdat, Moskva 1949.

Volevič, L. R., and S. G. Gindikin (Волевич, Л. Р., и С. Г. Гиндикин)
[1] Уравнение Винера-Хопфа в обобщенных функциях. Mosk. Math. Soc. **35** (1979), 165–216.

Vol'man, V. I. (Вольман, В. И.)
[1] Метод определения критических частот и собственных волн металлических волноводов со сложной формой поперечного сечения. Radiotehnika i elektronika XIX (1974) 7, 1368–1371.

Volterra, V.
[1] Sopra alcune condizioni caratteristiche della funzioni di una variabile complessa. Ann. Mat. Pura Appl., Ser. **2** (1883), 1–55.

Wacker, H. J.
[1] Eine Lösungsmethode zur Behandlung nichtlinearer Randwertprobleme. *In*: Collatz, L. (ed.): Iterationsverfahren, Numerische Mathematik, Approximationstheorie. Internat. Ser. Num. Math., vol. 15. Birkhäuser, Basel, Stuttgart 1970, 245–257.

van der Waerden, B. L.
[1] Algebra I (8th ed.), II (5th ed.). Springer, Berlin, Heidelberg, New York 1971, 1967.

Walk, H.
[1] Wachstumsverhalten zufälliger Potenzreihen. Z. Wahrscheinlichkeitsth. verw. Gebiete **12** (1969), 293–306.

Ward, J.
[1] A theory of analytic functions in linear associative algebras. Duke Math. J. 7 (1940), 233–248.

Warowna-Dorau, G.
[1] Application of the method of successive approximations to a non-linear Hilbert problem in the class of generalized analytic functions. Demonstratio Math. **2** (1970), 101–116.
[2] A compound boundary-value problem of Hilbert-Vekua type. Demonstratio Math. **7** (1974), 337–352.

Washizu, K., and M. Ikegawa
[1] Finite element technique in lifting surface problems. Internat. Symposium on Finite Element Methods in Flow Problems, University of Wales, Swansea 1974, 195–207.

Weitzman, A.
[1] A theorem on Nevanlinna deficiencies. Acta Math. **128** (1972), 41–52.

Wells, Jr., R. O.
[1] Differential analysis on complex manifolds. Prentice Hall, Englewood Cliffs (N.J.) 1973.
[2] Complex manifolds and mathematical physics. Bull. Amer. Math. Soc., New Series, **1** (1979) 2, 296–336.

Wen, Guo-Chun
[1] On Riemann-Hilbert boundary value problem of elliptic systems of linear partial differential equations of the first order. Acta Math. Sinica **15** (1965), 599–613 (Chinese).
[2] On Riemann-Hilbert problem for nonlinear elliptic systems of first order in the plane. Acta Math. Sinica **23** (1980), 244–255 (Chinese).
[3] Modified Dirichlet problem and quasiformal mappings for nonlinear elliptic systems of first order. Kexue Tongbao **25** (1980), 449–453.
[4] The continuous differential solutions for nonlinear elliptic complex equations of first order. Hebei Huagong Xueynan Xuebao, Shuxue Zhuanji, 1980, 62–83 (Chinese).
[5] The first boundary value problem for nonlinear elliptic equations of second order in the plane. Hebei Huagong Xueynan Xuebao, Shuxue Zhuanji, 1980, 84–103 (Chinese).

WENDLAND, W.
[1] An integral equation method for generalized analytic functions. *In*: COLTON, D. L., and R. P. GILBERT (eds.): Constructive and Computational Methods for Differential and Integral Equations. Lecture Notes in Mathematics, vol. 430. Springer, Berlin, Heidelberg, New York 1974, 414–452.
[2] On a class of semilinear boundary value problems for certain elliptic systems in the plane. *In*: Sb. Kompleksnyĭ analiz i ego priloženija. Nauka, Moskva 1978.
[3] On the imbedding method for semilinear first order elliptic systems and related finite element methods. *In*: WACKER, H. (ed.): Continuation Methods. Academic Press, New York, San Francisco, London 1978, 277–336.
[4] Elliptic systems in the plane. Pitman, London, San Francisco, Melbourne 1979.
[5] Zur Behandlung elliptischer Randwertaufgaben mit Integralgleichungen. Wiss. Schriftenr. Techn. Hochsch. Karl-Marx-Stadt, Tagungsberichte 7. Tagung über Probl. Meth. Math. Physik 1979, 143–153.
[6] On Galerkin collocation methods for integral equations of elliptic boundary value problems. *In*: ALBRECHT, J., and L. COLLATZ (eds.): Numerical Treatment of Integral Equations. Internat. Ser. Num. Math., vol. 53. Birkhäuser, Basel 1980, 244–275.

WENDLAND, W., and E. STEPHAN
[1] Remarks to Galerkin and least squares methods with finite elements for general elliptic problems. *In*: EVERITT, W. M., and B. D. SLEEMAN (eds.): Ordinary and Partial Differential Equations. Lecture Notes in Mathematics, vol. 564. Springer, Berlin, Heidelberg, New York 1976, 461–471.

WENDT, G.
[1] Statische Felder und stationäre Ströme. *In*: FLÜGGE, S. (ed.): Handbuch der Physik, vol. XVI. Springer, Berlin, Göttingen, Heidelberg 1958, 1–64.

WEYL, H.
[1] The method of orthogonal projection in potential theory. Duke J. **7** (1940), 411–444.
[2] Die Idee der Riemannschen Fläche. 3rd ed. Teubner, Stuttgart 1955.

WEYL, H., and J. WEYL
[1] Meromorphic functions and analytic curves. Princeton University Press, Princeton (N.J.) 1943.

WIDDER, D. V.
[1] The Laplace transform. Princeton University Press, Princeton (N.J.) 1946.

WINKLER, J.
[1] Bericht über Picardmengen ganzer Funktionen. *In*: LEHTO, O., I. S. LOUHIVAARA, and R. H. NEVANLINNA (eds.): Topics in Analysis. Lecture Notes in Mathematics, vol. 419. Springer, Berlin, Heidelberg, New York 1974, 384–392.

WITTICH, H.
[1] Ganze Lösungen der Differentialgleichung $w'' = f(w)$. Math. Z. **47** (1942), 422–426.
[2] Ganze transzendente Lösungen algebraischer Differentialgleichungen. Math. Ann. **122** (1950), 221–234.
[3] Bemerkung zu einer Funktionalgleichung von H. Poincaré. Arch. Math. **2** (1950), 90–95.
[4] Über das Anwachsen der Lösungen linearer Differentialgleichungen. Math. Ann. **124** (1951), 277–288.
[5] Eindeutige Lösungen der Differentialgleichung $w'' = P(z, w)$. Math. Ann. **125** (1953), 355–365.
[6] Zur Theorie der Riccatischen Differentialgleichung. Math. Ann. **127** (1954), 433–440.
[7] Einige Eigenschaften der Lösungen von $w' = a(z) + b(z) w + c(z) w^2$. Arch. Math. (Basel) **5** (1954), 226–232.
[8] Neuere Untersuchungen über eindeutige analytische Funktionen. Springer, Berlin, Heidelberg, New York 1955.
[9] Defektfreie Lösungen linearer Differentialgleichungen. Arch. Math. **7** (1956), 459 to 464.
[10] Eindeutige Lösungen der Differentialgleichung $w' = R(z, w)$. Math. Z. **74** (1960), 278–288.
[11] Über die Ordnung meromorpher Funktionen. Arch. Math. **16** (1965), 138–147.
[12] Zur Kennzeichnung linearer Differentialgleichungen mit konstanten Koeffizienten. *In*: KÜNZI, H. P., and A. PFLUGER (eds.): Festband zum 70. Geburtstag von Rolf Nevanlinna. Springer, Berlin, Heidelberg, New York 1966.

[13] Zur Theorie linearer Differentialgleichungen im Komplexen. Ann. Acad. Sci. Fenn. Ser. A I **379** (1966), 1–19.
[14] Anwendungen der Wertverteilungslehre auf gewöhnliche Differentialgleichungen. Ann. Acad. Sci. Fenn. A I **7** (1982), 89–97.

VON WOLFERSDORF, L.
[1] Optimale Steuerprobleme bei Noetherschen Operatorgleichungen. Math. Nachr. **61** (1974), 93–109.
[2] Necessary optimality conditions for control processes with singular integral equations and elliptic equations. *In*: KLUGE, R. (ed.) [1, 279–292].
[3] Necessary optimality conditions for optimal control problems with elliptic systems in the plane. Optimization Techniques; Part 1. Lecture Notes in Control and Information Sciences, vol. 22. Springer, Berlin, Heidelberg, New York 1980, 445–464.
[4] On some optimal control problems for elliptic systems in the plane. *In*: KLUGE, R. (ed.): Nonlinear Analysis. Theory and Applications. Abh. Akad. Wiss. DDR, Abt. Math., Nat.,Techn. **2N** (1981). Akademie-Verlag, Berlin 1981, 65–74.
[5] Monotonicity methods for class of first order semilinear elliptic systems. *In*: TUTSCHKE, W. (ed.): Komplexe Analysis und ihre Anwendungen auf partielle Differentialgleichungen. Martin-Luther-Univ. Halle-Wittenberg 1980, 369–373.
[6] Optimal control with semilinear first order complex differential equations of I. N. Vekua's type. Applicable Analysis **11** (1981), 259–277.
[7] On some optimal control problems for linear elliptic systems in the plane. Beitr. Anal. **17** (1981), 95–98.
[8] The linear-quadratic control problem for generalized analytic functions. Math. Nachr. **102** (1981), 201–216.
[9] A class of coefficients control problems with elliptic systems in the plane. Math. Nachr. **99** (1980), 285–300.

WOLSKA-BOCHENEK, J.
[1] On some generalized nonlinear problem of the Hilbert type. Zeszyty Naukowe Polytechniki Warszawskiej **183** (1968) 14, 15–32.
[2] A compound nonlinear boundary value problem in the theory of pseudo-analytic functions. Demonstratio Math. **4** (1972), 105–117.

WU, H.
[1] Mapping of Riemann surfaces (Nevanlinna theory). Entire functions and related parts of analysis. La Jolla (Calif.) 1966. Proc. sympos. pure math. **11** (1968), 480 to 532.
[2] Remarks on the first main theorem in equidistribution theory I–IV. J. Diff. Geom. **2** (1968), 197–202; 369–384; **3** (1969), 83–94; 433–446.
[3] The equidistribution theory of holomorphic curves. Ann. Math. Stud. 1970/Mir, Moskva 1973.

YANG, C. C.
[1] On deficiencies of differential polynomials. I, II. Math. Z. **116** (1970), 197–204; **125** (1972), 107–112.
[2] A note on Malmquist's theorem on first-order differential equations. *In*: WEISS, L. (ed.): NRL-MRC Conf. on Ordinary Differential Equations. Academic Press, New York 1972, 597–607; Yokohama Math. J. **20** (1972), 115–123.
[3] On meromorphic solutions of generalized algebraic differential equations. Ann. Mat. Pura Appl. (IV) **91** (1972), 41–52.

YANG, L., and KUAN-HEO CHANG
[1] Sur la distribution des direction des Borel des fonctions méromorphes. Sci. Sinica **16** (1973), 465–482.
[2] Sur la construction des fonctions méromorphes ayant des direction singulières données. Sci. Sinica **19** (1976), 445–459.
[3] Progress in value distribution theory of meromorphic functions. Kexue Tongbao **22** (1977), 375–380.

YANO, K.
[1] Differential geometry on complex and almost complex spaces. Pergamon Press, Oxford, London, Edinburgh, New York, Paris, Frankfurt/Main 1965.

YANO, K., and S. ISHIHARA
[1] Real hypersurfaces of a complex manifold and distributions with complex structure. Kodai Math. J. **1** (1978), 289–303.

Yosida, K.
[1] A generalization of a Malmquist's theorem. Japan J. Math. **9** (1932), 253–256.
[2] On algebroid solutions of ordinary differential equations. Japan J. Math. **10** (1933), 253–256.
[3] On the characteristic function of a transcendental meromorphic function of an algebraic differential equation of the first order and of the first degree. Proc. Math. Phys. Soc. Japan **15** (3) (1933), 337–338.
[4] A note on Riccati's equation. Proc. Math. Phys. Soc. Japan **15** (3) (1933), 227–237.
[5] A note on Malmquist's theorem on first order algebraic differential equations. Proc. Japan Acad. **53** (1977), 120–123.

Zabreĭko, P. P., A. I. Košelev, M. A. Krasnosel'skiĭ, S. G. Mihlin, L. S. Rakovščik, and V. Ja. Stecenko (Забрейко, П. П., А. И. Кошелев, М. А. Красносельский, С. Г. Михлин, Л. С. Раковщик, и В. Я. Стеценко)
[1] Интегральные уравнения (справочник). Nauka, Moskva 1968 (Engl. transl.: Integral equations — a reference text. Noordhoff, Leyden 1975).

Zolotarev, V. M. (Золотарев, В. М.)
[1] Несколько замечаний к статье Rossberg-Jesiak. Math. Nachr. **82** (1978), 301 to 305.

Zolotarevskiĭ, V. A. (Золотаревский, В. А.)
[1] О методе механических квадратур для систем сингулярных интегральных уравнений. Izv. vysš. učebn. Zaved., Mat. **4** (1976), 47-55.

Zuev, M. F. (Зуев, М. Ф.)
[1] О полианалитических функциях, равных по модулю. Litov. mat. sb. **5** (1968) 1, 31–36.

Žukova, N. I. (Жукова, Н. И.)
[1] Однородная задача Карлемана для правильного $2n$-угольника. Dokl. Akad. Nauk BSSR **21** (1977), 492–495.
[2] Задача Карлемана для многоугольника. Izv. vysš. učebn. zaved., Mat. 1979, **206** (1979) 7, 31–40.

Zverovič, È. I. (Зверович, Э. И.)
[1] Смешанная задача теории упругости для плоскости с прямолинейными разрезами. Annotaciĭ Dokl., Sem. Inst. priklad. Mat., Tbilis. Univ. **4** (1971), 39–43.

Zygmund, A.
[1] Trigonometric series, Vol. I, II. Cambridge University Press, Cambridge 1959.

Author Index

Subject Index

Bei Fragen zur Produktsicherheit wenden Sie sich bitte an:
If you have any questions regarding product safety,
please contact:

Walter de Gruyter GmbH
Genthiner Straße 13
10785 Berlin
productsafety@degruyterbrill.com